An Introduction to High-Pressure Science and Technology

An Introduction to High-Pressure Science and Technology

Edited by
J. Manuel Recio
Malta Consolider Team and
University of Oviedo, Spain

J. Manuel Menéndez
Malta Consolider Team and
University of Edinburgh, United Kingdom

Alberto Otero de la Roza
Malta Consolider Team and
National Institute for Nanotechnology, Canada

A MALTA-Consolider Initiative

CRC Press is an imprint of the
Taylor & Francis Group, an **informa** business

CRC Press
Taylor & Francis Group
6000 Broken Sound Parkway NW, Suite 300
Boca Raton, FL 33487-2742

First issued in paperback 2020

Version Date: 20150824

ISBN 13: 978-0-367-57539-7 (pbk)
ISBN 13: 978-1-4987-3622-0 (hbk)

Visit the Taylor & Francis Web site at
http://www.taylorandfrancis.com

and the CRC Press Web site at
http://www.crcpress.com

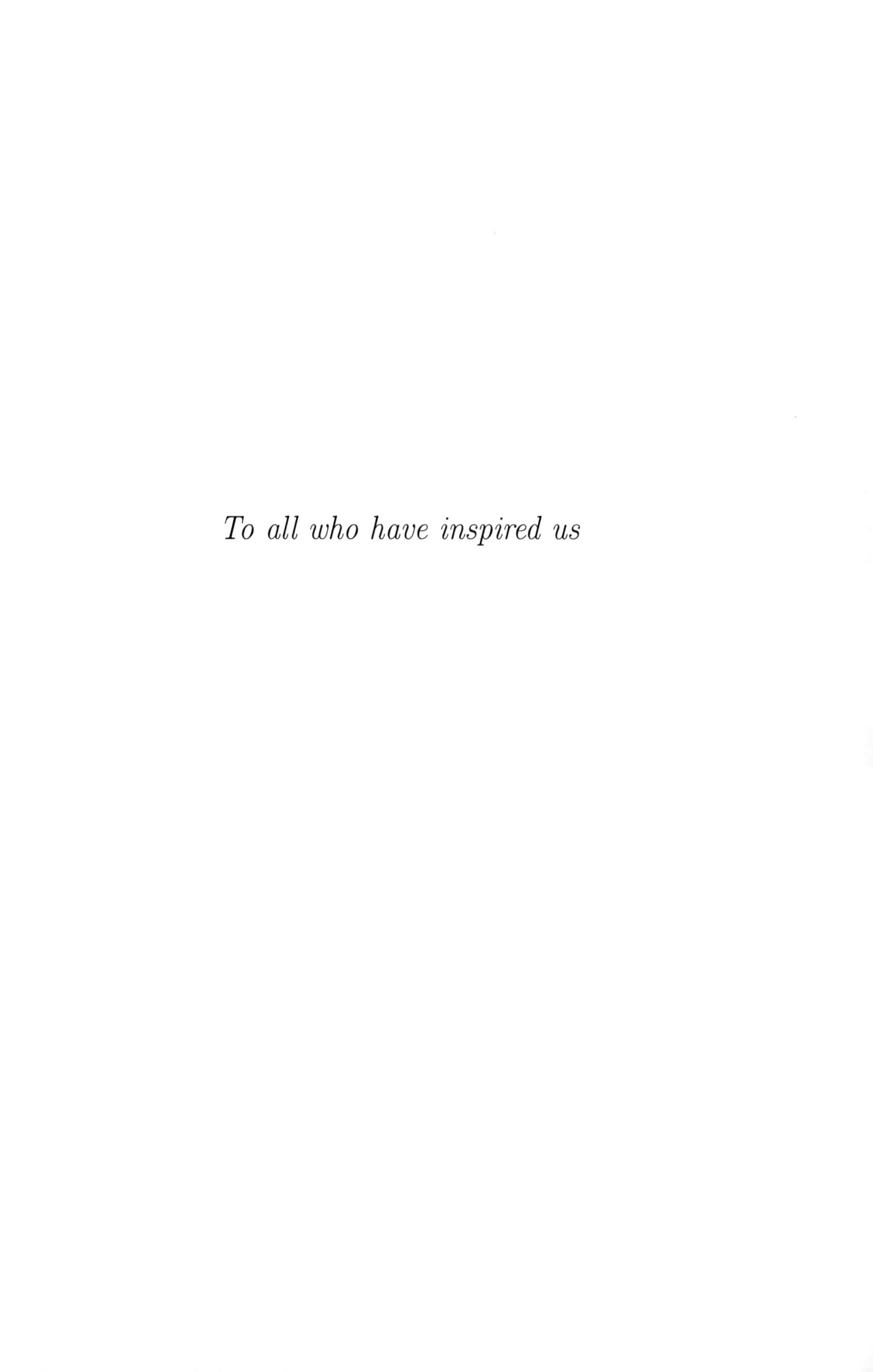

To all who have inspired us

Contents

Section II EXPERIMENTAL TECHNIQUES

Foreword

It is well known that when embarking on a new research subject, the first challenge is to find the convenient *entrance door*, i.e., a text introduces the topic at a basic level, using language that can be understood by newcomers—a concise text that gives an overview of the subject and contains the essential references for further reading. When José Manuel Recio first told me that the Spanish Malta consortium wrote an introductory textbook on high-pressure science and technology for the school that convenes regularly on this subject, I immediately encouraged him to translate it into English and publish it to make it available to a large audience. What I received a year later is truly impressive, and I find it a pleasure to read.

This book surveys the major current aspects of high-pressure research. It covers both modern theoretical approaches as well as state-of-the-art experimental high-pressure techniques with an emphasis on static high pressures using diamond anvil cells. About one-third of the book is devoted to *applications*, i.e., the use of high pressure for the synthesis of new materials, understanding the Earth's and other planetary interiors, and food processing and biotechnology. It reflects the profoundly interdisciplinary character of our research and our community and its close relationship with industrial applications. This text was written by 39 scientists who are internationally renowned experts on the topics they cover in this book. Many examples and results emanate directly from their own research interests, and are ideally suited to explaining basic concepts and methods in high-pressure research.

I highly recommend this book to those who want to become familiar with this topic, particularly postgraduate students. Last, but not least, this book demonstrates the vibrant state of the Spanish high-pressure community, and the undeniable success of the Malta consortium.

Stefan Klotz
Chairman: European High Pressure Research Group (EHPRG)
Université Pierre et Marie Curie
Paris, France

Preface

An *Introduction to High-Pressure Science and Technology* is the result of the work carried out by thirty-nine specialists from the high-pressure research field. This book came to being after six high-pressure school workshops celebrated throughout Spain: Valencia (2002), Madrid (2004), Barcelona (2007), Santander (2009), Tenerife (2011), and Oviedo (2013). Professors attending these courses wanted to record and organize their lectures with a structure similar to the one followed in the schools. Aware that basic aspects quickly lead to research results in class, they all agreed to use their one-hour lectures as starting points for writing their respective chapters. Writing allows a more thorough and reflexive presentation than an oral dissertation. The textbook *Materia a Alta Presión: Fundamentos y Aplicaciones*, first edition, was originally published in Spanish in 2011. The book was published by arrangement with the Oviedo University Press—Ediuno—and Cantabria University Press—EUC—. Now, we offer this improved and revised English version to the international community.

Perhaps the singularity of this book lies in the readers to whom it is targeted: postgraduate students and other scientists approaching the broad and interdisciplinary field of high pressure. Therefore, the book was conceived to guide its readers through the process of learning why pressure is considered a powerful scientific and technological tool, how pressure can be introduced into the laboratory, and which problems can be solved using this thermodynamic variable. We looked for the style and the tone of a textbook in which the natural sequence—theory, experiments and applications—divides the book into three sections. Its focus spans from basic thermodynamic equations and state-of-the-art computational tools to the responses of microorganisms, Earth constituents, and icy planets to pressure. This book includes many experimental techniques, which, when combined with pressure, reveal themselves as powerful sources of information for understanding many natural phenomena and reveal clear paths for the design of novel materials.

Although each chapter is relatively self-contained and can thus be read independently, the editors have made an effort to coordinate the contents of all chapters as coherently as possible. All in all, we expect this book to provide readers with an understanding of the connections between the different areas involved in the multidisciplinary science of high pressure. In that case, the main purpose of the book has been accomplished.

We would like to finish this preface with an acknowledgment to the people and institutions that have allowed this book to see the light. Stefan Klotz, EHPRG Chairman until September 2015, deserves special thanks for his continuous support and encouragement. It has been a real pleasure to have been guided by the CRC Press Physics Editor Francesca McGowan. Thank you very much, Francesca, for your clear advice, patience, and expertise. Also, thanks to our CRC Project Coordinator Ashley Weinstein for her help preparing the book before it was placed in production for editing. All the co-authors have understood that this book is not a simple sum of individual chapters but a common project in which the important goal is producing this book as a whole. Their understanding is gratefully acknowledged.

Finally, this book reflects in some way the enormous qualitative step forward given by the Spanish high-pressure science community in the past eight years. It has been possible thanks to the scientific harmony in which around one hundred scientists belonging to more than ten research groups have been working under the funding of the Spanish Consolider-Ingenio2010 CSD2007-00045 project, now called Matter at High Pressure (MALTA-Consolider). The fact that many of the group leaders from those groups are involved in this book compels us to acknowledge this financial support.

J. Manuel Recio
Oviedo, Spain

J. Manuel Menéndez
Edinburgh, United Kingdom

A. Otero de la Roza
Edmonton, Canada

Acknowledgments

Most of the contributing authors belong to the Malta Consolider Team, and the rest have been linked for years to our team in several ways, so we would like to express a general grateful acknowledgment for fruitful discussions to all the current and past members of the team and for financial support from the Spanish MALTA-Consolider Ingenio 2010 program under the project CSD2007-00045. We also want to acknowledge additional financial support from other projects and institutions, most of them linked to or resulting from the MALTA-Consolider endeavour. These are: CTQ2012-31174, CTQ2012-38599-C02, MAT2013-46649-C4-3-P, PrometeoII/2014/022, ACOMP/2014/270, CTQ2012-36253-C03-02, Junta de Castilla y León Grant CIP13/03, ANR-11-LABX-0037-01 (French National Research Agency), MAT2012-38664-C02-01, MAT2013-46649-C4-4-P, Alberta Innovates Technology Futures (AITF), C.E.I. Universidad de Oviedo and Mèrimèe, Comunidad de Madrid for Quimapres S2009/PPQ-1551, and Red Eléctrica Española (REE).

About the Editors

Professor J. Manuel Recio studied chemistry at the University of Oviedo and earned his master's degree and PhD in chemical sciences. He has been teaching physical chemistry for more than 25 years at undergraduate and postgraduate levels, including visiting professor positions at Michigan Technological University (where he was a Fulbright Fellow in 1991–92) and Université Pierre et Marie Curie (France). Dr. Recio leads a research group on theoretical and computational chemistry of solids under extreme conditions at the University of Oviedo and has supervised five PhD theses and ten MSc theses in this field. He is the author of more than one hundred scientific publications and has been invited for speaking engagements at a number of international meetings. Prof. Recio has also been involved as a coordinator of a cooperation project in Ethiopia setting up the first computational center in Addis Ababa University.

José Manuel Menéndez graduated with a bachelor's degree in chemistry from Universidad de Oviedo (Spain) in 2000 and with a MSc in physical chemistry two years later. After his PhD in chemical physics from Universidad de Valladolid (Spain) in 2006, he worked for two years in Nottingham (UK) as the main environmental/analytical chemist of the organic chemistry department of EIAG Ltd. In 2008 Dr. Menéndez rejoined academia as a research fellow of the quantum chemistry group at Universidad de Oviedo where he has focused on rationalizing the principles that govern the behavior of matter at extreme conditions. He was responsible for designing, upgrade planning and/or setting up several high-performance computing facilities tailored to specific needs in Spain, Ethiopia, and the UK, where he has been in charge of the computational center of the Centre for Neuroregeneration at the University of Edinburgh since August 2014. He has disseminated his research work through international conferences and prestigious research journals. He is the main coordinator of the first Spanish postgraduate book on high-pressure science and technology.

Alberto Otero de la Roza studied chemistry at the University of Oviedo and earned his PhD in 2011, under the supervision of Dr. Víctor Luaña, with a study of the thermodynamic and chemical bonding characteristics of solids, and the relations between them. After that, he moved to the University of California, Merced, to work with Prof. Erin Johnson in the development and application of density-functional theory methods for non-covalent interactions, his main research topic at the moment. After two years in Merced, Dr. Otero de la Roza accepted a position as research officer at the National Institute for Nanotechnology in Edmonton, where he has been continuing his work on computational approaches to non-covalent interactions.

Contributors

Fernando Aguado
Universidad de Cantabria
Santander, Spain

Miguel A. Alario-Franco
Universidad Complutense de Madrid
Madrid, Spain

Juan Andrés
Universidad Jaume I
Castellón, Spain

Valentín García Baonza
Universidad Complutense de Madrid
Madrid, Spain

Armando Beltrán
Universidad Jaume I
Castellón, Spain

Laura J. Bonales
CIEMAT
Madrid, Spain

Roberto A. Boto
Université Pierre et Marie Curie
Paris, France

Julia Contreras-García
Université Pierre et Marie Curie and CNRS
Paris, France

Antonio J. Dos santos-García
Universidad Politécnica de Madrid
Madrid, Spain

Daniel Errandonea
Universidad de Valencia
Valencia, Spain

Manuel Flórez
Universidad de Oviedo
Oviedo, Spain

Jesús González
Universidad de Cantabria
Santander, Spain

Lourdes Gracia
Universidad Jaume I
Castellón, Spain

Bérengère Guignon
Universidad Complutense de Madrid
Madrid, Spain

Stefan Klotz
Université Pierre et Marie Curie
Paris, France

Vanessa Labet
Université Pierre et Marie Curie and CNRS
Paris, France

Víctor Lavín
Universidad de La Laguna
Tenerife, Spain

Víctor Luaña
Universidad de Oviedo
Oviedo, Spain

Francisco Javier Manjón
Universitat Politècnica de València
Valencia, Spain

Miriam Marqués
Universidad de Valladolid
Valladolid, Spain

Jose Manuel Menéndez
University of Edinburgh
Edinburgh, UK

Oscar R. Montoro
Universidad Complutense de Madrid
Madrid, Spain

Ángel Morales
Charles University in Prague
Prague, Czech Republic

Emilio Morán
Universidad Complutense de Madrid
Madrid, Spain

Alfonso Muñoz González
Universidad de La Laguna
Tenerife, Spain

Victoria Muñoz-Iglesias
Centro de Astrobiología INTA-CSIC
Madrid, Spain

Alberto Otero de la Roza
National Institute for Nanotechnology
Edmonton, Alberta, Canada

Julia Pellicer-Porres
Universidad de Valencia
Valencia, Spain

Olga Prieto-Ballesteros
Centro de Astrobiología INTA-CSIC
Madrid, Spain

J. Manuel Recio
Universidad de Oviedo
Oviedo, Spain

Vanessa Riffet
Université Pierre et Marie Curie and CNRS
Paris, France

Fernando Rodríguez
Universidad de Cantabria
Santander, Spain

Plácida Rodríguez-Hernández
Universidad de La Laguna
Tenerife, Spain

Ulises R. Rodríguez-Mendoza
Universidad de La Laguna
Tenerife, Spain

Jordi Saldo
Universitat Autonòma de Barcelona
Barcelona, Spain

Javier Sánchez-Benítez
Universidad Complutense de Madrid
Madrid, Spain

David Santamaría
Universidad de Valencia
Valencia, Spain

Alfredo Segura
Universidad de Valencia
Valencia, Spain

Nadia A. S. Smith
National Physical Laboratory
Teddington, United Kingdom

Rafael Valiente
Universidad de Cantabria
Santander, Spain

I

THEORY

CHAPTER 1

Thermodynamics of Solids under Pressure

Alberto Otero de la Roza
MALTA Consolider Team and National Institute for Nanotechnology, National Research Council of Canada, Edmonton, Canada

Víctor Luaña
MALTA Consolider Team and Departamento de Química Física y Analítica, Universidad de Oviedo, Oviedo, Spain

Manuel Flórez
MALTA Consolider Team and Departamento de Química Física y Analítica, Universidad de Oviedo, Oviedo, Spain

CONTENTS

1.1 INTRODUCTION TO HIGH-PRESSURE SCIENCE

HIGH pressure research [1] is a field of enormous relevance both for its scientific interest and for its industrial and technological applications. Many materials undergo fascinating changes in their physical and chemical characteristics when subjected to extreme pressure. This behavior is caused by the involvement in bonding of electrons that would otherwise not be chemically active under zero-pressure conditions (the difference between zero pressure and atmospheric pressure is negligible from the point of view of its effect on materials properties, and we will use both concepts interchangeably). By and large, current chemical knowledge and the traditional rules for valence electrons are at a loss to explain most of the changes induced when materials are compressed, which makes chemical bonding under pressure an exciting research topic that has received much attention in recent years [1] (see Chapter 5).

The study of pressure effects can be approached in two mostly complementary ways: experimentally and computationally. In both cases, the basic object under study is the change in crystal structure a material undergoes when a given pressure and temperature are applied as shown in its thermodynamic pressure-temperature phase diagram. Experimentally, the application of temperature is relatively straightforward, but imposing high pressure on a sample requires specialized techniques that have been under development for the past 80 years. The field of experimental high-pressure physics was pioneered by Percy Bridgman, who received the Nobel Prize for his efforts in 1946 [2].

There are two main experimental high-pressure techniques: dynamic compression methods based on shock waves, and static compression methods which make use of pressure cells. In a shock-wave compression experiment [3], a strong shock is applied to the sample and the propagation velocities of the shock wave inside the material are measured. Very high pressures (up to 500–1000 GPa) and temperatures (tens of thousands of K) can be attained with this technique. Static compression techniques are applied using pressure cells, notably diamond anvil cells [4]. These methods are more accurate than dynamic techniques, but they are also limited by the pressure scale, with the mechanical resistance of a diamond (slightly above 300 GPa) being the ultimate upper pressure limit for the technique. High temperatures, in the range of thousands of K, can be attained by laser heating, and diamond anvil cells can be coupled to other techniques: spectroscopic (infrared, Raman, x-ray) and optical. Experimental high-pressure techniques are explored in Part II of this book.

From the computational point of view, in the framework of density functional theory (DFT), the application of pressure is relatively straightforward: one simply compresses the unit cell and calculates the applied pressure either analytically from the self-consistent one-electron states or from the volume derivative of the calculated total energy. The latter requires fitting an equation of state. Conversely, temperature effects are difficult to model because they involve collective atomic vibrations—called phonons—propagating throughout a crystal (see Chapter 3). The simplest computational approach to model the effects of temperature is to calculate the electronic energy, assuming the nuclear and electronic motions are decoupled (the adiabatic approximation), and then obtaining the phonon frequencies using the derivatives of said electronic energy with respect to the atomic movements. The phonon frequencies can be used in the harmonic approximation to calculate thermodynamic properties and phase stabilities at arbitrary pressure and temperature. Computational techniques for high-pressure studies are explored in this and the subsequent chapters in Part I.

Despite the difficulties in modeling temperature effects, theoretical approaches are necessary in cases for which experiments are either not feasible or difficult to interpret. An interesting example of synergy between computational and experimental approaches occurs in geophysics. Because the interior of the Earth is inaccessible, information about its composition and structure can only be inferred through indirect means [5, 6], primarily via seismic velocities measured during earthquakes. Calculated data (elastic moduli, longitudinal and shear velocities, thermal expansion and conduction, heat capacities, Grüneisen parameters, transition pressures, Clapeyron slopes, etc.) are very helpful in the interpretation of experimental data and in the construction of models of the Earth's interior [7, 8]. The geophysical and astrophysical applications of high-pressure science are presented in Chapters 15 and 16, respectively.

High-pressure research also has very relevant technological applications (Chapter 14). Materials that are thermodynamically stable at high pressure and temperature are often metastable at zero pressure and room temperature, so it is possible to access new material phases with unique properties by the application of appropriate pressure and temperature conditions. The application of high pressure to biological samples is also of technological relevance because it is known that the microorganism activity is diminished or canceled by application of high pressures, a process called pascalization. In contrast to the preceding applications, the pressures exerted in this case are in the order of the tenths of a GPa, much smaller than in previous examples. Pascalization can be used to increase the shelf lives of perishable foodstuffs: juice, fish, meat, dairy products, etc. This has been one of the research lines of MALTA [9–12], and is explored in Chapters 12 and 13.

To summarize, high pressure is a very active field of research and one whose ramifications affect many scientific and technological fields, from astrophysics and geophysics to materials physics and the food industry. From the fundamental view, the behavior of materials at high pressure is still poorly understood, and the usual textbook chemistry rules that apply to zero-pressure chemistry are essentially useless when molecules and materials are subjected to extreme compression. This makes high-pressure research an interesting and virtually unexplored field of study. In the remainder of the present chapter, we give an overview of the basic thermodynamic principles and mathematical tools that are needed to master the rest of the book. Namely, we review the thermodynamics fundamentals (Section 1.2), the equations of state used to describe the responses of materials to compression (Section 1.3), and the harmonic approximation, the basic theoretical framework in which temperature effects are modeled computationally (Section 1.4).

1.2 THERMODYNAMICS OF SOLIDS UNDER PRESSURE

1.2.1 Basic Thermodynamics

Hydrostatic pressure, together with its extensive associated variable volume, is a fundamental quantity covered in most thermodynamics textbooks [13–17]. Pressure can be defined by considering the mechanical work exerted on a closed system (in which there is no matter exchange with the environment) through a change in its volume:

$$\delta W = -P_E dV. \tag{1.1}$$

In this equation, P_E is the pressure applied on the system or environmental pressure [18] (also, see the subsection on irreversible pressure-volume work and references therein in Chapter 2 of Reference 15). When both system and environment are in mechanical equilibrium with each other, $P_E = p$, with p representing the system's pressure (constant throught the system). The definition of mechanical work is convenient to study, for instance, the behavior of gases and phase transitions in a closed system. Given that we are interested in the thermodynamic properties of solids under pressure, we start this chapter by presenting some fundamental quantities and the relations between them that will be used in the rest of the book.

Classical thermodynamics is a macroscopic theory based on a few axiomatic principles derived from empirical observations. These fundamental principles are called the laws of thermodynamics. The zeroth, first and second laws allow us to assert the existence of temperature T, internal energy U, and entropy S as state functions of the system (see below). The first law states that for any infinitesimal process in a closed system (at rest and in absence of external fields):

$$dU = \delta Q + \delta W. \tag{1.2}$$

In this equation, dU is the infinitesimal internal energy change undergone by the system, δQ is the infinitesimal heat flow into the system, and δW is the infinitesimal work done on the system during the process. For a finite process, Q and W depend on the intermediate states through which the process evolves, but ΔU depends only on the initial and the final states. This is symbolized in Equation (1.2) by using the exact differential notation (dU) for the internal energy but not for heat or work. Quantities like U that depend only on the state of the system are called state variables or state functions, and it is possible to define a function (whose changes can be empirically measured) that depends only on the variables that determine the state of the system, $U(p, T, \ldots)$. As we show below, the same thermodynamic quantity can act as a variable (if it is independent) or as a function (if its value is given as a function of other independent variables).

The first law establishes the conservation of energy principle in an isolated system, in which there is no energy or matter exchange with the environment. That is, in an isolated system, the internal energy is a constant. When a closed system receives energy, either in the form of heat or work, its internal energy increases and the environment loses energy by the same amount. If one considers a cyclic process in a closed system, i.e., a process in which the initial and final states of the system are the same, the internal energy change of the system is zero:

$$\oint dU = 0, \tag{1.3}$$

which makes building a machine whose only effect is the constant generation of energy impossible.

The second law of thermodynamics deals with the difference between heat and work. It originates from the realization that it is impossible to build a cyclic machine that transforms heat into work with 100 % efficiency. The second law, which was first proposed by Sadi Carnot through his analysis of steam engines, is therefore a principle of limited efficiency which, for a heat engine, is defined as the fraction of the heat input that is transformed into work. In the case in which heat is absorbed from a hot reservoir and released into a cold reservoir, the maximum efficiency of a heat engine depends only on the temperature of the two reservoirs, and can only be attained if the engine works reversibly (infinitely slowly and excluding dissipative effects) through an infinite series of equilibrium steps [15]. For any real (irreversible) process happening at a finite velocity and including dissipative effects, the efficiency is always less than the reversible upper bound.

The second law is formalized through the definition of entropy. Based on Carnot's work, Clausius proved that for a reversible process in a closed system (whose temperature T is constant throughout), the quantity $\delta Q_{\text{rev}}/T$ is an exact differential:

$$\oint \frac{\delta Q_{\text{rev}}}{T} = 0, \tag{1.4}$$

or, in other words, we can define a function S (the entropy) as:

$$dS = \frac{\delta Q_{\text{rev}}}{T}, \tag{1.5}$$

which depends only on the state of the system. The difference between the entropy in states A and B, $\Delta S = S_B - S_A$, can be calculated as the value obtained from Equation (1.5) in a hypothetical reversible process that goes from A to B. A corollary of the definition in Equation (1.5) is that the change in entropy of a closed system during a reversible adiabatic process is zero.

The observation that it is not possible in practice to achieve the maximum thermal efficiency of an engine can be recast in terms of entropy. For an irreversible process in a closed system, Equation (1.5) becomes,

$$dS > \frac{\delta Q}{T_E}, \tag{1.6}$$

with T_E being the temperature of the environment. Therefore, the effect of any process on an adiabatically isolated system is to always increase its entropy. A corollary of this statement is that in an isolated system, thermodynamic equilibrium is attained once the entropy of the system is maximized, and no spontaneous processes within the system can occur henceforth. Note that $T_E = T$, with T being the temperature of the system, only holds when both system and environment are in thermal equilibrium with each other. As we shall see below, the maximum entropy statement of the second law can be recast into minimum energy principles that affect certain functions when the state of the system is given by their associated variables.

For simplicity, let us now consider closed one-phase systems with pressure-volume work only undergoing a reversible process. Under these conditions, the mechanical work effected on the system is given by $\delta W = -pdV$ and the heat absorbed by the system is given by $\delta Q = TdS$. The change in internal energy is thus given by the fundamental equation:

$$dU = TdS - pdV, \tag{1.7}$$

which gives the expression of the exact differential dU in terms of two independent variables: entropy and volume. The internal energy $U(S, V)$ as a function of these variables permits the calculation of all the other thermodynamic properties of the system through its derivatives. U is called a thermodynamic potential and S and V are its natural variables.

Entropy and volume are difficult variables to work with because, in general, isochoric and isoentropic conditions are not easy to achieve experimentally. Thermodynamic potentials that depend on other variables can be defined by applying a Legendre transform to $U(S,V)$. In this way we define the enthalpy ($H(S,p)$), the Helmholtz free energy ($F(T,V)$), and the Gibbs free energy ($G(T,p)$) as:

$$H(S,p) = U + pV, \tag{1.8}$$
$$F(T,V) = U - TS, \tag{1.9}$$
$$G(p,T) = U + pV - TS = H - TS = F + pV. \tag{1.10}$$

H, F, and G are the thermodynamic potentials for the indicated independent natural variables.

If a thermodynamic potential can be determined as a function of its natural variables, then all other thermodynamic properties for the system can be calculated by taking partial derivatives. For instance, if G is known as a function of T and p, the thermodynamic behavior of the system is completely determined. The U, H, F, and G thermodynamic potentials can be used to express criteria for spontaneous change and equilibrium when the corresponding natural variables are held constant. For instance, G can only decrease during the approach to reaction equilibrium at constant T and p in a closed system doing pressure-volume work only, reaching a minimum at equilibrium.

The fundamental relation for U (Equation (1.7)) can be combined with the definitions for the enthalpy and the free energies (Equations (1.8) to (1.10)) to find analogous fundamental equations for the other thermodynamic potentials, valid for a reversible process:

$$dH = TdS + Vdp, \tag{1.11}$$
$$dF = -SdT - pdV, \tag{1.12}$$
$$dG = -SdT + Vdp. \tag{1.13}$$

In addition, a number of useful relations can be derived from these equations. For instance, Equation (1.12) can be compared to the exact differential for $F(T,V)$:

$$dF = \left(\frac{\partial F}{\partial T}\right)_V dT + \left(\frac{\partial F}{\partial V}\right)_T dV. \tag{1.14}$$

In this notation, the subscript in $\left(\frac{\partial F}{\partial T}\right)_V$ is used to express that the remaining independent variables other than T for the F function (in this case, only V) are held constant. Comparing Equation (1.14) and Equation (1.12), we arrive at:

$$S = -\left(\frac{\partial F}{\partial T}\right)_V \quad , \quad p = -\left(\frac{\partial F}{\partial V}\right)_T . \tag{1.15}$$

Likewise, for the rest of the thermodynamic potentials,

$$T = \left(\frac{\partial U}{\partial S}\right)_V, \qquad p = -\left(\frac{\partial U}{\partial V}\right)_S, \tag{1.16}$$

$$T = \left(\frac{\partial H}{\partial S}\right)_p, \qquad V = \left(\frac{\partial H}{\partial p}\right)_S, \tag{1.17}$$

$$S = -\left(\frac{\partial G}{\partial T}\right)_p, \qquad V = \left(\frac{\partial G}{\partial p}\right)_T. \tag{1.18}$$

The Maxwell relations can be obtained from the fundamental relations by calculating the mixed second derivatives of the thermodynamic potentials, which need to be the same regardless of the order in which they are taken. For instance,

$$\left(\frac{\partial^2 F}{\partial T \partial V}\right) = \left(\frac{\partial^2 F}{\partial V \partial T}\right) \tag{1.19}$$

leads, using Equation 1.15, to:

$$\left(\frac{\partial p}{\partial T}\right)_V = \left(\frac{\partial S}{\partial V}\right)_T. \tag{1.20}$$

Likewise, for the other state functions, one has:

$$\left(\frac{\partial p}{\partial S}\right)_V = -\left(\frac{\partial T}{\partial V}\right)_S \text{ (from U)}, \tag{1.21}$$

$$\left(\frac{\partial V}{\partial S}\right)_p = \left(\frac{\partial T}{\partial p}\right)_S \text{ (from H)}, \tag{1.22}$$

$$\left(\frac{\partial V}{\partial T}\right)_p = -\left(\frac{\partial S}{\partial p}\right)_T \text{ (from G)}. \tag{1.23}$$

These relations between thermodynamic potential derivatives are useful because, even though the potentials can not be measured directly, some of their variations correlate with quantities that are experimentally accessible. For instance, the volume derivative of the Helmholtz function (Equation (1.15)) gives the pressure as a function of volume and temperature, an object that is called the "volumetric equation of state", or simply the "equation of state", and which plays a key role in theoretical and computational studies of materials under pressure, as we shall see in the next sections. Other material properties and their relation to the corresponding thermodynamic potentials include the isothermal (κ_T) and adiabatic (κ_S) compressibilities:

$$\kappa_T = -\frac{1}{V}\left(\frac{\partial V}{\partial p}\right)_T = -\frac{1}{V}\left(\frac{\partial^2 G}{\partial p^2}\right)_T, \tag{1.24}$$

$$\kappa_S = -\frac{1}{V}\left(\frac{\partial V}{\partial p}\right)_S = -\frac{1}{V}\left(\frac{\partial^2 H}{\partial p^2}\right)_S, \tag{1.25}$$

with units of pressure^{-1}. The inverse of the compressibility, called bulk modulus, is often employed:

$$B_T = \frac{1}{\kappa_T} = -V\left(\frac{\partial p}{\partial V}\right)_T = V\left(\frac{\partial^2 F}{\partial V^2}\right)_T, \tag{1.26}$$

$$B_S = \frac{1}{\kappa_S} = -V\left(\frac{\partial p}{\partial V}\right)_S = V\left(\frac{\partial^2 U}{\partial V^2}\right)_S. \tag{1.27}$$

The bulk moduli have units of pressure. These intensive properties reflect the elastic response of a solid to hydrostatic compression. The bulk modulus is one of the basic quantities that determines the elasticity of a material (in the case of an isotropic material, together with the Poisson ratio, although one can use equivalently other measures such as the Young modulus or the shear modulus). Typical values for the bulk modulus at ambient conditions [19, 20] are: lithium (11 GPa), NaCl (24 GPa), aluminum (76 GPa), silicon (99 GPa), steel (160 GPa), and diamond (443 GPa). The difference between B_T and B_S is, in general, small compared to the value of the bulk modulus, so the terms are often used interchangeably.

The constant pressure (C_p) and constant volume (C_v) heat capacities are defined in terms of the heat necessary to raise reversibly the temperature of a closed system, and are defined as:

$$C_p = \left(\frac{\delta Q}{\delta T}\right)_p = \left(\frac{\partial H}{\partial T}\right)_p = T\left(\frac{\partial S}{\partial T}\right)_p = -T\left(\frac{\partial^2 G}{\partial T^2}\right)_p, \tag{1.28}$$

$$C_v = \left(\frac{\delta Q}{\delta T}\right)_V = \left(\frac{\partial U}{\partial T}\right)_V = T\left(\frac{\partial S}{\partial T}\right)_V = -T\left(\frac{\partial^2 F}{\partial T^2}\right)_V, \tag{1.29}$$

where we have used $\Delta H = Q_p$ and $\Delta U = Q_v$ in a closed system with pressure-volume work only undergoing constant pressure and constant volume processes, respectively. Heat capacities are extensive properties. Their equivalent intensive properties for pure substances, obtained by dividing the heat capacity by the amount of substance, are the constant pressure and constant volume molar heat capacities, usually represented in lower case (c_p and c_v). C_p and C_v measure a system's ability to store energy, and in consequence they are directly related to the available microscopic degrees of freedom. In a solid, their temperature dependence is characteristic: C_p and C_v go to zero for $T \to 0$ and to the Dulong–Petit limit ($3R = 3N_A k_B$ per mole of atoms, with R the gas constant, N_A the Avogadro's constant, and k_B the Boltzmann's constant, $1.3806488 \times 10^{-23}$ J/K) or slightly above it in the high-temperature limit.

The (volumetric) coefficient of thermal expansion or thermal expansivity measures the response of volume to changes in temperature. It is defined as:

$$\alpha = \alpha_V = \frac{1}{V}\left(\frac{\partial V}{\partial T}\right)_p = \frac{1}{V}\left(\frac{\partial^2 G}{\partial T \partial p}\right). \tag{1.30}$$

The thermal expansion coefficient is a quantity with great relevance in engineering and architecture. The vast majority of materials have positive α, that is, they expand when heated at constant pressure, but some materials present extensive temperature ranges with negative thermal expansivity. A characteristic example of this behavior is the cubic zirconium tungstenate (ZrW_2O_8), which has a negative α at all temperatures in its stability range [21]. Likewise, some composite materials like invar, a nickel-iron alloy, have very low thermal expansion coefficients and they are used in construction as well as in precision instruments.

The equations presented above can be used to derive thermodynamic relations between these material properties. For instance, the heat capacities are related by:

$$C_p - C_v = \frac{\alpha^2 V T}{\kappa_T}. \tag{1.31}$$

Likewise, the difference between compressibilities is:

$$\kappa_T - \kappa_S = \frac{\alpha^2 T V}{C_p}, \tag{1.32}$$

and the ratio between heat capacities is the same as between bulk moduli and compressibilities:

$$\frac{C_p}{C_v} = \frac{B_S}{B_T} = \frac{\kappa_T}{\kappa_S}. \tag{1.33}$$

An important quantity that will appear later in this chapter is the thermal Grüneisen parameter:

$$\gamma_{\text{th}} = \frac{V \alpha B_T}{C_v} = \frac{V \alpha B_S}{C_p}, \tag{1.34}$$

which represents the change in the vibrational characteristics caused by a change in volume. Because the product of the thermal coefficient with the bulk modulus is a simple derivative:

$$\alpha B_T = \left(\frac{\partial S}{\partial V}\right)_T = \left(\frac{\partial p}{\partial T}\right)_V, \tag{1.35}$$

the Grüneisen parameter can be rewritten as:

$$\gamma_{\rm th} = \frac{V}{C_v}\left(\frac{\partial S}{\partial V}\right)_T = \frac{V}{C_v}\left(\frac{\partial p}{\partial T}\right)_V. \tag{1.36}$$

More relations, and the proofs for the ones given above can be found in Wallace's book [16].

Finally, the Nernst–Simon statement of the third law of thermodynamics says that for any isothermal process that involves only substances in internal equilibrium, the entropy change ΔS goes to zero as T goes to zero. This statement emerged from the works by Richards, Nernst and Simon in the first decade of the past century. Based on the arbitrary choice that the entropy of each element is zero at 0 K, the Nernst–Simon statement of the third law is used to find conventional entropies of compounds: $S_0 = 0$ for each element or compound in internal equilibrium, where the zero subscript represents 0 K conditions [15].

Microscopically, the entropy is related to the number of microscopic states available to an isolated system for a given volume, internal energy, and number of particles. The relation between entropy and the available number of states compatible with the volume, energy, and number of particles of an isolated system Ω is given by Boltzmann's postulate:

$$S = k_B \ln \Omega(U, V, N). \tag{1.37}$$

At 0 K, only the system's lowest available energy level is populated and Ω becomes the degeneracy of this energy level. For this statistical-mechanical result to be consistent to the thermodynamic result $S_0 = 0$ for a substance, the degeneracy of the ground level of the one-component system would have to be 1. To make the statistical-mechanical and thermodynamic entropies agree, the convention of ignoring contributions to S coming from nuclear spins and isotopic mixing is adopted (see more details in Section 21.9 of Reference 15).

The Helmholtz free energy and the Gibbs free energy are related to different statistical ensembles whose macroscopic states are fixed by the values of their natural variables [17]. These are the canonical and the isothermal-isobaric ensembles, respectively. F and G are related to the partition functions of the corresponding ensemble by a simple relation:

$$F(N, V, T) = -k_B T \ln Z(N, V, T), \tag{1.38}$$

$$G(N, p, T) = -k_B T \ln \Delta(N, p, T). \tag{1.39}$$

The partition functions are calculated using the available energy levels of the system. For instance, the canonical partition function is:

$$Z(N, V, T) = \sum_i \exp\left(-\frac{E_i(N, V)}{k_B T}\right), \tag{1.40}$$

where the i sums over all states available to the system for the given volume and number of particles, E_i is the energy of that state, and T is the thermodynamic temperature. For a given temperature, if the energy levels are separated by differences much larger than k_BT, only the ground energy level (E_0) contributes to the sum, and $F(N, V, T) = E_0$ for a non-degenerate ground energy level. Conversely, if the separation of all available energy levels is very small compared to k_BT, the partition function reduces to the number of available states and F is a constant times k_BT, a result known as the equipartition theorem in classical statistical mechanics.

1.2.2 Principle of Minimum Energy

The second law of thermodynamics—that thermodynamic equilibrium in an isolated system is achieved when the entropy of the system is maximized—can be recast in several ways, depending on the independent variables we use. For every one of the thermodynamic potentials defined above (U, H, F, and G), and for a closed system (with pressure-volume work only) in which their corresponding independent variables are fixed, thermodynamic equilibrium is reached when the value of the corresponding potential is minimized.

A useful generalization of the Gibbs free energy is the *availability* or non-equilibrium Gibbs energy of the system, $G^\star$, a function that gives the Gibbs energy of the system away from equilibrium, either material—phase and chemical—or thermal and mechanical equilibrium [22, 23]. It can be shown from the second law that at a fixed environmental temperature T_E and pressure P_E, the most stable state of the system is that at which the function $G^\star = U + P_EV - T_ES$ has its lowest possible value. This function is only equal to the Gibbs free energy G when the system is at thermal and mechanical equilibrium with the environment ($T_E = T$, $P_E = p$). In the remaining chapters of this book, for simplicity, no distinction will be made between p and P_E or between G and $G^\star$, even for systems out of mechanical equilibrium. Which of those quantities is used will be clear from the context.

Let us consider now the particular case of a one-component crystal phase. For a solid, the more convenient variables, in the sense that they can be easily controlled experimentally, are pressure and temperature. In an infinite periodic solid, there are only two sources of energy levels that enter the isothermal-isobaric partition function: electronic and vibrational. For the majority of solids at the usual temperatures, the excited electronic levels are inaccessible, so the thermal contributions to $G^\star$ are mostly determined by the vibrational levels.

If a system is held at a fixed temperature $T_E = T$ and a constant hydrostatic pressure P_E (or simply P), the equilibrium state minimizes the non-equilibrium Gibbs energy of the crystal phase,

$$G^\star(\boldsymbol{x}, V; P, T) = F^\star(\boldsymbol{x}, V; T) + PV$$
$$= E_{\text{sta}}(\boldsymbol{x}, V) + F^\star_{\text{vib}}(\boldsymbol{x}, V; T) + PV, \tag{1.41}$$

with respect to all internal configuration parameters. In this equation, E_{sta} is the electronic (static) energy of the solid that is obtainable, for instance, through computational techniques. $F^\star_{\text{vib}}$ is the non-equilibrium vibrational Helmholtz free energy. The structure of a periodic infinite crystal is determined by its volume V and a number of coordinates, which include the variables that determine the geometry of the unit cell (angles and lattice parameters) as well as the crystallographic coordinates of the atoms whose values are not fixed by symmetry. All these variables are gathered in the $\boldsymbol{x}$ vector. Additional free energy terms can be used for other degrees of freedom that may represent accessible energy levels at the given temperature and that may contribute to the free energy of the solid, for instance, electronic contributions (in a metal), configurational entropy (in a disordered crystal), etc.

The principle of minimum energy for $G^\star$ says that thermodynamic equilibrium is reached at the $\boldsymbol{x}$ and V values for which $G^\star$ is a minimum at given T and P. That is:

$$G(p = P, T) = \min_{\boldsymbol{x}, V} G^\star(\boldsymbol{x}, V; P, T). \tag{1.42}$$

At the $\boldsymbol{x}$ and V for which $G^\star$ is a minimum, the value of the system pressure p is equal to the environmental pressure P.

1.2.3 Hydrostatic Pressure and Thermal Pressure

The minimum energy principle represented by Equation (1.42) can be used to predict $\boldsymbol{x}(p, T)$, the volume, $V(p, T)$, and the equilibrium Gibbs function $G(p, T)$ of a crystal at any given pressure and temperature. In the context of the computational determination of these quantities, however, finding the global minimum of $G^\star(\boldsymbol{x}, V; P, T)$, which involves finding all terms in Equation (1.41) for a large number of $\boldsymbol{x}$ and V values, is prohibitive.

A common approach to simplify this problem is to restrict the variables $\boldsymbol{x}$ to those resulting from a minimization of the electronic (static) energy at any given volume:

$$\boldsymbol{x}_{\text{opt}}(V) \quad \text{from} \quad E_{\text{sta}}(V) = \min_{\boldsymbol{x}} E_{\text{sta}}(\boldsymbol{x}, V). \tag{1.43}$$

If no crystal vibrations were present, then $F^\star_{\text{vib}}$ would disappear, and this approach would be strictly correct. Note, however, that $E(\boldsymbol{x}, V)$ is still a huge potential energy surface where *all* crystal phases compatible with that volume are represented. Despite this, crystal structure prediction or a judicious selection of candidate phases can render this problem tractable (see Chapter 4). In the presence of crystal vibrations, Equation (1.43) embodies the approximation that the variation of $F^\star_{\text{vib}}$ with $\boldsymbol{x}$ is negligible compared to that of E_{sta}. This "statically constrained" approximation [24–27] transforms Equation (1.41) into:

$$G^\star(V;P,T) = E_{\text{sta}}(\boldsymbol{x}_{\text{opt}}(V),V) + F^\star_{\text{vib}}(\boldsymbol{x}_{\text{opt}}(V),V;T) + PV, \tag{1.44}$$

where all degrees of freedom except for the volume have been eliminated (c.f. Equation (1.41)).

The minimum energy principle can now be applied to Equation (1.44). The equilibrium volume at P and T is found by making the volume derivative of $G^\star$ zero, which leads to the mechanical equilibrium condition:

$$\left(\frac{\partial G^\star}{\partial V}\right)_{P,T} = 0 = -p_{\text{sta}} - p_{\text{th}} + P. \tag{1.45}$$

In this equation, $p_{\text{sta}} = -dE_{\text{sta}}/dV$ is the static pressure, $p_{\text{th}} = -(\partial F^\star_{\text{vib}}/\partial V)_T$ is the thermal pressure, and P is the applied external pressure. At mechanical equilibrium, the sum of static and thermal pressure balances the external pressure applied on the solid:

$$P = p_{\text{sta}} + p_{\text{th}}. \tag{1.46}$$

The thermal pressure, which is usually negative, represents the expansion effect caused by the atomic vibrations in the crystal. It can be used, for instance, to estimate the thermal expansion of molecular crystals in order to compare static DFT with experimental x-ray crystal structures in molecular solids [28].

If the equilibrium is stable (or metastable) then $G^\star$ must be a minimum with respect to variations (or small variations) in the volume at constant P and T. In other words, the second derivative of $G^\star$ with respect to volume (at constant T and P) must be positive. This second derivative is equal to $B_T V$, thus obtaining the usual criterion for mechanical stability ($B_T > 0$).

In the statically constrained approximation, the thermal pressure depends only on volume and temperature. In reality, p_{th} depends on all variables $\boldsymbol{x}$ and particularly on the shape of the unit cell for a particular volume. The statically constrained approximation can be partially relaxed [26, 27, 29] by considering both volume and strain in Equation (1.43). The thermal pressure is then generalized to a thermal stress:

$$\sigma_{\text{th},ij}(\varepsilon_{ij},V;T) = \frac{1}{V}\frac{\partial F^\star_{\text{vib}}(\boldsymbol{x}_{\text{opt}}(\varepsilon_{ij},V),\varepsilon_{ij},V;T)}{\partial \varepsilon_{ij}}, \tag{1.47}$$

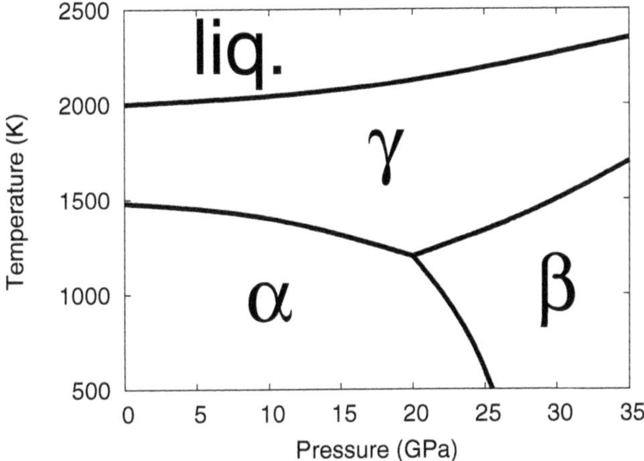

Figure 1.1 Pressure-temperature phase diagram for a one-component system, showing three solid phases (α, β, and γ) and the liquid phase.

where $\boldsymbol{\varepsilon}$ is the strain tensor—a measure of the deformation of the solid with respect to the equilibrium geometry at volume V (also see Chapter 2). In the statically constrained approximation, the thermal stress tensor $\boldsymbol{\sigma}$ would be diagonal and its elements equal to minus the thermal pressure. However, even this generalization leads to a computationally prohibitive iterative process involving an "excessive number of calculations" [26].

1.2.4 Phase Equilibria

At a given temperature T and pressure P, there are many different atomic arrangements of a crystal (corresponding to different values of V and $\boldsymbol{x}$ in Equation (1.41)) that are local minima of the non-equilibrium Gibbs function $G^{\star}$. A corollary of the minimum energy principle is that of all these minima, only the one with the lowest $G^{\star}$ (the global minimum) represents the thermodynamically stable phase, and the rest are metastable phases. The two-dimensional plot that represents the stable crystal phase as a function of temperature T and pressure p is called a (pressure-temperature) phase diagram. An example is shown in Figure 1.1.

The phase diagram of a solid determines the behavior of the material at arbitrary temperature and pressure conditions, and is therefore of great technological relevance. Even though only one of the phases is thermodynamically stable at a given pressure and temperature, metastable phases of pure crystalline solids, which are not usually represented in a phase diagram, are the rule rather than the exception. Carbon is the traditional example: graphite is the stable polymorph at zero pressure and room temperature, but diamonds are observed (metastable) because the transition from diamond to graphite is kinetically very unfavorable under room conditions (the energy barrier the material needs to overcome to go from one phase to the other is too high, see Chapter 2). As a matter of fact, the artificial generation of diamonds requires pressures and temperatures well beyond the thermodynamic limits of the graphite phase.

The boundaries between regions associated to different phases represent particular states of a system in which two phases may coexist. For a given temperature, the pressure at which one phase transforms reversibly into another is called the transition pressure, $p_{\mathrm{tr}}(T)$. Along the two-phase equilibrium curves, the value of $G(p,T)$ per mole is the same for both phases. By using this result, one can calculate properties of the curve such as its slope at a given point, which is given by the Clapeyron equation:

$$\frac{dp_{\mathrm{tr}}}{dT} = \frac{\Delta S_{\mathrm{tr}}}{\Delta V_{\mathrm{tr}}} = \frac{\Delta H_{\mathrm{tr}}}{T\Delta V_{\mathrm{tr}}}, \tag{1.48}$$

where ΔS_{tr} and ΔV_{tr} are the differences in entropy and volume between both phases at that point. During most phase transitions, energy is absorbed or released by the system and the volume increases or decreases without change in temperature and pressure. This "latent heat" (or the volume change) is used to drive the transition from one phase to the other. The latent heat is given by ΔH_{tr}.

From a thermodynamic point of view, phase transitions can be classified according to the derivatives of G that are discontinuous across the transformation [30]. First order phase transitions involve latent heat and are characterized by an infinite heat capacity at the transition temperature and pressure as well as by discontinuities in the properties that are derivatives of the Gibbs function: V and $-S$. Most phase transitions belong in this category; boiling water is an example of a first order phase transition.

In contrast, high order or continuous phase transitions do not involve latent heat. In second-order phase transitions, there is a discontinuity in the heat capacity, but it does not diverge. The isothermal compressibility and the thermal expansion coefficient, which contain also second derivatives of G with respect to T and/or p, are discontinuous too. The first derivatives of G, volume and minus entropy, show changes in their slopes, but they are continuous across the transition. In consequence, the Clapeyron equation cannot be applied. An example of a second-order phase transition is the onset of a superconducing regime in a material. See more details in Chapter 3 and also in References 15 and 30. Other phase transition classifications using different criteria are presented in Chapter 2.

1.3 EQUATIONS OF STATE

An equation of state (EOS) is a relationship between the thermodynamic variables relevant for a given mass of a solid phase under study at equilibrium. For a solid phase under pressure, the volumetric EOS is written in terms of pressure p, volume V, and temperature T:

$$F(p, V, T) = 0. \tag{1.49}$$

A simple equation of state, presented in Murnaghan's historical paper [31], is based on the principle of conservation of mass combined with Hooke's law for an infinitesimal variation of stress in a solid. The Murnaghan EOS can also be obtained by assuming a linear variation of the isothermal bulk modulus with pressure, $B_T(p) = B_0 + B_0'p$, and integrating $B_T = -V(\partial p/\partial V)_T$:

$$V(p) = V_0\left(1 + \frac{B_0'}{B_0}p\right)^{-1/B_0'}, \tag{1.50}$$

which can be inverted to:

$$p(V) = \frac{B_0}{B_0'}\left[\left(\frac{V_0}{V}\right)^{B_0'} - 1\right]. \tag{1.51}$$

The hydrostatic work equation, $\delta W = -pdV$, can then be used to integrate this equation of state to give:

$$E(V) = E_0 + \frac{B_0 V}{B_0'}\left[\frac{(V_0/V)^{B_0'}}{B_0' - 1} + 1\right] - \frac{B_0 V_0}{B_0' - 1}, \tag{1.52}$$

for the volume-dependent energy at zero temperature. In these and the subsequent equations in this section, the subscript "0" represents zero-pressure conditions. That is, V_0 is the (zero-pressure) equilibrium volume, E_0 is the energy at that volume, etc. In the rest of this section, we consider only static energies, although the same equations of state (or rather, their derivatives) can be used to fit experimental pressure-volume isotherms, which contain zero-point and temperature effects.

A collection of equations of state for solids can be obtained from the strain-stress (StSs) family of relations. We describe the energy of a crystal as a Taylor expansion in the strain f:

$$E = \sum_{k=0}^{n} c_k f^k. \tag{1.53}$$

Some of the equations of state arising from different strain definitions follow.

1.3.1 Birch–Murnaghan Family

The Birch–Murnaghan (BM) EOS family is based on the Eulerian strain, which is defined as:

$$f = \frac{1}{2}\left[(V_r/V)^{2/3} - 1\right], \tag{1.54}$$

where V_r, the reference volume, is usually the zero-pressure equilibrium volume of the phase under study (V_0).

The StSs series can be expanded to any order $n > 1$ (Equation (1.53)). The resulting n-degree polynomial EOS has coefficients that can be found by imposing the limiting conditions for the derivatives at the reference point (V_r):

$$\lim_{f\to 0}\left\{V; E; p = -\frac{dE}{dV}; B = -V\frac{dp}{dV}; B' = \frac{dB}{dp}; \dots\right\}$$
$$= \{V_0; E_0; 0; B_0; B_0'; \dots\}. \tag{1.55}$$

For instance, solving these equations for the fourth-order ($n = 4$) Birch–Murnaghan EOS gives:

$$V_r = V_0; \tag{1.56}$$
$$c_0 = E_0; \tag{1.57}$$
$$c_1 = 0; \tag{1.58}$$
$$c_2 = \frac{9}{2}V_0B_0; \tag{1.59}$$
$$c_3 = \frac{9}{2}V_0B_0(B_0' - 4); \tag{1.60}$$
$$c_4 = \frac{3}{8}V_0B_0\{9[B_0B_0'' + (B_0')^2] - 63B_0' + 143\}. \tag{1.61}$$

The polynomial strain EOS can be rearranged to produce the second-order Birch–Murnaghan equations, which are non-linear in the volume, and are typically used in solid-state simulations [32]:

$$E = E_0 + \frac{9}{2}B_0V_0f^2 = E_0 + \frac{9}{8}B_0V_0(x^{-2/3} - 1)^2, \tag{1.62}$$
$$p = 3B_0f(1+2f)^{5/2} = \frac{3}{2}B_0\left(x^{-7/3} - x^{-5/3}\right), \tag{1.63}$$
$$B = B_0(7f+1)(2f+1)^{5/2}, \tag{1.64}$$

with $x = (V/V_0)$.

Higher order BM EOS result in even more cumbersome non-linear expressions that are not always easy to converge. Furthermore, the Eulerian strain is just one of an infinite number of possible definitions [33]. Other options are

Table 1.1 The columns in order are: list of strain families ("Strain"), the associated equation-of-state names ("EOS"), definition of the strain (f), recurrence formula for higher order volume derivative of the strain ($f_{(n+1)V}$), and first volume derivative of the strain (f_{1V}). In these expressions, $x = V/V_r$ and $s = -1/(3V)$.

Strain	EOS	f	$f_{(n+1)V}$	f_{1V}
Eulerian	BM	$\frac{1}{2}\left(x^{-2/3}-1\right)$	$(3n+2)sf_{nV}$	$-\frac{x^{-2/3}}{3V}$
Lagrangian	Thomson	$\frac{1}{2}\left(x^{2/3}-1\right)$	$(3n-2)sf_{nV}$	$-\frac{x^{2/3}}{3V}$
Natural	PT	$\frac{1}{3}\ln x$	$3nsf_{nV}$	$\frac{1}{3V}$
Infinitesimal	Bardeen	$1-x^{-1/3}$	$(3n+1)sf_{nV}$	$\frac{(1-f)^4}{3V_r}$
x^3		x^3	$(3n-1)sf_{nV}$	$\frac{x^{1/3}}{3V}$
V		V	0	1

the infinitesimal [34], Lagrangian [35], and natural [36] strains. The question of fitting polynomial strain EOS to any order has been explored in one of our articles [24].

1.3.2 Polynomial Strain Families

StSs EOS can be arranged to produce complex nonlinear expressions but they are essentially polynomials in the strain for a given strain definition. It is possible to use linear polynomial fitting techniques to find the EOS coefficients. The transformations for the relevant strains are summarized in Table 1.1 [24, 25, 37].

Let us describe the polynomial fitting algorithm to a list of $E(V)$ pairs (such as would be obtained using computational techniques) using the BM family as an example. The first step is transforming the input volume data into the Eulerian strain:

$$f = \frac{1}{2}\left[\left(\frac{V_r}{V}\right)^{2/3} - 1\right], \quad V = V_r(2f+1)^{-3/2}, \tag{1.65}$$

where V_r is a reference volume. Essentially, any finite positive value can be used as V_r if the strain polynomial includes the $c_1 f^1$ linear term. We determine the equilibrium position by minimizing the polynomial with respect to f.

The second step is a linear fit of the $E(f)$ data to the polynomial form, Equation (1.53). To prevent numerical instabilities, it is important to use a robust and efficient numerical method [38] for solving the least squares equations.

Next, the minimum of the $E(f)$ curve must be determined to obtain the equilibrium properties of the crystal phase. In practice, this is a delicate step from the numerical view. Under some circumstances, the polynomial will have no minimum in the range of fitted volumes (for instance high-pressure phases) and this must be adequately detected and taken into account by the algorithm.

The derivatives of the energy with respect to volume are also required. For instance, values up to the third or fourth derivatives are used by thermal models (e.g., the Debye model, see below). The derivatives are calculated using the chain rule, which requires the derivatives of the energy with respect to the strain:

$$E_{mf} \equiv d^m E/df^m = \sum_{k=m}^{n} c_k k(k-1)\dots(k-m+1) f^{k-m} \quad 0 \leq m \leq n, \tag{1.66}$$

and the derivatives of the strain with respect to volume, $f_{mV} = d^m f/dV^m$, which are easily determined by a simple recursive procedure:

$$f_{(m+1)V} = -\frac{3m+2}{3V} f_{mV} \quad m = 1, 2, \dots, \tag{1.67}$$

where

$$f_{1V} = -\frac{1}{3V}\left(\frac{V_r}{V}\right)^{2/3}. \tag{1.68}$$

It must be noted that these last two equations are the only part of the procedure that is modified when the definition of the strain changes.

The derivatives $E_{nV} = d^n E/dV^n$ can now be written in terms of the E_{mf} and f_{mV} components:

$$E_{1V} = E_{1f} f_{1V}, \tag{1.69}$$

$$E_{2V} = f_{1V}^2 E_{2f} + f_{2V} E_{1f}, \tag{1.70}$$

$$E_{3V} = f_{1V}^3 E_{3f} + 3 f_{1V} f_{2V} E_{2f} + f_{3V} E_{1f}, \tag{1.71}$$

$$E_{4V} = f_{1V}^4 E_{4f} + 6 f_{1V}^2 f_{2V} E_{3f} + (4 f_{1V} f_{3V} + 3 f_{3V}^2) E_{2f} + f_{4V} E_{1f}. \tag{1.72}$$

Some of the more important properties of the solid can be obtained immediately in terms of the E_{mV} derivatives:

$$p = -E_{1V}; \tag{1.73}$$

$$B = V E_{2V}; \tag{1.74}$$

$$B' = -\frac{V E_{3V}}{E_{2V}} - 1; \tag{1.75}$$

$$B'' = E_{2V}^{-3}[V(E_{4V} E_{2V} - E_{3V}^2) + E_{3V} E_{2V}]; \dots \tag{1.76}$$

Table 1.1 shows the appropriate equations for arbitrary order Poirier–Tarantola (PT) [36], Thomson [35] and Bardeen [34] StSs families. Different implementations of the algorithm have already been provided [24, 25, 37]. The equations of state and their derivatives are used in the next section to calculate the pressure, bulk moduli and their derivatives and also in thermal models. Examples of the use of these EOS in the context of actual calculations are presented in Chapters 2 and 3.

1.4 LATTICE VIBRATIONS AND THERMAL MODELS

1.4.1 Lattice Vibrations in Harmonic Approximation

The non-equilibrium Gibbs function (Equation (1.41)) involves two terms that need to be determined to calculate the thermodynamic properties of a solid: the electronic energy and the vibrational Helmholtz free energy contribution. The electronic energy for the ground state of the crystal can be obtained by computational means with relative ease. The vibrational free energy, however, requires an approximate model for the atomic vibrations. A defining characteristic of a solid is that unlike liquids or gases, the atoms vibrate around their equilibrium positions in a potential well created by the rest of the atoms.

From the view of calculating thermodynamic properties, all vibrations contribute to the free energy of a crystal. In this sense, it is useful to define a phonon density of states (DOS) analogous to the density of states for the electronic Hamiltonian that gives the number of vibrational states per unit frequency range as a function of vibration frequency. Specifically, the phonon DOS is defined as:

$$G(\omega) = \int_0^\omega \sum_{\boldsymbol{k}\nu} \delta(\omega - \omega_{\boldsymbol{k}\nu}) d\omega, \tag{1.77}$$

$$g(\omega) = \frac{dG}{d\omega}, \tag{1.78}$$

$$\int_0^\infty g(\omega) d\omega = 3n, \tag{1.79}$$

where $\boldsymbol{k}$ runs over vectors in reciprocal space (also called $\boldsymbol{k}$-space, see Chapter 3) and ν is the phonon band index. $G(\omega)$ counts the number of vibrational states with frequencies lower than ω. $g(\omega)$ is the corresponding density function for the vibrational states. For a frequency interval $\Delta\omega = \omega_1 - \omega_0$, the integral of $g(\omega)$ between those frequencies gives the number of vibrational states in the corresponding frequency range. $g(\omega)$ is usually normalized to the number of phonon bands, $3n$, where n is the number of atoms in the primitive cell. Details on how the vibrational states are calculated are given in Chapter 3.

The calculation of the thermodynamic properties of a solid at a given $\boldsymbol{x}$, volume, and temperature starts with the equation for the canonical ensemble partition function (Equation (1.38)). At room temperature, the value of k_BT is 207.2 cm^{-1}, comparable to the typical vibrational energies in almost every solid. In consequence, all vibrational levels affect the thermodynamic properties under the usual temperature conditions. Each of the phonon normal modes determined by diagonalizing the dynamical matrix (see Chapter 3) behaves as an independent harmonic oscillator. The energy levels for these vibrations are quantized and the total vibrational energy $E_{\text{vib},j}$ for a certain quantum-mechanical state of the crystal is:

$$E_{\text{vib},j} = \sum_{\boldsymbol{k}\nu} (n_{\boldsymbol{k}\nu}(j) + 1/2)\hbar\omega_{\boldsymbol{k}\nu}, \tag{1.80}$$

where $n_{\boldsymbol{k}\nu}(j)$ is the vibrational quantum number of the $\boldsymbol{k}\nu$-th normal mode when the crystal is in the quantum-mechanical state j. $n_{\boldsymbol{k}\nu}$ is also referred to as the number of phonons with energy $\hbar\omega_{\boldsymbol{k}\nu}$ or the population of the normal mode $\boldsymbol{k}\nu$. In these equations, $\hbar = h/(2\pi)$ and h is Planck's constant ($h = 6.62606957 \times 10^{-34}$ $\text{m}^2\text{kg/s}$ in SI units, $\hbar = 1$ in atomic units). Different $\{n_{\boldsymbol{k}\nu}\}$ distributions give different crystal states j that enter the canonical partition function Equation (1.40). Phonons behave as bosonic quasiparticles because there is no restriction associated to the occupation of individual energy levels.

Equation (1.80) separates the vibration of the $3nN$ normal modes, where N is the number of primitive unit cells in the crystal in the order of Avogadro's number. Each normal mode is labeled by a different $\boldsymbol{k}\nu$, so the canonical partition function can be separated into a product of individual contributions, one from each normal mode:

$$Z = e^{-\beta E_0} \prod_{\boldsymbol{k}\nu} z_{\boldsymbol{k}\nu}, \tag{1.81}$$

where E_0 is the static zero-pressure energy and we have used the abbreviation $\beta = 1/(k_BT)$. The normal-mode partition functions are:

$$z = \sum_{n=0}^{\infty} e^{-\beta\hbar\omega(n+1/2)} = \frac{e^{-\beta\hbar\omega/2}}{1 - e^{-\beta\hbar\omega}}, \tag{1.82}$$

from which:

$$-\ln Z = \frac{E_0}{k_BT} + \sum_{\boldsymbol{k}\nu} \left[\frac{\hbar\omega_{\boldsymbol{k}\nu}}{2k_BT} + \ln\left(1 - e^{-\beta\hbar\omega_{\boldsymbol{k}\nu}}\right) \right]. \tag{1.83}$$

In consequence, the harmonic Helmholtz free energy is:

$$\begin{aligned} F &= -k_B T \ln Z \\ &= E_0 + \sum_{\boldsymbol{k}\nu} \left[\frac{\hbar\omega_{\boldsymbol{k}\nu}}{2} + k_B T \ln\left(1 - e^{-\hbar\omega_{\boldsymbol{k}\nu}/k_B T}\right) \right], \end{aligned} \tag{1.84}$$

or, in terms of the phonon density of states:

$$F = E_0 + \int_0^\infty \left[\frac{\hbar\omega}{2} + k_B T \ln\left(1 - e^{-\beta\hbar\omega}\right) \right] g(\omega) d\omega. \tag{1.85}$$

If $g(\omega)$ is normalized to $3nN$, this equation corresponds to the extensive quantity for N primitive cells. If, as is usual, $g(\omega)$ is normalized to $3n$, F is the free energy per primitive unit cell. In the following, we shall use values of the extensive quantities per primitive unit cell.

Note that the vibrational part (everything except E_0) of the free energy in Equation (1.84) can be separated in two parts: one that depends on the temperature and a zero-point contribution that is present even at zero kelvin:

$$F = E_0 + F_{\text{vib}} = E_0 + F_{\text{zp}} + F_{\text{T}}, \tag{1.86}$$

$$F_{\text{zp}} = \sum_{\boldsymbol{k}\nu} \frac{\hbar\omega_{\boldsymbol{k}\nu}}{2}, \tag{1.87}$$

$$F_{\text{T}} = \sum_{\boldsymbol{k}\nu} k_B T \ln\left(1 - e^{-\beta\hbar\omega_{\boldsymbol{k}\nu}}\right). \tag{1.88}$$

The harmonic approximation predicts the temperature dependence of the vibrational free energy, but it does not make any predictions regarding the volume variation. For that, we will need a way to make the phonon frequencies volume-dependent: the quasiharmonic approximation (see Section 1.4.3). However, Equation (1.84) allows us to calculate the thermodynamic properties that are related to the temperature derivatives of F, namely, the entropy (Equation (1.15)):

$$S = -\left(\frac{\partial F}{\partial T}\right)_V = -k_B \sum_{\boldsymbol{k}\nu} \left[\ln\left(1 - e^{-\beta\hbar\omega_{\boldsymbol{k}\nu}}\right) - \frac{\hbar\omega_{\boldsymbol{k}\nu}}{k_B T} \frac{1}{e^{\beta\hbar\omega_{\boldsymbol{k}\nu}} - 1} \right]. \tag{1.89}$$

The internal energy (Equation (1.9)) is:

$$U = F + TS = E_0 + \sum_{\boldsymbol{k}\nu} \frac{\hbar\omega_{\boldsymbol{k}\nu}}{2} + \sum_{\boldsymbol{k}\nu} \frac{\hbar\omega_{\boldsymbol{k}\nu}}{e^{\beta\hbar\omega_{\boldsymbol{k}\nu}} - 1}, \tag{1.90}$$

whose vibrational part can be rewritten in a manner analogous to Equation (1.80):

$$U_{\text{vib}} = \sum_{\boldsymbol{k}\nu} U_{\text{vib},\boldsymbol{k}\nu} = \sum_{\boldsymbol{k}\nu} \hbar\omega_{\boldsymbol{k}\nu} \left(\frac{1}{2} + \langle n_{\boldsymbol{k}\nu} \rangle \right), \tag{1.91}$$

but where $\langle n_{\boldsymbol{k}\nu} \rangle$ is the average population of the $\boldsymbol{k}\nu$ mode (that is, the average value of the vibrational quantum number for the $\boldsymbol{k}\nu$ mode) in the Bose–Einstein statistics:

$$\langle n_{\boldsymbol{k}\nu} \rangle = \frac{1}{e^{\hbar\omega_{\boldsymbol{k}\nu}/k_BT} - 1}. \tag{1.92}$$

The constant volume heat capacity (Equation (1.29)) is:

$$C_v = \left(\frac{\partial U}{\partial T}\right)_V = \sum_{\boldsymbol{k}\nu} C_{v,\boldsymbol{k}\nu} = \sum_{\boldsymbol{k}\nu} k_B \left(\frac{\hbar\omega_{\boldsymbol{k}\nu}}{k_BT}\right)^2 \frac{e^{\beta\hbar\omega_{\boldsymbol{k}\nu}}}{\left(e^{\beta\hbar\omega_{\boldsymbol{k}\nu}} - 1\right)^2}. \tag{1.93}$$

Both the internal energy and the constant volume heat capacity can be separated in individual contributions from each of the normal modes ($U_{\text{vib},\boldsymbol{k}\nu}$ and $C_{v,\boldsymbol{k}\nu}$).

1.4.2 Static Approximation

The volume dependence of the Helmholtz free energy Equation (1.84) is directly related to pressure effects on the crystal, but the harmonic approximation by itself cannot give any information about the evolution of E_{sta} and $F^{\star}_{\text{vib}}$ with volume. The primary source of free energy change with pressure is the electronic energy E_{sta} and we examine it in this section. In the simplest approach called the static approximation, one can simply disregard vibrational effects altogether ($F^{\star}_{\text{vib}} = 0$). This is not exactly the same as the behavior of the solid at zero kelvin because of the zero-point effects (Equation (1.87)).

However, the great advantage of the static approximation is that it is computationally much simpler because no phonon calculations are involved. The most popular method for calculating $E_{\text{sta}}(V)$ in a periodic solid is density-functional theory [39–43], particularly in the pseudopotentials/plane-waves approach. These computational methods are discussed in Chapter 3.

In the context of a computational calculation of the thermodynamic properties of a solid, $E_{\text{sta}}(V)$ is obtained on a discrete set of points (a uniform grid, typically) distributed in an interval encompassing the zero-pressure equilibrium volume (the "volume grid"). Alternatively, one can define a grid of pressures around zero and relax the crystal geometry at those target pressures to obtain $E_{\text{sta}}(p)$.

In the volume grid, the smallest volume determines the maximum pressure that can be modeled with the calculated data by the value of $-dE_{\text{sta}}/dV$ at that point. The largest volume corresponds to a negative pressure, and gives the maximum temperature that can be calculated at zero pressure. The pressure at individual points of the volume grid can be obtained by fitting an equation of state to the discrete $E_{\text{sta}}(V)$ data and then calculating the derivative, as discussed in the previous section. Using the same approach, the static bulk modulus as a function of pressure or volume and its pressure derivative (B') can also be obtained.

As a general rule, the relaxation of a crystal structure at a fixed volume gives the energy minimum in the local potential energy surface, which may be a metastable phase and not the (thermodynamic) global minimum for the crystal. If we consider several crystal phases and calculate $E_{\text{sta}}[(\boldsymbol{x}_{\text{opt}}(V), V] = E_{\text{sta}}(V)$ for each, it is possible to calculate the Gibbs energy of each phase in the static approximation, which is simply the enthalpy (c.f. Equations (1.44) to (1.46)):

$$G_{\text{sta}}(p_{\text{sta}}) = H_{\text{sta}}(p_{\text{sta}}) = E_{\text{sta}}(V(p_{\text{sta}})) + p_{\text{sta}}V(p_{\text{sta}}). \tag{1.94}$$

The stable phase at a given pressure in the static approximation is the one with the lowest enthalpy. Phase transitions can also be modeled; the static transition pressure between phases A and B is such that $H^{\text{A}}(p_{\text{tr}}) = H^{\text{B}}(p_{\text{tr}})$, where both H^i and p_{tr} are static quantities. Naturally, determining the correct thermodynamically stable phase requires a good selection of candidates or the use of crystal structure prediction techniques (see Chapter 4).

Figure 1.2 shows an example of a simple DFT calculation in sodium chloride. The $E_{\text{sta}}(V)$ plots show the volume grid we employed (the dots correspond to primitive cell volumes) and the fitted equation of state for two phases. At zero pressure, which corresponds to the bottoms of both curves, the B1 phase (NaCl-like) is more stable than the B2 (CsCl-like). The evolution of the enthalpy with pressure is also shown. According to our results, NaCl transforms from the B1 to the B2 phase at the point at which $\Delta H(p) = H^{\text{B2}} - H^{\text{B1}}$ crosses zero, 22.4 GPa. This transition is observed experimentally, and our prediction for the transition pressure is in fair agreement with the experimental value, 26.8 GPa [46]. The $V(p)$ curve shows that the B2 phase is more compact than the B1 phase, which is a reasonable observation since B2 is stable at high pressure.

The static transition pressure corresponds to a phase change observed at the bottom of the p (abscissa)-T (ordinate) phase diagram, neglecting zero point effects. This is a fair approximation as long as the phases involved in the transition do not have radically different bonding regimes. The Clapeyron slope dp_{tr}/dT (Equation (1.48)) at $T = 0$ K vanishes for ordered periodic solids (because $\Delta S = S^{\text{B2}} - S^{\text{B1}}$ is zero at $T = 0$ K by the third law), then the static transition pressures are also good approximations to phase stability at relatively low temperatures, all the more so in solids with high melting points. For instance, using a quasiharmonic Debye model (see below), the same phase transition at room temperature is predicted to occur at 22.5 GPa, only 0.1 GPa higher than the static value.

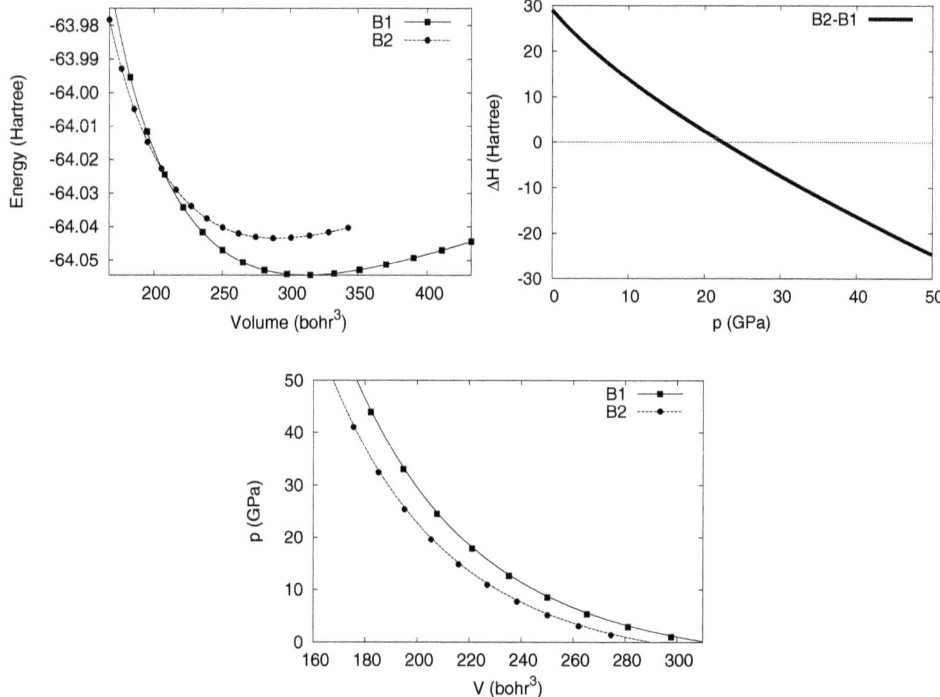

Figure 1.2 Calculated properties for the B1 (NaCl-type) and B2 (CsCl-type) phases of sodium chloride using DFT+PS/PW and the Perdew–Burke–Ernzerhof (PBE) exchange-correlation functional [44] as implemented in Quantum ESPRESSO [45]. The plots are: $E_{\text{sta}}(V)$ (top left), the difference between the B2 and the B1 enthalpies against pressure: $\Delta H(p) = H^{\text{B2}} - H^{\text{B1}}$(top right), and the static $p(V)$ curves (bottom). In the last plot, the continuous line is obtained by using the derivative of an equation of state fitted to the $E_{\text{sta}}(V)$ data and the pressure calculated from the Kohn–Sham wave functions at the grid volumes.

The accuracy of the comparison between calculated static and experimental transition pressures is affected by the quality of the underlying approximations in DFT+PS/PW. For instance, there is a sizable error in the prediction of the equilibrium geometries of many solids that varies depending on the exchange-correlation functional chosen [47–49]. Also, non-covalent interactions are poorly described by common DFT methods, requiring specialized exchange-correlation functionals incorporating dispersion corrections. The study of pressure effects in crystals whose lattice energy is dominated by dispersion interactions (e.g., molecular crystals) is still an unsolved problem [28, 50–53]. In addition, comparing calculated thermodynamic and observed transition pressures is affected by the usual kinetic (hysteresis) effects in the latter [46].

1.4.3 Quasiharmonic Approximation

The quasiharmonic approximation [22, 24, 25, 54, 55] (QHA) is one of the main computational approaches to access the p-T phase diagram and to calculate thermodynamic properties at arbitrary pressures and temperatures within the harmonic approximation. The QHA is based on the calculation of phonon dispersion relations at points other than zero-pressure equilibrium. In this approximation, the phonon frequencies depend on the volume (V) and on the cell shape and atomic coordinates ($\boldsymbol{x}$). The usual approach described in Sections 1.2.3 and 1.4.1 is the statically constrained QHA, in which the phonons only depend on the volume and $\boldsymbol{x}$ assumes the static equilibrium values.

For all the points in the static volume grid described in Section 1.4.2 the phonon dispersion frequencies are calculated as $\omega_{\boldsymbol{k}\nu}(V)$. The volume dependence of the phonons is given by the Grüneisen parameters (also called the "mode gammas"):

$$\gamma_{\boldsymbol{k}\nu} = -\frac{d\ln\omega_{\boldsymbol{k}\nu}}{d\ln V} = -\frac{V}{\omega_{\boldsymbol{k}\nu}}\frac{d\omega_{\boldsymbol{k}\nu}}{dV}. \tag{1.95}$$

The Grüneisen parameters capture the evolution of the crystal elasticity and its vibrational response under pressure. Neither the phonon frequencies nor the mode gammas depend directly on the temperature in QHA, although they depend indirectly through the thermal expansion effect caused by the thermal pressure. The volume dependence of $\omega_{\boldsymbol{k}\nu}$ transforms the vibrational Helmholtz free energy (Equation (1.84)) into:

$$F_{\text{vib}}(V,T) = \sum_{\boldsymbol{k}\nu}\left[\frac{\hbar\omega_{\boldsymbol{k}\nu}(V)}{2} + k_BT\ln\left(1 - e^{-\hbar\omega_{\boldsymbol{k}\nu}(V)/k_BT}\right)\right]. \tag{1.96}$$

The volume derivatives of F can be calculated, and they depend on the

Grüneisen parameters:

$$p(V,T) = -\left(\frac{\partial F}{\partial V}\right)_T$$
$$= -\frac{\partial E_{\text{sta}}}{\partial V} + \sum_{k\nu}\frac{\hbar\omega_{k\nu}\gamma_{k\nu}}{2V} + \sum_{k\nu}\frac{\hbar\omega_{k\nu}\gamma_{k\nu}/V}{e^{\hbar\omega_{k\nu}/k_BT}-1}, \tag{1.97}$$
$$p_{\text{sta}}(V) = -\frac{\partial E_{\text{sta}}}{\partial V}, \tag{1.98}$$
$$p_{\text{th}}(V,T) = \sum_{k\nu}\frac{\hbar\omega_{k\nu}\gamma_{k\nu}}{2V} + \sum_{k\nu}\frac{\hbar\omega_{k\nu}\gamma_{k\nu}/V}{e^{\hbar\omega_{k\nu}/k_BT}-1}$$
$$= \sum_{k\nu}\frac{\gamma_{k\nu}U_{\text{vib},k\nu}}{V} = \sum_{k\nu} p_{\text{th},k\nu}. \tag{1.99}$$

The individual mode contributions to the vibrational internal energy ($U_{\text{vib},k\nu}$) that enter the thermal pressure expression are defined in Equation (1.91). There is a zero-point contribution to the thermal pressure and a contribution that depends on the temperature. Therefore, the thermal expansion caused by vibrations is present even at zero kelvin, an important point to remember when comparing static and experimental (x-ray, neutron diffraction) structures [28]. Even if the experiments have been carried out at very low temperatures, zero-point vibrational effects still cause expansion.

The thermal Grüneisen parameter in Equation (1.34) can be rewritten in terms of mode gammas in the harmonic approximation. γ_{th} is an important quantity because it is used to extract data from shock experiments due to its approximately linear dependence with the pressure derivative of the bulk modulus (dB/dp) [33, 56]. A lot of empirical approximations to the relation between the thermal Grüneisen parameter and the pressure derivative of the bulk modulus have been proposed [33, 56], and will be reviewed in the next subsection.

In QHA, the thermal Grüneisen parameter can be calculated from Equation (1.36), using Equation (1.97) to evaluate $(\partial P/\partial T)_V$. This gives an expression that involves a $C_{\text{v},k\nu}$-weighted average over the individual mode Grüneisen parameters:

$$\gamma_{\text{th}} = \frac{\sum_{k\nu}\gamma_{k\nu}C_{\text{v},k\nu}}{Cv}, \tag{1.100}$$

where each mode gamma satisfies:

$$\gamma_{k\nu} = \frac{p_{\text{th},k\nu}V}{U_{\text{vib},k\nu}}. \tag{1.101}$$

If all the $\gamma_{k\nu}$ had the same value (which is the case for the approximate models in the next subsection), using Equation (1.97) and Equation (1.100) the thermal Grüneisen parameter can be calculated with:

$$\gamma_{\text{th}} = \frac{p_{\text{th}} V}{U_{\text{vib}}} = \gamma_{\boldsymbol{k}\nu} \text{ for all } \boldsymbol{k}\nu. \tag{1.102}$$

This expression is sometimes called the Mie–Grüneisen gamma or the "baric" Grüneisen parameter (γ_{bar}).

The γ_{th} quantity gives access to the thermodynamic properties in the crystal that depend on the volume derivatives of the free energy, namely, the thermal expansion coefficient, the constant pressure heat capacity, and adiabatic bulk modulus:

$$\alpha = \frac{\gamma_{\text{th}} C_v}{V B_T}, \tag{1.103}$$

$$C_p = C_v(1 + \gamma_{\text{th}} \alpha T), \tag{1.104}$$

$$B_S = B_T(1 + \gamma_{\text{th}} \alpha T). \tag{1.105}$$

The computational methods to apply the QHA from first principles data have been implemented in the Gibbs2 program [22, 24, 25]. A QHA calculation requires two sets of initial data provided by DFT calculations at every point in a volume grid around the equilibrium geometry: (i) the static energy obtained by relaxing the atomic coordinates and the cell shape at that volume, and (ii) the phonon density of states. Let us consider the case in which we are interested in the thermodynamic properties at pressure p and temperature T. The Helmholtz free energy at temperature T is generated on the same volume grid as the static energy. Then, $F(V;T)$ is fitted to an equation of state (using the techniques described in Section 1.3) and the equilibrium volume at pressure p and that temperature, $V(p,T)$, is calculated by minimizing $G^\star(V;T,P=p) = E_{\text{sta}}(V) + F^\star(V;T) + PV$. Once $V(p,T)$ is known, the rest of the equilibrium properties are obtained in the manner described above. Likewise, Gibbs2 can calculate the Gibbs free energy:

$$G(p,T) = F(V(p,T),T) + pV(p,T) \tag{1.106}$$

of more than one phase (each phase being in internal equilibrium) and compare them under the same pressure and temperature conditions to obtain the p-T phase diagram and the relevant thermodynamic transition pressures.

Figures 1.3, 1.4, 1.5, and 1.6 show token QHA calculations of several thermodynamic properties in MgO, diamond, and aluminum. These calculations were carried out using DFT+PS/PW with the PBE exchange-correlation functional [44] and empirical energy corrections to rectify the error in the static equilibrium volume [47]. The pressure- and temperature-dependent equation of state $V(p,T)$ in Figure 1.3 shows extraordinary agreement with the experimental values, even at temperatures close to the melting point (for MgO, 3124 K; Al, 933 K, near 4000 K for diamond). The same good agreement extends to B_S in Figure 1.4 (the data by Zouboulis show very high experimental noise and are not reliable), and to C_p, α, and C_v (Figures 1.5 and 1.6).

The limitations of QHA are: (i) the statically constrained approximation—thermal effects may not be captured by a simple thermal pressure p_{th} that induces a change in volume over the static curve, and (ii) anharmonic effects at high temperature (close to the melting point) may cause deviations from the experimental thermodynamic trends (for instance, see the calculated thermal expansion coefficient for MgO at high temperature in Figure 1.5).

1.4.4 Approximate Thermal Models

Despite its success, the application of the full quasiharmonic approximation as presented in the previous discussion is impractical for moderately large crystals because of the cost involved in the phonon frequency calculations. Under these circumstances, a possible way to estimate the temperature effects on the thermodynamic properties of a solid is to use an approximate thermal model to simplify the application of the quasiharmonic approximation. An approximate thermal model consists of two parts: (i) an approximation to the phonon DOS derived from data that is simpler to calculate than the full phonon spectrum itself, and (ii) an estimate of how that phonon DOS changes with volume.

The Debye model [19, 20, 88] is a thermal model that approximates the phonon DOS using the elastic moduli of the solid. In the Debye model, the solid behaves like a continuum with no atomic structure. This is the same as assuming that all the bands behave as acoustic bands, which, at low temperature, contribute the most to the heat capacity and the other thermodynamic properties. The acoustic wavelengths are long compared to the interatomic separations. Despite the drastic nature of this approximation, the Debye model offers a reasonable quantitative estimate for the thermodynamic properties of a solid, particularly if it is relatively compact. In addition, the model correctly describes the low temperature ($C_v \to 0$ with a T^3 dependence) and high temperature limits ($3k_B$ per atom in the harmonic approximation) of the constant volume heat capacity.

The vibrational states for a continuous solid are composed of standing (acoustic) phonon waves. It is possible to show that the phonon density of states is a simple parabola [19, 20, 55]:

$$g_{\text{Debye}}(\omega) = \frac{3\omega^2 V}{2\pi^2 v^3}, \tag{1.107}$$

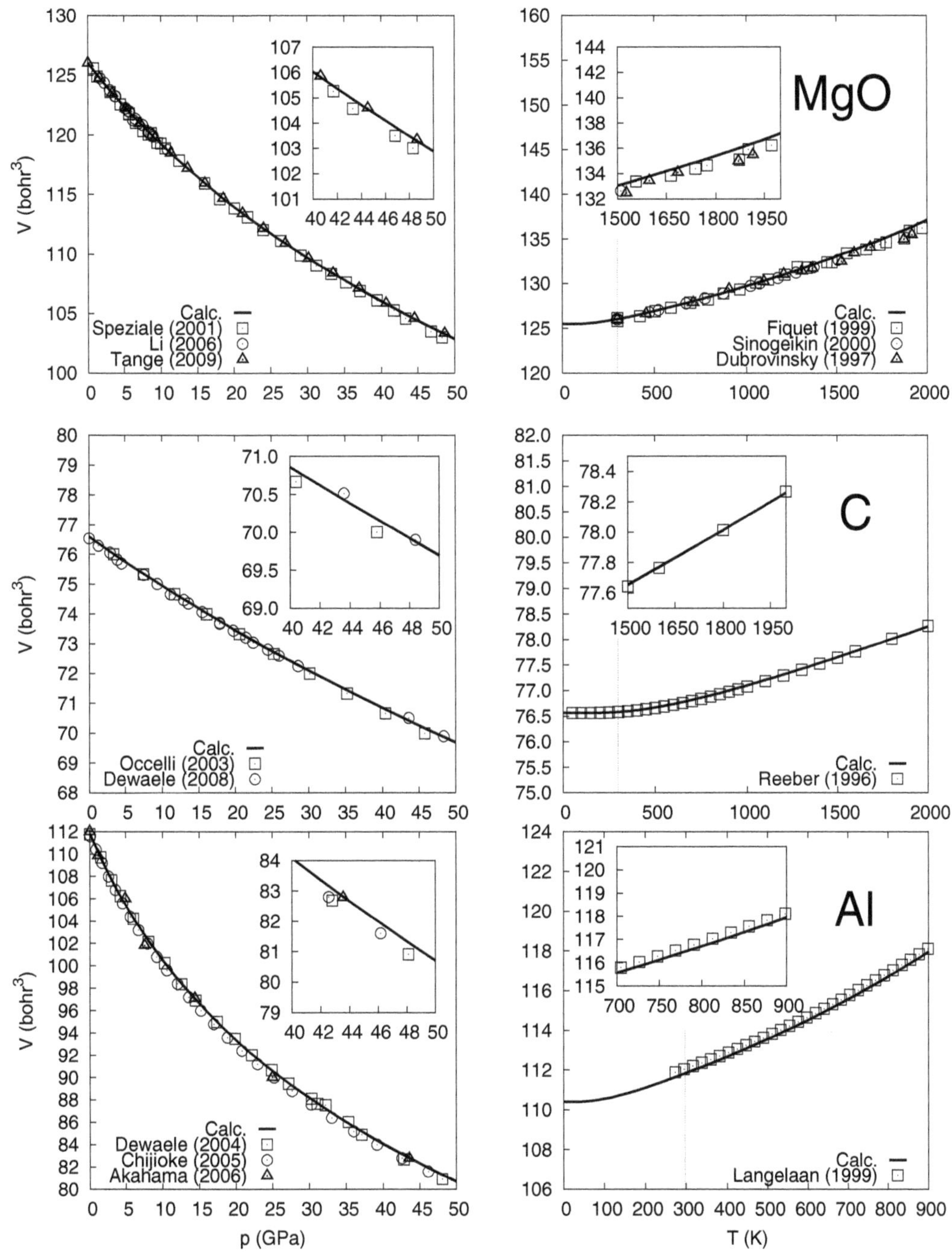

Figure 1.3 Zero-pressure equation of state for MgO, diamond (C), and aluminum. Adapted, with permission, from Reference 47. The panes on the left show the room temperature isotherm. The plots on the right represent $V(T)$. Experimental data from the literature is also given (Speziale [57], Li [58], Tange [59], Fiquet [60], Sinogeikin [61], Dubrovinsky [62], Occelli [63], Dewaele (2004) [64], Dewaele (2008) [65], Reeber [66], Chijioke [67], Akahama [68], and Langelaan [69]). The high-temperature high-pressure regions of the graph are shown in the insets. Room temperature is represented by the verticals line in the $V(T)$ plots.

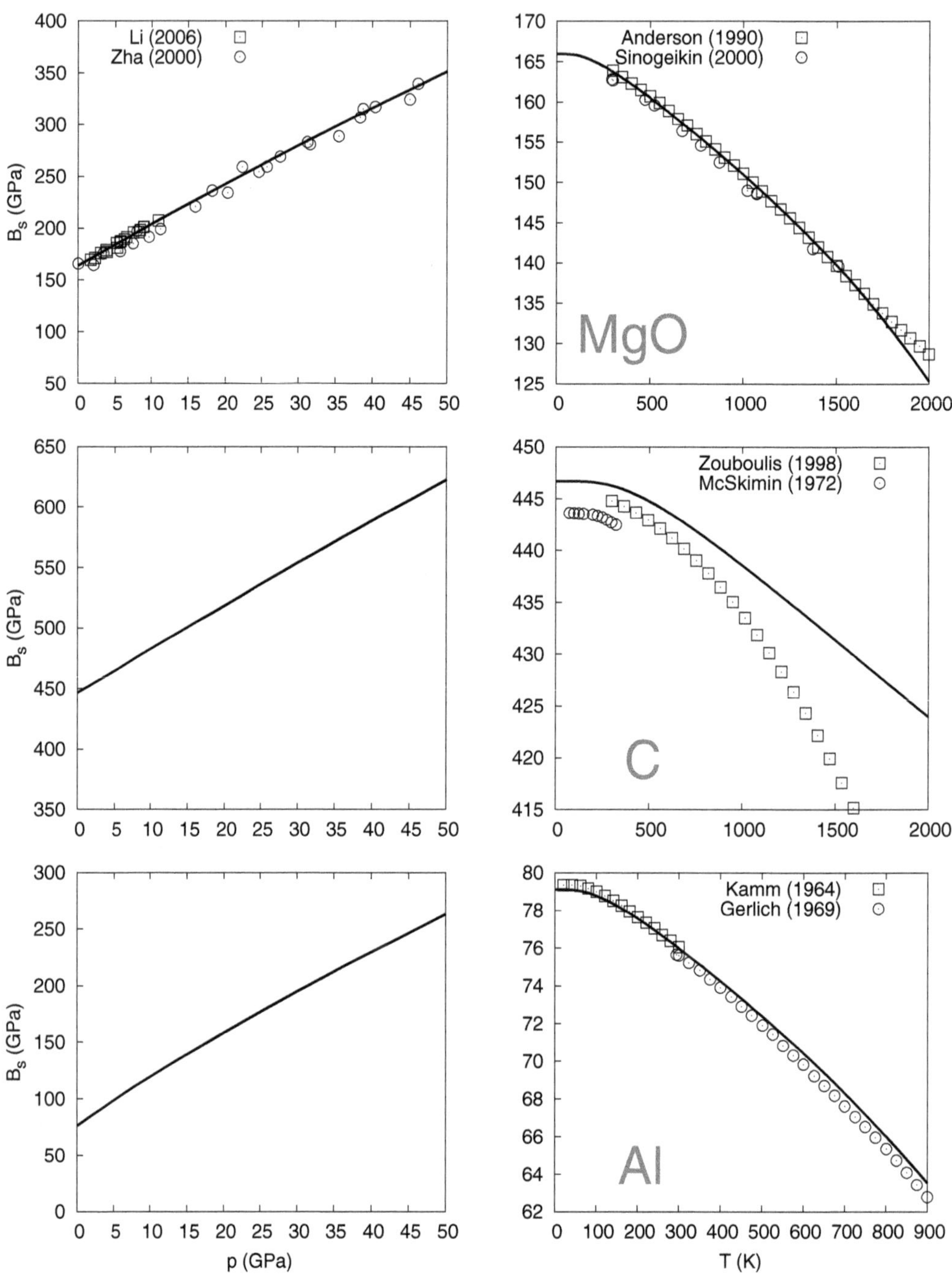

Figure 1.4 Calculated adiabatic bulk modulus (B_S) against pressure at room temperature (left) and against temperature at zero pressure (right) for MgO, diamond, and Al. Adapted, with permission, from Reference 47. The points correspond to experimental results (Li [58], Zha [70], Anderson [71], Sinogeikin [61], Zouboulis [72], McSkimin [73], Kamm [74], and Gerlich [75]).

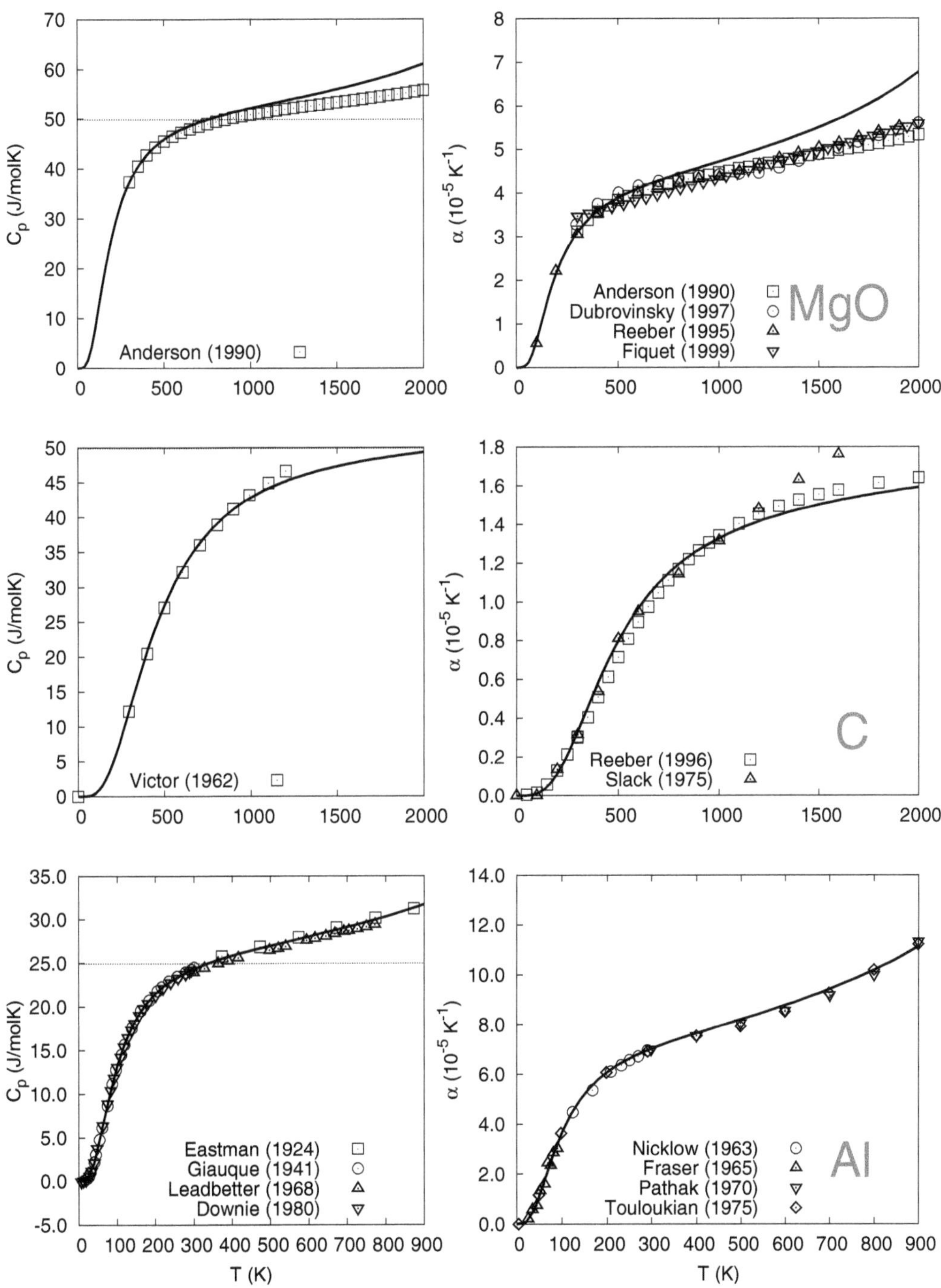

Figure 1.5 Calculated constant pressure heat capacity (C_p, left) and thermal expansion coefficient (α, right) of MgO, diamond, and Al against temperature at zero pressure. Adapted, with permission, from Reference 47. The points correspond to experimental results (Anderson [71], Dubrovinsky [62], Reeber (1995) [76], Fiquet [60], Reeber (1996) [66], Slack [77], Victor [78], Eastman [79], Giauque [80], Fraser [81], Leadbetter [82], Downie [83], Pathak [84], Nicklow [85], and Touloukian [86]).

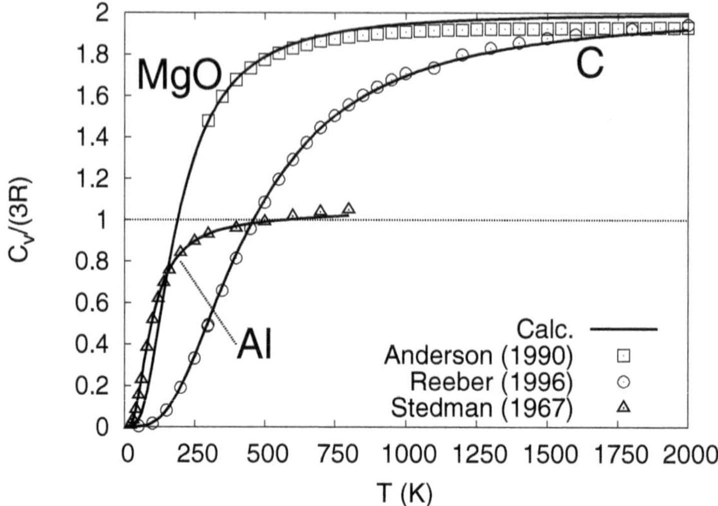

Figure 1.6 Calculated constant volume heat capacity (C_v) of MgO, diamond, and Al against temperature at zero pressure. Adapted, with permission, from Reference 47. The points correspond to experimental results (Anderson [71], Reeber [66], and Stedman [87]).

with v the average sound velocity and V the volume of the crystal (or the volume of the primitive unit cell). In a periodic solid, the velocity averaged over all possible directions is used instead:

$$\frac{3}{\bar{v}^3} = \frac{1}{4\pi} \int d\Omega \left(\frac{2}{v_t^3(\Omega)} + \frac{1}{v_l^3(\Omega)} \right), \tag{1.108}$$

where v_l and v_t are the direction-dependent sound velocities corresponding to longitudinal and transverse waves. If the volume in Equation (1.107) corresponds to the primitive unit cell volume, then the Debye phonon DOS needs to be normalized to $3n$. The normalization of $g_{\text{Debye}}(\omega)$ gives:

$$g_{\text{Debye}}(\omega) = \begin{cases} \frac{9n\omega^2}{\omega_{\text{D}}^3} & \text{if } \omega < \omega_{\text{D}} \\ 0 & \text{if } \omega \geq \omega_{\text{D}} \end{cases}, \tag{1.109}$$

with ω_{D} a parameter in the model called the Debye frequency, which is related to the Debye temperature by:

$$\Theta_{\text{D}} = \frac{\hbar\omega_{\text{D}}}{k_B} = \frac{\hbar}{k_B} \left(\frac{6\pi^2 n}{V} \right)^{1/3} \bar{v}. \tag{1.110}$$

The Debye temperature can be calculated in several ways. The exact value of the Debye temperature would be obtained from the average of the transversal and longitudinal sound velocities (Equation (1.108)). However, this requires the calculation of the elastic constants at every point in the volume grid, which, once more, involves a high computational cost.

An approximation to $\Theta_D(V)$, originally proposed by Slater [22, 24, 25, 89, 90], can be made by assuming that the solid is isotropic. In an isotropic solid, the elastic response is determined by any two of its elastic moduli: the bulk modulus (B), Young's modulus (E), the shear modulus (G), and Poisson's ratio (σ). This is convenient because the adiabatic bulk modulus can be approximated from the $E_{sta}(V)$ curve alone. Poisson's ratio is a parameter in this model and to simplify even more, it can be assumed to be the same as in a Cauchy solid ($\sigma = 0.25$). Common values for Poisson's ratio do not, in general, deviate too much from the Cauchy values: MgO (0.18), steel (0.28), aluminum (0.33), diamond (0.2). Under the isotropic approximation (and approximating B_S by B_{sta}), the Debye temperature can be rewritten as:

$$\Theta_D = \frac{\hbar}{k_B}\left(6\pi^2 V^{1/2} n\right)^{1/3} f(\sigma)\sqrt{\frac{B_{sta}}{M}}, \tag{1.111}$$

where M is the molecular mass per primitive unit cell, B_{sta} is the static bulk modulus, and $f(\sigma)$ is a function of the Poisson ratio:

$$f(\sigma) = \left\{3\left[2\left(\frac{2}{3}\frac{(1+\sigma)}{(1-2\sigma)}\right)^{3/2} + \left(\frac{1}{3}\frac{(1+\sigma)}{(1-\sigma)}\right)^{3/2}\right]^{-1}\right\}^{1/3}. \tag{1.112}$$

The Poisson's ratio can also be derived from any of the other elastic moduli that may be experimentally available and the calculated bulk modulus. Figure 1.7 shows the phonon density of states in the Debye approximation compared to the DOS calculated using the density functional perturbation theory (DFPT) for MgO, diamond, and Al (see Chapter 3 for a brief description of DFPT). The approximation is crude, but the essential behavior of the DOS is captured, and the agreement is good considering the Debye DOS was obtained from the static bulk modulus alone.

The second piece necessary for the implementation of the Debye model is the volume evolution of the phonon density of states (Equation (1.107)). The simplest approach is to use the natural volume dependence of the bulk modulus in Equation (1.111) and to assume that Poisson's ratio does not change with pressure. This results in a volume-dependent Debye temperature $\Theta_D(V)$ that can be quantified by the Debye–Grüneisen ratio:

$$\gamma_D = -\frac{\partial \ln \Theta_D}{\partial \ln V}. \tag{1.113}$$

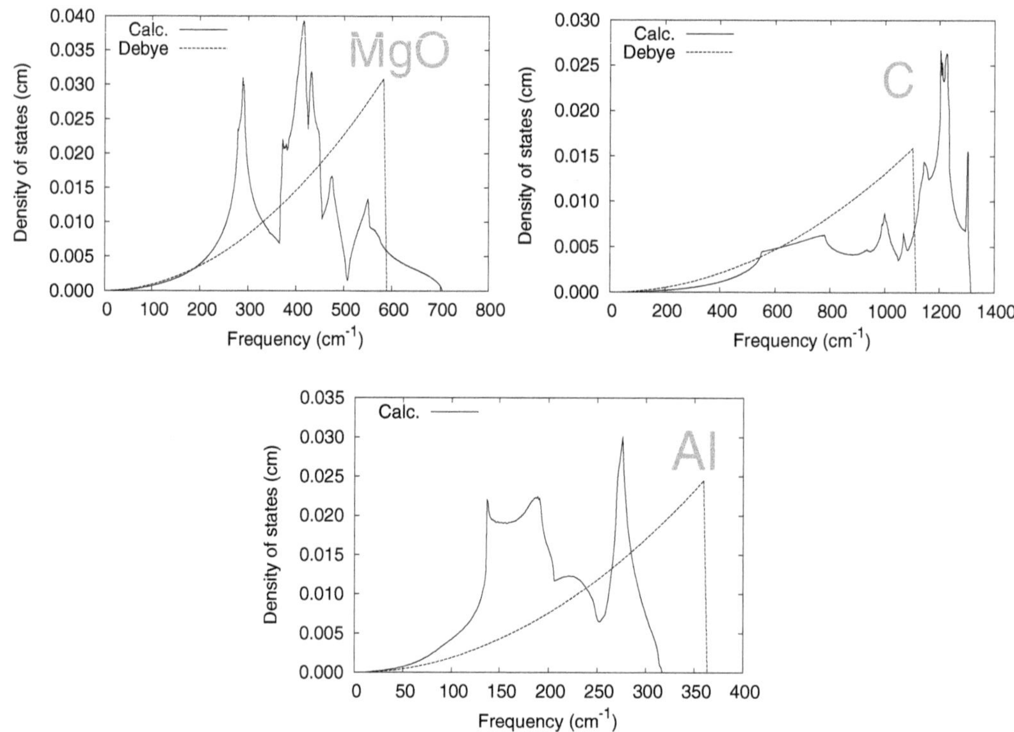

Figure 1.7 Calculated phonon density of states at the zero static pressure equilibrium volume for MgO, diamond (C), and Al, compared to the approximate Debye phonon density of states. Adapted, with permission, from Reference 25.

Inserting the phonon DOS for the Debye model in the quasiharmonic formulas of the previous subsection, it is possible to obtain all the relevant thermodynamic quantities (per primitive unit cell in the case of extensive quantities):

$$F = E_{\text{sta}} + \frac{9}{8}nk_B\Theta_{\text{D}} + 3nk_BT\ln\left(1 - e^{-\Theta_{\text{D}}/T}\right) - nk_BTD(\Theta_{\text{D}}/T), \quad (1.114)$$

$$S = -3nk_B\ln\left(1 - e^{-\Theta_{\text{D}}/T}\right) + 4nk_BD(\Theta_{\text{D}}/T), \quad (1.115)$$

$$U = E_{\text{sta}} + \frac{9}{8}nk_B\Theta_{\text{D}} + 3nk_BTD(\Theta_{\text{D}}/T), \quad (1.116)$$

$$C_v = 12nk_BD(\Theta_{\text{D}}/T) - \frac{9nk_B\Theta_{\text{D}}/T}{e^{\Theta_{\text{D}}/T} - 1}, \quad (1.117)$$

$$p_{\text{th}} = \frac{\gamma_{\text{D}}}{V}\left[\frac{9}{8}nk_B\Theta_{\text{D}} + 3nk_BTD(\Theta_{\text{D}}/T)\right] = \frac{\gamma_{\text{D}}U_{\text{vib}}}{V}, \quad (1.118)$$

in which D is the Debye integral:

$$D(x) = \frac{3}{x^3}\int_0^x \frac{y^3e^{-y}}{1 - e^{-y}}dy. \quad (1.119)$$

In the Debye model, the thermal Grüneisen parameter (Equation (1.34)) is the same as the baric Grüneisen parameter (Equation (1.102)) and equal to γ_{D}. All mode gammas in the Debye model are also equal to γ_{D}. The volume derivative of the Debye temperature in Equation (1.111) leads to:

$$\gamma_{\text{D}} = -\frac{1}{6} + \frac{1}{2}\frac{dB_{\text{sta}}}{dp}. \quad (1.120)$$

This expression is called the Slater gamma [56, 91].

The evolution of the vibrational properties embodied in the Slater gamma is affected by the assumption that the Poisson's ratio in the crystal does not change with pressure, a shortcoming that was already pointed out by Slater [89]. A generalization of the Debye–Slater model can be proposed by generalizing Equation (1.120):

$$\gamma = a + b\frac{dB_{\text{sta}}}{dp} = a - b\frac{d\ln B_{\text{sta}}}{d\ln V}, \quad (1.121)$$

resulting in what is called the Debye–Grüneisen model [92]. In this approach, the Debye temperature at the static equilibrium volume is obtained in the usual way (Equation (1.111)), but its volume evolution is given by:

$$\Theta_{\text{D}}(V) = \Theta_{\text{D}}(V_0)\frac{(B_{\text{sta}}(V)/B_0)^b}{(V/V_0)^a}, \quad (1.122)$$

where V_0 and B_0 are the volume and bulk modulus at zero static pressure equilibrium and a and b are the parameters in Equation (1.121). Several sets of values for the a and b parameters have been proposed in the literature, resulting in a collection of approximate models (see References 56 and 91 and citations therein): Dugdale–McDonald ($a = -1/2, b = 1/2$), Vaschenko–Zubarev ($a = -5/6, b = 1/2$), and the mean free volume gamma model ($a = -0.95, b = 1/2$). Each of these models introduces a different dependence of the Poisson ratio with volume.

For solids that are not compact, the Debye model has clear shortcomings. For instance, in molecular solids, intramolecular vibrations play a major role in determining the vibrational behavior of a solid [28]. These intramolecular vibrations have a definite frequency that is barely affected when a molecular crystal is formed, and are almost independent of the wavevector $\boldsymbol{k}$. In consequence, they appear as thin but very high peaks in the high frequency region of the phonon DOS. Naturally, representing these features of the phonon DOS with a parabola is not ideal. A more accurate model for these systems is to use the Debye model only for the acoustic branches or for the low-frequency intermolecular vibrations. The rest of the branches are represented by a single Dirac delta function located at the frequency of the intramolecular vibration. The treatment of the optical modes is similar to what is done in the Einstein model [19, 20] and hence this approximation is called the Debye–Einstein model [25, 93, 94].

The Debye model and variants, including Debye–Einstein, are also implemented in the Gibbs2 program [22, 24, 25] and can be used to estimate the thermodynamic properties of a periodic solid with a limited amount of information in cases in which computational cost for the full QHA is prohibitive. Thermal models are also widely used to extrapolate experimental data to ranges of pressure and temperature outside the experimentally available particularly in the mineralogy community [55, 93, 95].

1.4.5 Anharmonicity and Other Approaches

The harmonic approximation relies on the assumption that atomic vibration lengths are small compared to interatomic spacings and that the atoms sit at the bottom of a harmonic potential. The harmonic approximation alone does not explain many observed experimental phenomena, and in that case we need to resort to anharmonic corrections [19, 20, 55]. For instance, the harmonic approximation alone fails to predict any thermal expansion. In a way, the quasiharmonic approximation is a simple way to introduce anharmonicity and overcome these limitations.

QHA, however, fails at high temperature, particularly close to the melting point of the solid, and it does not provide a way to estimate transport properties, such as thermal conduction. Evidence of the failures of QHA can be seen in Figures 1.5 and 1.6. The constant volume heat capacity in Figure 1.6 tends to $3k_B$ per atom in the harmonic and quasiharmonic approximations, but in reality most crystals go above this value if the temperature is close enough to the melting point. In addition, the high-temperature behavior of the thermal expansion, which should be linear, is misrepresented by QHA in Figure 1.5.

The simplest way to introduce anharmonicity on top of QHA is to consider cubic and quartic terms as perturbations on the harmonic Hamiltonian. In the harmonic approximation, phonons behave independently of each other. Anharmonic terms couple the motion of different phonons and are described as phonon-phonon interactions. These interactions are subject to a number of conservation laws: momentum, energy, and symmetry. They can be used to estimate, for instance, the thermal conductivity of a crystal, the evolution of the harmonic phonon frequencies with temperature (in addition to and with an opposite effect to the thermal expansion effects caused by temperature, already modeled by QHA), and the width in the vibrational absorption peaks. A full treatment of anharmonic effects is given elsewhere [19, 55].

An alternative to using QHA plus anharmonic corrections, particularly when we are interested in pressures and temperatures close to a phase transition, is molecular dynamics (MD) simulations [55, 96–98]. Other methods such as path-integral Monte Carlo can also be used, and are described elsewhere [99, 100]. In MD, an approximate interatomic potential (also called a force field) is proposed that reproduces the potential energy surface that would be obtained quantum mechanically by solving the electronic Hamiltonian for the given system. Force fields model the relevant physical interactions using ball-and-spring terms. For instance, a covalent bond may be represented by a harmonic potential between the two bonded atoms with a minimum at the correct bond length. More complex MD force fields may use coarse graining, in which groups of atoms (whole functional groups, monomers, etc.) are represented as a unit.

MD simulations solve the classical Newton's equations of motion using the force field as the potential energy component. In this way, one can simulate the motion of atoms in a system quantitatively. Thermodynamic properties are calculated as averages over time of the corresponding phase function. These functions depend on the position of the phase point in the (position, momentum) phase space. Likewise, one can determine phonon frequencies and other dynamical properties using autocorrelation functions.

MD has the advantage over QHA that temperature and pressure are introduced in a very natural way and phase transitions can be observed explicitly. One limitation of MD is that it is usually complex and computationally expensive, which limits the system size and time length that may be simulated. Another limitation is that the results will be as reliable as the underlying force field. For instance, accurately representing bond breaking and formation or even atomic polarizabilities is still a problem under study. The application of Car–Parrinello molecular dynamics [98] in which a combination of DFT and MD is used can alleviate the latter.

1.5 CONCLUSIONS

In this chapter, we have given a brief introduction to the fundamentals and applications of high-pressure science. Basic thermodynamic concepts that will be used in the rest of the book have been presented. The basic thermodynamic quantity in studies at high pressure and temperature is the Gibbs function of a crystal phase at thermal and mechanical equilibrium with the environment, $G(p, T)$, and its non-equilibrium (thermal and mechanical) variant $G^{\star}(V, \boldsymbol{x}; P_E, T_E)$, also called availability. For a given environment pressure and temperature, the principle of minimum energy applies: $G^{\star}$ is a minimum at thermal and mechanical equilibrium (or, equivalently, at internal thermodynamic equilibrium). This allows us to obtain the thermal pressure for a solid through which thermal expansion is modeled and also the pressure-temperature phase diagram for the solid.

A basic experimental observable is the volume of a solid as a function of pressure and temperature, $V(p, T)$, called the equation of state. We have seen in this chapter a number of analytical expressions for the equation of state and their relation to relevant elastic moduli such as the bulk modulus or its pressure derivatives. The equation of state can also be derived from calculated $E_{\text{sta}}(V)$ in what is called the static approximation.

In order to correctly represent thermal effects, it is necessary to treat crystal vibrations, and the simplest model to do so is the harmonic approximation. The basic assumption is that the atoms are constrained to move around (but not too far from) their equilibrium position by a harmonic potential. The harmonic approximation leads directly to the definition of a phonon: a quasiparticle representing a wave of collective atomic translations that carries energy and momentum across a crystal. Phonon frequencies and normal modes can be calculated from first principles with a reasonable computational cost.

The simple calculation of thermodynamic properties requires the use of the quasiharmonic approximation (QHA). In QHA, phonons are made to depend on the crystal volume, and temperature effects are modeled via a thermal pressure that represents the expansion effect caused by the crystal vibrations. QHA is feasible computationally and gives quantitatively good results for all temperatures except those close to the melting point. In larger systems, where even QHA is computationally unfeasible, one can resort to approximate thermal models (Debye, Debye–Einstein, etc.) that circumvent the necessity of calculating the full phonon spectrum at every point of a volume grid. Anharmonic effects may be added *a posteriori* as a correction on top of the harmonic approximation—the phonon-phonon interactions that allow, for instance, the calculation of thermal conductivity or the band widths in the infrared spectrum—or explicitly modeling a system via molecular dynamics simulations.

Bibliography

[1] J. M. Menéndez, F. Aguado, R. Valiente, *et al.*, eds. *Materia a alta presión. Fundamentos y aplicaciones.* Universidad de Oviedo, Servicio de Publicaciones; Universidad de Cantabria, Servicio de Publicaciones, Oviedo (2011).

[2] Bridgman's Nobel prize lecture can be downloaded from `http://www.nobelprize.org/nobel_prizes/physics/laureates/1946/bridgman-lecture.pdf`.

[3] J. W. Forbes. *Shock wave compression of condensed matter: a primer.* Springer, Berlin (2012).

[4] A. Jayaraman. *Rev. Mod. Phys.*, 55, 65 (1983).

[5] H. Mao and R. Hemley. In R. Hemley, ed., *Ultrahigh-pressure mineralogy: physics and chemistry of the Earth's deep interior*, Volume 37 of Reviews in Mineralogy and Geochemistry, 1–32. Mineralogical Society of America, Chantilly, VA (2010).

[6] M. J. Gillan, D. Alfè, J. Brodholt, *et al.* *Rep. Prog. Phys.*, 69, 2365 (2006).

[7] R. M. Wentzcovitch, Y. G. Yu, and Z. Wu. In *Theoretical and computational methods in mineral physics: geophysical applications*, Volume 71 of Reviews in Mineralogy and Geochemistry, 59–98. Mineralogical Soc. Amer., Chantilly, VA, USA, Mineralogical Society of America, Chantilly, VA (2010).

[8] R. M. Wentzcovitch, Z. Wu, and P. Carrier. In *Theoretical and computational methods in mineral physics: geophysical applications*, Volume 71 of Reviews in Mineralogy and Geochemistry, 99–128. Mineralogical Society of America, Chantilly, VA (2010).

[9] A. Delgado, C. Rauh, W. Kowalczyk, *et al.* *Trends Food Sci. Tech.*, 19, 329 (2008).

[10] A. Hereu, P. Dalgaard, M. Garriga, *et al. Innov. Food Sci. Emerg.*, 16, 305 (2012).

[11] M. Zimmermann, D. W. Schaffner, and G. M. Aragão. *LWT-Food Sci. Technol.*, 53, 107 (2013).

[12] B. Guignon, I. Rey-Santos, and P. D. Sanz. *Food Res. Int.*, 64, 336 (2014).

[13] H. B. Callen. *Thermodynamics and an introduction to thermostatistics.* Wiley, West Sussex, UK (1985).

[14] D. Kondepudi. *Introduction to modern thermodynamics.* Wiley, West Sussex, UK (2008).

[15] I. Levine. *Physical chemistry.* McGraw-Hill, New York (2008).

[16] D. C. Wallace. *Thermodynamics of crystals.* Courier Corporation, New York (1998).

[17] D. A. McQuarrie. *Statistical thermodynamics.* Harper & Row, New York (1973).

[18] M. Mau and J. W. McIver. *J. Chem. Educ.*, 63, 880 (1986).

[19] N. W. Ashcroft and N. D. Mermin. *Solid state physics.* Thomson Learning, Stamford (1976).

[20] C. Kittel. *Introduction to solid state physics.* Wiley, New York, 8th edition (1996).

[21] T. Mary, J. Evans, T. Vogt, *et al. Science*, 272, 90 (1996).

[22] M. A. Blanco, E. Francisco, and V. Luaña. *Comput. Phys. Commun.*, 158, 57 (2004).

[23] J. R. Waldram. *The theory of thermodynamics.* Cambridge University Press, Cambridge, UK (1985).

[24] A. Otero-de-la Roza and V. Luaña. *Comput. Phys. Commun.*, 182, 1708 (2011).

[25] A. Otero-de-la Roza, D. Abbasi-Pérez, and V. Luaña. *Comput. Phys. Commun.*, 182, 2232 (2011).

[26] P. Carrier, R. M. Wentzcovitch, and J. Tsuchiya. *Phys. Rev. B*, 76, 064116 (2007).

[27] P. Carrier, R. Wentzcovitch, and J. Tsuchiya. *Phys. Rev. B*, 76, 189901 (2007).

[28] A. Otero-de-la Roza and E. R. Johnson. *J. Chem. Phys.*, 137, 054103 (2012).

[29] P. Carrier, J. F. Justo, and R. M. Wentzcovitch. *Phys. Rev. B*, 78, 144302 (2008).

[30] J. C. Tolédano, ed. *Geometry and thermodynamics. Common problems of quasi crystals, liquid crystals, and incommensurate systems.* Springer, Berlin (1990).

[31] F. D. Murnaghan. *Proc. Natl. Acad. Sci. USA*, 30, 244 (1944).

[32] K. Lejaeghere, V. Van Speybroeck, G. Van Oost, *et al. Crit. Rev. Solid State Mater. Sci.*, 39, 1 (2014).

[33] O. L. Anderson. *Equations of state for solids in geophysics and ceramic science.* Oxford University Press, Oxford, UK (1995).

[34] J. Bardeen. *J. Chem. Phys.*, 6, 372 (1938).

[35] L. Thomson. *J. Phys. Chem. Solids*, 31, 2003 (1970).

[36] J.-P. Poirier and A. Tarantola. *Phys. Earth Planet. Int.*, 109, 1 (1998).

[37] A. Otero-de-la Roza and V. Luaña. *Comput. Phys. Commun.*, 182, 1708 (2011).

[38] L. N. Trefethen and D. Bau, III. *Numerical linear algebra.* SIAM, Philadelphia, PA (1997).

[39] R. G. Parr and W. Yang. *Density functional theory of atoms and molecules.* Oxford University Press, New York (1989).

[40] R. M. Dreizler and E. K. V. Gross. *Density functional theory.* Springer, Berlin (1990).

[41] W. Koch and M. C. Holthausen. *A chemist's guide to density functional theory.* Wiley-VCH, Weinheim (2001).

[42] S. Cottenier. *Density functional theory and the family of (L)APW-methods: a step-by-step introduction.* `http://www.wien2k.at/reguser/textbooks`, Instituut voor Kern- en Stralingsfysica, Katholieke Universiteit Leuven, Belgium (2002).

[43] C. Fiolhais, F. Nogueira, and M. A. L. Marques. *A primer in density functional theory.* Springer, Berlin (2003).

[44] J. Perdew, K. Burke, and M. Ernzerhof. *Phys. Rev. Lett.*, 77, 3865 (1996).

[45] P. Giannozzi, S. Baroni, N. Bonini, *et al.* *J. Phys.: Condens. Matter*, 21, 395502 (2009).

[46] X. Li and R. Jeanloz. *Phys. Rev. B*, 36, 474 (1987).

[47] A. Otero-de-la Roza and V. Luaña. *Phys. Rev. B*, 84, 184103 (2011).

[48] A. Otero-de-la Roza and V. Luaña. *Phys. Rev. B*, 84, 024109 (2011).

[49] J. Perdew, A. Ruzsinszky, G. Csonka, *et al.* *Phys. Rev. Lett.*, 100, 136406 (2008).

[50] G. A. DiLabio and A. Otero-de-la Roza. In K. B. Lipkowitz, ed., *Reviews in computational chemistry.* Wiley-VCH, Hoboken, NJ (2014).

[51] A. Otero-de-la Roza and E. R. Johnson. *J. Chem. Phys.*, 136, 174109 (2012).

[52] A. Otero-de-la Roza, B. H. Cao, I. K. Price, *et al.* *Angew. Chem. Int. Ed.*, 53, 7879 (2014).

[53] A. Otero-de-la Roza, V. Luaña, E. R. T. Tiekink, *et al.* *J. Chem. Theory Comput.*, 10, 5010 (2014).

[54] M. Born and K. Huang. *Dynamical theory of crystal lattices.* Oxford University Press, Oxford, UK (1988).

[55] M. T. Dove. *Introduction to lattice dynamics.* Cambridge University Press, Cambridge, UK (1993).

[56] J.-P. Poirier. *Introduction to the physics of the Earth's interior.* Cambridge University Press, Cambridge, UK, 2nd edition (2000).

[57] S. Speziale, C. S. Zha, T. S. Duffy, *et al.* *J. Geophys. Res.*, 106, 515 (2001).

[58] B. Li, K. Woody, and J. Kung. *J. Geophys. Res.*, 111, B11206 (2006).

[59] Y. Tange, Y. Nishihara, and T. Tsuchiya. *J. Geophys. Res.*, 114, B03208 (2009).

[60] G. Fiquet, D. Andrault, J. P. Itié, *et al.* *Phys. Earth Planet. Int.*, 95, 1 (1996).

[61] S. V. Sinogeikin, J. M. Jackson, B. O'Neill, *et al.* *Rev. Sci. Instrum.*, 71, 201 (2000).

[62] L. S. Dubrovinsky and S. K. Saxena. *Phys. Chem. Miner.*, 24, 547 (1997).

[63] F. Occelli, P. Loubeyre, and R. LeToullec. *Nature Mater.*, 2, 151 (2003).

[64] A. Dewaele, P. Loubeyre, and M. Mezouar. *Phys. Rev. B*, 70, 094112 (2004).

[65] A. Dewaele, F. Datchi, P. Loubeyre, *et al.* *Phys. Rev. B*, 77, 094106 (2008).

[66] R. Reeber and K. Wang. *J. Electron. Mater.*, 25, 63 (1996).

[67] A. D. Chijioke, W. J. Nellis, and I. F. Silvera. *J. Appl. Phys.*, 98, 073526 (2005).

[68] Y. Akahama, M. Nishimura, K. Kinoshita, *et al.* *Phys. Rev. Lett.*, 96, 045505 (2006).

[69] G. Langelaan and S. Saimoto. *Rev. Sci. Instrum.*, 70, 3413 (1999).

[70] C. S. Zha, H. K. Mao, and R. J. Hemley. *Proc. Natl. Acad. Sci. USA*, 97, 13494 (2000).

[71] O. Anderson and K. Zou. *J. Phys. Chem. Ref. Data*, 19, 69 (1990).

[72] E. S. Zouboulis, M. Grimsditch, A. K. Ramdas, *et al.* *Phys. Rev. B*, 57, 2889 (1998).

[73] H. J. McSkimin and P. Andreatch. *J. Appl. Phys.*, 43, 2944 (1972).

[74] G. N. Kamm and G. A. Alers. *J. Appl. Phys.*, 35, 327 (1964).

[75] D. Gerlich and E. S. Fisher. *J. Phys. Chem. Solids*, 30, 1197 (1969).

[76] R. R. Reeber, K. Goessel, and K. Wang. *Eur. J. Miner.*, 7, 1039 (1995).

[77] G. A. Slack and S. F. Bartram. *J. Appl. Phys.*, 46, 89 (1975).

[78] A. C. Victor. *J. Chem. Phys.*, 36, 1903 (1962).

[79] E. D. Eastman, A. M. Williams, and T. F. Young. *J. Am. Chem. Soc.*, 46, 1178 (1924).

[80] W. F. Giauque and P. F. Meads. *J. Am. Chem. Soc.*, 63, 1897 (1941).

[81] D. B. Fraser and A. C. H. Hallett. *Can. J. Phys.*, 43, 193 (1965).

[82] A. J. Leadbetter. *J. Phys. C*, 1, 1481 (1968).

[83] D. B. Downie and J. F. Martin. *J. Chem. Thermodyn.*, 12, 779 (1980).

[84] P. D. Pathak and N. G. Vasavada. *J. Phys. C*, 3, L44 (1970).

[85] R. M. Nicklow and R. A. Young. *Phys. Rev.*, 129, 1936 (1963).

[86] Y. Touloukian, ed. *Thermophysical properties of matter.* IFI/Plenum, New York (1970).

[87] R. Stedman, L. Almqvist, and G. Nilsson. *Phys. Rev.*, 162, 549 (1967).

[88] P. Debye. *Ann. Physik*, 39, 789 (1912).

[89] J. Slater. *Introduction to chemical physics.* McGraw-Hill, New York (1939).

[90] M. Álvarez Blanco. *Métodos cuánticos locales para la simulación de materiales iónicos. Fundamentos, algoritmos y aplicaciones.* Tesis doctoral, Universidad de Oviedo (1997).

[91] L. Vocadlo, J. Poirer, and G. Price. *Am. Mineralog.*, 85, 390 (2000).

[92] V. L. Moruzzi, J. F. Janak, and K. Schwarz. *Phys. Rev. B*, 37, 790 (1988).

[93] S. W. Kieffer. *Rev. Geophys.*, 17, 35 (1979).

[94] J. L. Fleche. *Phys. Rev. B*, 65, 245116 (2002).

[95] W. B. Holzapfel. In J. Loveday, ed., *High pressure physics.* CRC Press, Boca Raton, FL (2011).

[96] D. Marx and J. Hutter. *Ab initio molecular dynamics: basic theory and advanced methods.* Cambridge University Press, Cambridge, UK (2009).

[97] M. P. Allen and D. J. Tildesley. *Computer simulation of liquids.* Clarendon Press, New York (1989).

[98] R. Car and M. Parrinello. *Phys. Rev. Lett.*, 55, 2471 (1985).

[99] J. Barker. *J. Chem. Phys.*, 70, 2914 (1979).

[100] B. Militzer and D. Ceperley. *Phys. Rev. Lett.*, 85, 1890 (2000).

CHAPTER 2

Mechanisms of Pressure-Induced Phase Transitions

Manuel Flórez
MALTA Consolider Team and Departamento de Química Física y Analítica, Universidad de Oviedo, Oviedo, Spain

J. Manuel Recio
MALTA Consolider Team and Departamento de Química Física y Analítica, Universidad de Oviedo, Oviedo, Spain

CONTENTS

2.1 INTRODUCTION

POLYMORPHISM is one of the most genuine phenomena in high-pressure science. The existence of different crystalline structures for the same chemical compound and the possibility of keeping some of them as metastable phases at ambient conditions arouse great interest within both the scientific and the technological realms. We can think of many examples, graphite and diamond carbon allotropes are probably the most common and representative examples. The key to understand metastability in pressure-induced phase transformations lies in the kinetics of the process. This is a difficult topic to deal with, as we will show in this chapter, mainly due to the complexity of describing the time evolution of a transition. Fortunately, mechanistic aspects, the other part inherent to the kinetics of the transition, can be described with approximate models providing valuable information. Thus, we can understand at a microscopic level the energetics, structural and bonding changes associated with the atomic reorganization involved in the transformation.

Concerning the mechanism, most of what we know comes from phenomenological models, theoretical formalisms, and computational simulations. During the 1990s and the first decade of this century, this topic attracted the attention of a number of research laboratories and became very popular. Some good references can be found in the works of those groups that were specifically involved in the development and application of the so-called martensitic approach. The authors of this chapter participated in the formalization of this approximation by providing a chemical perspective of phase transitions. See for example References 1–4 and references therein. Space symmetry has always played an essential and clarifying role, and powerful tools as those available at the Bilbao Crystallographic Server [5] deserve to be recognized. We would like to emphasize the contributions of Sowa [6], Stokes and Hatch [7], Catti [8], Tolédano and Dmitriev [9], Lambretch and co-workers [10], and Zahn and Leoni [11]. The last group introduced variable time in their dynamic simulations and described the nucleation and growth steps of some simple transformations. Nevertheless, not much progress has been made since the traditional papers of Johnson, Avrami and Mehl (see Reference 12 and references therein) concerning formalisms describing the time evolution of the transformation.

This chapter is divided in two parts. In the first one, we present fundamentals and concepts of the kinetics of solid-solid pressure-induced phase transitions. Under applied pressure, a crystalline system progressively reduces its volume in such a way that, eventually, it can undergo a transformation into another different denser polymorph with a more compact structure. The atomic reorganization associated with this transition may involve the formation and/or the breaking of bonds, and can be described by means of a mechanism following a similar procedure to that used to describe chemical reactions. The main approach to this view is the martensitic conception of the transformation in which the translational symmetry is maintained along the transition path. This martensitic approximation will be the central topic of this chapter. Theoretical concepts needed to follow its contents can be found in many graduate textbooks on thermodynamics, kinetics, quantum chemistry, chemical bond theory, and crystallography. This first part of the chapter also includes a summary of some key thermodynamic results presented in more detail in Chapter 1 and a discussion of different classifications of solid-solid phase transitions.

The second part of the chapter is devoted to the analysis of selected martensitic simulations that allow us to illustrate general aspects related to these transformations: structural changes, universal behaviors, atomic displacements, energy barriers, bond reorganization, etc. Although the language of some of these topics can be difficult to grasp, the chapter is written in such a way that *a priori* knowledge is not required to understand the examples and to learn from their results. We follow a sequence of increasing complexity in the cases of study. Starting with the well known rock salt → cesium chloride (B1 → B2) transition in the alkali halides, we introduce transformations exhibited by other binary compounds, silica, and zircon.

2.2 KINETICS OF SOLID–SOLID PHASE TRANSITIONS

2.2.1 Some Thermodynamic Aspects of Phase Transitions

In Chapter 1, basic intra- and inter-phase thermodynamic concepts are presented in detail and, therefore, that chapter should be read beforehand. Here, we only want to stress three meaningful results from Chapter 1. The first one is that at constant temperature and pressure, the thermodynamic potential that controls the equilibrium and stability of a pure substance is the Gibbs free energy (G). Second, the thermodynamic stability (metastability) condition of one phase with respect to a volumetric isotropic strain leads to a decreasing volume when pressure increases at constant temperature. The stability with respect to other non-isotropic strains of the lattice (not discussed in Chapter 1) can be expressed by inequalities involving appropriate elastic constants relationships (see Chapter 3 for definition of elastic constants). Since volume is the first derivative of G with respect to pressure at constant temperature, the $(\partial V/\partial p)_T < 0$ condition leads to:

$$\left(\frac{\partial^2 G}{\partial p^2}\right)_T < 0. \tag{2.1}$$

This means that in the common pressure ranges explored experimentally or in computer simulations, G-p curves show monotonous increasing sublinear behaviors. Finally, a discussion on the relative values of G among different phases is also presented in Chapter 1. Let us imagine that for a given pure substance, phase α is stable at low pressures and another phase, say β, has higher G at those conditions. If G in the β phase increases with pressure at a lower rate than G in the α phase does, then a crossing point between the two curves will eventually appear. This point defines the thermodynamic transition pressure p_{tr} (see Figure 2.1). The first derivative of G with respect to p at p_{tr} is different for the α and β phases and therefore a change in the volume associated with the transition occurs. This and a change in the entropy are the signatures of first-order phase transitions.

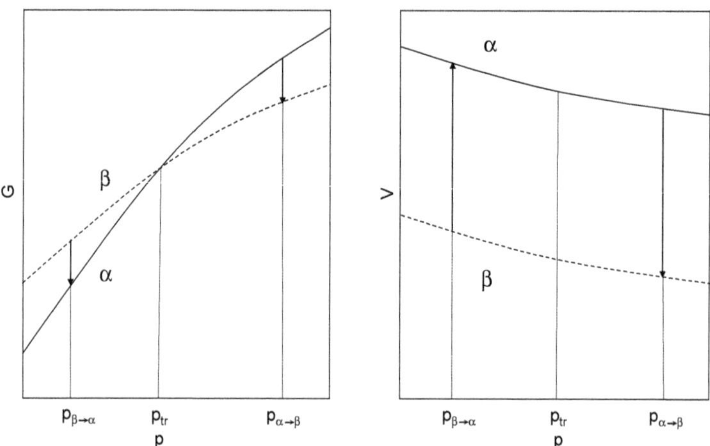

Figure 2.1 Hysteresis cycle in a G-p diagram (left) and in a V-p diagram (right).

Usually, a pressure-induced phase transition is observed in the laboratory or in nature at a pressure different from the thermodynamic transition pressure. Using the same example, if we look at the process of loading pressure on the α phase, its transformation to β will be experimentally observed at a pressure above p_{tr}. On the other hand, starting at high pressure with the β phase and decreasing pressure, we will find the $\beta \rightarrow \alpha$ transformation at a lower value than p_{tr}. The loading and unloading pressure process can be seen as a hysteresis cycle in a G-p or in a volume-pressure (V-p) diagram as depicted in Figure 2.1.

This mismatch between the observed and the thermodynamic transition pressures is due to the finite rate of conversion between phases, which is not high enough to avoid the appearance of metastable phases. In fact, a metastable phase (α) can be defined as a polymorphic structure that has higher Gibbs energy than other polymorphs (β) of the same compound at given T and p, but with such a low rate of $\alpha \rightarrow \beta$ conversion that α can be detected during a meaningful observable period. The extremely important role that metastability plays in materials science indicates that the kinetics of pressure-induced phase transformations deserve detailed study.

2.2.2 Nucleation and Growth Models

As we anticipated in Section 2.1, two basic aspects can be distinguished in the kinetic study of a solid-solid phase transition (these two aspects may be found in many other kinetic phenomena too). The first one concerns changes in the atomic ordering as the transformation is taking place. These changes usually involve formation and breaking of bonds, thus introducing energetic barriers. The description of these atomic movements is called the mechanism or the transition path. The second aspect refers to the time evolution of this atomic reorganization. In this chapter, the focus is on the mechanism, although information on models for the time evolution of the transition is also summarized.

The classical models of solid-solid phase transitions were developed in the 1940s by Johnson, Mehl, Avrami, and others (see Reference 12 and references therein), and were used in other branches of chemistry and physics such as electrochemistry. To simplify, we consider that the $\alpha \rightarrow \beta$ transformation occurs in two steps. First, an embryo or domain of the β phase emerges in the matrix of the initial α phase. This is the nucleation step with a characteristic rate defined by I. Then, in the second step, β domains grow at the expense of the α matrix with rate u. The transformation $\alpha \rightarrow \beta$ takes place only at the interfaces between β domains and α. The fraction of the α phase, $x(t)$ that transforms into β increases with time at a rate involving both I and u rates.

Assuming that u is time-independent (growing is not controlled by long-range diffusion), the basic kinetic law can be written as:

$$x(t) = 1 - \exp\left(-\int_0^t vI d\tau\right), \tag{2.2}$$

where $v = gu^3(t-\tau)^3$; v is the volume of a β domain growing at a constant linear rate before other domains or free surfaces interfere with it; g is a shape factor; t is the total time of the transformation; and τ is the time since the beginning of the transformation at which the β domain nucleates. For example, if I is time-independent then

$$x(t) = 1 - \exp\left(-gu^3It^4/4\right). \tag{2.3}$$

Other approximations lead to the following expressions for I and u:

$$I = A\exp\left(-\frac{\Delta G^\star + \Delta G^\dagger}{k_BT}\right), \tag{2.4}$$

$$u = B\exp\left(-\frac{\Delta G^\dagger}{k_BT}\right)\left(1 - \exp\left(\frac{\Delta G + V\epsilon}{k_BT}\right)\right), \tag{2.5}$$

where T is temperature, k_B is the Boltzmann constant, and $\Delta G^\star$ is the energy barrier for the nucleation that in the case of a spherical nucleus has the expression:

$$\Delta G^\star = \frac{16\pi(\sigma - \sigma_h)^3}{3(V^{-1}\Delta G + \epsilon)^2}. \tag{2.6}$$

$\Delta G^\dagger$ is the energy barrier for the short-range diffusion of the particles through the interphase, ΔG is the change of Gibbs energy of the transformation, σ is the surface free energy per unit area of the nucleus, and ϵ is the deformation energy per unit volume associated with the slow relaxation rate of the internal stresses produced by the volume difference between the two phases in the vecinity of domain walls. In this process, the heterogeneous nucleation is characterized by the presence of a surface energy, σ_h, associated with the existence of defects and/or impurities in the structure. We refer to the classical works cited above for a more thorough explanation of the kinetics of solid-solid transformations that is beyond the scope of this chapter. However, it is to be emphasized that some energy terms as ΔG and $\Delta G^\dagger$ (and other parameters such as ϵ and σ) can be related and/or estimated with mechanistic models, as we will see below, thus providing indications on how the kinetics of the process is influenced by these properties.

Nucleation and growth steps are simultaneously present in many transitions where experiments and calculations emphasize the key role that defects and impurities play in the transformation rate. Nevertheless, we have evidence of sudden and fast phase transitions in which the crystal as a whole changes from one phase to another. Usually this happens when large and ultra-pure crystals are involved. For these last transformations, the martensitic approximation presented in Section 2.2.4 provides a more rigorous description of the phase transition.

2.2.3 Classification of Solid–Solid Phase Transitions

Here we briefly discuss some of the criteria used to classify solid-solid phase transitions. Firstly, as already seen in Chapter 1, from the thermodynamic view we can distinguish between first- and high-order transitions. In first-order phase transitions, the first derivatives of G with respect to T and p, $-S$ and V, respectively, are discontinuous and, therefore, ΔS_{tr}, ΔV_{tr} and ΔH_{tr} are not zero. Clapeyron's equation provides the slope, $\mathrm{d}p_{\text{tr}}/\mathrm{d}T$, of the equilibrium boundary between two phases in a T-p phase diagram:

$$\frac{dp_{\text{tr}}}{dT} = \frac{\Delta H_{\text{tr}}}{T\Delta V_{\text{tr}}}. \tag{2.7}$$

In second- or higher-order transitions, ΔS_{tr}, ΔV_{tr} and ΔH_{tr} are zero. In second-order phase transitions, the second derivatives of G (related to heat capacity at constant pressure C_p, compressibility κ, and thermal coefficient α) are discontinuous, whereas the first derivatives are continuous. For these last transitions, Clapeyron's equation lacks meaning.

Regarding the mechanism, transitions can be divided into diffusive and displacive. The former take place by nucleation and growth driven by diffusion if temperature is high enough. Displacive transitions involve cooperative atomic reorganizations due to elastic or phonon softening because, for example, the frequency of a specific vibrational mode goes to zero. Atomic displacements follow the direction of the eigenvector associated with that specific normal mode that usually shows a T or p dependence. We say that there is a softening in a lattice oscillation or that the corresponding phonons are frozen.

Buerger [13, 14] proposed a classification of structural transitions in two groups, reconstructive and displacive. In reconstructive transitions, there is a substantial modification of the unit cell, which is accompanied by breaking and formation of primary chemical bonds (those involving atomic first coordination spheres). These transitions usually show heterogeneous nucleation and are first order with coexistence of both phases at equilibrium. They also present metastability and hysteresis phenomena. In displacive transitions, there is neither breaking nor reconstruction of primary chemical bonds, and the new phases can nucleate homogeneously. These transitions can be of second order or they can show a weak first order character. In the case of weakly first order displacive transitions, the close relationship between the initial and final structures allows visualizing the transformation in a continuous manner using few structural parameters. Obviously, there is not a clear and unambiguous way to decide to which group of the Buerger's classification a particular transition belongs. Among the reasons we can cite the elusive character of the chemical bonding concept or the fact that so-called martensitic transitions may involve large atomic displacements with changes in coordination numbers (reconstructive) but they are not effectively diffusive.

Finally, a less ambiguous classification is proposed based on symmetry criteria [15]. In this classification, there are: (i) transitions with a group-subgroup relationship between the initial and final phases, (ii) transitions without a group-subgroup relationship, and (iii) transitions involving intermediate steps of the two former types. The following features are inherent to type (i) transitions: structural changes in which interatomic distances are slightly modified, atoms of the high symmetry structure that move from their special positions, breaking the local symmetry; the low symmetry phase gradually approaching the situation where new symmetry elements appear; and lack of phase metastability and hysteresis. These transitions can be modeled using the phenomenological formalism of Landau [9]. Here, the differences between the high and low symmetry structures can be described using an order parameter η (η

= 0 for the high symmetry structure and $\eta = 1$ for the low symmetry one) that transforms as a non-totally symmetric irreducible representation of the high symmetry group and belongs to the totally symmetric irreducible representation of the low symmetry group. This η parameter presents in a good approximation a linear dependence on the cell strains and atomic displacements. The change in the relevant thermodynamic potential (G, for example) between the initial and final phases can be expressed with few terms as a sum of powers of the η parameter. These transitions are second order or even first order from the thermodynamic point of view. The order of the transition depends on the coefficients of the thermodynamic potential expansion (they are functions of pressure and temperature). In particular, the presence of cubic terms leads to intermediate states with higher energies than the initial and final structures at equilibrium conditions and therefore to first-order transitions. Displacive phase transitions are commonly of this type.

Transitions without group-subgroup relationship between the initial and final phases (type (ii)) usually involve a great atomic reorganization with formation and breaking of bonds. Both diffusion processes and cooperative movements of the atoms may take place simultaneously. In the case of long-range diffusionless transitions, it is possible to establish a transition path based on an intermediate structure with a spatial group being a common subgroup of the initial and final phases. Then it is possible to define a transformation coordinate to follow the transition using the intermediate structure. In these cases, it is possible to model the transition using a modification of Landau's approach: the corresponding order parameter is defined as a non-linear periodic function of particular atomic displacements, strains, etc. involved in the transformation [9]. Reconstructive transitions are of this type.

2.2.4 Martensitic Description of Transition Paths

The whole kinetic description of pressure-induced phase transitions is a formidable task due to the inherent complexity of the nucleation and growth steps and the time dependence involved in the atomic reorganization during the process. Computer simulations deal with the problem by means of two strategies. The first one is the static or martensitic approximation whereas the second one takes time evolution into account. Both can be connected with the two types of fundamental aspects of the kinetic study presented above. In what follows, we focus on the martensitic description. It only covers the mechanistic view of the transformation with an almost exclusive reference to the description of the transition path. This is useful for the simulation of the nucleation step of the transformation. In spite of the reduced scope at first impression, transition paths proposed by means of the martensitic approximation are also very helpful in the dynamic simulations of these transformations, and some energetic properties (as we anticipated above) can be calculated from the static approach and related later to the kinetic models.

Let us review now the main ideas of this approach. The martensitic or static approximation has to do with a military concept of the atomic movements: atoms form a well organized and disciplined army that, although suffering local distortions or macroscopic strains, never loses the global order. It is assumed that domains do not appear in the transformation, but the crystal as a whole transforms from one phase to the other. Along the process, atoms move simultaneously in a concerted manner while keeping translational symmetry. In this way, we can choose a unit cell to describe the transition path. Thus, the first step requires us to find a common space subgroup of both the initial and final structures. This martensitic description follows the formalism of the generalized Landau theory and allows us to select a transformation coordinate and to define a transition path as in the case of chemical reactions (see below).

Under this approximation, it is possible to visualize cell strains and atomic displacements along the transition. Quantum mechanical calculations can be carried out to simulate this martensitic approach of the transition path and provide valuable data describing some of the energetic properties appearing in the classical models discussed above, for example, $\Delta G^{\dagger}$ and ΔG. The last one is non-zero if pressure is different from p_{tr}, and its negative value leads to an increase in the transformation rate. For this reason, ΔG is usually known as the *driving force* of the transformation.

2.2.4.1 Modeling and Computational Strategies

In practice, the study and determination of possible transition mechanisms involves the following steps: (i) a preliminary election based on symmetry criteria, (ii) application of structural criteria, (iii) evaluation of energetic profiles for the chosen transition paths, and (iv) structural and chemical bonding analysis along the pathways.

Symmetry criteria help to define a hierarchy of transition paths using the space group (common subgroup) and the size (number of formula units) of the unit cell connecting the initial and final structures. The method requires the same number of atoms in the common subgroup unit cell for the two structures as well as a correspondence between the Wyckoff positions of the atoms of both structures in the reference unit cell. The representation of the initial and final phases in the basis of the common subgroup allows their direct comparison with the possibility of visualizing all the changes in the lattice parameters and atomic coordinates associated with the transformation. Structural criteria, mainly the unit cell strain (S) and the atomic displacements ($\delta(i-j)$) involved in the transformation, lead to a second hierarchical classification of those paths that are compatible by symmetry. Low values of S and $\delta(i-j)$ suggest an *a priori* competitive transition path.

It should be recalled that S is an increasing function of the strain tensor eigenvalues, and that the strain tensor is related to a change in the metric tensor $\boldsymbol{G}-\boldsymbol{G}_0$, *i.e.*, in the geometry of the unit cell corresponding to a homogeneous deformation of the crystalline structure that fixes the fractional atomic coordinates [5]. If these coordinates change, an internal deformation appears. The total deformation of the crystalline structure contains both contributions. Incidentally, the changes in the interatomic distances that are only due to the internal deformation can be used to establish atom-atom correspondences between the two structures. It is logical to choose the atomic correspondence that provides the shortest distances between atoms of the same type in both structures: atom i in one structure and atom j in the other. Usually, distances (atomic displacements $\delta(i-j)$) are calculated taking as the reference the lattice parameters of the initial structure. The `TRANPATH` utility from the Bilbao Crystallographic Server [5] is a practical tool to carry out these calculations.

Obviously, for a phase transition at a given pressure and temperature, a rigorous conclusion on the most favorable martensitic mechanism requires the comparison of calculated Gibbs energetic profiles for different transition paths and, in particular, the calculation of the corresponding energy barriers. The evaluation of this energetic profile follows the next procedure. Here, we select T and p and a transformation coordinate connecting the initial and final structures. For each of the grid values of the transformation coordinate, the structure is optimized by minimizing the Gibbs energy with respect to the remaining degrees of freedom. Since *ab initio* calculations can be computationally expensive when the number of atoms is high, a reduction of the dimensionality in the energy hyper-surface is desirable. It can be achieved by following approximate schemes in which the lattice parameters and/or atomic positions change along the transition path with linear trends. It is also convenient to follow the criterion of minimum atomic displacements to identify and label the atoms for each of the values of the transformation coordinate [3]. Most of these computational strategies will be illustrated in the examples presented in the second part of this chapter.

Although it is known that nucleation and growth mechanisms are usually studied by means of molecular dynamics simulations (see Reference [11]), continuous models as the martensitic approximation can be of great help in the prediction of intermediate phases. An example is found in the well-known rock salt $\rightarrow$ cesium chloride phase transition in NaCl. Molecular dynamics calculations show the preference for a nucleation and growth process instead of a concerted mechanism [16]. These simulations indicate the formation

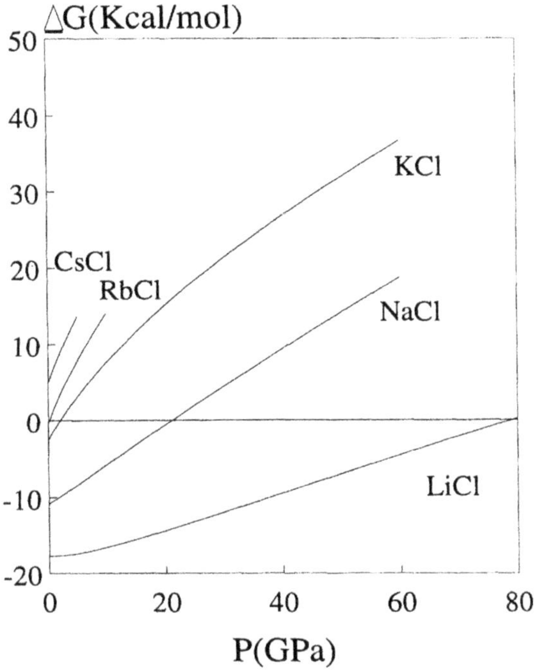

Figure 2.2 Dependence with pressure of ΔG for the alkali chlorides [19]. Adapted, with permission, from Reference 19.

of an interphase between the rock salt and cesium chloride regions with a well defined structure (α-TlI-type) that was previously anticipated in periodic quantum mechanical calculations [17] and also proposed from symmetry arguments [9, 18]. In general, many pressure-induced phase transitions are expected to occur first after a competition of different transition mechanisms, and second following complex mechanisms involving martensitic (cooperative) and diffusive (nucleation and growth) phenomena. Finally, we need to stress that transition pathways involving common high symmetry subgroups become important from the kinetic view at high pressure and are often important at least in the nucleation steps of many of these transformations.

2.3 ILLUSTRATIVE EXAMPLES

2.3.1 Equation of State and B1–B2 Phase Transition in Alkali Halides

Before studying the transition mechanism in a given compound, the starting step is to determine the thermodynamic stability pressure range of its different polymorphs. In Figure 2.2, we show the ΔG-p curves of the alkali chlorides from static quantum mechanical calculation [19]. Notice that, in this context, static means that calculations are performed considering $T = 0$ K and

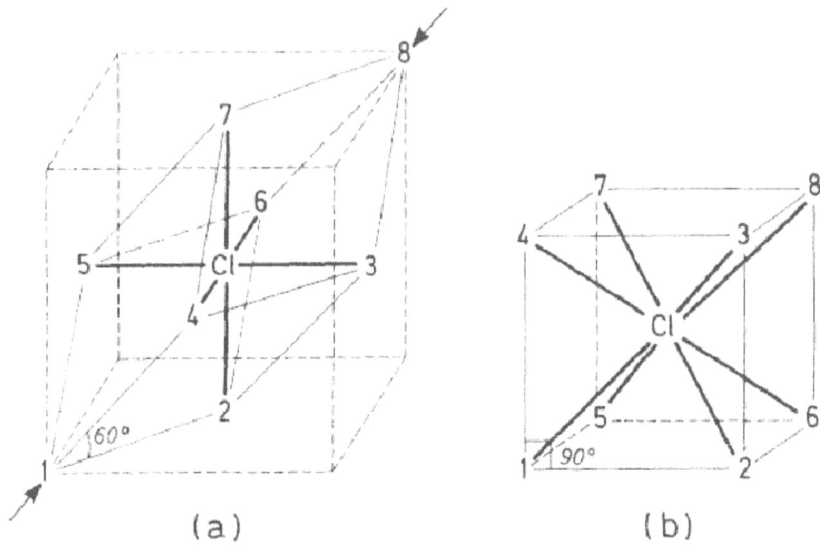

Figure 2.3 Rhombohedral cells for the B1 (a) and B2 (b) phases. Arrows show the cell deformation according to Buerger's mechanism [13]. Numbers help to establish a correspondence between atoms in both structures. Adapted, with permission, from Reference 1.

neglecting zero point vibrational contributions. $\Delta G = G(\mathrm{B2}) - G(\mathrm{B1})$, and thus a negative value of this property means that the B1 phase is the thermodynamically stable one. The crossing point with the $\Delta G = 0$ line represents the (static) thermodynamic transition pressure, p_{tr}.

The most interesting results are: (i) at ambient conditions, the B1 phase is the stable structure for all the compounds except for CsCl, in agreement with the experimental observations, (ii) ΔG values at zero pressure increase as the cation size increases, (iii) p_{tr} decreases as the cation size increases, which is also in agreement with the experimental data, (iv) the slope of the ΔG-p curves at p_{tr} (*i.e.*, the transition volume at p_{tr}) increases with the cation size, again in concordance with the observed trend. We notice that for LiCl the transition pressure has not been experimentally determined yet. The calculated value is around 80 GPa (see Figure 2.2). LiCl is the example we use below to illustrate the B1-B2 transition mechanism.

This pressure-induced rock salt (B1) $\rightarrow$ cesium chloride (B2) phase transition in the alkali halide crystal family has been frequently studied since the seminal works of Slater [20] and Bridgman [21]. For the B1 $\rightarrow$ B2 mechanism, the most popular reference is Buerger [13]. The transition path and the visualization of the transformation are easy to follow in this case. As depicted in Figure 2.3, both structures can be described by a rhombohedral cell with an angle of 60° in the B1 phase and 90° in the B2 phase. The common space subgroup is $R\bar{3}m$ and the atomic positions are fixed in the origin and the center of the cell along the transition path. The unit cell angle continuously changes from 60° to 90°, whereas the lattice parameter evolves also in a continuous way from the value in the B1 phase to the value in the B2 phase as the atomic coordination increases from 6 to 8.

In Figure 2.4, G static profiles along transition paths for LiCl at different pressures are plotted. The cell angle α is taken as the transformation coordinate. At each pressure, G is referred to the B1 phase value to better demonstrate the relative stability of both phases. We observe topological differences in the profiles, depending on pressure. At zero pressure, B1 is the thermodynamically stable phase and it is not possible to find B2 even as a metastable structure since the G-α curve presents a maximum at 90°. Between $p = 0$ and the transition pressure ($p_{\rm tr}$ = 80 GPa), both B1 and B2 phases show minima at 60° and 90°, though B1 is thermodynamically stable (absolute minimum) whereas B2 is metastable (relative minimum). At $p_{\rm tr}$ = 80 GPa, the two minima are obtained with $G(\text{B1}) = G(\text{B2})$ and therefore both phases are at equilibrium. Barrier heights for the B1 → B2 and B2 → B1 transitions are different at $p \neq p_{\rm tr}$ and only display the same value around 30 kJ/mol at the transition pressure. Above $p_{\rm tr}$, both phases again show minima but now the B2 phase is thermodynamically stable (B1 is metastable). At very high pressures (above 300 GPa), B1 is not even metastable since the minimum at 60° disappears. The presence of high energy barriers with asymmetry and pressure dependence explains the observations related to the hysteresis phenomena. This example also illustrates the difference between absolute and internal stability (metastability). These results also show that along the B1-B2 pathway there is a coupled movement in which the cell angle opens as the lattice parameter decreases below or above the transition pressure. A more detailed description of this example can be found in Reference 1.

Another interesting issue related to this transition is that many features of the rhombohedral transition path are common to the entire alkali halide crystal family, and probably to other I-VII binary compounds too. The corresponding state principle is the well known basis for universal equations of the traditional aggregate states of matter. In solids, the volume at zero pressure is often used as the reference property and state for many analytical equations of state (EOS, see details in Chapter 1). An example of a popular universal EOS is the one proposed by Vinet *et al.* [22]:

$$\ln H = \ln B_0 + \frac{3}{2}(B_0' - 1)(1 - x), \tag{2.8}$$

where $x = V/V_0$ and

$$H = \frac{px^2}{3(1 - x)}. \tag{2.9}$$

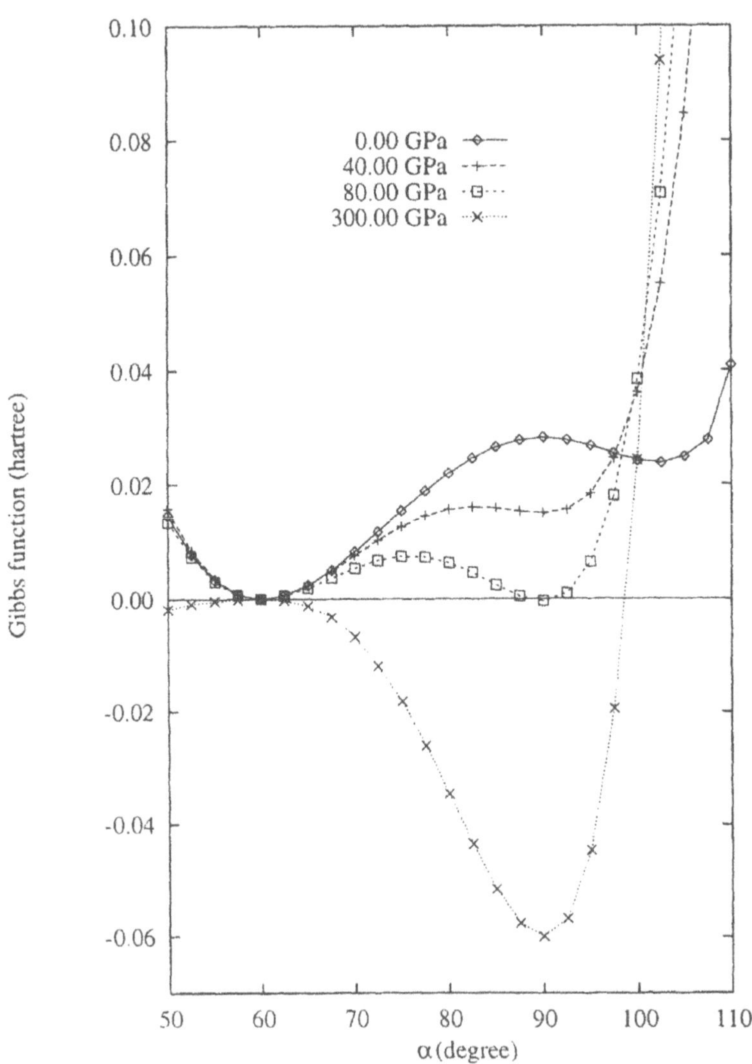

Figure 2.4 Energetic profiles for the $R\bar{3}m$ B1-B2 transition path of LiCl at different pressures [1]. Adapted, with permission, from Reference 1.

When applied to the alkali chloride crystal family, $\ln H$-$(1-x)$ calculated curves show a linear behavior with B_0 and B_0' static values in agreement with 0 K experimental data. Besides the quantitative concordance, theoretical simulated solids behave as Vinet solids fulfilling the general corresponding state principle in all the analyzed pressure ranges. This kind of analysis can be seen as a requisite to the application of quantum mechanical methodologies to the simulation of transition paths in solid-solid phase transformations.

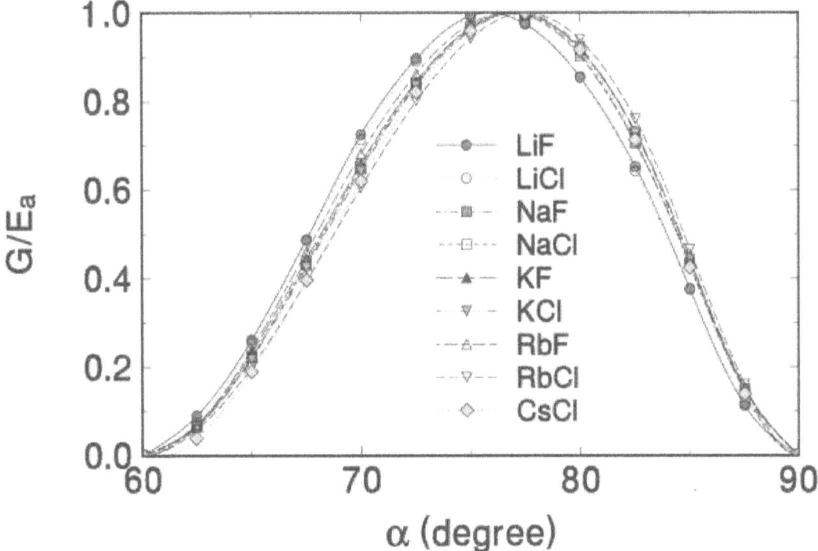

Figure 2.5 Universal energy profile for the $R\bar{3}m$ path of the B1-B2 transition phase of several alkali halides. Adapted, with permission, from Reference 23.

We present here an extension of this principle to interphase phenomena for the case of the B1-B2 transition mechanism [23] in alkali halide crystals. The scaling parameter is the barrier height of the transition $\Delta G^{\dagger}$ or E_a. The idea is to normalize the energetic profile of the transition path using E_a. If we plot this normalized profile for the B1-B2 pathways of many alkali halides, we obtain almost the same curve with the transition state located at practically the same value of the α coordinate (see Figure 2.5). This behavior has been also found for other II-VI binary compounds along the zinc blende B3-B1 transition path [24] that we will discuss below. We can conclude that universal equations are not only valid for intraphase p-V-T relationships but also for interphase mechanisms under the martensitic approximation.

2.3.2 B3–B1 Phase Transition in Binary Compounds

This transition is very interesting because it is exhibited by a number of binary semiconductors involving noticeable changes in their electronic

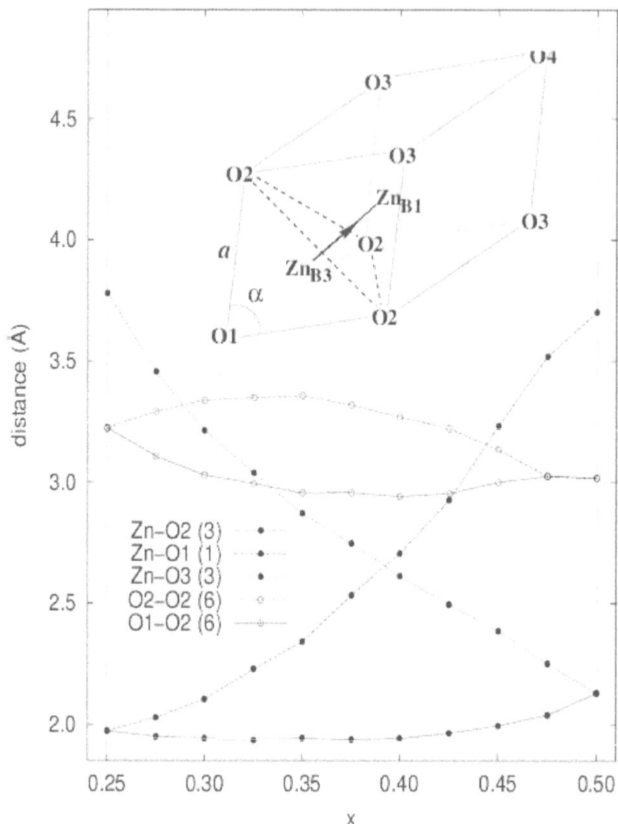

Figure 2.6 Rhombohedral $R3m$ path for the B3-B1 transition in ZnO. Unit cell, atomic environment, and interatomic distances are based on Reference 2. Adapted, with permission, from Reference 2.

properties (see Chapter 5): II-VI (ZnS), III-V (GaAs) and IV-IV (3C-SiC polytype) compounds show this pressure-induced phase transition. It constitutes a more complex example in terms of the application of the martensitic approximation. Fortunately, it has been widely studied by several groups [2, 24–27]. In essence, we will discuss two transition paths connecting the conventional unit cells, with symmetries $F\bar{4}3m$ (B3) and $Fm\bar{3}m$ (B1).

In the first path, the highest symmetry common subgroup leading to a single molecule mechanism, the rhombohedral $R3m$ space group, is considered. The cell angle is 60° for the initial and final structures. One of the atoms is at the origin and the other atom is at $(0.25, 0.25, 0.25)$ (B3) or $(0.5, 0.5, 0.5)$ (B1). There are three degrees of freedom: a, α, and the $x = y = z$ coordinate of the second atom. Results for the B3 $\rightarrow$ B1 transition path in ZnO using this rhombohedral mechanism have been previously discussed in detail [2]. Considering the oxygen atom at the origin, the natural transformation coordinate to visualize the atomic displacements along the transition path is the $x = y = z$ coordinate of Zn that evolves from 0.25 (B3) to 0.50 (B1). This is displayed in the inset of Figure 2.6.

At a value of $x = y = z = 0.375$, Zn has to cross the center of a triangle of three equivalent oxygen atoms (O2) as it moves away from the oxygen at the origin (O1) and approaches the three equivalent oxygens with the label O3 (see Figure 2.6). Although the cell angle is the same in the initial and final structures, the cell has to open to facilitate the atomic movement of Zn, specially at the 0.375 crossing point; a cell angle of around 70° is reached in the transition path at this point. The change in the coordination from 4 to 6 is easily followed using Figure 2.6: one Zn-O distance increases (O1), three keep the same approximate value (O2), and the other three decrease (O3) along the transition path (multiplicities are in brackets in Figure 2.6). We can conclude that one Zn-O primary bond is broken and three are formed along this mechanism.

A more favorable transition path might involve only the formation of two bonds without breaking others. A detailed study of the changes in the chemical bonding for this B3-B1 transition in BeO is presented in Chapter 5. Here we recall the mechanistic aspects reported by Sowa [6], Catti [25], and Lambretch and co-workers [10, 24], and more recently described by Cai *et al.* [26] in their atomistic simulations of the B3-B1 phase transformation in GaAs. Although there are competitive pathways of lower symmetries and this issue was a matter of debate in the last decade, the consensus view is that an orthorhombic cell of symmetry $Imm2$ provides the adequate transition mechanism for the B3-B1 transformation. The number of formula units in this common cell is two. Now, we have four degrees of freedom: the three lattice parameters (a, b, c) and the z coordinate of one of the atoms that changes from 0.25 (B3) to 0.50 (B1). Therefore, z is the natural transformation coordinate for this path. Barrier heights up to 20 kJ/mol lower than those for the $R3m$ mechanism were found under the $Imm2$ symmetry for several compounds such as ZnS and GaAs. This is a clear manifestation of the lower chemical reorganization involved along the orthorhombic path.

2.3.3 α-Cristobalite–Stishovite Phase Transition in Silica

This is an example of a phase transition mechanism with initial and final structures displaying a group-subgroup relationship. Although this feature is often linked to a displacive character for the transition, we will illustrate that it is not the case here. α-cristobalite belongs to the tetragonal $P4_12_12$ space group. Its conventional unit cell contains 4 Si and 8 O atoms with Si at $(x, x, 0)$ and O at general (x, y, z) positions. Stishovite also belongs to a tetragonal

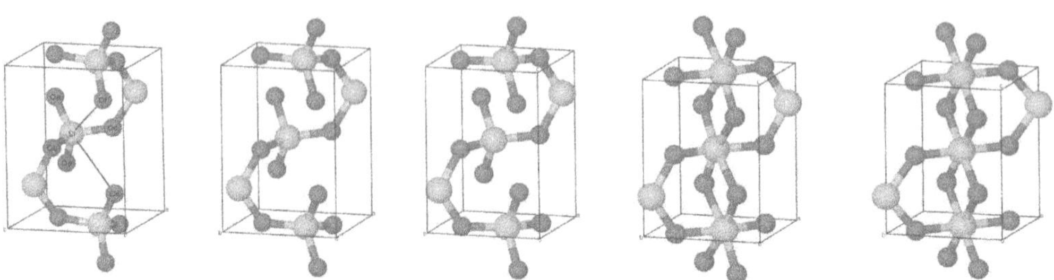

Figure 2.7 Evolution of the local environment of Si and O in the α-cristobalite-to-stishovite transformation along the $P4_12_12$ path. ξ is the normalized transition coordinate. Black lines are prefigured new Si-O bonds. Adapted, with permission, from Reference 4.

space group, the $P4_2/mnm$ in this case. Its conventional unit cell contains 2 Si and 4 O atoms with Si at $(0,0,0)$ and O at $(x,x,0)$. Since the space group of α-cristobalite is a subgroup of the stishovite space group, we can check whether a $P4_12_12$ unit cell with four formula units can be used to describe the mechanism of the α-cristobalite-to-stishovite transformation. Fortunately, in this space group, a double unit cell for stishovite contains Si and O atoms at the same Wyckoff positions as in α-cristobalite with $x_{\text{Si}} = 0.5$, $y_{\text{O}} = x_{\text{O}}$, and $z_{\text{O}} = 0.25$.

The parameter x_{Si} is a good transformation coordinate to walk along the transition path. The optimized value of x_{Si} in α-cristobalite at the calculated transition pressure is 0.33 [4]. Therefore, to accomplish the transition, x_{Si} has to move from this value to 0.50 (stishovite). For convenience, we can define a normalized parameter, ξ, that varies from 0 (cr, α-cristobalite) to 1 (st, stishovite):

$$\xi = \frac{x_{\text{Si}} - x_{\text{Si}}^{\text{cr}}}{x_{\text{Si}}^{\text{st}} - x_{\text{Si}}^{\text{cr}}}. \tag{2.10}$$

At selected values of ξ (selected values of x_{Si}), the cell parameters and the oxygen coordinates (five degrees of freedom in total) have to be optimized by minimizing the Gibbs energy of the system. As in the previous examples, we can follow the energetic profile and the structure along the transition path. An activation barrier of around 1000 kJ/mol is obtained for $\xi = 0.55$. At the transition state, the volume collapse with respect to the unit cell volume of the α-cristobalite phase is close to 24 %, compared to the total volume reduction of 34 %. These values confirm the reconstructive character of the transition. In particular, we can visualize how the Si coordination number increases from 4 to 6 and that of O from 2 to 3 (see the sequence of Figure 2.7). Not until the last stages of the transformation (according to this ξ coordinate) do new bonds emerge, once the distance between Si and the two oxygen atoms becoming new nearest neighbors in the final structure is as close as 2.0 Å [4]. We can conclude that, in spite of the group-subgroup relationship between the two phases, a reconstructive phase transition is expected due to the increasing atomic coordination accompanying this transformation, which can be successfully monitored using the symmetry of the space subgroup. Besides, it is striking to see an activation barrier as high as 100 kJ/mol for a transition without breaking chemical bonds. We can explain this value as arising from distortions of the rigid O-Si-O angles, the increasing of Si-O shortest distances, and other local chemical and structural atomic reorganizations involved in the transformation.

2.3.4 Zircon–Reidite Phase Transition in $ZrSiO_4$

Silicates arouse great interest in geophysics and materials science. In particular, the $ZrSiO_4$ silicate is used as a host of radioactive atoms (U, Pu) and for dating rocks in which it is found. It exhibits a pressure-induced transformation from the ambient condition stable zircon phase to a high pressure phase known as reidite at around 8–10 GPa. In addition to the group-subgroup relationship between zircon and reidite space groups (as in the previous example), what makes this transition peculiar is that the coordination numbers of Si (4) and Zr (8) are the same in both low- and high-pressure phases. For transitions displaying these two features, a displacive character without a great bonding reorganization along the transition path is expected. However, we will see that this transition is reconstructive and breaking and formation of bonds do occur.

Zircon and reidite conventional unit cells belong to the tetragonal $I4_1/amd$ and $I4_1/a$ space groups, respectively, the second being a subgroup of the first. If the common subgroup ($I4_1/a$) is used for the transition pathway (trying to lead the transformation through a displacive route), a convenient transformation coordinate is the c/a ratio that evolves from a value lower than 1 (zircon) to a value greater than 2 (reidite) (see Reference 28 for details). Results are quite conclusive: the high value of the calculated energy barrier (more than 230 kJ/mol) allows us to rule out the consideration of zircon-reidite as a displacive transformation and invite us to investigate other transition pathways. Although in this example the main focus is not on the mechanism itself (as in the previous ones), a symmetry as low as monoclinic involving 12 degrees of freedom has been proposed to successfully describe the transformation [3]. This monoclinic pathway involves the rupture of two Zr-O bonds and the subsequent formation of two new Zr-O bonds. For this reason, Smirnov *et al.* have used the term "bond switching" to name the mechanism of the zircon-reidite phase transition [29]. This bond reorganization leads to an activation barrier of around 80 kJ/mol, in good agreement with estimates based on the

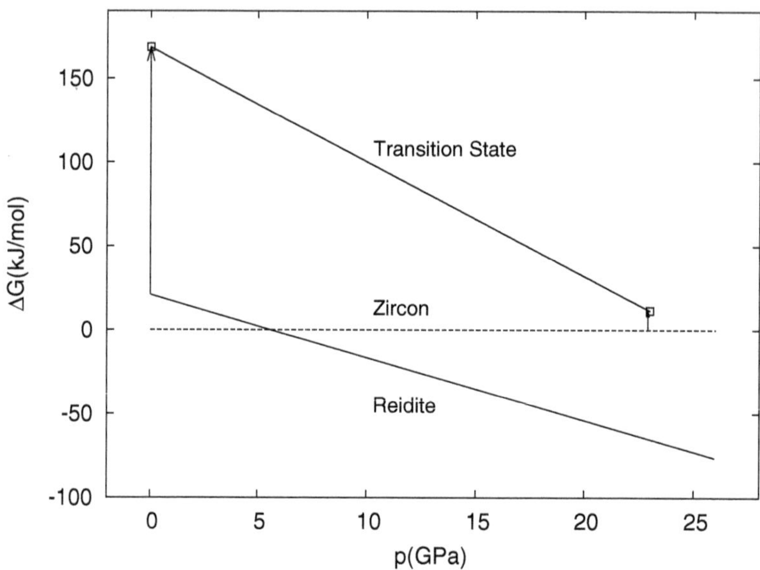

Figure 2.8 Estimated thermal barrier at the thermodynamic transition pressure (central arrow) for the zircon → reidite transition of $ZrSiO_4$. Adapted, with permission, from reference 28.

experimental observations analyzed in the next paragraph, and considering that surface and/or diffusion contributions to the barrier are not taken into account in our simulations.

This example is also interesting to illustrate thermal effects on the transition. At room temperature, it is possible to observe the zircon → reidite transition by increasing pressure up to 23 GPa. Starting with reidite at 23 GPa, a progressive pressure decrease at room temperature does not produce the conversion back to zircon, *i.e.*, reidite can exist as a metastable phase at zero pressure. Only if the temperature is increased up to 1300 K is it possible to recover zircon at zero pressure. These facts indicate the existence of activation barriers associated with the transition mechanism. With the above experimental data and using a simple Debye thermal model (see Chapter 1), it is possible to estimate the energy of this barrier at the thermodynamic pressure. To do that, a plot of the pressure dependence of the reidite Gibbs energy with respect to the zircon phase is needed (see Figure 2.8). Obviously, the straight line for zircon always has a zero value. For reidite, ΔG values above zero are obtained from quantum mechanical calculations at pressures below $p_{\rm tr} = 5.3$ GPa and negative for $p > p_{\rm tr}$ [28].

The line for the transition state can be drawn as follows. We assume a linear trend for the evolution of the activation barrier with pressure. Therefore, only two points are needed. The point at 23 GPa (small arrow) is estimated from the available vibrational energy of zircon at 300 K temperature at which the zircon → reidite transformation was observed at 23 GPa [30]. It amounts to 12 kJ/mol according to the simple Debye model, once the zero point contribution was removed. At 0 GPa, the reidite phase transforms back to zircon at 1273 K [31]. The corresponding available thermal energy of reidite is 148 kJ/mol at this temperature. The straight line for the transition state energy provides an activation barrier at the thermodynamic transition pressure of 133 kJ/mol. This vibrational energy can be converted back into temperature and leads to a value around 1200 K, in the range of the experimental data of Ono *et al.* [32]. Their static experiments showed that the zircon → reidite phase transition is observed at pressures in the range of 10–15 GPa when temperature is increased up to 1000–1500 K, thus validating the hypotheses behind the estimation of the transition barrier.

2.4 CONCLUSIONS

The kinetics of solid-solid transformations involve time evolution and mechanistic aspects. The former are more difficult to deal with using realistic phenomenological models or quantum mechanical simulations. A global description of the nucleation and growth steps of pressure-induced transformations requires consideration of simplified approaches. The martensitic approximation conceives the system as constituted by a periodic assembly of atoms that do not lose the translational symmetry. All atoms move in a concerted way from the initial to the final structure, following a transition pathway with a given space group symmetry. A transformation coordinate can be proposed, as in gas-phase chemical reactions, with an associated energetic profile. The corresponding activation barrier is a key property used to discriminate between competitive mechanisms. The examples presented in this chapter have been selected to illustrate how the martensitic approximation is applied and what kind of microscopic information can be derived from it.

Bibliography

[1] A. M. Pendás, V. Luaña, J. M. Recio, *et al.* *Phys. Rev. B*, 49, 3066 (1994).

[2] M. A. Blanco, J. M. Recio, A. Costales, *et al.* *Phys. Rev. B*, 62, R10599 (2000).

[3] M. Flórez, J. Contreras-García, J. M. Recio, *et al.* *Phys. Rev. B*, 79, 104101 (2009).

[4] M. A. Salvadó, P. Pertierra, A. Morales-García, *et al.* *J. Phys. Chem. C*, 117, 8950 (2013).

[5] E. Kroumova, M. Aroyo, J. Pérez-Mato, *et al.* *Phase Transitions*, 76, 155 (2003).

[6] H. Sowa. *Zeitsch. Kristall.*, 215, 335 (2000).

[7] H. T. Stokes and D. M. Hatch. *Phys. Rev. B*, 65, 144114 (2002).

[8] M. Catti. *Phys. Rev. Lett.*, 87, 035504 (2001).

[9] P. Tolédano and V. Dmitriev. *Reconstructive phase transitions.* World Scientific, Singapore (1996).

[10] M. S. Miao, M. Prikhodko, and W. R. L. Lambrecht. *Phys. Rev. B*, 66, 064107 (2002).

[11] D. Zahn and S. Leoni. *J. Phys. Chem. B*, 110, 10873 (2006).

[12] D. Turnbull. *Solid State Phys.*, 3, 225 (1956).

[13] M. Buerger. *Phase transformations in solids.* Wiley, New York (1951).

[14] A. R. West. *Solid state chemistry and its applications.* Wiley, Chichester (1985).

[15] A. G. Christy. *Acta Crystallogr., Sect. B: Struct. Sci.*, 49, 987 (1993).

[16] D. Zahn and S. Leoni. *Phys. Rev. Lett.*, 92, 250201 (2004).

[17] M. Catti. *Phys. Rev. B*, 68, 100101 (2003).

[18] H. Sowa. *Acta Crystallogr., Sect. A: Found. Crystallogr.*, 56, 288 (2000).

[19] J. M. Recio, A. M. Pendás, E. Francisco, *et al. Phys. Rev. B*, 48, 5891 (1993).

[20] J. C. Slater. *Phys. Rev.*, 23, 488 (1924).

[21] P. Bridgman. In *Proceedings of the American Academy of Arts and Sciences*, 19–38 (1929).

[22] P. Vinet, J. H. Rose, J. Ferrante, *et al. J. Phys.: Condens. Matter*, 1, 1941 (1989).

[23] A. M. Pendás, J. M. Recio, E. Francisco, *et al. Phys. Rev. B*, 56, 3010 (1997).

[24] M. S. Miao and W. R. L. Lambrecht. *Phys. Rev. Lett.*, 94, 225501 (2005).

[25] M. Catti. *Phys. Rev. B*, 65, 224115 (2002).

[26] J. Cai, N. Chen, and H. Wang. *J. Phys. Chem. Solids*, 68, 445 (2007).

[27] G.-R. Qian, X. Dong, X.-F. Zhou, *et al. Comput. Phys. Commun.*, 184, 2111 (2013).

[28] M. Marqués, J. Contreras-García, M. Flórez, *et al. J. Phys. Chem. Solids*, 69, 2277 (2008).

[29] M. B. Smirnov, A. P. Mirgorodsky, V. Y. Kazimirov, *et al. Phys. Rev. B*, 78, 094109 (2008).

[30] H. P. Scott, Q. Williams, and E. Knittle. *Phys. Rev. Lett.*, 88, 015506 (2001).

[31] K. Kusaba, T. Yagi, M. Kikuchi, *et al. J. Phys. Chem. Solids*, 47, 675 (1986).

[32] S. Ono, Y. Tange, I. Katayama, *et al. Am. Mineral.*, 89, 185 (2004).

CHAPTER 3

Ab Initio High-Pressure Simulations

Alfonso Muñoz González

MALTA Consolider Team and Departamento de Física, Instituto de Materiales y Nanotecnología, and Universidad de La Laguna, Santa Cruz de Tenerife, Spain

Plácida Rodríguez-Hernández

MALTA Consolider Team and Departamento de Física, Instituto de Materiales y Nanotecnología, and Universidad de La Laguna, Santa Cruz de Tenerife, Spain

CONTENTS

3.1 INTRODUCTION

QUANTUM mechanics can be applied to successfully calculate the total energy of a system of nuclei and electrons: atoms, molecules, clusters, and solids. Many properties of these systems are related to the total energy or to total energy differences between configurations. For instance, the cell parameters of a crystal are those that minimize the total energy of the structure. If this structure is also determined by internal parameters, these parameters minimize the energy of the structure and the forces on the atoms. Furthermore, quantum mechanics techniques allow the calculation not only of the optimized geometry of a system, but also of other properties related to the total energy such as bulk modulus, elastic constants, phonon spectra, phase transition pressures between different structures, etc.

The study of material properties on the atomic level is a complicated problem because it involves complex interactions between the basic constituents (nuclei and electrons). Nowadays, it is possible to deal with this problem using computational simulation. In this process a computer is used to numerically solve the equations that describe a particular situation. Simulations are used in all fields of sciences: mathematics, physics, chemistry, biology, geology, etc. Computational methods are very useful when experiments are impossible or extremely difficult, for example, to study materials at very high pressure or to analyze metastable phases. The simulation of materials allows the study from an atomistic point of view that is not possible experimentally. Combined with experiments, computer simulations help to reveal new chemical and physical properties of matter under extreme conditions. Moreover, computational experiments can be performed, since simulations allow control of physical parameters (cell parameters, length and bonds angles, etc.) in a way that is not possible experimentally, to provide insight into the physical processes and their properties.

The simulation of materials at the atomic scale has experienced a large increase in popularity in recent decades for two reasons: (i) the great progress in the underlying theory, which allowed the development of new and more efficient algorithms (software) that improve and simplify the calculations, and (ii) the advances in power of the new generations of computers (hardware). Computational simulation in physics and chemistry (and other scientific fields such as biology) has become a third way of research, between "traditional" theory and experiments.

To perform a simulation, in addition to a sufficiently powerful computer, one needs three more components: a model for the interactions between the components of the material (this model is described by a set of equations that must be solved), an algorithm to solve the equations that describe the interaction between the constituents of the material numerically, and finally, a set of tools to analyze the numerical results obtained by the simulation.

Today it is possible to solve the Schrödinger equation by methods that only require the atomic numbers of the atoms and their positions in the material under study without experimental input. These are known as *ab initio* (or first principles) methods. In addition, the atomic positions can be obtained from the chemical formulas alone using crystal structure prediction methods, see Chapter 4.

One of the most popular methods to perform quantum total energy calculations *ab initio* (first principles) is density functional theory (DFT). The DFT model defines a set of equations to be solved in a self-consistent manner: the

Kohn–Sham equations. There are various algorithms to solve these equations; each of these algorithms is based on different approximations and implemented on different simulation codes. Readers interested in learning more about *ab initio* methods can find information in References 1–4.

In this chapter, we give an overview of density functional theory with the pseudopotentials method. We describe the application of this technique to the study of structural, dynamical and elastic properties of solids under high pressure and to the calculation of pressure-driven phase transitions. We also provide information about computational codes to perform these studies and software to analyze the results.

3.2 AB INITIO SIMULATIONS: DENSITY FUNCTIONAL THEORY

The aim of a computational simulation is to predict the electronic and geometric structure of a material and its properties. To perform this task, it is necessary to solve the Schrödinger equation to obtain the total energy and then minimize that energy with respect to the electronic and nuclear positions.

We can describe a material as a set of atomic nuclei and electrons interacting via electrostatic forces; the Hamiltonian of such a system has the form:

$$\hat{H} = -\sum_I \frac{\hbar^2}{2M_I}\nabla_I^2 - \frac{\hbar^2}{2m_e}\sum_i \nabla_i^2 + \frac{1}{2}\sum_{I,J(I\neq J)} \frac{Z_I Z_J e^2}{|\mathbf{R}_I - \mathbf{R}_J|} + \frac{1}{2}\sum_{i,j(i\neq j)} \frac{e^2}{|\mathbf{r}_i - \mathbf{r}_j|} - \sum_{i,I} \frac{Z_I e^2}{|\mathbf{R}_I - \mathbf{r}_i|}, \tag{3.1}$$

where $\mathbf{R}_I$ is the set of nuclear coordinates and $\mathbf{r}_i$ is the set of electronic coordinates; M_I and Z_I are the mass and charge of nucleus I; m_e is the mass of the electron; and e is the electronic charge.

In principle, all the properties of the system can be obtained by solving the time-independent Schrödinger equation:

$$\hat{H}\Psi(\mathbf{R}_I;\mathbf{r}_i) = E\Psi(\mathbf{R}_I;\mathbf{r}_i), \tag{3.2}$$

where $\Psi(\mathbf{R}_I;\mathbf{r}_i)$ is the wave function that describes the state of the system and E is its energy. This many-body problem is formidable, and the calculation is too complicated to actually carry out. To obtain a solution to this problem, it is necessary to use some approximations to develop a theory that is still *ab initio*, but makes the problem tractable. Therefore, an *ab initio* simulation has no experimental input but it is not free of approximations.

The first approximation we can use is the adiabatic or Born–Oppenheimer approximation. Materials are composed of nuclei bound together by electrons.

The forces on electrons and nuclei due to their electric charges are of the same order of magnitude, but since the nuclei are much more massive than the electrons, they must accordingly have much smaller velocities. On the typical time scale of nuclear motion, the electrons will relax to their ground state configuration instantaneously as the nuclei move. We assume that the nuclei can be treated adiabatically and look for a solution of the dynamical problem of the electrons moving in the potential created by the frozen ionic configuration.

The most widely used approach to find the ground state of the electrons in solids from "first principles", is density functional theory. This theory, first formulated by Hohenberg and Kohn [5], is used to describe a system of many electrons with mutual interactions moving in an external potential (V_{ext}). The Hamiltonian is:

$$\hat{H} = -\frac{\hbar^2}{2m_e}\sum_i \nabla_i^2 + \sum_i V_{\text{ext}}(\mathbf{r}_i) + \frac{1}{2}\sum_{i,j(i\neq j)} \frac{e^2}{|\mathbf{r}_i - \mathbf{r}_j|}. \tag{3.3}$$

DFT is based on the two Hohenberg–Kohn theorems that can be summarized as follows:

1. For any system of particles interacting in an external potential $V_{\text{ext}}(\mathbf{r})$, this potential is determined uniquely by the ground state particle density $n_0(\mathbf{r})$ (up to an additive constant). Since $\hat{H}$ is determined by $V_{ext}(\mathbf{r})$, it follows that the properties of the system, including the total energy, are determined only by $n_0(\mathbf{r})$.

2. For any $V_{\text{ext}}(\mathbf{r})$, the energy is a unique functional, $E[n]$, of the particle density $n(\mathbf{r})$. Minimizing this functional with respect to variations in $n(\mathbf{r})$, one finds the energy and the density of the ground state. The ground state energy of the system is the minimum of this functional and the density that minimizes it is the ground state density $n_0(\mathbf{r})$.

We can express any physical property of the system in the ground state as a functional of the electron density. In DFT, instead of dealing with the many-body Schrödinger equation and the corresponding wave function, the problem is formulated in a way that involves the electron density $n(\mathbf{r})$. The Hohenberg–Kohn theorems do not provide a way to construct the functional. However, Khon and Sham [6] used this formalism to derive a procedure that provides a set of equations that can be solved with a self-consistent method.

The ground state energy of the system can be written as a functional of the electron density (and the following atomic units are used: $e = \hbar = m_e = 1/(4\pi\varepsilon_0) = 1$):

$$E[n] = \int V_{\text{ext}}(\mathbf{r})n(\mathbf{r})d\mathbf{r} + F[n], \tag{3.4}$$

under the condition $\int n(\mathbf{r})d\mathbf{r} = N$, where N is the number of electrons in the system. The universal (i.e., independent of $V_{\text{ext}}(\mathbf{r})$) functional $F[n]$ includes the kinetic energy and all the electron-electron interactions, including the Hartree energy:

$$\frac{1}{2}\int \frac{n(\mathbf{r})n(\mathbf{r}')}{|\mathbf{r}-\mathbf{r}'|}d\mathbf{r}d\mathbf{r}', \tag{3.5}$$

but the form of $F[n]$ is unknown.

Kohn and Sham noted than in a system of N non-interacting electrons in an external potential, $F[n]$ is simply the kinetic energy functional, $T_s[n]$. In this case the one-electron Schrödinger equations to be solved are:

$$\left\{-\frac{1}{2}\nabla^2 + V_{\text{ext}}(\mathbf{r})\right\}\psi_i(\mathbf{r}) = \varepsilon_i\psi_i(\mathbf{r}) \quad ; \quad i = 1\ldots N, \tag{3.6}$$

and the density is given by $n(\mathbf{r}) = \sum_{i=1}^{N}|\psi_i(\mathbf{r})|^2$.

Following this approach, the energy for an interacting system can be written as:

$$E[n] = \int V_{\text{ext}}(\mathbf{r})n(\mathbf{r})d\mathbf{r} + \frac{1}{2}\int \frac{n(\mathbf{r})n(\mathbf{r}')}{|\mathbf{r}-\mathbf{r}'|}d\mathbf{r}d\mathbf{r}' + T_s[n] + E_{\text{xc}}[n], \tag{3.7}$$

where $E_{\text{xc}}[n]$ is the exchange-correlation energy.

The ground-state density of the N interacting electrons system is found by solving in a self-consistent manner the following set of one-particle equations:

$$\left\{-\frac{1}{2}\nabla^2 + V_{\text{eff}}(\mathbf{r})\right\}\psi_i(\mathbf{r}) = \varepsilon_i\psi_i(\mathbf{r}) \quad ; \quad i = 1\ldots N, \tag{3.8}$$

with the density given by $n(\mathbf{r}) = \sum_{i=1}^{N}|\psi_i(\mathbf{r})|^2$ with the sum extended only to the first N lowest eigenstates of the single particle equation set. The effective potential for a system of electrons moving in the external potential of the ions is given by:

$$V_{\text{eff}}(\mathbf{r}) = V_{\text{ext}}(\mathbf{r}) + \frac{1}{2}\int \frac{n(\mathbf{r})}{|\mathbf{r}-\mathbf{r}'|}d\mathbf{r}' + \mu_{\text{xc}}(n(\mathbf{r})), \tag{3.9}$$

with

$$\mu_{\text{xc}}(n(\mathbf{r})) = \frac{\delta E_{\text{xc}}[n(\mathbf{r})]}{\delta n(\mathbf{r})}. \tag{3.10}$$

If the functional $E_{\text{xc}}[n]$ were known, the Kohn–Sham (KS) equations would give the ground state density and energy for the interacting-particles system. However, the exact exchange-correlation functional is unknown. It is necessary to adopt a reasonable form for the exchange-correlation energy, ($E_{\text{xc}}[n]$), that can be approximated as a local or nearly local functional. If the density $n(\mathbf{r})$ varies slowly then, for practical uses, the local density approximation (LDA) can be adopted:

$$E_{\text{xc}}[n] = \int n(\mathbf{r})\varepsilon_{\text{xc}}(n(\mathbf{r}))d\mathbf{r}. \tag{3.11}$$

In this approximation, the exchange-correlation energy density at each point is the same as in a homogeneous electron gas with density, $\varepsilon_{\text{xc}}(n(\mathbf{r}))$. The LDA approximation works remarkably well and has been extensively used.

Regarding the implementation of LDA, no exact analytical expressions for the exchange-correlation of a homogeneous gas are known. A widely used expression is the one obtained by Ceperley and Alder [7] performing quantum Monte Carlo simulations. The corresponding values were parameterized by Perdew and Zunger [8].

Although there is not a systematic way of improving the exchange-correlation functional, there are approximations for the $E_{\text{xc}}[n]$ that improve in some cases the results for solids and molecules obtained with LDA. For example, different flavors of generalized gradient approximations (GGAs) that include the gradient of the density [2]. In general all the GGA functionals lead to cohesion energies lower than LDA. In materials with localized and strongly interacting electrons, the LDA+U and GGA+U methods allow to improve the descriptions of some properties as the electronic band gap [2].

DFT can be generalized to the case of solids with magnetic order and molecules with net spin [2].

3.3 HOW TO SOLVE KOHN–SHAM EQUATIONS

To solve the KS equations that describe a particle moving in an effective potential, the wave functions are expanded in basis sets. There are three basic approaches. Each of the methods has advantages and disadvantages, and is most appropriate for a certain range of problems:

- Plane waves basis set. Plane waves are very simple and easy to use with *ab initio* methods. Mathematically, the use of plane waves implies the use of Fourier transform between real and reciprocal space, and there are many efficient algorithms to realize this task. The inconvenience is the large number of plane waves needed to develop the wave functions and potentials reasonably.

- Localized atomic-like orbitals (LCAOs). These orbitals are only important in some regions of the space, near the atoms. The number of orbitals needed is low for each atom but it is difficult to achieve good convergence of the basis set. LCAOs are widely used in chemistry.

- Atomic sphere methods. The basic idea is to divide the space in regions near each nucleus where all the magnitudes present atomic-like features and regions between the atoms where all the magnitudes change in a smooth way. These basis sets combine the good features of the localized functions and the plane waves, but they are difficult to implement and the computational cost is high.

In this chapter we will concentrate in the solution of the Kohn–Sham equations employing a set of plane waves and the pseudopotential method. A material is composed of nuclei bound together by electrons. However, electrons in the solid experience two different kinds of potentials. Electrons near the cores remain localized around the atoms because they feel a strong atomic potential, so their wave functions resemble those of individual atoms and oscillate rapidly. The remaining electrons, the valence electrons, determine the majority of solid properties in which we are interested. These electrons experience a smooth potential.

To facilitate calculations and reduce the computational cost, it is possible to replace the strong Coulomb potential created by the nucleus and the effect of the tightly bound core electrons by an effective *ab initio* potential acting on the valence electrons. This pseudopotential is constructed in such a way that the pseudowave function has no radial nodes in the core region and the pseudopotential and the pseudowave function are identical to the real wave function and the real potential in a region outside the core determined by a cutoff radius R_c [1–4].

Pseudopotentials can be constructed in several ways [1–4] but they must describe the salient feature of the valence electrons and be transferable to a wide variety of systems. Among the most employed pseudopotentials are the "*ab initio* norm conserving pseudopotentials" [9–12]. Recently, new pseudopotential forms have been utilized such as the ultrasoft pseudopotentials [13] and the projector-augmented wave pseudopotentials (PAWs) that greatly improve the accuracy of simulation and reduce the computer time devoted to them [14].

In a crystal, the positions of the nuclei are repeated periodically in space. The crystal is built by the repetition of one unit cell where the atomic positions are fixed and applying cell translations in the three directions of space. Therefore, electrons feel a periodic potential. This periodicity simplifies enormously the problem we want to solve (the KS equations). As we must perform calculations on a periodic system, we can apply Bloch's theorem to each electronic wave function. Bloch's theorem states that in a periodic solid each wave function can be written as the product of a plane wave and a function with the periodicity of the crystal. However, indeed in *a priori* non-periodic systems like atoms, molecules, point defects or surfaces it is possible to impose periodic conditions employing a periodic supercell that contains the system under study.

Using the periodicity of the crystal, each electronic wave function can be written as a sum of plane waves:

$$\psi_{n,\mathbf{k}}(\mathbf{r}) = \sum_{\mathbf{G}} C_{n,\mathbf{k}+\mathbf{G}} e^{i(\mathbf{k}+\mathbf{G})\cdot\mathbf{r}}, \tag{3.12}$$

where the sum runs over the reciprocal lattice vectors $\mathbf{G}$, n is the band number and $\mathbf{k}$ refers to a vector in the first Brillouin zone that identifies the state.

In principle, to correctly obtain the electron density and the contributions to the total energy, all the $\mathbf{k}$ points corresponding to the occupied states must be taken into account. However, the electronic wave functions at $\mathbf{k}$ points that are close to each other will be almost identical. Several methods have been devised to obtain accurate results using only a finite small number of $\mathbf{k}$ points, named "special $\mathbf{k}$ points" within the Brillouin zone. The most popular among them is the Monkhorst–Pack scheme [15]. A few $\mathbf{k}$ points suffice to yield accurate results for semiconductors and insulators. In the case of metals, dense meshes of $\mathbf{k}$ points are needed. The error due to $\mathbf{k}$ point sampling can be always reduced by employing a denser set of $\mathbf{k}$ points. Indeed there are techniques that allow reducing the number of needed $\mathbf{k}$ points [1–4].

As stated above, the electronic wave functions at each $\mathbf{k}$ point can be expanded in term of a discrete plane wave basis set, although, this base should in principle be infinite. In practice, this basis is finite due to the fact that the coefficients $C_{n,\mathbf{k}+\mathbf{G}}$ for plane waves with small kinetic energy, $\frac{\hbar^2}{2m_e}|\mathbf{k}+\mathbf{G}|^2$, are more important than those with large kinetic energy, so the plane wave basis set can be truncated to include only those plane waves that have kinetic energies lower than some particular cutoff energy, E_{cutoff}:

$$\frac{\hbar^2}{2m_e}|\mathbf{k}+\mathbf{G}|^2 < E_{\mathrm{cutoff}}. \tag{3.13}$$

The truncation of the plane wave basis leads to an error in the computed total energy. This error can be reduced by increasing the value of the cutoff energy until convergence is reached. Hence, the cutoff energy must be chosen so that the total energy of the system and the other properties in which we are interested are well converged. It should be noted that the pseudofunctions and pseudopotentials that substitute the real functions and potentials can be expanded easily with a lower number of plane waves functions than the real ones.

Finally, to obtain the quantities we are interested in, we must solve in a self-consistent way the Kohn–Sham equations. That is, obtain the effective potential consistent with the electron density for the ground state. In order to get self-consistent solutions, first we select the system we want to study. The atoms and their positions determine the pseudopotential, $V_{\text{ext}}^{\text{pseudo}}$. We first obtain an initial guess for the electronic charge density, $n^{\text{in}}(\mathbf{r})$, from which the effective potential can be calculated. This allows solving the KS equations and obtaining the eigenstates to calculate a new electronic charge density $n^{\text{out}}(\mathbf{r})$. If the solution is consistent, i.e., the input and output potential agree and the input and output densities also agree, then the magnitudes of interest can be obtained: the self-consistent $n(\mathbf{r})$, total energy, forces on the atoms, etc. If the solution is not consistent, then the new $n(\mathbf{r})$ must be employed as input to construct a new effective potential and a new Hamiltonian that provides new eigenstates, which produce in turn a new $n(\mathbf{r})$. The process continues until self-consistency is obtained. Care must be taken to ensure the total energy is converged as a function of the number of **k**-points and as function of the cutoff energy for the plane wave basis set.

After solving the KS equations, it is necessary to be sure that we have obtained the optimized relaxed geometry for the structure under study. The energy of the structure can be determined as a function of the atomic positions and all the cell parameters. For the optimized structure, the forces on the atoms must be almost zero. The forces on the atoms can be determined with the Hellmann–Feynman theorem [16]. The stress theorem [17] allows us to obtain the stress tensor for each atomic configuration and hence the pressure on the system. In the case of hydrostatic pressure, the stress tensor is diagonal. Therefore, in a total energy calculation we obtain not only the energy but also the pressure in the system and the forces on the atoms. If, for a particular atomic configuration, the forces on the atoms are not zero or the stress tensor is not diagonal, a minimization algorithm would change the atomic positions and the process described above must start again. This process continues until we obtain the optimized relaxed structure: the positions of the atoms in the unit cell and the size and shape of the cell that minimize the energy.

3.4 GEOMETRY OPTIMIZATION

To perform a first principles simulation of a crystal, the computer code will need only:

- The number of atoms, their atomic numbers, and their positions in the cell.
- The cell parameters, that determine the cell volume, V.

From these quantities, the equilibrium structure is obtained, i.e., the atomic positions and the cell parameters that minimize the energy, and hence the energy and all the physical properties of the system.

Many relevant magnitudes of the system can be obtained as derivatives of the energy respect to the volume. For instance, the pressure:

$$p = -\frac{dE}{dV}, \tag{3.14}$$

the bulk modulus

$$B = V\frac{d^2E}{dV^2}, \tag{3.15}$$

and other higher-order energy derivatives.

It is important to emphasize that the equilibrium configuration of a system is obtained only if the calculations are well converged respect to the number of plane waves, the energy cutoff, and the number of special **k** points employed in the integration over the first Brillouin zone. If this is the case, the forces on the atoms obtained using the Hellmann–Feynman theorem are approximately zero (in practice, usually fewer than 0.006 eV/Å). If the stress tensor, (σ_{ij}), is diagonal (hydrostatic pressure), then the external pressure:

$$p = -\frac{dE}{dV} = \frac{1}{3}(\sigma_{xx} + \sigma_{yy} + \sigma_{zz}) \tag{3.16}$$

is almost zero (~0.1 GPa) for the ground state.

Once the energy is obtained for a fixed volume, the calculation of the total energy of the system for different volumes (this means, in practice, different pressures) can be performed. Note that the volume, V, for a particular cell can be fixed easily: it suffices to choose the parameters corresponding to a certain crystal symmetry. Once the calculations are done, the energy-volume $E(V)$ curve is obtained. The minimum of this curve corresponds to the equilibrium configuration of the system (the configuration with lowest energy). The pressure at this point is zero and the volume, V_0, is the equilibrium volume of the system. This volume can be compared with available experimental data.

The energy-volume data can be fitted with an equation of state (for example, the Murnaghan equation of state [18]) and this fit can be used to obtain the energy at equilibrium, E_0, the equilibrium volume, V_0, the bulk modulus, B_0, and its pressure derivative, B_0'. When the equilibrium volume or the bulk modulus is known experimentally, we can test the accuracy of our calculations; otherwise our results can be taken as predictions. As our calculations are within the adiabatic approximation, the atoms are at fixed positions in the cell and the thermal vibrations of the system are ignored. This means that the simulation is performed in the static approximation: at zero temperature, $T = 0$ K, and the zero-point vibrational contribution is not included.

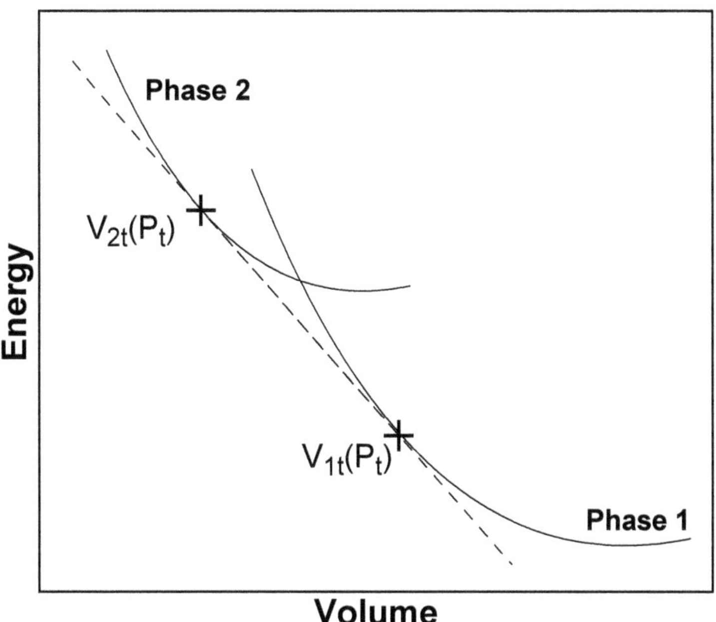

Figure 3.1 Plot of energy versus volume for two crystal phases, showing the common tangent method used to obtain the transition pressure between them.

3.5 PHASE TRANSITIONS INDUCED BY PRESSURE

It is possible to compare the relative stabilities of different phases of a system (the same compound in different crystal structures) to either predict new phases which are not known experimentally or to calculate metastable phases not accessible by experiments. For this task, we only need to obtain the $E(V)$ curves for the phases we want to compare. By analyzing these curves, the relative stability of each phase can be known. In addition, to obtain the thermodynamic transition pressure between two phases at a temperature T, we must to take into account that the transformation takes place when the Gibbs free energy, $G = E + pV - TS$, is the same in both phases (S is the entropy of the system). See Chapter 1 for more details.

In a static simulation ($T = 0$ K and no zero-point vibrations), the relevant thermodynamic magnitude is the enthalpy:

$$H = E + pV. \tag{3.17}$$

The transition between two phases occurs when the enthalpy is the same for both. It is sufficient to calculate the enthalpy for phase 1 and phase 2, as a function of pressure; the enthalpy of both phases will be equal at the transition pressure, p_t. Alternatively, the thermodynamic transition pressure, p_t, can also be obtained by the construction of "the common tangent" (see

Figure 3.1). Taking into account that p_t is the same for both phases:

$$H_{1t} = E_{1t} + p_t V_{1t} = H_{2t} = E_{2t} + p_t V_{2t}, \tag{3.18}$$

where V_{1t} and V_{2t} are the volumes of phase 1 and phase 2 at the transition pressure and E_{1t} and E_{2t} are the corresponding energies, it follows that:

$$p_t = -\frac{E_{1t} - E_{2t}}{V_{1t} - V_{2t}}. \tag{3.19}$$

The transition pressure is the tangent of the $E(V)$ curves at V_{1t} and V_{2t} (as illustrated in Figure 3.1). The volume reduction at the transition can be calculated easily: $\Delta V_t = V_{2t} - V_{1t}$. Of course, the thermodynamic transition pressure may be different from experiment results when, for example, kinetic barriers are present during the phase transition.

Using this method, it is not possible to predict new structures because one can only compare the phases considered in the simulation. The truly high pressure structure may not be one of the tested structures. However, it is possible test a set of candidate structures taking into account their structural relations with the equilibrium structure, and previous theoretical and experimental results. Some simulation techniques (random search, genetic algorithms, metadynamics, etc.) can be used to try to find the most stable structure, but the computational cost is often high and there are limitations in the number of atoms per cell; see Chapter 4 for more details.

3.5.1 Illustrative Example

DFT has been applied to study the phase transitions at high pressure in a variety of systems with great success [19]. To illustrate the methods described above, we present in the following an example of an *ab initio* simulation of the behavior of GaN under pressure.

Although most III-V semiconductors have a zincblende structure (space group 216, $F\bar{4}3m$), the III-nitrides have a wurtzite structure (space group 186, $P6_3mc$) at zero pressure. GaN belongs to this family, and has four atoms in the unit cell. The Ga atoms are located at Wyckoff positions 2b $(1/3, 2/3, 0)$ and the N atoms are at 2b $(1/3, 2/3, u)$. The high pressure behavior of GaN was predicted by Muñoz and Kunc [20]. They found a theoretical phase transition from the wurtzite to a rock-salt (NaCl) type structure. This transition was corroborated in an experimental study by Perlin *et al.* [21]. Both studies reported for the first time a phase transition of III-V compound to a six-fold coordinated rock-salt structure. The rock-salt structure was observed previously as a high pressure phase in II-VI semiconductors. Here, we show how to carry out an *ab initio* simulation of GaN in the wurtzite, zincblende and rock-salt structures under pressure (see Figure 3.2).

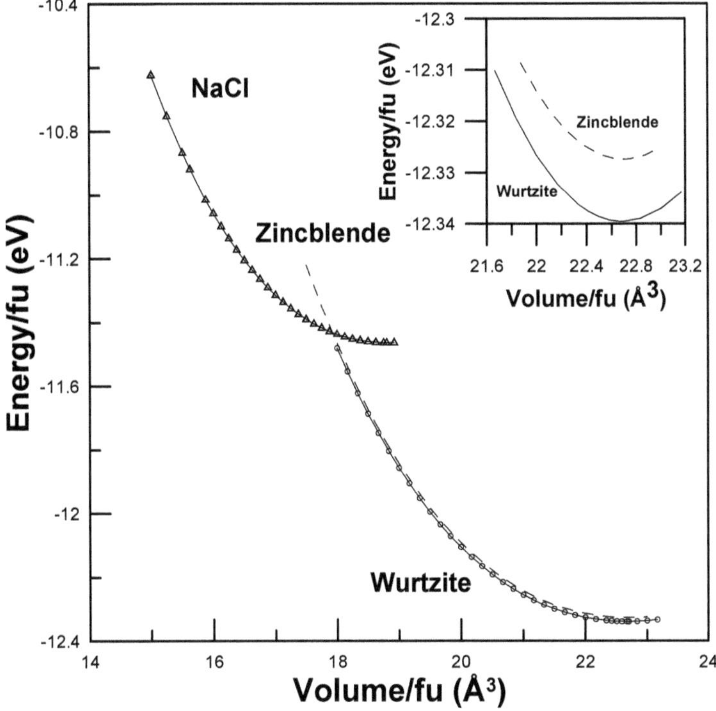

Figure 3.2 Energy versus volume for wurtzite (circles), zincblende (dashed lines) and rock-salt (triangles) phases of GaN. The inset shows the small energy difference between the wurtzite and zincblende structures. Both structures are quite similar except for the stacking of the crystal planes.

All calculations were performed within the framework of the density-functional theory [5] using the GGA exchange-correlation functional PBEsol [22] (PBE for solids). The static approximation was used ($T = 0$ K and no vibrational zero-point motion). This scheme is known to give an accurate enough description of the structural phase diagrams of groups IV, III-V, and II-VI semiconductors [19].

A plane-wave basis set was used to solve the KS equations in the pseudopotential implementation. Only the outermost electrons of each atom were explicitly considered in the calculation and the effects of the inner electrons and the nucleus were described within a pseudopotential scheme. The projector augmented wave (PAW) scheme [14] was employed to take into account the full nodal character of the all-electron charge density distribution in the core region. The semicore 3*d* electrons of Ga were treated explicitly. Even using smooth PAW pseudopotentials, the hardness of the *p* component of the N pseudopotential and that of the *d* component of the Ga considered in the valence required the use of a large kinetic energy cutoff to produce well converged results. We have found that a cutoff of 520 eV is enough to have the total energy converged to an accuracy of about 1 meV per formula unit.

The reciprocal space integrations were done in the Monkhorst–Pack scheme [15]. The **k** point meshes used were $6 \times 6 \times 6$ for the wurtzite, rock-salt, and zincblende semiconducting phases. As explained previously, the forces on the atoms were calculated through the Hellmann–Feynman theorem [16] and the pressure with the stress theorem [17]. In the optimized configurations, the forces on the atoms are fewer than 0.006 eV/Å and the deviation of the stress tensor from a diagonal hydrostatic form is fewer than 0.1 GPa. The simulation was performed with the code and pseudopotentials provided by the Vienna *ab initio* simulation package, VASP [23].

For each of the structures considered (wurtzite, rock-salt, and zincblende), we made a set of calculations at different volumes (equivalently, at different pressures). Figure 3.2 shows the energy as function of volume for the three phases. Although the wurtzite and zincblende phases are close in energy, it is clear that the energy is lower for the wurtzite structure. At zero pressure and temperature, the wurtzite is the most stable phase; the minimum of the $E(V)$ curve corresponds to the optimized geometry at zero pressure. From the optimized geometry at each pressure we get the cell parameters and their evolution under pressure; see Figure 3.3. The internal parameter for the atomic position, $u = 0.3765$ and the cell parameters at equilibrium ($p = 0$) are $a_0 =$ 3.1789 Å, $c_0 = 5.1798$ Å and are in good agreement with the experimental data [24]: $u = 0.375 \ldots 0.378$, $a_0 = 3.186$ Å, and $c_0 = 5.176$ Å.

The $E(V)$ data for the wurtzite structure were fitted with a third-order Birch–Murnaghan equation of state [25] to obtain the volume, $V_0 = 45.372$ Å^3,

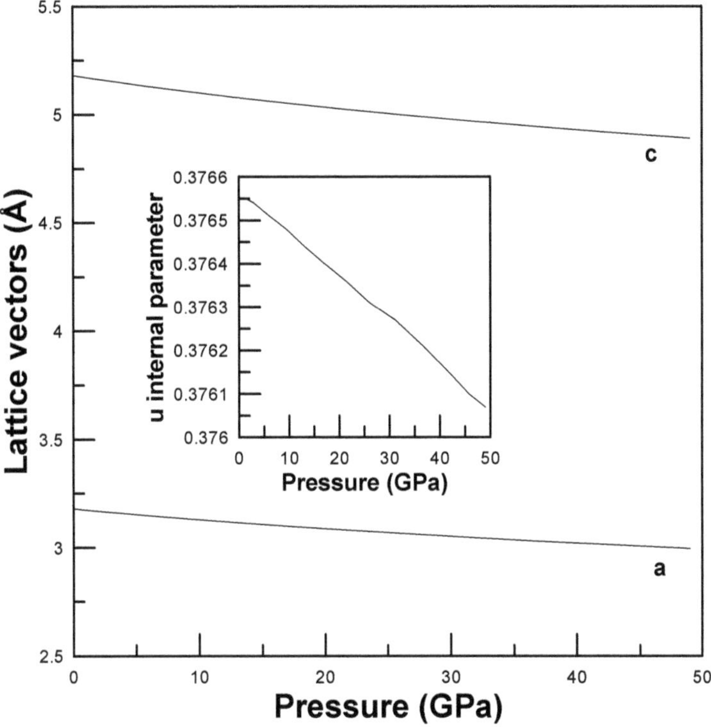

Figure 3.3 Pressure evolution of the wurtzite lattice parameters a and c in GaN. The inset shows the evolution of the internal parameter u.

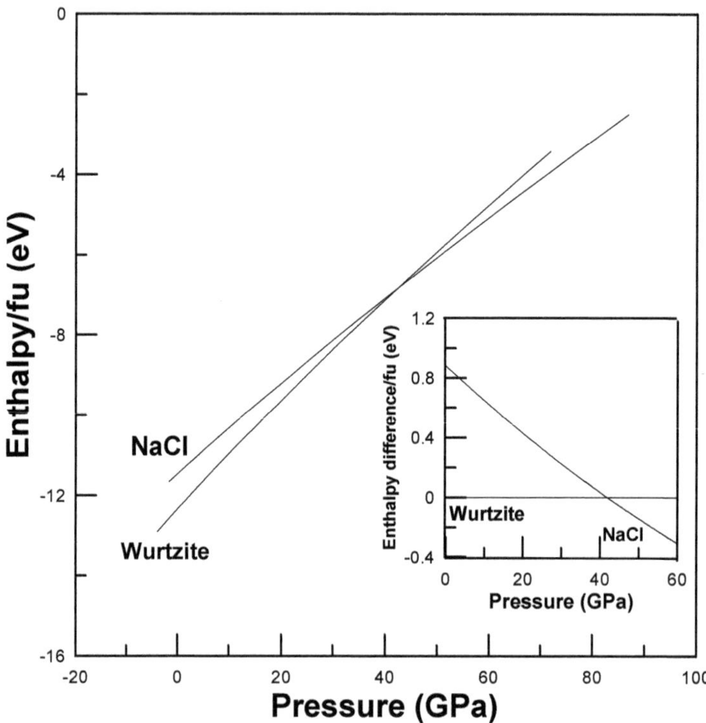

Figure 3.4 Enthalpy per formula unit versus pressure for the wurtzite and rocksalt phases of GaN. The inset shows the enthalpy difference with the wurtzite phase as reference.

the bulk modulus $B_0 = 189.7$ GPa, and its first derivative $B_0' = 4.25$, at zero pressure. The calculated values are in very good agreement with previous theoretical and experimental results [26].

By applying hydrostatic pressure to the wurtzite structure (i.e., calculating the energy and the optimized structure for increasingly small volumes), a transition to the rock-salt phase takes place, as can see in Figure 3.2. The enthalpy versus pressure curves for the wurtzite and rock-salt structures are plotted in Figure 3.4. It is clear that at a pressure over 42 GPa, the enthalpy of the rock-salt structure is the lowest. The calculated transition pressure is therefore $p_t = 42$ GPa. The experimental result is a hysteresis behavior for the transition pressure: 48 GPa on upstroke and 30 GPa on downstroke [21]. The construction of the common tangent or the $p(V)$ data allow calculation of the volume contraction at the transition, 15 % in the present case.

3.6 PHONONS

The vibrational properties of solids determine a wide variety of physical properties: specific heat, thermal expansion, heat conduction, resistivity of metals, superconductivity, sound velocity, infrared and Raman absorptions, and elasticity. In some cases, the modes with low frequencies can be associated with phase transformations, because the appearance of imaginary frequencies can be an indication that the structure under study is dynamically unstable.

Let us consider an infinite periodic crystal with N unit cells (N in the order of Avogadro's number) and n atoms per cell. The atom j in the cell given by lattice vector $\boldsymbol{L}$ (represented by the lattice index L) vibrates around its equilibrium position $\boldsymbol{r}_0(jL)$, and we define its displacement as:

$$\boldsymbol{u}_{jL} = \boldsymbol{r}_{jL} - \boldsymbol{r}_{0,jL}. \tag{3.20}$$

There are a total of $3nN$ atomic coordinates. The static energy of the crystal E can be expanded in a Taylor series using the atomic displacements around the equilibrium geometry. Since the static energy is minimum at equilibrium, the energy in the neighborhood of this point does not contain a linear term, and reads:

$$E = E_0 + \frac{1}{2} \sum_{jj'LL'} \boldsymbol{u}_{jl}^T \boldsymbol{D}_{jLj'L'} \boldsymbol{u}_{j'L'} + \ldots, \tag{3.21}$$

where E_0 is the static energy at zero-pressure (equilibrium), $\boldsymbol{u}_{jl}^T$ is the transpose of $\boldsymbol{u}_{jl}$, and the sum goes over all pairs of atoms in the crystal. $\boldsymbol{D}_{jLj'L'}$ is the 3×3 matrix of second derivatives with respect to the atomic displacements of atoms jL and $j'L'$:

$$(\boldsymbol{D}_{jLj'L'})_{\alpha\beta} = \frac{\partial^2 E}{\partial(\boldsymbol{u}_{jL})_\alpha \partial(\boldsymbol{u}_{j'L'})_\beta}, \tag{3.22}$$

where α and β run over Cartesian coordinates (x, y, and z).

In the harmonic approximation [27–30], the Taylor expansion of the crystal energy in Equation (3.21) is truncated to second order. The harmonic approximation assumes that the atoms in the crystal vibrate inside a harmonic well around fixed positions and that their vibrational lengths are small compared to interatomic spacings. The equation of motion for each atom is given by:

$$m_j \ddot{\boldsymbol{u}}_{jL} = \boldsymbol{F}_{jL} = -\sum_{j'L'} \boldsymbol{D}_{jLj'L'} \boldsymbol{u}_{j'L'}, \tag{3.23}$$

where m_j is the mass of atom j, and $\boldsymbol{F}_{jl}$ is the force on the atom at the jL position. There are $3nN$ equations of motion, one for each Cartesian component of every atom. Their solutions have the form of plane waves [2, 3, 27, 29, 31]:

$$\boldsymbol{u}_{jL} = \boldsymbol{\varepsilon}_{\boldsymbol{k}\nu j} e^{i(\boldsymbol{k}\cdot\boldsymbol{L} - \omega_{\boldsymbol{k}\nu} t)}. \tag{3.24}$$

The $\boldsymbol{k}$ vector index arises from imposing periodic boundary conditions on the crystal [27]. For a periodic system with N primitive cells, there are N permissible $\boldsymbol{k}$ vectors that can be used as solutions of Equation (3.23) (the $\boldsymbol{k}$ vectors in the first Brillouin cell). ν is an index that counts the number of solutions for a given $\boldsymbol{k}$ vector, and runs from 1 to $3n$, where n is the number of atoms in the primitive unit cell. The vector $\boldsymbol{\varepsilon}_{\boldsymbol{k}\nu j}$ is called the displacement vector, and gives the propagation direction for the wave of collective atomic displacements. $\omega_{\boldsymbol{k}\nu}$ is the frequency of the wave. The atomic displacement waves represented by Equation (3.24) (or linear combinations thereof) are called phonons. They carry vibrational energy across the crystal. Phonons possess momentum and they follow conservation of momentum rules in their interaction with other waves (like photons during a spectroscopy experiment or other phonons).

Substitution of Equation (3.24)) in the equation of motion for the atomic displacements Equation (3.23) gives:

$$m_j \omega_{\boldsymbol{k}\nu}^2 \boldsymbol{\varepsilon}_{\boldsymbol{k}\nu j} = \sum_{j'} \left(\sum_{L'} \boldsymbol{D}_{j0j'L'} e^{i\boldsymbol{k}\cdot\boldsymbol{L}'} \right) \boldsymbol{\varepsilon}_{\boldsymbol{k}\nu j'}, \tag{3.25}$$

with $\boldsymbol{L} = \boldsymbol{0}$ as it refers to the reference unit cell. The quantity in parentheses is the Fourier transform of the matrix formed by the second derivatives of the energy. In order to simplify this equation and eliminate the dependence on the atomic masses, we define $\boldsymbol{\eta}_{\boldsymbol{k}\nu j}$, which is simply $\boldsymbol{\varepsilon}_{\boldsymbol{k}\nu j}$ but weighted by the square root of the mass of atom j:

$$\boldsymbol{\varepsilon}_{\boldsymbol{k}\nu j} = \frac{1}{\sqrt{m_j}} \boldsymbol{\eta}_{\boldsymbol{k}\nu j}. \tag{3.26}$$

This allows us to define the matrix on the right hand side of Equation (3.25) as:

$$\boldsymbol{D}_{jj'}(\boldsymbol{k}) = \frac{1}{\sqrt{m_j m_{j'}}} \sum_{L'} \boldsymbol{D}_{j0j'L'} e^{i\boldsymbol{k}\cdot\boldsymbol{L}'}. \tag{3.27}$$

Equation (3.25)) now reads:

$$\omega_{\boldsymbol{k}\nu}^2 \boldsymbol{\eta}_{\boldsymbol{k}\nu j} = \sum_{j'} \boldsymbol{D}_{jj'}(\boldsymbol{k}) \boldsymbol{\eta}_{\boldsymbol{k}\nu j'}. \tag{3.28}$$

All the 3×3 matrices for the n atoms in the unit cell can be collected in a square $3n \times 3n$ matrix $D(\boldsymbol{k})$, called the dynamical matrix. The Fourier transform of the dynamical matrix to real space gives a matrix similar to the second derivative matrix $\boldsymbol{D}_{j0j'L'}$ Equation (3.22), but weighted with the atomic masses. The real-space transform of the dynamical matrix is called the force-constant matrix:

$$\boldsymbol{C}(\boldsymbol{L})_{j\alpha j'\beta} = \frac{1}{\sqrt{m_j m_{j'}}} \frac{\partial^2 E}{\partial(\boldsymbol{u}_{j0})_\alpha \partial(\boldsymbol{u}_{j'L'})_\beta}, \tag{3.29}$$

also with dimension $3n \times 3n$.

Merging all the atomic coordinates in the displacement vector $\boldsymbol{\eta}_{\boldsymbol{k}\nu}$ transforms Equation (3.28) into an eigenvalue problem:

$$\omega^2_{\boldsymbol{k}\nu} \boldsymbol{\eta}_{\boldsymbol{k}\nu} = \boldsymbol{D}(\boldsymbol{k}) \boldsymbol{\eta}_{\boldsymbol{k}\nu}. \tag{3.30}$$

Diagonalization of the dynamical matrix allows the calculation of the phonon frequencies ($\omega_{\boldsymbol{k}\nu}$) and the phonon eigenvectors ($\boldsymbol{\eta}_{\boldsymbol{k}\nu}$). The latter are called the polarization vectors or normal modes. Since the dynamical matrix is Hermitian ($\boldsymbol{D}_{ij} = \boldsymbol{D}^{\star}_{ji}$), the squares of the frequencies are real and the phonon eigenvectors can be chosen to form an orthonormal set.

It is now possible to perform reliable theoretical calculations of the dynamical properties of a system using *ab initio* quantum mechanical techniques. Vibrational mode frequencies and their symmetry can be obtained along with the static dielectric constant, piezoelectric constant, effective charges, and other information that is not directly (or easily) available from an experiment.

To calculate the phonon mode frequencies of a crystal, two basic methods are used: the frozen-phonon (direct method) and a method based on the density functional perturbation theory (DFPT) [32]. In both methods the study of the dynamical properties of the system requires us to obtain first the equilibrium state for the system (as explained above).

The frozen-phonon method allows the calculation of frequencies for selected modes from energy differences or from the forces acting on the atoms produced by the finite, periodic displacement of a few atoms in a perfect crystal at equilibrium. The atoms are displaced by small quantities, u, to obtain a system configuration representing a frozen-phonon (a snapshot of the system undergoing vibration). The total energy and the atomic forces and stresses are calculated for that configuration. In others words, if we have a system with N atoms in each cell, the system has $3N$ phonon branches. By making a series of small displacements and evaluating the total energy and the forces (through the Hellman-Feynman theorem) on the atoms, we obtain the interatomic force constants and the dynamical matrix of the system. The diagonalization of the dynamical matrix provides the phonon frequencies. It is interesting to emphasize that, in principle, as the number of atoms per cell is N, the number of displacement to be made is $3N$. However, in practice, the number of displacements required to build the matrix constant is lower and is determined by the crystal symmetry.

It is important to note that we are working under the harmonic approximation. The displacement of the atoms must be small to ensure we are indeed in the harmonic regime, but large enough to ensure that energy differences between configurations of the system will be significant and larger than the numerical noise. The frozen-phonon method allows the calculation of the phonons at the zone center Γ ($\mathbf{k} = \mathbf{0}$) with minimal computational effort. An analysis of the eigenvectors of the phonons at Γ with group theory allows us to identify the Raman-active, infrared-active, and silent phonons and the associated displacement patterns of the atoms, which can help assign the phonon modes.

Phonon dispersion along high symmetry lines requires the use of supercells ($2\times2\times2$, $3\times3\times3$, etc.) to find exact solutions for the **k** points commensurable with the supercell size. The rest of the phonons are obtained by interpolation. The phonon density of states $g(\omega)$ (DOS) and the partial density of states (PDOS) projected onto a particular atom can be obtained with the direct method.

Phonon calculations can be carried at different pressures, as explained before, simply by varying the volume of the cell. The evolution of the vibrational properties with pressure can be studied in this way. The Grüneisen parameters (the volume derivatives of the phonon frequencies) can be obtained, along with the evolution of the Raman and infrared spectra under pressure. The appearance of phonon branch softening with increasing pressure can be associated with a phase transition and the analysis of the phonon eigenvectors can determine structural relations with the high-pressure phase. The vibrational contributions to the free energy, the heat capacity, and other thermodynamic properties can be calculated using the harmonic formulas in Chapter 1.

The direct method does not allow the study of the LO-TO splitting at Γ [31]. The LO-TO splitting is related to the non-analytical energy term associated to the electric fields in polar crystals. However, there are some approximations that introduce this effect. By using Born effective charges and the dielectric constant, it is possible to model the effect of the electric field and to obtain the LO-TO splitting.

The study of the dynamical properties of a crystal with the frozen-phonon approach does not require any special code; it suffices to use a standard total energy code. The principal limitation is the requirement for large supercells, and the computational cost can be high.

Another approach for the calculation of phonon modes is based on response functions. The density functional perturbation theory DFPT [32] is designed to calculate the response of the system using perturbation theory, which describes the properties of a system as an expansion in powers of the perturbation. With DFPT, it is now possible to calculate phonon frequencies at arbitrary wave vectors and avoid the use of supercells. The phonon frequencies can be obtained on a fine grid of wave vectors in the Brillouin zone and several properties of the system can be calculated, such as heat capacities, thermal expansion coefficients, etc. (see Chapter 1). Further information on this method can be found in the review of Baroni *et al.* [32] and online at the Scuola Internazionale Superiore di Studi Avanzanti (Trieste) website.

3.6.1 Illustrative Example

We present the case of GaN as an example of a practical study of the dynamical properties using the direct approach. The simulation was performed for the GaN in the wurtzite structure as explained previously. Group theory predicts two E_2, two A_1, two E_1, and two B_1 modes. One A_1 mode and one E_1 mode are acoustic and the other two are both Raman and infrared active. The E_2 modes are Raman active and the B_1 modes are silent. The phonons obtained at Γ were fitted with a second order polynomial $\omega = \omega_0 + ap + bp^2$, where ω is given in cm^{-1} and p in GPa . The calculated frequency for the phonons at zero pressure, ω_0, and the Grüneisen parameters, $\gamma = (B/\omega)(\partial\omega/\partial p)$, are:

$\mathbf{E_2}$: $\omega_0 = 138.3\ \text{cm}^{-1}$, $a = -0.27$, $b = -0.003$, and $\gamma = -0.370$.

$\mathbf{A_1}$: $\omega_0 = 517.2\ \text{cm}^{-1}$, $a = 3.99$, $b = -0.015$, and $\gamma = 1.463$.

$\mathbf{E_1(TO)}$: $\omega_0 = 541\ \text{cm}^{-1}$, $a = 3.92$, $b = -0.017$, and $\gamma = 1.374$.

$\mathbf{E_2}$: $\omega_0 = 551\ \text{cm}^{-1}$, $a = 4.26$, $b = -0.018$, and $\gamma = 1.466$.

These values were calculated without including the LO-TO splitting of the Raman and infrared-active modes. They do not provide the E_1(LO) or the A_1(LO) phonons, but are in agreement with the experimental values reported by Perlin *et al.* [21] The silent B_1 modes are difficult to study experimentally. The *ab initio* results are:

$\mathbf{B_1(low)}$: $\omega_0 = 328.4\ \text{cm}^{-1}$, $a = 1.60$, $b = -0.008$, and $\gamma = 0.92$.

$\mathbf{B_1(high)}$: $\omega_0 = 678.6\ \text{cm}^{-1}$, $a = 4.37$, $b = -0.017$, and $\gamma = 1.22$.

The pressure evolution of the calculated phonon frequencies is displayed in Figure 3.5. The phonon dispersion in the Brillouin zone shown in Figure 3.6 was calculated using a 2×2×2 supercell. This figure shows the LO-TO splitting of E_1 and A_1 phonons obtained by including the Born effective charges and the dielectric constant as in Reference [33]. The total and partial density of states, Figure 3.7, were obtained from the *ab initio* data and processed using the PHONON [33] utility software. As shown in Figure 3.7, the acoustic modes are mainly related to the Ga atoms, and the optical modes are related to the

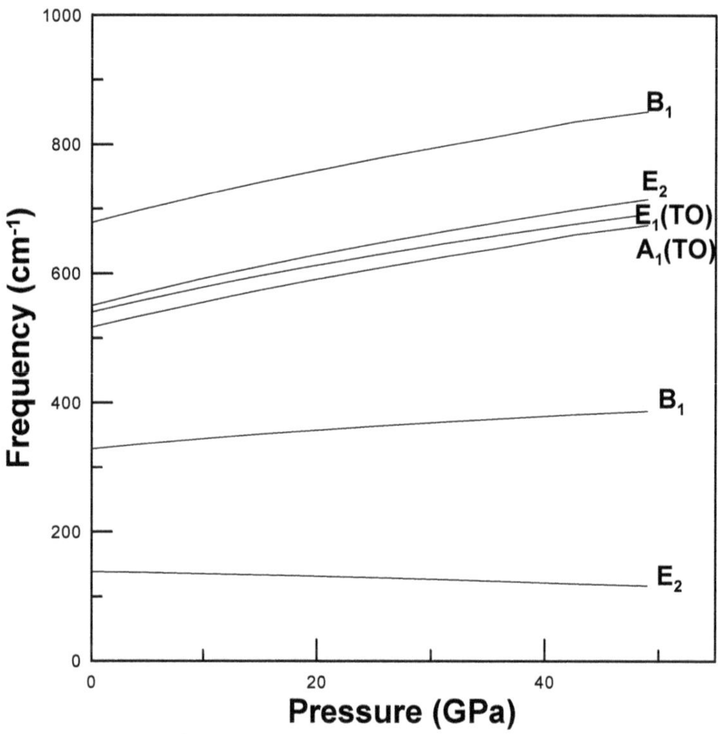

Figure 3.5 Pressure evolution of the wurtzite phonon frequencies in GaN at Γ in absence of LO-TO splitting. A_1 and E_1 are both Raman- and infrared-active. Acoustic phonons, which are zero at Γ, are not shown. B_1 is a silent mode.

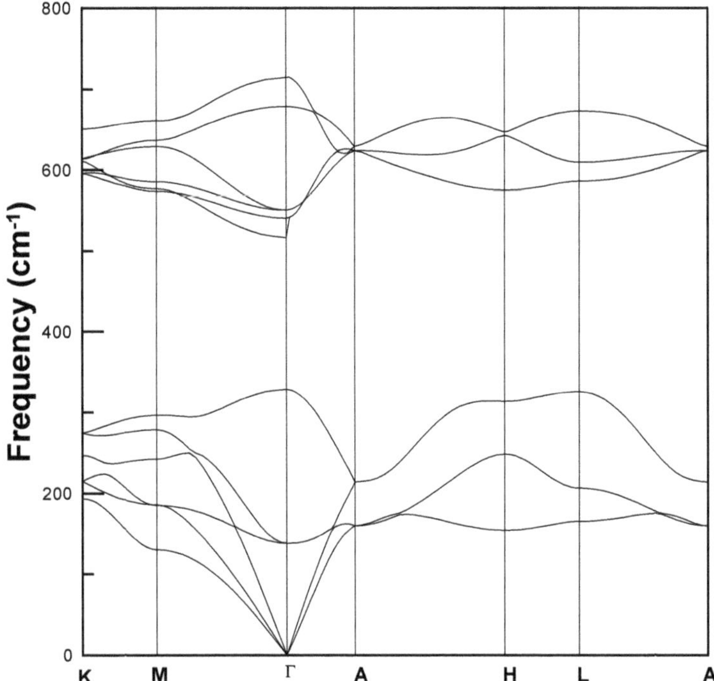

Figure 3.6 Calculated phonon dispersion of GaN wurtzite at zero pressure including the LO-TO splitting.

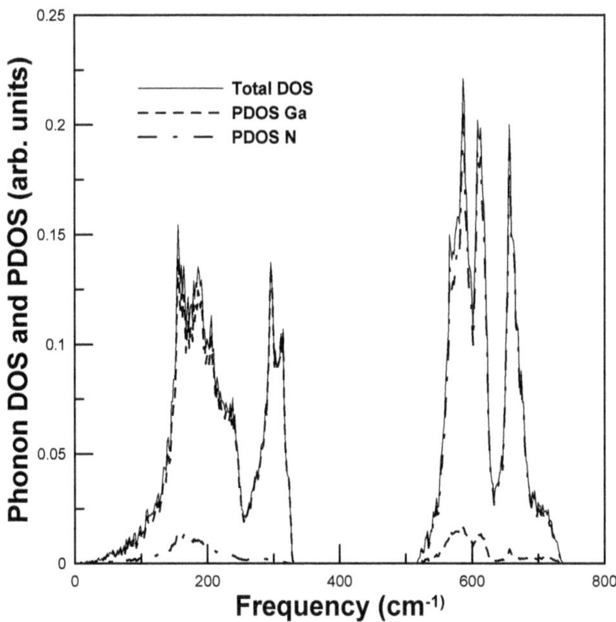

Figure 3.7 Phonon density of states and projected density of states of GaN wurtzite at zero pressure.

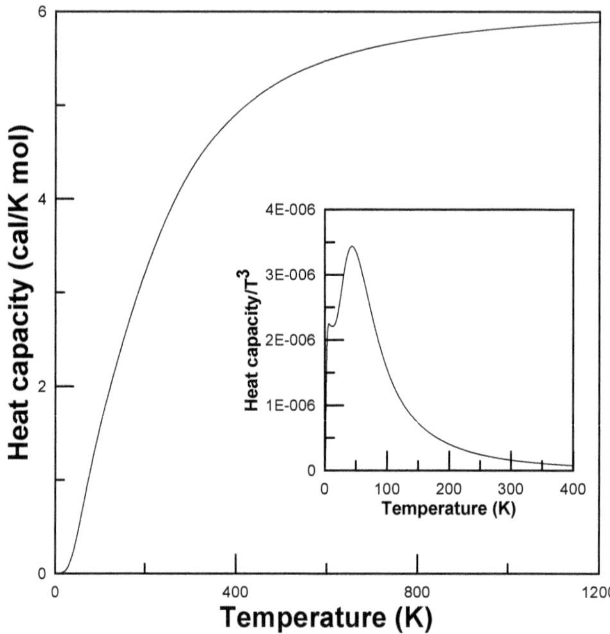

Figure 3.8 Temperature dependence of heat capacity in GaN wurtzite. The inset shows the heat capacity divided by T^3, emphasizing the deviations from the regular Debye behavior at low temperatures.

N atoms (N atoms have lower masses than Ga atoms). Figure 3.8 shows the evolution of the heat capacity of wurtzite GaN as function of the temperature computed using the DOS and the harmonic approximation formula.

3.7 ELASTIC CONSTANTS

The elastic properties of a material can be described by the stress-strain relations [30]. To a linear order, the elastic constants of a system (second derivative of the energy with respect to the strain tensor, per unit volume) are given by:

$$C_{ijkl} = \frac{\sigma_{ij}}{\varepsilon_{kl}}, \tag{3.31}$$

where ε_{kl} represents the components of the strain tensor and σ_{ij} denotes the components of the stress tensor defined as in Reference [34]. Using symmetry and the tensor notation, the C_{ijkl} elements can be expressed as a 6×6 matrix of C_{ij} elements [34]. The number of independent elastic constants, C_{ij}, depends on the crystal symmetry [34].

To perform elastic constant calculations, the crystal must be at equilibrium for the configuration considered. The positions of the atoms in the unit cell are relaxed at every strain. For any configuration corresponding to a particular strain, the stress can be calculated as the derivative of the energy with respect to the strain. Therefore, the stress can be calculated as a function of the strain.

Interesting properties of the crystal such as bulk modulus, Poisson's ratio, Zener anisotropy, sound velocity, etc. can be expressed as combinations of the elastic constants. Furthermore, the mechanical stability of the system can be studied trough the Born stability conditions [30]. The study of the evolution of the elastic properties and the crystal stability under pressure is possible by changing the volume of the cell. Care must be taken when studying the crystal stability under pressure as the requirements for stability that apply in this case are the generalized stability conditions [35].

Hexagonal crystals such as GaN wurtzite are characterized by an elastic constant tensor with six independent elastic constants. The calculated elastic constants obtained for the GaN are:

$$C_{11} = 349.5 \text{ GPa} \qquad C_{12} = 128.7 \text{ GPa} \qquad C_{13} = 91.6 \text{ GPa},$$
$$C_{33} = 384.7 \text{ GPa} \qquad C_{44} = 93.1 \text{ GPa} \qquad C_{66} = 110.3 \text{ GPa},$$

which are in good agreement with other theoretical and experimental results for the same system [36, 37]. The bulk modulus can be calculated from the elastic constants

$$B = \frac{1}{9}(2(C_{11} + C_{12}) + C_{33} + 4C_{44}) \tag{3.32}$$

and equals 190.2 GPa. This value is in very good agreement with the value obtained from the total energy calculations using the equation of state. The coincidence of both results indicates the quality and consistency of the simulation.

3.8 SOME CODES AND UTILITY PROGRAMS

We conclude this chapter with a list of useful programs for the study of the materials properties under pressure. They are all programs with many capabilities and we highlight a few to give readers an idea of their potential use. This list is not intended to be exhaustive; it was designed as a reference for searching utility programs to suit particular needs.

ABINIT The main component of this package allows the calculation of the total energies, charge densities and electronic structures of molecules and periodic solids within DFT using pseudopotentials and plane wave or wavelet basis sets. ABINIT also implements the calculation of other properties using DFPT. Additional utility programs are provided. This program is distributed under the terms of the GNU General Public License. `http://www.abinit.org/`

Quantum ESPRESSO Quantum ESPRESSO is an integrated suite of open-source computer codes for electronic structure calculations and materials modeling at the nanoscale. It uses DFT, plane waves, and pseudopotentials. It also allows the study of response properties using DFPT. `http://www.quantum-espresso.org/`

VASP The Vienna *Ab initio* simulation package (VASP) was developed at Vienna University. VASP is a computer program for electronic structure calculations and quantum mechanical molecular dynamics from first principles within DFT or Hartree-Fock (HF) concepts. VASP makes use of ultrasoft pseudopotentials and the projector-augmented wave method (PAW). `https://www.vasp.at`

CASTEP CASTEP is a code for calculating the properties of materials from first principles using DFT, plane waves, and pseudopotentials. It can simulate a wide range of materials properties. The code is available under a free-of-charge license to all UK academic research groups. `http://www.castep.org/`

WIEN2k This package performs electronic structure calculations of solids using DFT. It is based on the full-potential (linearized) augmented plane wave plus local orbitals method. WIEN2k is an all-electron scheme including relativistic effects. `http://www.wien2k.at/`

SIESTA SIESTA is a computer program for performing electronic structure calculations and *ab initio* molecular dynamics simulations that uses localized basis sets and linear-scaling algorithms. `http://departments.icmab.es/leem/siesta/`

Bilbao Crystallographic Server The server is a crystallographic site at the University of the Basque Country Condensed Matter Physics Department (Bilbao, Spain). The server is accessible online and provides crystallographic programs and databases to analyze and visualize problems in structural and mathematical crystallography, solid state physics and structural chemistry. `http://www.cryst.ehu.es/`

ICSD The Inorganic Crystal Structure Database contains data on inorganic compounds and related structures. `http://cds.dl.ac.uk/cds/datasets/crys/icsd/llicsd.html`.

PHON This program was developed by Dario Alfè to calculate force constant matrices and phonon frequencies in crystals. The procedure uses the direct method and can be used in combination with any program that calculates forces on the atoms of a crystal. It also calculates thermodynamic quantities like the Helmholtz free energy, the entropy, the specific heat and the internal energy in the harmonic approximation. The program is available on the web from the author and can be used free of charge. `http://www.homepages.ucl.ac.uk/~ucfbdxa/`

PHONON This software is for calculating phonon dispersion curves and phonon DOS of periodic crystal, crystals with defects, surfaces, etc. from a set of force constants or from Hellmann–Feynman forces calculated with an external *ab initio* program. The program allows users, among other many utilities, to build a crystal structure and to find the polarization vectors and the irreducible representations (Γ point) of phonon modes. For polar crystals, the LO-TO mode splitting can be calculated. `http://wolf.ifj.edu.pl/phonon/`

PHONOPY This open source package for phonon calculations is based on the direct method. It allows the calculation of phonon dispersion curves, LO-TO splitting, phonon DOS, PDOS, thermal properties and other features. `http://phonopy.sourceforge.net/`

ISOTROPY This software was developed by Harold T. Stokes, Dorian M. Hatch, and Branton J. Campbell at the Department of Physics and Astronomy, Brigham Young University (Utah, USA). The ISOTROPY suite is a collection of software that applies group theoretical methods to the analysis of phase transitions in crystals. The program is available for Linux online. `http://stokes.byu.edu/iso/isotropy.php`

More information about free and commercial software can be found at: `http://electronicstructure.org/`

Bibliography

[1] M. C. Payne, M. P. Teter, D. C. Allan, *et al.* *Rev. Mod. Phys.*, 64, 1045 (1992).

[2] R. M. Martin. *Electronic structure: basic theory and practical methods.* Cambridge University Press, Cambridge (2004).

[3] E. Kaxiras. *Atomic and electronic structure of solids.* Cambridge University Press, Cambridge (2003).

[4] J. Kohanoff. *Electronic structure calculations for solids and molecules: theory and computational methods.* Cambridge University Press, Cambridge (2006).

[5] P. Hohenberg and W. Kohn. *Phys. Rev.*, 136, B864 (1964).

[6] W. Kohn and L. J. Sham. *Phys. Rev.*, 140, A1133 (1965).

[7] D. M. Ceperley and B. J. Alder. *Phys. Rev. Lett.*, 45, 566 (1980).

[8] J. P. Perdew and A. Zunger. *Phys. Rev. B*, 23, 5048 (1981).

[9] D. R. Hamann, M. Schlüter, and C. Chiang. *Phys. Rev. Lett.*, 43, 1494 (1979).

[10] G. P. Kerker. *J. Phys. C: Solid State Phys.*, 13, L189 (1980).

[11] G. B. Bachelet, D. R. Hamann, and M. Schlüter. *Phys. Rev. B*, 26, 4199 (1982).

[12] N. Troullier and J. L. Martins. *Phys. Rev. B*, 43, 1993 (1991).

[13] D. Vanderbilt. *Phys. Rev. B*, 32, 8412 (1985).

[14] P. E. Blöchl. *Phys. Rev. B*, 50, 17953 (1994).

[15] H. J. Monkhorst and J. D. Pack. *Phys. Rev. B*, 13, 5188 (1976).

[16] R. P. Feynman. *Phys. Rev.*, 56, 340 (1939).

[17] O. H. Nielsen and R. M. Martin. *Phys. Rev. B*, 32, 3780 (1985).

[18] F. D. Murnaghan. *Proc. Natl. Acad. Sci. U.S.A.*, 30, 244 (1944).

[19] A. Mujica, A. Rubio, A. Muñoz, *et al. Rev. Mod. Phys.*, 75, 863 (2003).

[20] A. Muñoz and K. Kunc. *Phys. Rev. B*, 44, 10372 (1991).

[21] P. Perlin, C. Jauberthie-Carillon, J. P. Itie, *et al. Phys. Rev. B*, 45, 83 (1992).

[22] J. P. Perdew, A. Ruzsinszky, G. I. Csonka, *et al. Phys. Rev. Lett.*, 100, 136406 (2008).

[23] G. Kresse and D. Joubert. *Phys. Rev. B*, 59, 1758 (1999).

[24] D. Bimberg. *Physics of group IV elements and III-V compounds*, Volume III/17. Springer, Berlin (1982).

[25] F. Birch. *Phys. Rev.*, 71, 809 (1947).

[26] J. Serrano, A. Rubio, E. Hernández, *et al. Phys. Rev. B*, 62, 16612 (2000).

[27] N. W. Ashcroft and N. D. Mermin. *Solid state physics.* Thomson Learning Inc. (1976).

[28] C. Kittel. *Introduction to solid state physics.* John Wiley & Sons, New York, 8th edition (1996).

[29] M. T. Dove. *Introduction to lattice dynamics.* Cambridge University Press, Cambridge (1993).

[30] M. Born and K. Huang. *Dynamical theory of crystal lattices.* Clarendon Press/ Oxford University Press, New York (1988).

[31] P. Yu and M. Cardona. *Fundamentals of semiconductors: physics and materials properties.* Springer, Berlin (2010).

[32] S. Baroni, S. de Gironcoli, A. Dal Corso, *et al. Rev. Mod. Phys.*, 73, 515 (2001).

[33] K. Parlinsky. Computer code PHONON (2008). http://wolf.ifj.edu.pl/phonon.

[34] J. F. Nye. *Physical properties of crystals: their representation by tensors and matrices.* Clarendon Press Oxford University Press, New York (1985).

[35] D. C. Wallace. *Thermodynamics of crystals.* Courier Corporation, North Chelmsford MA (1998).

[36] S. Q. Wang and H. Q. Ye. *Phys. Stat. Solidi B*, 240, 45 (2003).

[37] A. Polian, M. Grimsditch, and I. Grzegory. *J. Appl. Phys.*, 79, 3343 (1996).

CHAPTER 4

Structure Prediction at High Pressures

Miriam Marqués

MALTA Consolider Team and Departamento de Física Teórica, Atómica y Óptica. Universidad de Valladolid, Valladolid, Spain

Ángel Morales

MALTA Consolider Team and Department of Physical and Macromolecular Chemistry, Charles University in Prague, Prague, Czech Republic

José Manuel Menéndez

MALTA Consolider Team and Centre for Clinical Brain Sciences, University of Edinburgh, Edinburgh, United Kingdom

CONTENTS

4.1 INTRODUCTION

THE pressure range in the universe spans more than 60 orders of magnitude from the remotest vacuum of space to the centers of neutron stars.

Pressure causes dramatic changes in the behaviors of materials. It converts common liquids into spectacular crystals and turns common gases into exotic metals by the dramatic modification of interatomic distances and bond angles. Pressure provides an extremely powerful means of probing the relationship between structure and properties, which is necessary for a better fundamental understanding of the underlying phenomena and also for the improved design of materials [1]. In crystals, the most dramatic effects of pressure are phase transitions, which result in new polymorphs or different chemical species. The new crystal structures, sometimes with unusual stoichiometries, may present unexpected chemical and physical properties. For example, some chemical elements become superconductors under pressure [2] and simple metals such as lithium and sodium [3, 4], transform into insulators.

Thermodynamically, only one crystal structure is stable at given pressure and temperature conditions. However, the transformation from one crystal structure to another may be associated with a high free energy barrier that precludes the transformation. The transition between diamond and graphite is the archetypal example. At ambient conditions, only graphite, the thermodynamically stable phase, should exist. However, diamond (stable above 4.5 GPa) does not convert spontaneously to graphite because of its strong covalent C-C bonds. Therefore, in some cases, high-pressure phases can remain metastable at atmospheric pressure, which paves the way for new chemistry and physics even at ambient conditions.

Pressure also facilitates chemical reactions by lowering the kinetic barriers. For instance, hydride phases of most transition metals can be obtained [5] through applied pressure. The study of the phase transitions and the associated polymorphism is important in condensed matter physics and materials science and, also in geology because matter in the deep interiors of planets is under high pressure and/or temperature. An example is the interpretation of the seismic anisotropy of the Earth's D″ layer as due to the post-perovskite structure for $MgSiO_3$ [6].

Experimentally, diamond anvil cell devices [7], combined with x-ray and neutron diffraction, as well as other experimental techniques such as Raman spectroscopy, have been able to determine crystal structures in the megabar region (pressure in the Earth's inner core reaches up to 3.6 Mbar). More recently, pressures beyond 10 Mbar have been reached due to advances in dynamic compression techniques. However, these methods can be expensive and difficult (use of toxic chemicals, reactions between light elements and diamond, etc.), x-ray powder diffraction data may be of poor quality or incomplete and pressures within giant planets such as Jupiter are not experimentally accessible. Therefore, a theoretical approach is necessary to interpret and complement experiments and also predict novel materials as targets for synthesis.

As noted by John Maddox, editor of *Nature*, in 1988, "One of the continuing scandals in the physical sciences is that it remains in general impossible to predict the structure of even the simplest crystalline solids from a knowledge of their chemical composition" [8]. The challenge of crystal structure prediction (CSP), in other words, to find the stable crystal structure of a compound at given pressure and temperature conditions from the chemical formula only, remains nowadays. However, there have been significant advances toward this goal with the increase in computing power, and the development of first-principles quantum mechanical programs and global searching methods. In this chapter, we present an overview of the different methodologies illustrated by recent examples.

4.2 POTENTIAL ENERGY SURFACES

Thermodynamically, crystal structure prediction refers to the problem of minimizing the Gibbs free energy $G = U + pV - TS$, with U being the internal energy, p the pressure, V the volume, T the temperature, and S the entropy of an atomic arrangement (see Chapter 1 for a detailed discussion of the thermodynamics of solids under pressure). Local and global minima of the Gibbs free energy correspond to metastable and stable crystal structures, respectively. In general, the vibrational contributions are expensive to calculate and are often neglected. The enthalpy, which equals the Gibbs free energy at $T = 0$ and in absence of vibrations (the static regime), is minimized instead.

The main difficulty in CSP stems from the multidimensionality of the potential energy surface (PES) given by the enthalpy. The PES can be easily visualized as a collection of mountains and valleys. The bottoms of the valleys are the local minima and the hillsides of the mountains enclosing the valleys their corresponding attraction basins. Since atomic positions and lattice parameters are the variables, the dimensionality is represented by $d = 3N + 3$, where N is the number of atoms. For this high-dimensional problem, the number of local minima was estimated to increase exponentially with size. In consequence, locating the global minimum and local minima of a PES is considered as a non-deterministic polynomial-time hard problem. For these problems, it is assumed that the solution cannot be found in polynomial time.

Because an exhaustive exploration of the entire PES is impossible, solving the CSP problem requires both an efficient search of the PES and an accurate evaluation of it. Some of the known properties of the PES that can be exploited to solve the CSP problem are:

- A large part of the PES corresponds to very high energy structures with unphysically short distances, fragmented units, etc. and contains almost no minima.

- As a consequence of the Bell–Evans–Polanyi principle, which states that highly exothermic chemical reactions have low activation energies, going from a minimum to another minimum with lower energy is more likely if the barrier between their basins is small. In addition, low energy minima are likely to occur near other low energy minima, generating a funnel.

- Attraction basins corresponding to lower energy minima tend to be larger.

- Minima with very low or very high energy usually correspond to symmetrical structures.

The simplest theoretical approach to CSP involves calculating the energy for a collection of candidate structures. These can be guessed following topological approaches, structural analogies between chemically similar compounds [9], phenomenological phase diagrams [10], and heuristic models such as the Zintl-Klemm concept [11] or the anions-in-metallic-matrices model [12]. Experimental information or, at least, chemical intuition are, therefore, necessary for proposing suitable candidate structures.

A more elaborate approach consists of combining data mining techniques with *ab initio* calculations [13]. *Ab initio* energies are calculated for a given number of crystal structures (usually experimentally observed) in similar compounds. The patterns (or correlations) that control the phase stability for this database are statistically recorded and used to build probabilistic or regression models. From these models, the stable structure of a different compound can be inferred. The energy of the candidate structure suggested using this method is then calculated and added to the previous database. As the database size increases, a better crystal structure prediction can be expected. The main problem with this technique lies in the limited size of the database. Unfortunately, due to the limited availability of high-pressure (or high-temperature) data, this database is normally far from complete. Moreover, only structures present in the database can be predicted. Therefore, predictions of novel crystal structures using this technique are impossible.

Ideally, the CSP problem should be solved with unbiased stochastic methods without relying on prior knowledge or assumptions about the system. Some of these methods are discussed below.

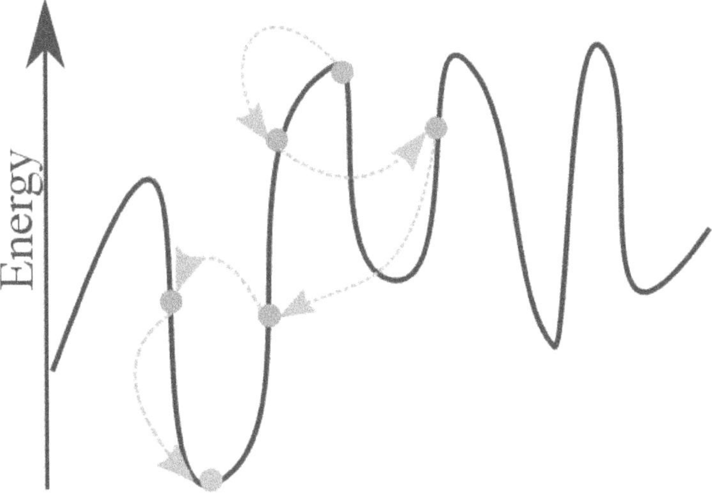

Figure 4.1 Simulated annealing algorithm in a 1D PES.

4.3 SIMULATED ANNEALING

The simulated annealing (SA) approach [14] (Figure 4.1) mimics the procedure of the same name applied in metallurgy to reduce defects and obtain larger microcrystals. The technique is based on heating a metal and then cooling it in a controlled way.

When a metal melts, the atoms are in a disordered state in which atomic diffusion and kinetic barrier crossing are allowed. If the temperature decreases abruptly (the system is quenched), the resulting crystal usually contains defects such as vacancies or dislocations. In contrast, if the cooling process is slow, the atoms are likely to rearrange in a configuration with lower free energy. In SA, the atomic configuration (candidate structure) is continuously perturbed through random atomic displacements, atomic mutations, and changes of lattice parameters using a Monte Carlo scheme. The random movements are accepted or rejected according to a Metropolis criterion. That is, if the energy of the new configuration is lower than the original one, the change in the structure is accepted. Otherwise, the probability $P = \exp(-\Delta E/k_B T)$ is calculated, where ΔE is the energy difference between the two configurations, k_B is the Boltzmann constant, and T is the simulation temperature. Only if $\varepsilon < P$, with ε being a random number between 0 and 1, the new structure is accepted and the procedure repeated.

A simulated annealing calculation starts at high temperature where most of the candidate higher energy configurations are accepted , that is, the kinetic barriers are easily overcome. By decreasing the temperature, fewer high energy configurations are accepted and, if the cooling is slow enough, the system

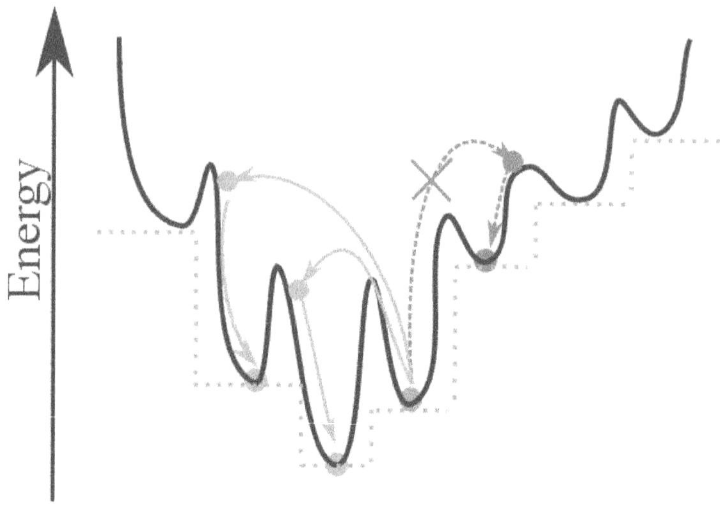

Figure 4.2 Basin hopping algorithm in a 1D PES.

will crystallize at the global minimum. Unfortunately, the system often gets stuck in a local minimum or a wrong funnel, especially if these local minima are surrounded by high energy barriers. The efficiency of this method depends strongly on the types and magnitudes of the perturbations. They need to be large enough to let the system jump out of local minima, but small enough to enable learning. Normally, the temperature is reduced linearly or exponentially, but more complicated annealing procedures in which the temperature is raised during the run can also improve the efficiency. Moreover, multiple runs with different starting structures and different temperature schedules can be performed. This method has been used to predict the structures of several inorganic compounds, including LiF, BN [15], and PbS [16].

4.4 BASIN HOPPING

A related technique based on the Monte Carlo approach is basin hopping [17] (Figure 4.2). Basin hopping starts at a random configuration, which is relaxed to its local minimum. After applying a random displacement, the new configuration is also relaxed to its local minimum and the Metropolis criterion is applied on the local minimum energies of the old and new structures. The temperature can be maintained during the run or constantly reduced. The original PES is transformed into a multi-dimensional stepwise function representing the energies of the local minima of the corresponding attraction basins. Importantly, the energy differences between local minima are smaller than the barriers between points on the corresponding attractor basins so jumping out of a local minimum is easier than in SA. As in the SA approach, the efficiency

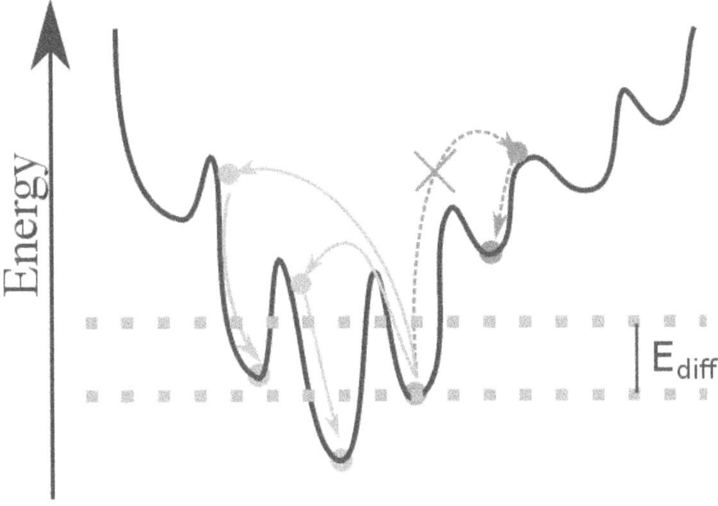

Figure 4.3 Minima hopping algorithm in a 1D PES.

of this method depends on the choice of temperature and on the magnitudes of the perturbations. Basin hopping has been mainly used to predict the structures of clusters but to our knowledge has not been used for crystal prediction under high pressure.

4.5 MINIMA HOPPING

The minima hopping method [18] (Figure 4.3) uses molecular dynamics (MD) instead of a Monte Carlo method to overcome barriers between local minima. The MD simulation starts from a local minimum. Once the trajectory reaches a different potential energy minimum, the MD simulation is stopped and the geometry is optimized to the nearest local minimum. The new configuration is accepted if $E(\mathrm{M}) < E(\mathrm{M}_i) + E_{\mathrm{diff}}$, where $E(\mathrm{M})$ and $E(\mathrm{M}_i)$ are the energies of the new and the old minimum, respectively, and E_{diff} is a positive energy threshold. E_{diff} is dynamically adjusted during the run so that on average half of the proposed minima are accepted. In particular, E_{diff} decreases (increases) by a factor of $\alpha < 1$ ($\alpha > 1$) if the new configuration is accepted (rejected).

The kinetic energy used in the MD simulation needs to be large enough to escape from the current minimum and enter a new basin of attraction. According to the Bell–Evans–Polanyi principle, excessive kinetic energy will result in visiting many undesirable local minima. On the other hand, insufficient kinetic energy will require a large number of MD steps to escape the current basin of attraction. Therefore, in analogy with E_{diff}, the kinetic energy of the system E_{kin} is adjusted dynamically to ensure that on average half of the proposed attraction basins are accepted. E_{kin} is decreased if a new minimum is found

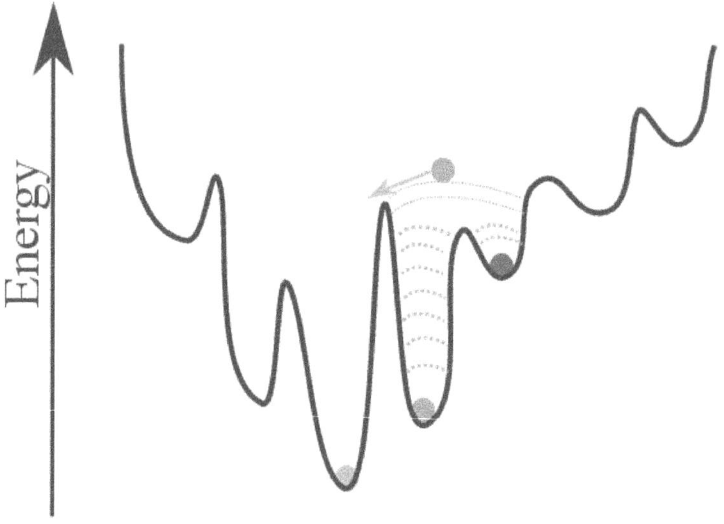

Figure 4.4 Metadynamics algorithm in a 1D PES.

but is increased if the escape trial is unsuccessful or if the current minimum has been visited previously. The minima hopping method has been employed to predict the high-pressure structures of Si_2H_6 [19], C [20], $LiAlH_4$ [21], and $Zn(BH_4)_2$ [22].

4.6 METADYNAMICS

An alternative approach to CSP is the simulation of phase transitions in crystals. This approach not only predicts the stable crystal structures at given p-T conditions, but also provides insight into the kinetics by unveiling the mechanisms involved in the transformation between structures. The standard Parrinello–Rahman variable-cell constant-pressure molecular dynamics method [23] and its generalizations are the most straightforward for phase transition prediction. However, these methods face a time-scale problem, since only transitions with activation barriers in the order k_BT or less are observable on time scales reachable in MD simulations. Solid-solid phase transitions are often first-order and proceed via nucleation and growth (see Chapter 2), which are usually slow and need to overcome high energy barriers. Moreover, within the periodic boundary conditions typical of MD simulations, the simulation of heterogeneous nucleation at a surface or defect is not possible. Instead, the transition is assumed to proceed in a collective and concerted way, resulting in a larger energy barrier for the transformation. This barrier can be overcome by increasing the pressure, but this overpressurization may lead to unrealistic kinetics and some phases that exist only at a small range of pressures may be overlooked as a result.

The metadynamics technique [24] (see Figure 4.4) proposed by Laio and Parrinello is based on modifying the original Gibbs energy landscape by adding a suitable potential to overcome the energy barrier associated to a phase transformation. The simulation of the phase transition is formulated as the search for the lowest energy path connecting two minima in the space of a suitable order parameter. A natural choice for this order parameter is a supercell box matrix $h = (a, b, c)$ where a, b, and c are the lattice vectors of the supercell. A steepest-descent-like dynamics is applied in the space of the h vector, with forces derived from a modified Gibbs free energy which includes a history-dependent term. This term is a sum of Gaussian functions placed at the visited points in the h space to discourage the dynamics from visiting them again. The attraction basin of the initial crystal structure is gradually filled with Gaussian functions that increase the Gibbs energy. As the dynamics progresses, the Gibbs energy becomes high enough to overcome the barrier between local minima and the system enters into the attraction basin of a new structure.

Instead of the difficult direct calculation of $G(h)$, the algorithm only requires the calculation of the first derivative of G with respect to the order parameter. Taking into account that a change in the simulation cell can be considered a deformation (strain) relative to the original structure, the derivative of the free energy with respect to the order parameter is easily obtained from the stress tensor, which can be evaluated employing a short NVT molecular dynamics simulation in which the atomic coordinates are equilibrated.

One of the challenges of the metadynamics method is the choice of the width and amplitude of the Gaussian functions. Choosing too large Gaussians may lead to wrong phase transitions, but too small Gaussians may require an impractical number of metadynamics steps (metasteps). The method has been successfully applied to a number of systems including SiO_2 [25], CO_2 [26], and Ca [27].

4.7 RANDOM SEARCHING

Although unbiased, metadynamics requires a good guess for the initial structure to avoid excessively long simulations. If no knowledge about the system is available, the simplest approach to explore to whole PES is calculating the Gibbs free energy of a set of random structures, a random search (Figure 4.5). The procedure is simple: generate a trial structure at random, relax the configuration to the local minimum, and repeat these two steps until no minima with lower energy are found. In spite of its simplicity, some of the properties of the PES such as the correlation between the size of the basin of attraction and the stability of the minimum make a blind random search successful for systems with up to 12 atoms and even more.

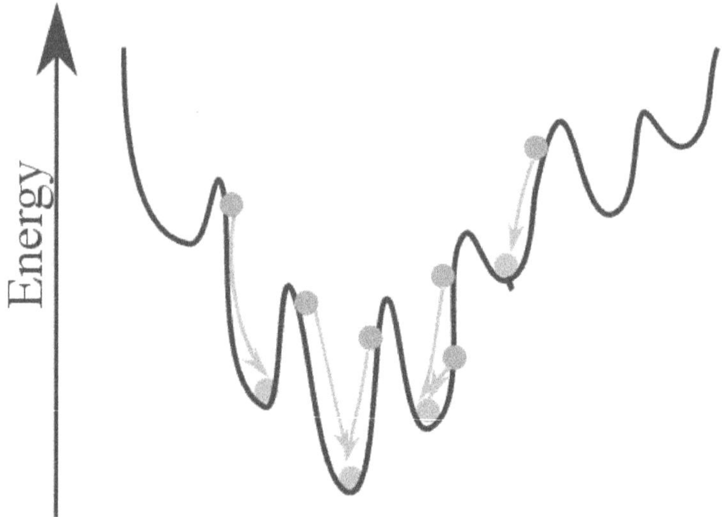

Figure 4.5 Random search algorithm in a 1D PES.

Larger systems can be treated by imposing chemical constraints on the search. For instance, to avoid unphysical trial structures, it is useful to sample structures with reasonable volumes. The volume can be easily estimated from other structures with the same atomic species. In any case, the atomic bond lengths lie between 0.75 and 3 Å, and the volume depends mainly on the number of atoms, and only slightly on their identities and the p-T conditions (from this argument, it is clear that imposing a constraint on the bond lengths can also be used). For molecular solids or systems likely to contain structural units (e.g., SiO_4 tetrahedra), it is desirable to consider the molecules or clusters as building blocks to bias the search towards the correct bonding pattern. Because the low-energy minima tend to correspond to symmetrical structures, imposing symmetry by randomly selecting space groups can also be exploited to generate good trial structures.

Although no experimental information is needed, available information about the lattice parameters or the highest space group compatible with the experimental data can be used to accelerate the search. All these constraints are employed by the powerful *ab initio* random structure searching (AIRSS) method developed by Pickard and Needs [28]. The main criticism to the method is its lack of learning from previously discarded configurations. To deal with this shortcoming and considering that low energy structures are close to even lower energy structures, the AIRSS method allows random atomic and cell displacements (shaking) to escape the attraction basin of the current structure. The criterion to end the search is that the same low-energy structure is found several times. A summary of AIRSS calculations on a variety of systems, such as SiH_4, AlH_3, H_2, N_2, and H_2O is presented in Reference 28.

4.8 EVOLUTIONARY ALGORITHMS

Evolutionary algorithms are stochastic methods inspired by the natural evolution of a population, in which the offspring resemble their parents and procreation is a reward for success. Different computational codes based on evolutionary algorithms have been developed to approach the CSP problem: USPEX [29], XtalOpt [30], EVO [31], GASP [32], algorithms by Zunger [33], Abraham and Probert [34], Fadda [35], and the adaptive GA of Wentzcovitch *et al.* [36]. In all of them, an initial population of structures is generated, and after optimization to the local minimum, children are obtained from combinations of two parents (crossovers) or mutations of a single parent. The methods differ on how crossover and mutation are defined, how the candidates are selected for procreation and survival, and how the population is updated.

USPEX is one of the most commonly used methods. The first population of structures is randomly generated for systems with up to 200 atoms under the constraints of an estimated volume and minimum bond lengths. The lattice parameters for the search can be fixed if they are experimentally known, but simulations are usually performed with variable cell shapes. In analogy with the AIRSS method, other chemical constraints can be imposed (using molecules or coordination polyhedra as building blocks, including symmetry, etc.). It is crucial to ensure the diversity in large systems, in which the structures will be highly disordered and almost identical (fluid-like). Seeding the first generation with plausible candidate structures is also possible.

After relaxing the initial structures to the local minimum, a fingerprint function based on the interatomic distances is calculated for all the structures. In addition, a fitness value (usually the enthalpy) is also assigned to each local minimum. Identical structures, identifiable by the similarity between their fingerprints, are eliminated to prevent cancer growth, in which a structure proliferates, creating its own clones and overwhelming the population. Cancer growth occurs when the method is trapped in the energy funnel of one of the local minima and new, possibly better structures are not allowed to appear.

In the next step, a fraction of the low-enthalpy structures is chosen to create the children for the next generation. The algorithm "learns" the energetically favorable atomic arrangements and, by creating new structures as a combination of these, effectively zooms in on the most promising parts of the PES. The following variation operators are applied to the selected structures in order to generate the children:

- Heredity: spatially coherent slabs of two parents' atoms are selected and joined in the fractional coordinate system. Similar parents are chosen after comparison of the fingerprint functions to create "good" children.

- Lattice mutation: a random deformation of the lattice vectors of a parent structure.
- Permutation: atoms of different species are exchanged within a parent structure.
- Special atomic mutations: displacements of atoms of a parent structure based on chemical ideas (larger displacements for low-order atoms, following of a soft mode, etc.).

In every generation, only a few of the more stable structures (usually, one) survive unchanged into the next generation ("survival of the fittest"). This enhances the learning power and ensures the variational principle on the enthalpies: the minimum enthalpy in each generation can only decrease or remain constant with respect to the previous generation. A percentage of random structures is also added in each generation to preserve diversity. The structures in the new generation are relaxed to the local minimum and the procedure is repeated until the enthalpy does not decrease further after a given number of generations (Figure 4.6). Examples include the high-pressure structures of Ca, Na, B, SnH_4, GeH_4, and LiH_x [37].

4.9 PARTICLE SWARM OPTIMIZATION

The particle swarm optimization (PSO) method (Figure 4.7) is a population-based stochastic optimization technique inspired by the collective behavior of bird flocking or fish schooling. It shares many similarities with the evolutionary methods, but no evolution operators such as crossover and mutation are present. Ma *et al.* implemented this algorithm for crystal structure prediction in the CALYPSO code [38]. A first population (swarm) of structures (particles) is generated randomly, imposing constraints on volume, interatomic distances, and symmetry. The initial structures are optimized to the local minimum and identical structures are discarded. To measure structural similarity, matrices listing the interatomic distances and number of first and second nearest neighbors for the different bond types of each structure are built. If a newly generated structure shares the same number of bonds with a previous structure and the deviation of the corresponding bond lengths is within a given threshold, the new structure is discarded. A fraction of this random swarm formed by the structures with lower enthalpies is used to produce 60 % of the structures in the next generation. The remaining 40 % is generated randomly to preserve the population diversity.

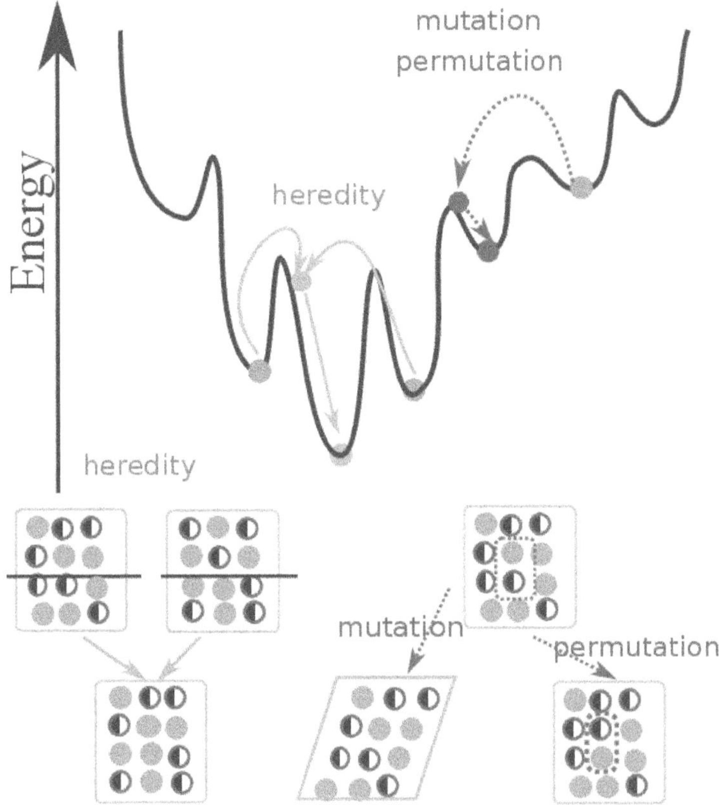

Figure 4.6 Evolutionary algorithm in a 1D PES.

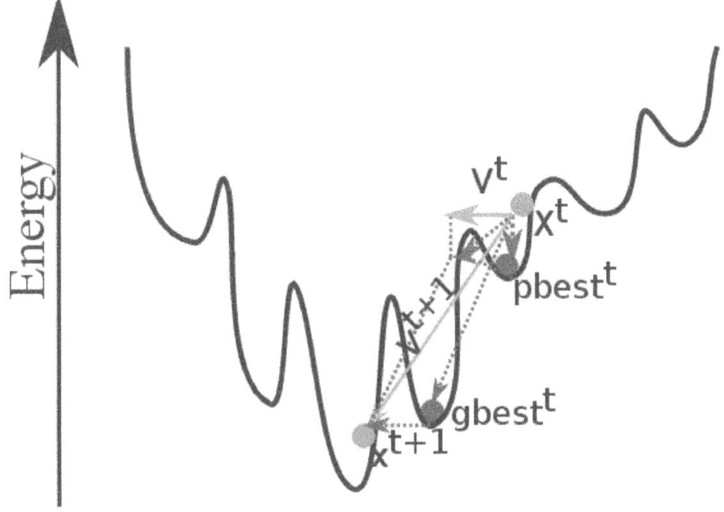

Figure 4.7 Particle swarm optimization technique in a 1D PES.

Within the PSO scheme, the position of each particle (x^t) is updated according to

$$x^{t+1} = x^t + v^{t+1}, \tag{4.1}$$

where the velocity of the particle is calculated via

$$v^{t+1} = \omega v^t + c_1 r_1 (p^t_{\text{best}} - x^t) + c_2 r_2 (g^t_{\text{best}} - x^t), \tag{4.2}$$

with ω being an inertia weight (between 0.4 and 0.9) that controls the momentum of the particle; v^t is the velocity at step t; c_1 and c_2 are factors that measure, respectively, how much the particle trusts its own position and that of the swarm, and are empirically fixed to 2; r_1 and r_2 are random numbers determined using a uniform distribution between 0 and 1 to ensure a non-deterministic exploration of the configuration space and prevent the swarm from becoming trapped in a local minimum; x^t is the initial position of the particle (before the local optimization); p^t_{best} is the position with the lowest enthalpy that the particle has already visited (after the local optimization), and g_{best} is the position of the leading particle (global minimum for a given population). Since the movement of the particles is clearly influenced by their individual past experience (p^t_{best}, v^t) and that of the swarm (g^t_{best}), this method is also able to learn from history and is more likely to zoom in on the most promising region of configurational space. The PSO method has been applied to predict the high-pressure structures of systems such as Li, Mg, Bi_2Te_3, CaH_x, and CO_2 [38].

4.10 APPLICATIONS

4.10.1 Lithium

The light alkali metals Li and Na are often considered "simple", because their electronic properties at ambient conditions are well described by the free-electron model. This is because the motion of the conduction electrons is only weakly perturbed by the interactions with the atomic cores that adopt the highly symmetric body-centered-cubic (bcc) structure, leaving a single valence electron per primitive unit cell. However, a deviation from the "ideal" metallic behavior is observed under pressure.

Lithium, in spite of being the least complex metallic element, has been until recently the least studied under compression. In 1999, Neaton and Ashcroft [39] suggested that Li should transform at high pressures to less symmetric structures with decreasing metallic character, leading to a nearly insulating phase (α-Ga structure) with Li_2 units at about 100 GPa. One year later, diffraction studies performed by Hanfland *et al.* found a transition from the ambient pressure bcc to the face-centered cubic (fcc) and then to the complex cubic cI16 structure [40].

A constant-volume *ab initio* molecular dynamics SA method was employed to verify the stability of the α-Ga structure at 100 GPa [41]. Both the internal coordinates and the shape of a $3 \times 2 \times 2$ supercell of the α-Ga unit cell under those conditions were allowed to relax. The temperature was fixed to 500 K for 1 ps, and then slowly reduced to 10 K for 1 additional ps. The resulting structure was found to be orthorhombic with 24 atoms and $Cmca$ space group. From enthalpy calculations, a transition from the cI16 structure to this new oC24 structure was predicted at 88 GPa. Evolutionary searches with the USPEX code including 4, 6, 8, 12, 16, and 24 atoms in the simulation cell also found oC24 to be the most stable structure from 100 to 300 GPa [42].

A subsequent study using diffraction and electrical resistance measurements at 25 K along with visual observations revealed two structural transitions at $\sim$70 and $\sim$80 GPa, first to a pseudo-metallic Li-VI phase and then to a semiconducting Li-VII phase [43]. Structural searches with the AIRSS method including 12, 16, and 24 atoms in the simulation cell proposed two structures with symmetry $Pbca$ and $C2cb$ as candidates for the Li-VI and Li-VII phases, respectively [44].

Additional searches with the USPEX method, taking as input the 80 lower-enthalpy structures found with the AIRSS code (supercells containing 8, 12 and 16 atoms) proposed the previous $C2cb$ structure as Li-VII, and a $C2$ structure as Li-VI [45]. All these structures have small band gaps (of the order of 0.3 eV) in agreement with the observed semiconducting behavior, but they are only marginally favored with respect to the $Cmca$ structure (by less than 1 meV). Recently, the phase diagram of lithium was mapped experimentally up to 130 GPa and from 77 K to 300 K [46]. At 200 K, the study located transitions from cI16 to Li-VI (at 60 GPa) and then to Li-VII (at 65 GPa), and also a further transition to a previously unreported Li-VIII phase at 95 GPa. Li-VI, VII, and VIII were all found to have C-face-centered orthorhombic symmetries, with probable numbers of atoms per unit cell of 88, 40, and 24, respectively, referred to as oC88, oC40, and oC24 in Pearson notation (see Figure 4.8).

The high-quality x-ray diffraction data for Li-VIII confirmed it as the previously proposed $Cmca$ structure. Single crystals of Li-VII were of poorer quality than those obtained for Li-VIII, resulting in broader, weaker reflections. Still, it was possible to determine the space group as $Cmca$ or $C2cb$ and least-squares fitting of reflection d-spacings provided the lattice parameters. Taking as input the experimentally determined symmetry, lattice paramcters, and the known number of atoms in the unit cell (40), the AIRSS method was applied to determine the structure of oC40 Li-VII [3]. Searches for possible oC40 structures were initially conducted using a 20-atom primitive cell with space group $P2_1/c$ (a subgroup of $Cmca$). The initial lattice parameters

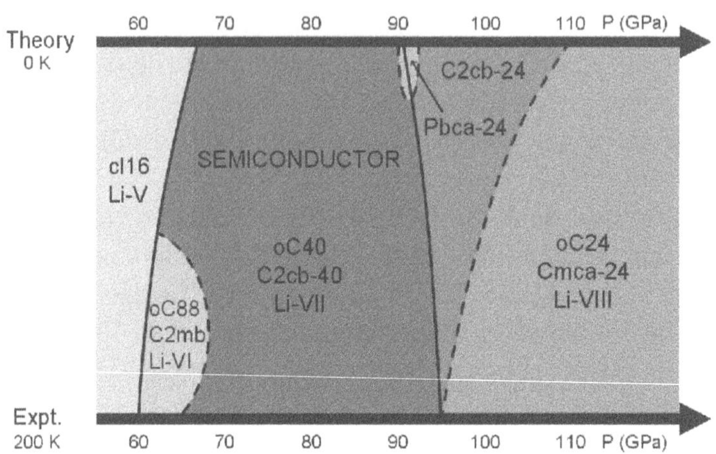

Figure 4.8 Stable phases of lithium under pressure. Solid lines indicate the connection between the experimental transition pressures at 200 K ("Expt." at bottom of the plot) and the theoretical static transition pressures ("Theory" at top). Dashed lines show phases identified as stable either experimentally or theoretically, but not both.

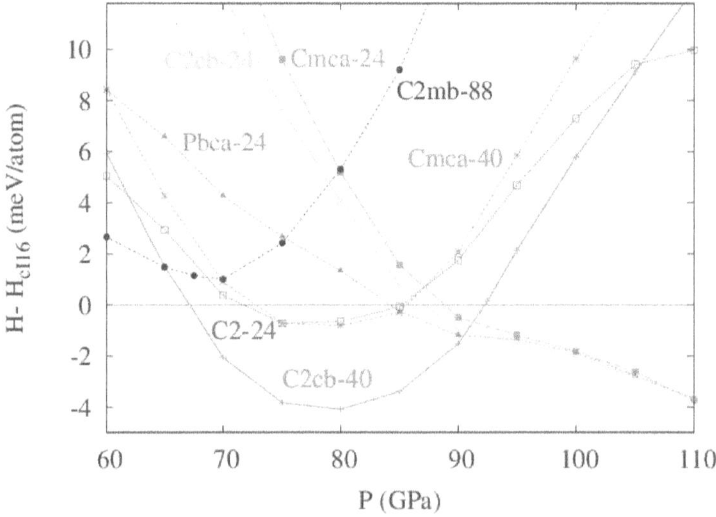

Figure 4.9 Enthalpy differences (relative to the cI16 structure of Li-V) as a function of pressure for the 24-atom structures of Li-VIII (*Cmca*-24, *C*2*cb*-24), the 40-atom structures of Li-VII (*Cmca*-40 and *C*2*cb*-40), the calculated *C*2*mb*-88 structure of Li-VI, and the previously-proposed structures *Pbca*-24 and *C*2-24. Adapted, with permission, from reference 3.

were chosen as the experimental values, but allowed to change during the optimizations. After searching through 250 possible structures, a low enthalpy candidate structure with *Cmca* symmetry was found. However, further trials with four additional batches of 150 20-atom structures and using space groups *Pc* (subgroup of both *Cmca* and *C*2*cb* space groups), *P*1, and two randomly-selected space groups with 4 and 2 symmetry operations were conducted. The exploration led to a structure with space group *C*2*cb* and significantly lower enthalpy than the previously-proposed *C*2, *Pbca* and *C*2*cb* structures (Figure 4.9). The electronic density of states confirmed that this new phase is a semiconductor with an indirect band gap that increases from 0.82 to 1.15 eV between 75 and 95 GPa. This gap is clearly larger than that of any other candidate structure (Figure 4.10). At the same time, structure predictions with the CALYPSO code using simulation sizes up to 48 atoms per cell also found the *C*2*cb* structure [47].

The resolution of the Li-VII and Li-VIII structures left Li-VI, stable between 60 GPa and 65 GPa at 200 K, as the only unknown structure in high-density Li. Data collected from two single crystals of Li-VI at 65 GPa enabled the determination of the lattice parameters and highest symmetry space groups compatible with the data (*Cmma* or *C*2*mb*). Constrained AIRSS searches using the experimental space group and lattice parameters were performed [3]. The lowest enthalpy structure had the *C*2*mb* space group and was in reasonable agreement with the diffraction data. Although the structure does not have an enthalpy lower than cI16, it was more stable than the rest of the high-pressure phases of lithium in its experimental range of appearance (from 60 to 65 GPa). In fact, the inclusion of both zero-point and finite-temperature thermal effects was shown to stabilize this phase compared to the cI16 structure [48].

The story of the crystal structures of lithium under pressure underlines the usefulness of a combined experimental-theoretical approach to CSP in which experimental information can guide theoretical searches. The inclusion of the correct number of formula units in the simulation cell was also shown to be crucial. For instance, the oC40 structure was hidden in previous searches because they did not include 20 or 40 atoms in the simulation cells. Ideally, the structural searches should be performed at finite temperature. In spite of this, most of the crystal structure prediction methods are restricted to static conditions since it is computationally expensive to perform phonon calculation of all the structures. However, the inclusion of phonon free energies can affect the phase diagrams at high pressure, even for systems that are not usually considered quantum solids at ambient conditions. The stabilization of the *oC*88 phase confirms this observation.

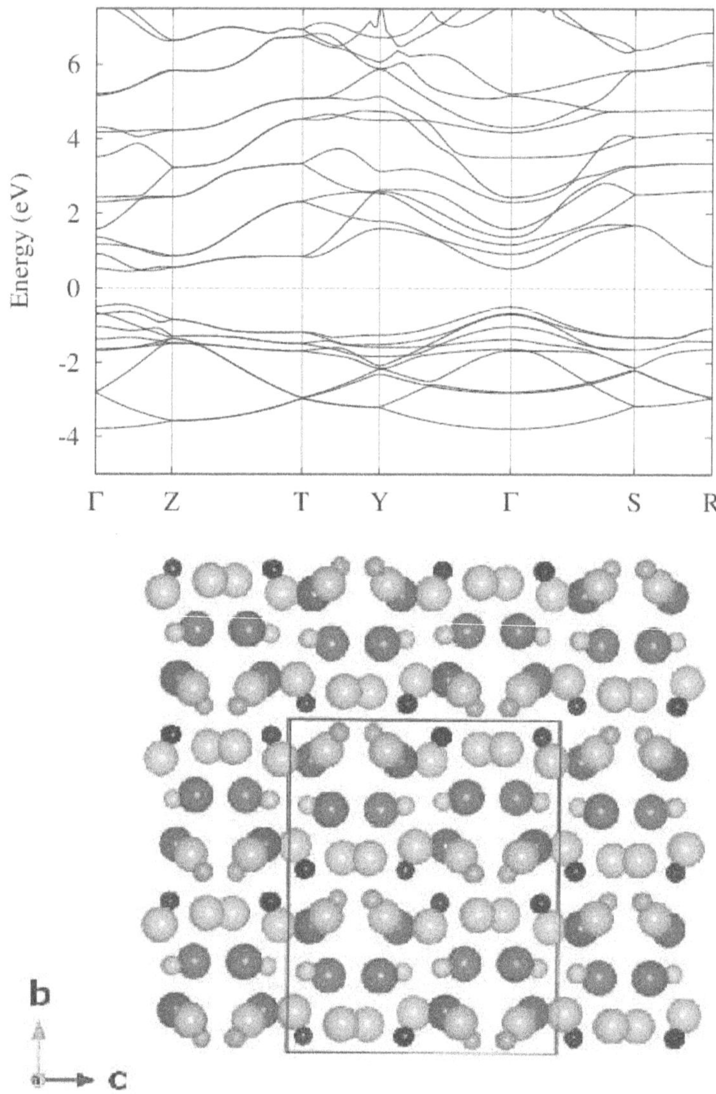

Figure 4.10 Electronic band structure of the *C*2*cb* structure of lithium at 78 GPa (top), showing an indirect band gap of 0.90 eV along the Γ-Z line (Λ_1 $(0, 0, 0.152)$–Λ_2 $(0, 0, 0.217)$), similar in magnitude to the direct band gap of 1.03 eV at Γ. The crystal structure with atoms and electron localization function (ELF) attractors represented by big and small spheres, respectively, is also shown (bottom). Adapted, with permission, from reference 3.

4.10.2 Iridium Trihydride

Noble metals adopt close-packed structures at ambient pressure and are fairly unreactive at ambient conditions. However, chemical reactions can be induced by high pressure. In particular, their reactivity with hydrogen to form hydride phases is interesting based on their potential hydrogen storage capabilities and hydrogen-mediated superconductivity.

Synchrotron x-ray diffraction studies were performed on iridium in a hydrogen medium at pressure up to 125 GPa [49]. Iridium adopts the fcc crystal structure at ambient conditions. At 55 GPa, a new cubic phase (space group $Pm\bar{3}m$) appears with a drastically increased volume that signals the formation of a metal hydride. This phase is stable up to 125 GPa. In the $Pm\bar{3}m$ space group with Ir atoms located at the corners of the cubic unit cell, there are three potential stoichiometries for the IrH_n, $n = 1$–3 hydride. Different cubic structures were built for the experimentally observed volume at 81 GPa. From the calculated pressures, it was concluded that hydrogen atoms most probably occupy the centers of the faces, making this phase an iridium trihydride (IrH_3). Unfortunately, this structure is dynamically unstable.

Structural searches were performed with the USPEX code in the pressure range of the experimentally detected phase (from 40 to 140 GPa), at 20 GPa intervals. The initial IrH_3 structures were randomly generated and contained up to four formula units. Each subsequent generation was created using 60 % of the lower enthalpy structures of the preceding generation through heredity, lattice mutation, and permutation of atoms. In addition, the structure with lowest enthalpy in each generation was carried over unchanged into the next generation. The searches at 40 and 60 GPa led to a competition between the structures Cc and $Pna2_1$, which are thermodynamically stable below 68 GPa. Interestingly, both structures are distortions of the P$m\bar{3}m$ structure, showing decreased distortion amplitudes with increasing pressure and, more importantly, the structures are dynamically stable.

At pressures above 68 GPa, a different phase with the $Pnma$ space group emerges as energetically preferred and dynamically stable. In this phase, although the metallic atoms are hexacoordinated as in the (distorted) cubic structures, the motif of iridium octahedra with hydrogens located close to the centers of the edges transforms into triangular prisms with interstitial H_3 clusters, with short distances in between (see Figure 4.11). No experimental evidence for this phase has been observed up to 125 GPa. Its formation may be hindered by a large energy barrier.

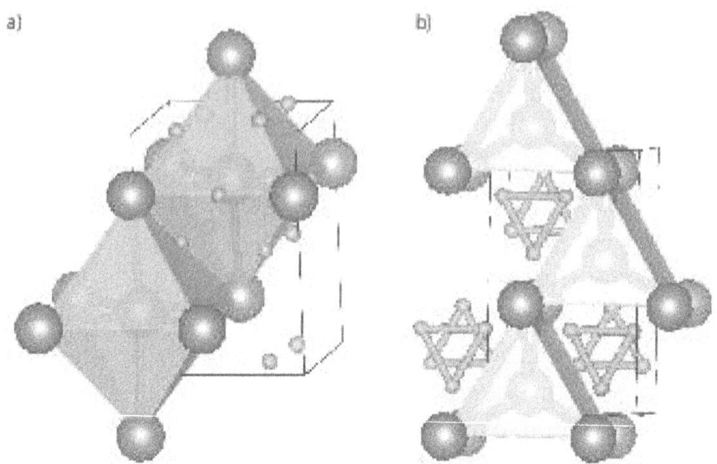

Figure 4.11 Polyhedral representation of the $Pna2_1$ (a, left) and $Pnma$ (b, right) structures of IrH_3 at 125 GPa. Iridium and hydrogen atoms are represented by large and small spheres, respectively. Adapted, with permission, from reference 49.

4.10.3 Xenon Binary Compounds

The abundance (or scarcity) of noble gases such as Ar, Kr, and Xe constrain the models for Earth's formation and, therefore, their reactivity with planetary materials at high pressure and temperatures is fundamental to our understanding of Earth's geological history. It is believed that a class of meteorites (carbonaceous chondrites) are the raw materials that made up the Earth. However, the proportion of Xe in the Earth's atmosphere is much lower than it is in these meteorites. A possible explanation for this "missing Xe paradox" is that gaseous Xe escaped from the atmosphere into space. But most geophysicists believe that Xe may be trapped in the interiors of the Earth. A recent experiment has reported the reactivity of Xe with water ice under conditions found in the interior of giant planets, such as Uranus and Neptune (above 50 GPa and at 1500 K) [50]. It has been also suggested that Xe may be incorporated into silicates by substitution of Si [51]. A subsequent experiment reported the synthesis of Xe oxides [52]. The Earth's core has been considered a potential Xe reservoir. The hypothesis is based on the ability of Xe to form stable compounds with Fe and Ni (main constituents of the Earth's core). However, no evidence of chemical reactions between Xe and Fe has been experimentally found up to 155 GPa [53].

Structure searches were performed with the CALYPSO code at 150, 250, and 350 GPa for $XeFe_x$ (x = 0.5, 1–6) and at 100, 200, and 350 GPa for $XeNi_x$ (x = 0.5, 1–6), respectively, with one to four formula units in the simulation cells [54]. Each generation contained 40 structures, and the first

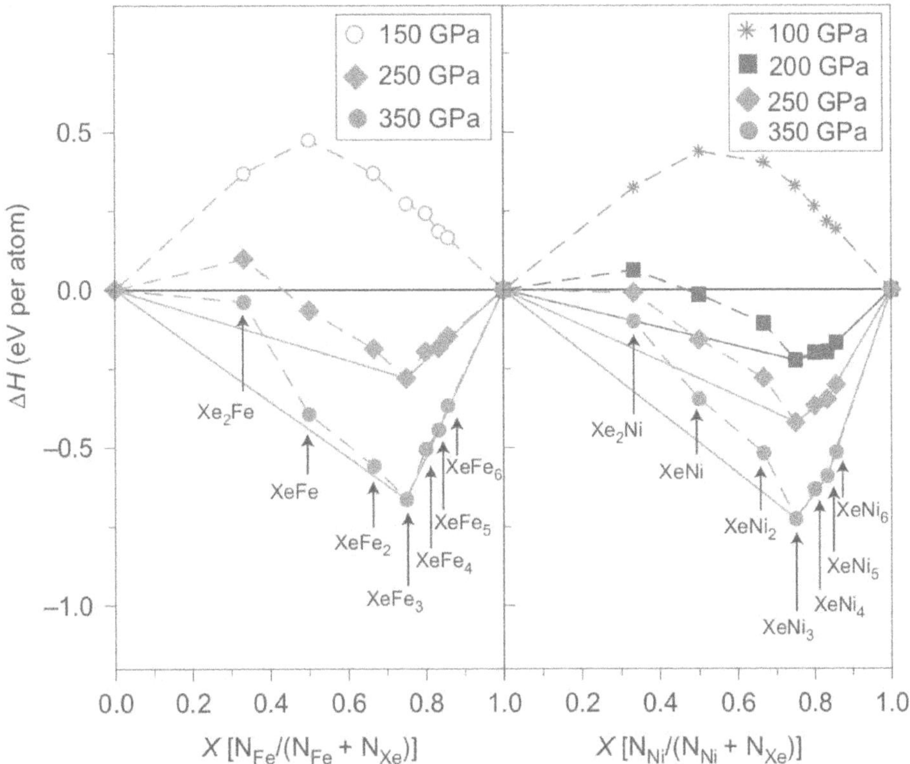

Figure 4.12 Calculated formation enthalpies of Xe-Fe (left) and Xe-Ni (right) phases with respect to the elemental decomposition into Xe and Fe/Ni. Dashed lines connect data points and solid lines represent the convex hulls at different pressures. Adapted, with permission, from reference 54.

generation was randomly created under symmetry constraints. The 60 % low enthalpy structures of each generation were used to produce the structures in the next generation, and the remaining 40 % structures were randomly generated. For each simulation, the search was stopped after formation of 1000 to 1200 candidate structures (about 20 to 30 generations). At each pressure, the formation enthalpy of XeM_x (M=Fe,Ni) was calculated according to:

$$\Delta H_f(\mathrm{XeM}_x) = \frac{H(\mathrm{XeM}_x) - H(\mathrm{Xe}) - xH(\mathrm{M})}{1+x}, \tag{4.3}$$

where $H(\mathrm{XeM}_x)$ is the enthalpy of the most stable structure for the XeM_x formula. For Xe, Fe, and Ni, the enthalpies of the known elemental structures under those conditions, hexagonal close-packed (hcp), ϵ-phase, and fcc, respectively, were used. A negative enthalpy of formation indicates that a mixture is more stable than its elements.

The enthalpies of formation with respect to the atomic ratio x are plotted in Figure 4.12. At each pressure, the convex hull connecting the enthalpically stable phases (stable with respect to the decomposition into other binary stoichiometries or the elements) are also shown in the figure. The convex hull is an elegant way of presenting the stability of a binary mixture and is also called the global stability line. At 150 GPa, the formation enthalpies of all stoichiometries are positive, which is in agreement with the reported lack of reactivity between Xe and Fe up to 155 GPa. Stable Xe-Ni and Xe-Fe compounds emerge at 200 GPa and 250 GPa, respectively. The most stable stoichiometries are $XeNi_3$ and $XeFe_3$, which adopt the hcp structure (space group *Pmmn*, two formula units per cell) and a fcc Cu_3Au-type structure, respectively. Thermal effects included after computation of the phonon spectra further stabilize the XeM_3 (M: Ni, Fe) stoichiometry for both compounds.

In addition, metadynamics calculations were performed at 350 GPa and 6000 K to verify the phase stabilities of these crystals; $2 \times 2 \times 2$ supercells of the Cu_3Au-type structure for $XeFe_3$ and the *Pmmn* structures for $XeNi_3$ were built. Extensive metadynamics simulations with more than 200 and 100 metasteps for $XeFe_3$ and $XeNi_3$, respectively, were conducted. Each metastep included 600 time steps of 1 fs. The canonical NVT ensemble was used for molecular dynamics runs. The two structures remained stable relative to their component elements, supporting the stability of these compounds under the Earh's inner core conditions. Although it is not clear that these findings solve the missing Xe paradox (doubts about how Fe and Ni trapped Xe in the core and where the xenon was between the Earth's formation and the core formation), they unveil unexpected novel chemistry of Xe under extreme pressure.

4.11 CHALLENGES IN CRYSTAL STRUCTURE PREDICTION

Crystal structure prediction requires both an efficient sampling and an accurate evaluation of the PES. Materials with defects, nanomaterials or biomaterials, for example, present large unit cells. A completely random sampling of the PES would require an impractical number of calculations and lead to low structural diversity.

Biasing the searches by including symmetry constraints can improve the sampling efficiency, but the problem of the accurate evaluation of the PES still remains. Searches are usually performed with first-principles DFT codes such as VASP [55] or CASTEP [56], which use a plane wave basis set and pseudopotentials (details about computational techniques, with emphasis on DFT simulations, are given in Chapter 3). However, *ab initio* local optimizations of large systems are computationally very expensive. Faster orbital-free DFT and force field methods can be used, but at the expense of the accuracy.

The application of DFT presents other problems, especially at high pressure. One of the main difficulties is the choice of the approximate exchange-correlation functional. It is known that standard local density approximation (LDA) and generalized gradient approximation (GGA) functionals do not provide proper descriptions of van der Waals interactions and electride behaviors under pressure. Transition and metallization pressures are usually predicted incorrectly. Most sophisticated hybrid and Hubbard-corrected DFT functionals may be necessary, but the computational cost increases for the former. The choice of adequate pseudopotentials is also essential. They must have sufficiently small core radii and the appropriate number of valence electrons need to be treated explicitly. A rule of thumb is that the distance between neighboring atoms must be greater than the sum of the core radii of the atoms. The inclusion of thermal effects is also a bottleneck for DFT-based searches. Although it is possible to estimate the free energy with quasi-harmonic lattice dynamics models (see Chapter 1), it is still computationally very expensive to compute the phonon dispersion curves of all the structures, and the correct inclusion of anharmonic effects remains a challenge. Metadynamics and molecular dynamics simulations are alternatives to CSP at finite temperatures, but they usually require long simulation runs and it is difficult to verify that the global minimum has been found.

4.12 CONCLUSIONS

In this chapter, we discussed several available methods for structure prediction, such as simulated annealing, basin and minima hopping, metadynamics, random structure searches, evolutionary algorithms, and the particle swarm optimization technique. Recent examples of high-pressure systems (lithium, iridium hydrides and xenon binary compounds) have been presented to portray the features of the different codes and the technical details. A combined experimental-theoretical approach is shown to be efficient for structure characterization. However, even in absence of experimental information, the CSP methods are able to predict material structures with unusual stoichiometries and novel properties. The development of efficient computational methods and the systematic increase of computing power can help solve some of the problems that CSP faces, such as dealing with large simulation cells, using hybrid DFT methods, and including finite temperature effects.

Bibliography

[1] W. Grochala, R. Hoffman, J. Feng, *et al.* *Angew. Chem. Int. Ed.*, 46, 3620 (2007).

[2] C. Buzea and K. Robbie. *Supercond. Sci. Technol.*, 18, R1 (2005).

[3] M. Marqués, M. I. McMahon, E. Gregoryanz, *et al.* *Phys. Rev. Lett.*, 106, 095502 (2011).

[4] M. Marqués, M. Santoro, C. L. Guillaume, *et al.* *Phys. Rev. B*, 83, 184106 (2011).

[5] V. E. Antonov. *J. Alloys Compounds*, 330, 110 (2002).

[6] A. R. Oganov and S. Ono. *Nature (London)*, 430, 445 (2004).

[7] A. Jayaraman. *Rev. Mod. Phys.*, 55, 65 (1983).

[8] J. Maddox. *Nature (London)*, 335, 201 (1988).

[9] A. Y. Liu and M. L. Cohen. *Phys. Rev. B*, 41, 10727 (1990).

[10] D. G. Pettifor. *J. Phys. C*, 19, 285 (1986).

[11] E. Zintl and W. Dullenkopf. *Z. Phys. Chem. Abt. B*, 16, 195 (1932).

[12] M. Marqués, M. Flórez, J. M. Recio, *et al.* *J. Phys. Chem. B*, 110, 18609 (2006).

[13] C. C. Fischer, K. K. Tibbetts, D. Morgan, *et al.* *Nat. Mater.*, 5, 641 (2006).

[14] S. Kirkpatrick, C. D. Gellat, and M. P. Vecchi. *Science*, 220, 671 (1983).

[15] J. C. Schon, K. Doll, and M. Jansen. *Phys. Status Solidi B*, 247, 23 (2010).

[16] D. Zagorac, K. Doll, and M. Jansen. *Phys. Rev. B*, 84, 045206 (2011).

[17] D. J. Wales and J. P. K. Doye. *J. Phys. Chem. A*, 101, 5111 (1997).

[18] S. Godecker. *J. Chem. Phys.*, 120, 9911 (2004).

[19] J. A. Flores-Livas, M. Amsler, T. J. Lenosky, *et al. Phys. Rev. Lett.*, 108, 117004 (2012).

[20] T. D. Huan, M. Amsler, M. A. L. Marques, *et al. Phys. Rev. Lett.*, 110, 135502 (2013).

[21] M. Amsler, J. A. Fores-Livas, T. D. Huan, *et al. Phys. Rev. Lett.*, 108, 205505 (2012).

[22] T. D. Huan, M. Amsler, V. N. Tuoc, *et al. Phys. Rev. B*, 86, 224110 (2012).

[23] M. Parrinello and A. Rahman. *Phys. Rev. Lett.*, 45, 1196 (1980).

[24] A. Laio and M. Parrinello. *Proc. Natl. Acad. Sci. U.S.A.*, 99, 12562 (2002).

[25] R. Martonak, D. Donadio, A. R. Oganov, *et al. Nat. Mater.*, 5, 623 (2006).

[26] J. Sun, D. D. Klug, R. Martonak, *et al. Proc. Natl. Acad. Sci. U.S.A.*, 106, 6077 (2009).

[27] Y. Yao, D. D. Klug, J. Sun, *et al. Phys. Rev. Lett.*, 103, 055503 (2009).

[28] C. J. Pickard and R. J. Needs. *J. Phys.: Condens. Matter*, 23, 053201 (2011).

[29] C. W. Glass, A. R. Oganov, and N. Hansen. *Comput. Phys. Commun.*, 175, 713 (2011).

[30] D. C. Lonie and E. Zurek. *Comput. Phys. Commun.*, 182, 372 (2011).

[31] S. Bahmann and J. Kortus. *Comput. Phys. Commun.*, 184, 1618 (2013).

[32] W. W. Tipton, C. R. Bealing, K. Mathew, *et al. Phys. Rev. B*, 87, 184114 (2013).

[33] G. Trimarchi and A. Zunger. *Phys. Rev. B*, 75, 104113 (2007).

[34] N. L. Abraham and M. I. J. Probert. *Phys. Rev. B*, 73, 224104 (2006).

[35] A. Fadda and G. Fadda. *Phys. Rev. B*, 82, 104105 (2010).

[36] S. Q. Wu, M. Ji, C. Z. Wang, *et al. J. Phys. Condens. Matter*, 26, 035402 (2014).

[37] A. R. Oganov, A. O. Lyakhov, and M. Valle. *Acc. Chem. Res.*, 44, 227 (2011).

[38] Y. Wang and Y. Ma. *J. Chem. Phys.*, 140, 040901 (2014).

[39] J. B. Neaton and N. W. Ashcroft. *Nature (London)*, 400, 141 (1999).

[40] M. Hanfland, K. Syassen, N. E. Christensen, *et al. Nature (London)*, 408, 174 (2000).

[41] R. Rousseau, K. Uehara, D. D. Klug, *et al. Chem. Phys. Chem.*, 6, 1703 (2005).

[42] Y. Ma, A. R. Oganov, and Y. Xie. *Phys. Rev. B*, 78, 014102 (2008).

[43] T. Matsuoka and K. Shimuzu. *Nature (London)*, 458, 186 (2009).

[44] C. J. Pickard and R. J. Needs. *Phys. Rev. Lett.*, 102, 146401 (2009).

[45] Y. Yao, J. S. Tse, and D. D. Klug. *Phys. Rev. Lett.*, 102, 115503 (2009).

[46] C. L. Guillaume, E. Gregoryanz, O. Degtyareva, *et al. Nat. Phys.*, 7, 211 (2011).

[47] J. Lv, Y. Wang, L. Zhu, *et al. Phys. Rev. Lett.*, 106, 015503 (2011).

[48] F. A. Gorelli, S. F. Elatresh, C. L. Guillaume, *et al. Phys. Rev. Lett.*, 108, 055501 (2012).

[49] T. Scheler, M. Marqués, Z. Konôpková, *et al. Phys. Rev. Lett.*, 111, 215503 (2013).

[50] C. Sanloup, S. A. Bonev, M. Hochlaf, *et al. Phys. Rev. Lett.*, 106, 265501 (2013).

[51] C. Sanloup, B. C. Schmidt, E. C. Perez, *et al. Science*, 310, 1174 (2005).

[52] D. S. Brock and G. J. Schrobilgen. *J. Am. Chem. Soc.*, 133, 6265 (2011).

[53] D. Nishio-Hamane, T. Yagi, N. Sata, *et al. Geophys. Res. Lett.*, 37, L04302 (2010).

[54] L. Zhu, H. Liu, C. J. Pickard, *et al. Nat. Chem.*, 6, 644 (2014).

[55] G. Kresse and J. Furthmuller. *Phys. Rev. B*, 54, 11169 (1996).

[56] S. J. Clark, M. D. Segall, C. J. Pickard, *et al. Z. Kristallogr.*, 220, 567 (2005).

CHAPTER 5

Chemical Bonding under Pressure

Roberto A. Boto

Laboratoire de Chimie Théorique, Université Pierre et Marie Curie and CNRS, Paris, France

Miriam Marqués

MALTA Consolider Team and Departamento de Física Teórica, Atómica y Óptica, Universidad de Valladolid, Valladolid, Spain

Armando Beltrán

MALTA Consolider Team and Departamento de Química Física y Analítica, Univesidad Jaume I, Castellón, Spain

Lourdes Gracia

MALTA Consolider Team and Departamento de Química Física y Analítica, Univesidad Jaume I, Castellón, Spain

Juan Andrés

MALTA Consolider Team and Departamento de Química Física y Analítica, Univesidad Jaume I, Castellón, Spain

Vanessa Riffet

Laboratoire de Chimie Théorique, Université Pierre et Marie Curie and CNRS, Paris, France

Vanessa Labet

Laboratoire MONARIS, Université Pierre et Marie Curie and CNRS, Paris, France

Julia Contreras-García

MALTA Consolider Team and Laboratoire de Chimie Théorique, Université Pierre et Marie Curie and CNRS, Paris, France

CONTENTS

5.1 INTRODUCTION

As defined by Linus Pauling, "Chemistry is the science of substances: their structure, their properties, and the reactions that change them into other substances" [1]. The first two aspects, structure and properties, are clearly associated to the arrangement of atoms in a molecule, i.e., the chemical bond. These bonds determine chemical reactivity —the third aspect— and their visualization allows chemists to understand how atoms or molecules bond at a most fundamental level. A mechanistic understanding of chemical and biological functions and the structures of solid materials depends on knowing the geometric structures and the natures of bonds. But, despite the fact that the chemical bond is a fundamental concept in chemistry, "what is a chemical bond?" still remains a critical question for the chemical community because of the lack of a unique definition and inadequate understanding of its physical nature.

Visualization of bonding interactions between atoms and molecules is a long-standing quest in theoretical and computational chemistry. In recent years, it has become possible for most chemists to calculate molecular structures and stabilities based on quantum chemistry approaches. The main interest lies in creating a tool that enables researchers to see the interactions, and also interpret their characters and properties. Successful numerical solution of the Schrödinger equation has yielded energies and properties of atoms and molecules, but not yet a clear physical explanation of chemical bonding. There is even a controversy on the mechanistic origin of the simplest chemical bond: covalent bonding, as it was noted by Burdett in his classical book [2].

Chemical bonds together with other concepts such as atomic orbitals, electron shells, lone pairs, aromaticity, atomic charges, (hyper-) conjugation, strain, etc. do not correspond to physical observables. Such concepts therefore cannot be unambiguously defined in pure quantum theory, but constitute a rich set of "fuzzy", yet invaluably useful concepts [3–5]. They lead to constructive ideas and developments when appropriately used and defined.

To overcome this "fuzziness", several approaches have been developed to reveal the microscopic electronic natures of solids, both in reciprocal and in real space. As far as the reciprocal space is concerned, we have to keep in mind that the underlying idea behind the electronic structure of solids is that the valence electrons from the atoms involved spread throughout the entire structure, i.e., molecular orbitals are generally extended over all the constitutive atoms. A large number of overlapping atomic orbitals lead to molecular orbitals very similar in energy over a certain range. They form almost continuous bands. These bands are separated by band gaps, which are the energy values where there are no orbitals.

Within real space approaches, topological analysis is one of the most useful tools to characterize intra- and intermolecular interactions. Different types of bonding can be revealed by various topological methods, each based on a different scalar field [6]. Probably the best-known approaches are the electron density [7, 8], the Laplacian of the electron density [9, 10], the Electron Localization Function (ELF) [11, 12], and the molecular electrostatic potential. By topological analysis of the three-dimensional electron density and the definition of surfaces of zero electronic flux, the Quantum Theory of Atoms In Molecules (QTAIM) [10, 13] divides direct space into discrete atomic basins, which provides self-consistent atomic properties such as charges and volumes as well as a topological interatomic bond paths motif, which is often associated to the molecular structure. QTAIM is a powerful theory because all of these concepts and properties are derived from the empirically observable electron density [10, 14, 15].

High-pressure studies provide us with a new data concerning phenomena in condensed-matter physics and materials science, they help test fundamental theories and simulations [16, 17] and yield new materials with previously unknown structures and innovative chemical and physical properties [18, 19]. Compression induced by pressures at the gigapascal-scale can yield dramatic structural and electronic changes in solids [20], including semiconductor-to-metal transitions of both low-band-gap organic [21] and inorganic solids [22], polymerization of conjugated organic groups [23], spin-polarization reversal [23], metal-to-metal charge transfer [24], and induced conductivity and piezochromism [25]. Recently, morphology-tuned high pressure phase transitions have been found in several nanomaterials with various morphologies [26–28], providing important insight for the potential development of a method for synthesizing nanomaterials with high pressure structures.

In this context, the present chapter displays a working methodology to determine the chemical bonds under pressure both in real and reciprocal spaces.

This chapter is structured in three parts. First, we introduce real space analysis: chemical functions (electron density, ELF, and NCI) and their topological analysis. Illustrative examples are presented, from systems in which only changes in coordination take place to cases with changes in chemical bonding (e.g. from molecular to metallic or covalent). In a second part, reciprocal space analyses will deal with phonons, band structures and DOS. The importance of the local coordination environments of atoms in the lattices of crystalline matter and their responses to pressure in demonstrating physical and chemical property relations is emphasized through several examples, where phase competition becomes the big issue. In the third part, we will combine analyses in reciprocal and/or real space to reveal unexpected chemical changes upon pressurization.

5.2 ANALYSIS IN REAL SPACE: ELECTRON DENSITY, ELF, AND NCI

A thorough understanding of microscopic aspects of chemical transformations requires a knowledge of the behavior of electronic structures under pressure and its effect on the system's valence electrons. In this section, we focus on the analysis of chemical bonds (and valence electrons in general) and their changes from the real space view, working under a basic hypothesis: chemical bonds are three dimensional objects. Quantum chemistry should be able to associate chemical entities with different regions of space. This idea takes us to the topological partition of real space: a function with chemical meaning is defined, i.e., a chemical function (CF), whose topology confirms its chemical meaning. The chemical entities defined by such CFs, may be easily visualized by the representation of the CF isosurfaces and contour lines. In what follows, we present the most common CFs that can reveal atoms, bonds, and weak interactions of systems.

5.2.1 Chemical Functions

The *electron density*, $\rho(\boldsymbol{r})$, is a key property of any electronic system. As the first theorem of Hohenberg and Kohn states [29], it is possible to relate the ground state energy of any non-degenerate system with its electron density. Similarly, the main properties of the system may be quantitatively calculated from $\rho(\boldsymbol{r})$, providing valuable information about atomic and bonding property, stability, and chemical reactivity. The electron density is defined by the integration of the wave function to all the electrons of the system but the reference formula:

$$\rho(\boldsymbol{r}) = N \int \ldots \int |\Psi(\boldsymbol{x}_1, \boldsymbol{x}_2, \ldots, \boldsymbol{x}_N)|^2 ds_1 d\boldsymbol{x}_2 \ldots d\boldsymbol{x}_N, \qquad (5.1)$$

where s_i and $\boldsymbol{x}_i$ represent spin and spin-spatial coordinates, respectively. QTAIM defines atoms from the topological partition induced by the electron density. Chemical bonds and the underlying molecular graph may be traced by the analysis of critical points of the partition [10]. Among the achievements of QTAIM, we may outline the identification of the natures of functional groups and the transferability of their properties among systems.

The *Electron localization function* is a powerful tool to identify regions of space where electrons are localized in molecular and solid systems: atomic shells, bonds, and lone pairs [30, 31]. The first formulation of ELF was derived from the spin conditional probability defined from the wave function of the system. Alternatively, it may understood in terms of the excess of local kinetic energy density due to Pauli repulsion, t_P:

$$\chi = \frac{t_P}{c_F \rho^{5/3}} \tag{5.2}$$

$$\eta = \frac{1}{1+\chi^2}, \tag{5.3}$$

where $c_F = 2(3\pi^2)^{1/3}$ is the Fermi constant. ELF provides elegant and unambiguous explanations of many paradigmatic situations present in the molecular structure and chemical reactivity fields [32–34].

The *non-covalent interactions index* (NCI) [35] localizes interatomic surfaces where weak interactions occur. Regions with low reduced density gradient (RDG) and low electron density are identified where the reduced density gradient is given by:

$$\text{RDG} = \frac{1}{c_F}\frac{|\nabla\rho|}{\rho^{4/3}}. \tag{5.4}$$

The resulting regions provide an intuitive picture of weak bonding (hydrogen bonding), non-bonding (van der Waals) and anti-bonding (steric clashes) interactions. The small variations of the density in weak interaction regions make this index almost independent of the quality of the electron density making this technique useful even with promolecular densities (addition of non-relaxed atomic densities). Thanks to this property it is possible to apply NCI to large systems for moderate computational times.

For the sake of comparison, Figure 5.1 shows the different chemical images of diamond and graphite obtained with the three CFs. It is easy to notice how the electron density in both cases recovers the carbon atoms, ELF shows the covalent C-C bonds, and NCI reveals the steric clashes between close carbon atoms and the van der Waals interactions in graphite. The three analyses are complementary because each one recovers a different chemical concept from the wave function.

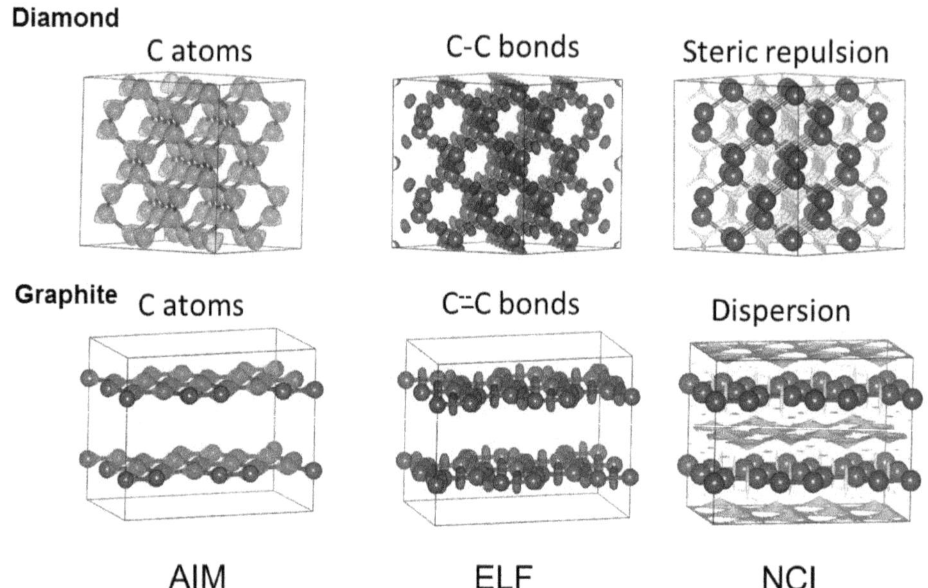

Figure 5.1 From left to right: density, ELF and NCI of diamond (top) and graphite (bottom).

5.2.2 Topological Analysis

To obtain insight into the transformations undergone by a system upon pressurization, we need a CF able to identify chemical entities and their interactions and also quantify their characteristics. Such information can be obtained by combining the above chemical functions with topological concepts based on the analysis of the gradient vector flux associated to CFs. This strategy enables a discrete partition of real space and the properties of the system. Since each point of space has only one associated flux vector, the space may be divided into discrete regions following the maximum at which the flux vectors finish. By definition, such regions contain only one maximum. Each of them is separated from the other regions by zero flux surfaces of the associated gradient vector field. It is possible to find an analogy between this topological partition and the division of a mountain chain into its constitutive mountains by observing the valleys.

This type of topological analysis has very important properties:

- *Chemical meaning.* The partition associated to a CF inherits its chemical meaning. For instance, the partition provided by the topology of the ELF gradient recovers regions associated with Lewis entities: cores, bonds, etc. We are able to recover chemically meaningful regions from the wave function of the system.

- *Exhaustiveness.* Each of the points in the system belongs to only one region. This means that it is possible to split the properties of the system into contributions associated with each region, $\langle A \rangle_\Omega$. These properties are additive:

$$\langle A \rangle = \sum_\Omega \langle A \rangle_\Omega. \tag{5.5}$$

The contribution of each region may be calculated by integration of the property density operator $\rho_A(\boldsymbol{r})$ over the volume of the region:

$$\langle A \rangle_\Omega = \int_\Omega \rho_A(\boldsymbol{r}) d\boldsymbol{r}. \tag{5.6}$$

As an example, if the electron density is integrated over the region associated with a bond, we obtain the charge of this bond. If the same procedure is repeated for all the regions of the system, the addition of all the integrals recovers the number of electrons of the system. In addition, it has been possible to recover the aufbau principle from ELF integrations over atomic shells along the periodic table [36]. The volumes of each of the regions also have chemical meaning; for example, the VSEPR principles may be easily recovered with this analysis.

5.2.3 Phase Transition with Changes in Atomic Coordination

In the next two subsections we will apply the concepts developed above to describe the changes in the electronic structure induced by pressure in solid-solid phase transitions. First, we will deal with transformations where the main source of phase reorganization under pressure is the achievement of a more efficient atomic packing that does not involve a change in the chemical bonding type.

Upon pressurization of a sample, a shortening of nearest neighbor distances occurs. This shortening is accompanied by a raise in the Gibbs free energy (identified by the chemical potential) of the solid. Since the free energy always increases with pressure (at constant temperature), the volume decreases to keep the free energy as low as possible. The process continues until the increase in connectivity becomes energetically competitive with the shortening of distances, and eventually a phase transition to a denser structure occurs.

It is interesting to know whether the high pressure coordination appears from the beginning or emerges at a later stage. The main goals here are to gain insight into the process of creation of these new atomic coordinations and investigate how the bond reconstruction correlates with the energetic profile and atomic displacements involved in the transformation. We will review the case of the phase transition from the B3 to the B1 phase in BeO, where a change in coordination for both Be and O takes place from 4 to 6 while keeping the ionic nature of the chemical bonds. Details of the martensitic approach applied to this transition are given in Chapter 2. We will show here the ability of ELF to track the chemical changes in the bonds and in the valence shells.

Analysis of the atomic coordinations in a covalent solid is straightforward. It suffices to count the number of bonds surrounding a given core. However, coordination counting can sometimes become a difficult task, especially when we are dealing with low symmetry ionic patterns. In these cases, establishing the limit of what belongs and what does not to the active sphere of coordination might become a matter of choosing a threshold. Even though ELF does not show basins for ionic bonding, it can still be a very useful tool in revealing the coordination of the valence maxima. Known as OCM (for outer core maxima), these maxima reflect in an indirect and subtle way the coordination of the solid. The OCM of the soft ions (usually anions) are disposed along the nearest neighbor directions and reflect the polarization of the ions. If we analyze the outer core of a hard ion instead, the OCM are disposed to occupy the interstitial voids between the bonding basins.

One example of the use of OCM to explain the changes in coordination in ionic crystals is provided by the B3 $\rightarrow$ B1 phase transition in BeO. Since both phases are ionic, there are no bonding basins and all electron pairs are distributed in both B1 and B3 structures, forming closed atomic electronic shell around the nuclei: one shell for beryllium (the core K(Be)-shell) and two for oxygen (the core K(O)-shell and the valence L(O)-shell). Following the changes in OCM arrangements of the anion also sheds light on the polarization reorganization along the transition: from four-fold tetrahedral (B3) to six-fold octahedral (B1) at around 100 GPa [37–39]. Using the orthorhombic $Imm2$ unit cell [40] discussed in Chapter 2, we can use z_{Be} to monitor the transition mechanism since the cation is displaced from the $(0, 0, 1/4)$ crystallographic position in the B3 phase to $(0, 0, 0)$ in the B1 phase [38]. We can normalize this change in the [0,1] range by means of the transformation:

$$\zeta = \frac{z_{Be} - z_{Be}(B3)}{z_{Be}(B1) - z_{Be}(B3)}. \tag{5.7}$$

The change in ionic coordination is reflected within the ELF approach by an increase in oxygen OCM from 4 to 6. Two new Be atoms from the second coordination sphere approach each oxygen and give rise to the corresponding new OCM (Figure 5.2). The topological change proceeds across three domains of structural stability:

- $z_{Be} > 0.18$: 4-fold coordination
- $0.18 > z_{Be} > 0.10$: 5-fold coordination

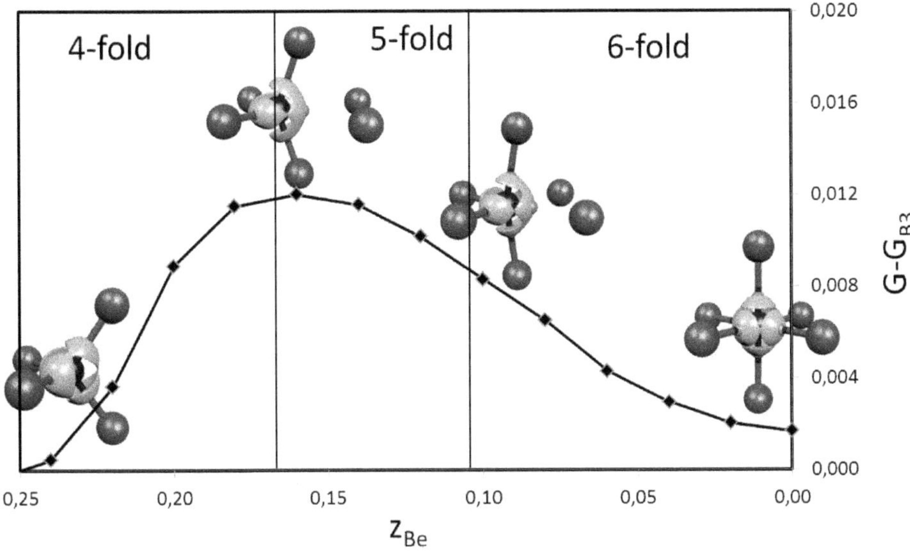

Figure 5.2 Gibbs energy profile versus z_{Be} across the B3 $\rightarrow$ B1 transition in BeO. Characteristic chemical patterns for each topological region are shown along the Gibbs profile with oxygen at the center of the polyhedron formed by Be atoms. One new oxygen basin emerges at $z_{Be} = 0.18$ associated with the transition state and at z_{Be}=0.10 when the final coordination is reached.

- $z_{Be} < 0.10$: 6-fold coordination

A new OCM first appears close to the transition state (z_{Be}=0.18) induced by the two approaching Be atoms (see Figure 5.2). This new OCM is located along the plane that bisects the angle formed by the O atom and the two approaching Be atoms. At $z_{Be} = 0.10$, the initial OCM divides into two, one for each new O-Be interaction. Along this last step, the new oxygen L-shell basins grow in volume and charge until they are equivalent to the other four at $\zeta = 1$.

The ELF topology has enabled us to identify the bonding nature in the initial and final phases, to track the change from one phase to the other, and even to understand the microscopic changes at the transition state (the appearance of the 5-fold coordination).

5.2.4 Phase Transition with Changes in Chemical Bonding

Transitions with changes in chemical bonding are characterized by the appearance of new electronic features in the solid. Metallization is a typical example. The analysis of solid-solid phase transitions involving dramatic changes in the localization pattern has become a major topic in recent years due to the availability of new high-pressure techniques [41, 42]. From the chemical view,

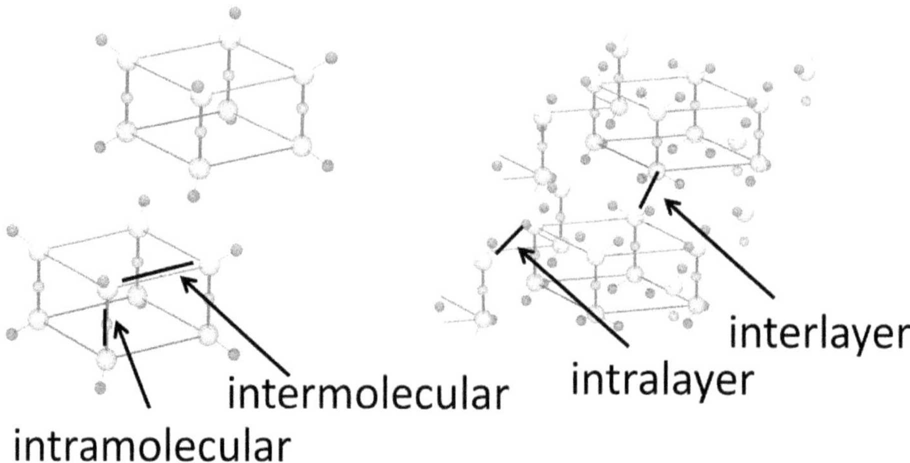

Figure 5.3 ELF maxima for the ϵ phase (left) and ζ phase (right) of crystalline oxygen. New intercluster connections in the ζ phase are highlighted.

these transitions reveal an incredibly rich potential in aiding our understanding of chemical bonding. A unifying principle is of general applicability to the microscopic factors determining the progression of these polymerizations. The minimization of the chemical potential is clearly facilitated by the Lewis entity, which changes its nature to provide a more efficient packing of the structure. Lone pairs and multiple bonds (more voluminous than the underlying σ-bonds) are good examples of voluminous Lewis entities that enable the formation of new bonds following this compacting principle. Throughout the next two sections we will see some examples of the application of ELF topology to chemical transformations:

- From molecular to metallic: by means of ELF we will review the phase transformation ϵ-O_2 $\rightarrow$ ζ-O_2, where O_2 molecules form clusters that preserve their molecular nature, and become metallic under pressure.

- From molecular to covalent: bonds are formed between the molecules of a solid to yield a covalent solid. We will use NCI to present a model system for hydrogen polymerization under pressure.

Molecular to metallic: ϵ-O_2 $\rightarrow$ ζ-O_2. The ϵ phase of oxygen is formed by $(O_2)_4$ clusters (see Figure 5.3) in a monoclinic C_2/m cell [43, 44]. At pressures over 95 GPa, the ϵ phase undergoes an isostructural transition towards the metallic ζ phase [45]. Raman studies of the pressure-induced metallization of oxygen ($\epsilon \rightarrow \zeta$) unveil the persistence of the O_2 vibron across the transition. In other words, O_2 molecules persist after the transition. In spite of the fact that the cluster geometry is maintained, the intrinsic properties and the chemical bond change completely.

Topological analysis enables us to understand this change: it suffices to analyze the intramolecular bonding in the O_2 molecules and the intermolecular bond between clusters. In the ϵ phase, O_2 molecules are observed with a single bond character. The rest of the electrons (in comparison with the double bonds observed in *in vacuo* molecules) are lodged in lone pairs that hold a population of 5.03 electrons each. We are thus witnessing a stabilization of the ionic resonant forms. There are no ELF maxima between the O_2 molecules, although high values of ELF are observed, indicating high delocalization within the cluster.

At around 95 GPa, a discontinuity in the cell parameters introduces a change in the electronic distribution: an increasing electronic delocalization across the layers connecting the different clusters appears (see Figure 5.3). Indeed, taking into account that the homogeneous electron gas has an ELF value of 0.5, valence electrons of metals should deviate very slightly from this quantity and flatter surfaces are expected for metallic compounds. After the transition, this is the case for ζ-O_2. We have a flatter profile arising from intercluster interactions that may be related to the metallicity of this compound.

Molecular to covalent: hydrogen polymerization. We study now the changes in the cohesion of a solid from purely long-range electrostatic and van der Waals interactions between fragments (molecular phases) to covalency. Following the compacting principle during this polymerization, some intermolecular interactions are expected to strengthen as the intramolecular bond weakens. As an example, we will track the pressure-induced polymerization of hydrogen.

Hydrogen is the simplest of elements; nonetheless its phase diagram is still only poorly understood despite tremendous experimental and theoretical work [46], driven in part by the remarkable properties predicted for this material under extreme conditions (high-temperature superconductivity [47], low- or zero-temperature quantum fluidity [48, 49]).

At low pressures and low temperatures, hydrogen is well known to exist as a molecular solid whose cohesion is ensured by van der Waals interactions between H_2 molecules characterized by the prototypical 2e-2c covalent bond seen in freshman chemistry textbooks. If the pressure is increased enough, hydrogen is expected to metallize like any other material; it is also expected to increase its coordination number, i.e., to polymerize. Will the two phenomena occur concomitantly or independently? The question remains open.

The existence of at least three molecular phases for solid hydrogen at low temperatures has been evidenced experimentally although for two of them there is no consensus on the associated structure. Particularly interesting is phase III, whose thermodynamic stability range extends from $p \sim 150$ GPa to $p > 300$ GPa [50]. The most promising candidate for this phase is a structure of

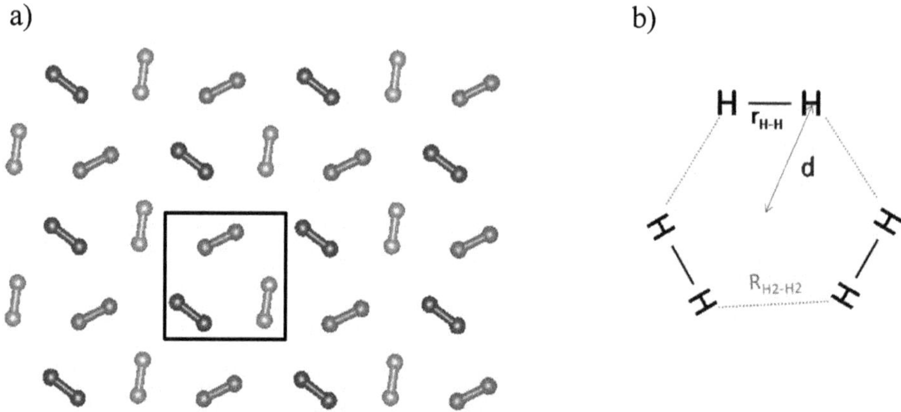

Figure 5.4 a) A layer of the $C2/c$ structure in hydrogen; b) isolated "three H_2" ring model system constrained to D_{3h} symmetry, with the d distance adjusted to model the effect of pressure.

$C2/c$ symmetry consisting of rings of three H_2 molecules (see Figure 5.4a) [51]. We have cut out this motif from the crystal to study the correlation between the intermolecular H_2-H_2 separation and the intramolecular H-H bond length as the pressure increases (simulated by constraining the distance d between the center of array and each H_2 molecule in an imposed D_{3h} symmetry; see Figure 5.4b). This model has shown that as the intermolecular distance decreases under pressure, the intramolecular bond length first slightly decreases and then increases, both characteristic distances going towards equalization [52–55]. At the same time, the Wiberg bond index (bond order measure) [56] for the intermolecular interaction increases from 0 to 0.438 while that of the intramolecular interaction decreases from 1.000 to 0.438.

Besides the geometrical aspects, an NCI analysis of the system gives us insight on the pressure regime where H_2 fragments can be considered true H_2 molecules and the system can be considered completely polymerized. For a long distance (e.g., $d = 250$ pm) corresponding to a low pressure, the three H_2 fragments of the three H_2 system interact through van der Waals interactions as depicted in Figure 5.5. Moreover, the density at the bond critical point coincides with that of an isolated H_2 molecule, $\rho = 0.255$ a.u., indicating that in the three H_2 system, the three H_2 fragments are true H_2 molecules. All these observations are consistent with Wiberg's bond order analysis.

As the value of d is decreased to model the effect of compression ($d = 250 \rightarrow 160 \rightarrow 100 \rightarrow 89$ pm), the NCI analysis reveals (i) a strengthening of the van der Waals interactions between the H_2 fragments, as shown by the increased density in the interaction region, and (ii) the weakening of the

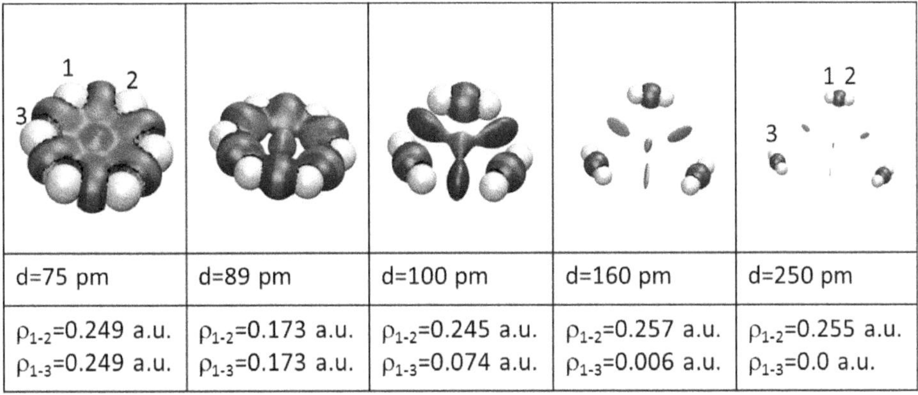

1
2
3
1 2
3
d=75 pm
d=89 pm
d=100 pm
d=160 pm
d=250 pm
$\rho_{1\text{-}2}$=0.249 a.u. $\rho_{1\text{-}3}$=0.249 a.u.
$\rho_{1\text{-}2}$=0.173 a.u. $\rho_{1\text{-}3}$=0.173 a.u.
$\rho_{1\text{-}2}$=0.245 a.u. $\rho_{1\text{-}3}$=0.074 a.u.
$\rho_{1\text{-}2}$=0.257 a.u. $\rho_{1\text{-}3}$=0.006 a.u.
$\rho_{1\text{-}2}$=0.255 a.u. $\rho_{1\text{-}3}$=0.0 a.u.

Figure 5.5 NCI for d equal to 250, 160, 100, 89, and 75 pm. Values for the density at the NCI peaks are shown: 1-2 corresponds to the initial (i.e., low pressure) intramolecular H-H interaction and 1-3 to the intermolecular one.

covalent bonds of the three H_2 fragments (ρ going from 0.255 a.u. at $d =$ 250 pm to 0.173 a.u. at $d = 89$ pm; see Figure 5.5).

For $d = 100$ pm, the density at the bond critical point of the cluster molecules no longer coincides with that of an isolated H_2 molecule, suggesting that the H_2 fragments of the three H_2 system can no longer be considered as true H_2 molecules. In other words, the system is polymerizing. For $d = 89$ pm and $d = 75$ pm, only one type of covalent bond is found. The six H-H bonds can be considered equivalent and the model system is completely polymerized.

It is possible to see a correspondence between the value of d in the three H_2 system and the pressure p imposed on solid hydrogen by computing and comparing an equalization index (measuring the degree of equalization of intra- and inter-molecular H-H separations) in both systems [52]. NCI analysis reveals that in the pressure range where the $C2/c$ structure seems to be thermodynamically stable (p = 125–250 GPa, corresponding to d = 95–110 pm), solid hydrogen has already begun its transition from true H_2 molecules to completely polymerized atomic hydrogen.

5.3 ANALYSIS IN RECIPROCAL SPACE: PHONONS, BAND STRUCTURE, AND DOS

5.3.1 Band Structure and Total Energy

Reciprocal space is also called Fourier space, $\boldsymbol{k}$-space, or momentum space in contrast to real space or direct space. Therefore it is possible to have separate s and p bands. Whether they form two distinct bands with a band gap or overlap depends on the separation of the orbitals and how strong the interaction is. Strong interaction means wide bands and a greater chance of overlap. The distinction between metallic and non-metallic solids comes from the way the orbitals are filled.

The highest energy associated with a populated state of a crystal occurs at the Fermi level, the analogue of the highest occupied molecular orbital (HOMO) energy levels in molecules. Metallic behavior arises from partially occupied bands, in which there is no energy gap between the Fermi level and the lowest unoccupied band. However, a non-metallic solid has a completely filled level (the valence band, VB) and an empty one (the conduction band, CB). These two bands are separated by a band gap of energy E_g. For conduction to occur, electrons must be excited up to the conduction band by overcoming an activation energy and hence the conduction of these compounds increases with temperature.

An Insulator has a completely filled valence band separated from the empty conduction band by a large gap. A good example is diamond: this compound has a large band gap ($E_g \sim 6$ eV), which makes it an excellent insulator. Very few electrons have sufficient energy to be promoted and the conductivity is negligible.

Thus, from band analysis we can differentiate:

- *Ionic solids*: bonding due to charge transfer from one atom to another. Energy bands formed from the atomic orbitals of anions and cations.
- *Covalent solids*: bonding due to orbital overlap and sharing of electrons. Bands formed from bonding molecular orbitals (filled bands) and anti-bonding orbitals (empty bands).
- *Metallic solids*: bonding due to orbital overlap forming a delocalized cloud of electrons. Overlap of atomic orbitals can be so strong that bands formed are much broader than the original energy separation of the orbitals. Orbitals lose their individuality and electrons can be considered as moving freely.

5.3.2 Electronic Density of States Example

In this subsection, we will review two examples in which the band structure, phonons, and electronic density of states (DOS) concepts will be applied to pressure-induced solid polymorphism and connected with their electronic structures. Hence, special attention will be paid to the changes in the nature of the chemical bonding through the DOS and the electronic bands. In this first example, we choose ZnS to illustrate the relationship between structural and electronic properties in a prototypical semiconductor material.

ZnS presents three polymorphs: cubic zinc blende (ZB), hexagonal wurtzite (W), and the rarely observed cubic rock salt (RS) obtained at relatively high

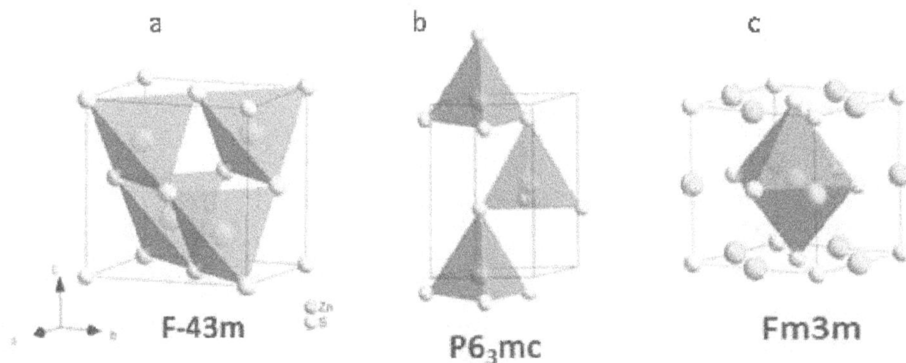

Figure 5.6 Unit cells for ZnS phases: (a) Zinc blende, (b) wurtzite, and (c) rock salt. Adapted, with permission, from reference 57.

pressures (Figure 5.6) [58]. In the W and ZB structures, each Zn atom is surrounded by four S atoms forming tetrahedral clusters [ZnS_4] that share one or two corners, respectively. It is interesting to note that these two phases show similar ionic environments, the difference lying in the different arrangements of the ionic layers. The W phase is more compact than the ZB phase and lacks a center of symmetry that generates a residual polarization in the [ZnS_4] clusters. ZB is the thermodynamically stable ZnS phase at room conditions, while the W polymorph is stable above 1293 K [59]. ZB and W phases have industrial applications, and due to their size- and shape-dependent properties both materials may be obtained in different ways to tune their properties to specific needs [60, 61]. Thus, understanding this polymorphism at an atomic level is very important for technological applications of ZnS-based materials [57].

As far as the band structure is concerned, both the ZB and W phases have direct band gaps at the Γ points, while the RS phase has an indirect band gap at the $L \rightarrow X$ points. The calculated band gaps for ZnS polymorphs are 4.10, 4.14, and 1.45 eV for ZB, W, and RS, respectively [62, 63].

Following the procedure presented in Chapter 3, we computed energy versus volume curves for the three different structures of ZnS and the subsequent enthalpy curves as a function of pressure. As expected from the crystalline environments, both W and ZB phases present similar enthalpies. Upon further compression, we find that RS becomes the thermodynamically stable phase around 15 GPa. The calculated W $\rightarrow$ RS and ZB $\rightarrow$ RS transition pressures are in excellent agreement with the experimental values in the 15.0–16.0 GPa range.

The atomic contributions of the VB and CB in the DOS analysis, known as atoms present at the active site (APAS), have been obtained following

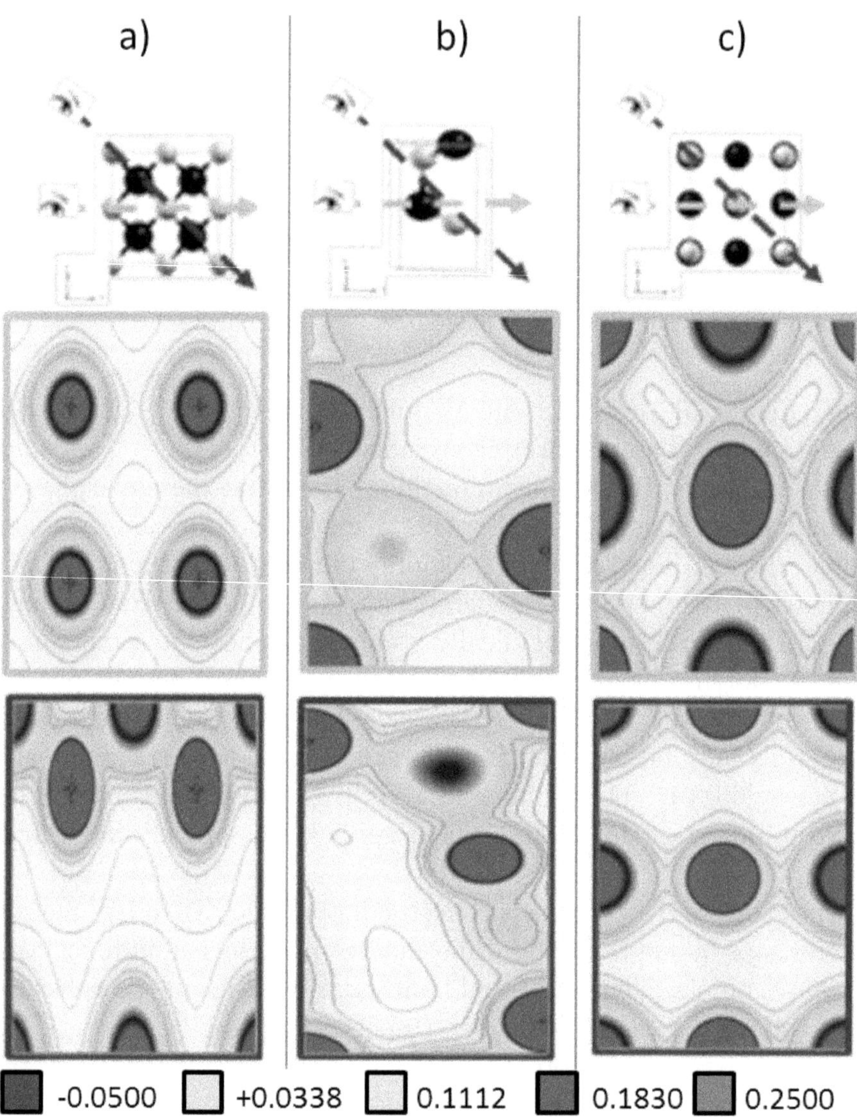

Figure 5.7 Charge density maps in ZnS polymorphs: a) zinc blende, b) wurtzite and c) rock salt across two different planes. The maps in the bottom row correspond to the diagonal plane. The maps in the middle row are parallel to the crystallographic a-b plane. Adapted, with permission, from reference 57.

Table 5.1 APAS contributions (%) to densities of states in ZnS polymorphs. Adapted, with permission, from reference 57.

Polymorph	Atom	APAS contribution %	
		VB	CB
ZB	Zn	30.16	63.05
	S	69.84	36.95
W	Zn	29.90	61.65
	S	70.10	38.35
RS	Zn	31.90	71.58
	S	68.10	28.42

a recently developed protocol [64, 65]. Results for the ZnS polymorphs are presented in Table 5.1. These results reveal the fundamental relationship between the nature of chemical bonding and bulk properties in ZnS polymorphs, showing the presence of mixed ionic-covalent bonding for these materials. The bonding is predominantly covalent for the ZB and W phases while it is more ionic for the RS structure. The Zn-S distances are 2.308 Å, 2.313–2.316 Å, and 2.510 Å for ZB, W, and RS, respectively. In Figure 5.7, the charge density contours are depicted for bulk ZnS polymorphs. A detailed analysis of these results points out that a larger contribution of covalent bonds in the ZnS clusters causes an energetic stabilization of the crystal structure.

The transition to RS involves a larger volume collapse. For the RS phase, the bulk modulus is approximately 25 % larger than the value for the ZB and W structures ($B_0 = 155.4$ GPa and $B_0' = 5.1$). This decrease of the bulk compressibility is caused by a rearrangement of the polyhedral units that takes place at the transition. The W $\rightarrow$ RS phase transition is a reconstructive phase transition and the transformation mechanism is associated with large atomic displacements and strain. The volume reduction during the phase transitions for both ZB and W phases are predicted to be about 13.9 and 14.3 %, respectively. In other words, the transition pressures of the W $\rightarrow$ RS and ZB $\rightarrow$ RS phase changes are very close to each other. From enthalpy calculations, we can state that no phase transition between the bulk ZB and W phases occurs with the application of pressure alone because their enthalpies are always parallel to each other.

Regarding the polyhedral changes taking place as the pressure is applied, calculations show that the distortion of the ZnS_4 tetrahedral clusters is reduced in the W phase, and at 35 GPa there is only one Zn-S distance. Due to the increase in coordination, the average Zn-S bond distances are increased at the transition pressure: 8.8 % for W $\rightarrow$ RS (at 15.0 GPa) and 9.0 % for ZB $\rightarrow$

RS (at 15.5 GPa). The computed linear compressibilities κ_c and κ_a show a slightly anisotropic behavior for the W structure, which is not significantly compressed along the c-direction.

5.3.3 Band Structure Example

A more complex material for exploring pressure-induced transformations on the electronic structure is zinc stannate, Zn_2SnO_4. This is an important semiconductor material (E_g = 3.6 eV) with a typical inverse spinel structure (I-ZTO). Recently, Zn_2SnO_4 has attracted attention as an alternative to solid-state devices due to its low fabrication cost and relatively high efficiency, as an anode material for dye-sensitized solar cells. From a fundamental view, it is interesting to study Zn_2SnO_4 as a model ternary oxide to understand how the optical and electronic properties are controlled by composition and structure. High-pressure studies on the cubic spinels indicate that a cubic-to-orthorhombic phase transformation takes place under pressure.

The calculated band structures for Zn_2SnO_4 along the adequate symmetry lines of the cubic and orthorhombic Bravais lattices are 3.46 eV, 2.48 eV, and 2.64 eV for inverse-ZTO (I-ZTO), titanite-type (T), and Ferrite-type (F), respectively. These values are typical for the wide band gap semiconductors. Figure 5.8 shows the band structure, the total DOS, and the DOS projected on atoms for these Zn_2SnO_4 polymorphs.

I-ZTO has a band gap of 3.6–3.7 eV with a forbidden direct transition from the VB to the CB [67, 68]. The transition is direct because the CB minimum and VB maximum are both located at the Γ point for both normal (N-) and I-ZTO spinels. Due to the inversion symmetry, the states at the CB minimum and VB maximum have the same parity, which makes the transition forbidden. For the I-ZTO, the calculations show that the VB maximum is derived mostly from O-2p with a minor contribution from Zn-3d states; whereas in the CB, the Zn-4s and 3p states (tetrahedral sites) dominate over the Sn-5s contribution (octahedral sites). Therefore, a significant dependence of the Zn and Sn CB DOS on the local coordination is observed.

The mixing (or hybridization) between the O and Zn states can be an indication of the directional character of the bonds; consequently the ionicity of Zn_2SnO_4 is lower than that of other inverse spinel phase $A^{IV}B_2^{II}O_4$ compounds (like Cd_2SnO_4) and a higher pressure is needed for the phase transition, as reported in previous works [69].

T and F orthorhombic phases also present direct band gap values. For both, the maximum VB is derived mostly from O-2p with a minimal contribution from Zn-3d states. Unlike I-ZTO, there is a first CB with dominance of Sn-5s over Zn-4s and O-2p contributions. A second CB is governed by Zn-4s.

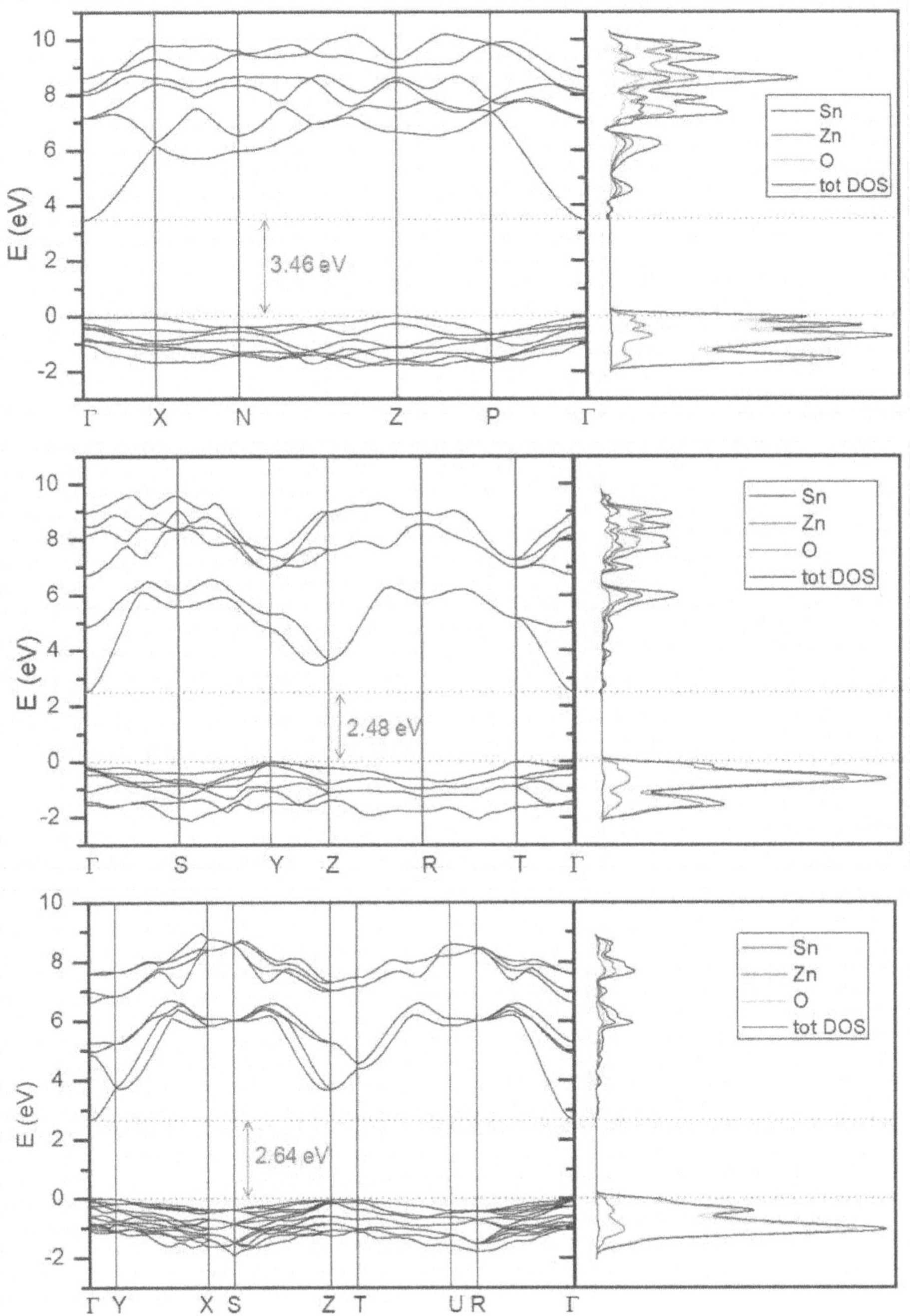

Figure 5.8 Band structure, DOS, and projected DOS on atoms for the Zn_2SnO_4 polymorphs: inverse spinel (top), titanite-type (middle), and ferrite-type (bottom). Adapted, with permission, from reference 66.

In the case of the F structure, the Zn-3p also participates in the second CB. Therefore, the energy separation between the first and second CBs is also large for these compounds in their respective ground states. According to quantum mechanics, a material is transparent to certain a wavelength when its energy level diagram indicates no energy difference corresponding to that wavelength. Thus, there can be no difference in energy of the order of visible light. However, a material can absorb, for example, part of the infrared or ultraviolet radiation. The energy gaps between the VB and CB correspond to UV transitions while the difference between the first and second levels of the CB corresponds to infrared light, so the material does not absorb the visible light. This explains the transparency observed in these degenerate n-type conducting oxides.

5.4 COMBINING REAL AND RECIPROCAL SPACE ANALYSIS

It is becoming increasingly clear that the free-electron model of metals is not applicable when metals are compressed. A deviation from the ideal metallic behavior is manifested by the appeareance of open and incommensurate structures with unusual properties such as decreased conductivity and melting points as well as superconductivity [70]. This state of matter has been explained in terms of Peierls distortions, Fermi surface-Brillouin zone interactions, $s \rightarrow p$, $s \rightarrow d$ electronic transitions, and more recently in terms of an increase of valence electrons in interstitial regions leading to pseudo-ionic materials [71]. We will analyze K and Na through ELF, band structure, and DOS analysis.

5.4.1 K-fcc to K-hP4

Potassium, like the rest of alkali metals, adopts the body-centered-cubic (bcc) structure at ambient pressure. It transforms to the face-centered-cubic (fcc) structure at 11 GPa. Recently, several host-guest structures have been identified from 20 to 40 GPa. What is more interesting is that a commensurate phase (hP4) is also observed between 25 and 35 GPa. This phase is hexagonal, space group $P6_3/mmc$, with four atoms per unit cell located on the $2a$ and $2c$ Wyckoff sites, respectively [72]. Interestingly, the double hexagonal closely packed (DHCP) structure, which has the same space group and atomic positions, has been proposed as the stable structure of K above 250 GPa [73]. However, the proposed DHCP structure has a c/a ratio close to the ideal close packing value (3.266), very different from the experimental c/a ratio of 1.36.

Curiously, this atomic arrangement with similar c/a ratios appears as the cation sublattice in several dialkali metal monochalcogenides and their corresponding oxides. It can be explained through the anions in metallic matrices model (AMM) [74]. According to this model, the metallic matrix acts as a host

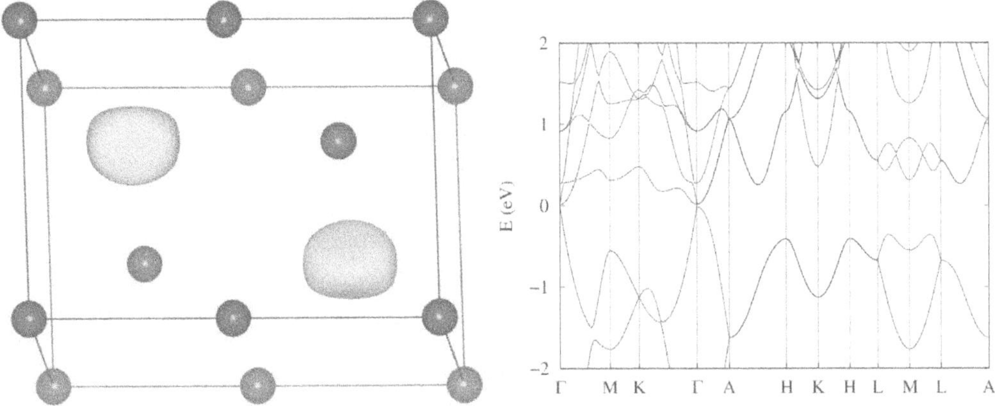

Figure 5.9 ELF isosurface $\eta = 0.9$ (left) and band structure for hP4 potassium at $p = 27$ GPa (right). Adapted, with permission, from reference 72.

lattice for the non-metallic atoms, the formation and localization of the anions being determined by the geometric and electronic structure of the metallic lattice. In fact, the analysis of ELF for elemental potassium at 25 GPa reveals maxima with high ELF value (~ 0.95) at the interstitial positions on the $2d$ sites. Their basins form well defined chemical entities holding approximately two electrons each, but with no core inside (see Figure 5.9 left). In reciprocal space, this electronic distribution leads to the opening of a gap in the band structure and a semiconductor character for this phase (see Figure 5.9 right). Therefore, the validity of the AMM is clear by considering potassium as an electride in which the perfectly localized valence electrons act as pseudoanions. The pseudo-ionic character is highlighted by the fact that the pseudoanions are located at the same sites as the anions in the Ni_2In-type compounds.

5.4.2 Na-cI16 to Na-oP8 to Na-hP4

The negative melting curve of sodium is one of the archetypal examples of unusual properties of elements under pressure [75]. Although negative melting curves have been previously reported, the broad non-monotonic decrease from 35 GPa to more than 100 GPa is unique. Moreover, at the minimum of the melting curve, seven different crystalline phases appear with slight changes of pressure and/or temperature [76]. Interestingly, the clustering of these phases sharing stability conditions at the minimum of the melting curve is related to the appearance of a boundary between delocalization and localization represented by the cI16 and oP8 structures, respectively [77].

Na-cI16 shows a very disordered electronic distribution with coexisting attractor regions of high and low ELF, with approximate values of 0.75 and

0.25, respectively. The existence of the attractors with very low ELF values, which also decrease upon increasing pressure, is likely to be related to weaker bonding and therefore lower melting temperature. When passing to the oP8 phase, the low ELF attractors disappear and only high ELF attractors, with ELF values close to unity are present. This indicates a change to an electride (i.e. pseudo-ionic) behavior, which leads to high melting temperatures. At around 240 GPa, this phase is predicted to transform to the hP4 structure observed in Na. Experimentally, this phase appears at 179 GPa. The valence electrons localize even more strongly and there is almost no exchange between them, as the very low ELF value between basins shows (regions become more independent from one another).

Traditionally, the closure of the gap between the valence and conduction bands is the signature of the metallic state. Within an ELF perspective, a typical metallic system shows a very flat ELF profile where the difference in the ELF values of the ELF maxima and saddle points is very low. However, when a metal turns into an electride (and possibly an insulating state) very localized units (pseudoanions) with almost no connections between them appear. In fact, the hP4 phase is a direct band gap insulator, with a band gap of 2.1 eV measured by optical transmission spectroscopy. Unusually strong Raman activity is also detected in these electride phases. In particular, the very intense E_{2g} mode of the hP4 phase corresponds to the stretching along the metal-pseudoanion direction. It provides experimental proof of the pseudoanionic ELF attractors.

5.5 SUMMARY AND CONCLUSIONS

In this chapter, we have reviewed the most important tools available to the theoretical chemist to demonstrate the chemical patterns of solids under pressure and changes induced by pressurization. We have seen that there are mainly two types of tools: those working in real space and those based on the reciprocal space. Among those methods that exploit the real space, we focused on topological tools. By the use of scalar fields such as the electron density, NCI or ELF, we can visualize the chemical pattern of a system in three dimensions. We have used these tools in the descriptions of coordination and chemical changes under pressure. As an example, we followed the polymerization of hydrogen using the reduced density gradient and we revisited the increase in coordination of an ionic solid by looking at its valence distribution. As far as reciprocal space is concerned, we reviewed important concepts such as bands and DOS projections. We applied these concepts to three different cases in which the polymorphisms induced by external pressures were analyzed in detail.

The novel message of this chapter is that pressure can change chemical bonding nature in a profound manner. This can involve entirely different structures and symmetries and, most importantly, new energetics and mechanisms of chemical bonding. This conclusion is supported by a mosaic of complementary experimental results. It should also help to stimulate searches for similar phenomena in other systems while keeping in mind that understanding the microscopic organization allows for the control, enhancement and optimization of the physical and chemical properties of materials under pressure.

Bibliography

[1] L. Pauling. *General Chemistry.* Dover Publications, New York (1947).

[2] J. K. Burdett. *Chemical Bonds: A Dialog.* Wiley, Chichester (1997).

[3] P. v. R. Schleyer. *Chem. Rev.*, 105, 3433 (2005).

[4] P. L. A. Popelier. *Faraday Discuss.*, 135, 3 (2007).

[5] I. V. Alabugin, K. M. Gilmore, and P. W. Peterson. *WIREs Comput. Mol. Sci.*, 1, 109 (2011).

[6] P. L. Ayers, R. J. Boyd, *et al. Comp. Theor. Chem.*, 1053, 2 (2015).

[7] K. Collard and G. G. Hall. *Int. J. Quantum Chem.*, 12, 623 (1977).

[8] R. F. W. Bader and M. E. Stephens. *J. Am. Chem. Soc.*, 97, 7391 (1975).

[9] R. F. W. Bader. *Acc. Chem. Res.*, 8, 34 (1975).

[10] R. F. W. Bader. *Atoms in Molecules: A Quantum Theory.* Clarendon Press, Oxford (1990).

[11] A. D. Becke and K. Edgecomb. *J. Chem. Phys.*, 92, 5397 (1990).

[12] B. Silvi and A. Savin. *Nature*, 371, 683 (1994).

[13] R. F. W. Bader. *The Encyclopedia of Computational Chemistry.* Wiley, Chichester (1998).

[14] R. F. W. Bader. *Phys. Rev. B*, 49, 13348 (1994).

[15] C. F. Matta and R. J. Boyd. *The Quantum Theory of Atoms in Molecules from Solid State to DNA and Drug Design.* Wiley-VCH, Weinheim (2007).

[16] R. J. Hemley and N. W. Ashcroft. *Physics Today*, 51, 26 (2008).

[17] W. Grochala, R. Hoffmann, J. Feng, *et al. Angew. Chem., Int. Ed.*, 46, 3620 (2007).

[18] V. Schettino and R. Bini. *Phys. Chem. Chem. Phys.*, 5, 1951 (2002).

[19] P. F. McMillan. *Nat. Mater.*, 1, 19 (2002).

[20] J. C. Tan and A. K. Cheetham. *Chem. Soc. Rev.*, 40, 1059 (2011).

[21] J. W. L. Wong, A. Mailman, and K. Lekin. *J. Am. Chem. Soc.*, 136, 1070 (2014).

[22] A. P. Nayak, S. Bhattacharyya, and J. Zhu. *Nat. Commun.*, 5, 3731 (2014).

[23] C. Murli and Y. J. Song. *J. Phys. Chem. B*, 114, 9744 (2010).

[24] Z. Huang, J. E. Auckett, P. E. R. Blanchard, *et al. Angew. Chem., Int. Ed.*, 53, 3482 (2014).

[25] A. Jaffe, Y. Lin, W. L. Mao, *et al. J. Am. Chem. Soc.*, 137, 1673 (2015).

[26] Y. J. Wang, J. Z. Zhang, J. Wu, *et al. Nano Lett.*, 8, 2891 (2008).

[27] L. H. Shen, X. F. Li, Y. M. Ma, *et al. Appl. Phys. Lett.*, 89, 141903 (2006).

[28] I. Zardo, S. Yazji, C. Marini, *et al. ACS Nano*, 6, 3284 (2012).

[29] P. C. Hohenberg and W. Kohn. *Phys. Rev. B*, 136, 864 (1964).

[30] C. Gatti. *Z. Kristallogr.*, 220, 399 (2005).

[31] A. Savin, R. Nesper, S. Wengert, *et al. Angew. Chem., Int. Ed. Engl.*, 36, 1808 (1997).

[32] J. C. Santos, J. Andrés, A. Aizman, *et al. J. Phys. Chem. A*, 109, 3687 (2005).

[33] J. Berski, S Andrés, B. Silvi, and L. R. Domingo. *J. Phys. Chem. A*, 110, 13939 (2006).

[34] J. Polo, V Andrés and B. Silvi. *J. Comput. Chem.*, 28, 857 (2007).

[35] E. R. Johnson, S. Keinan, P. Mori-Sáchez, *et al. J. Am. Chem. Soc.*, 132, 6498 (2010).

[36] M. Kohout and A. Savin. *J. Comput. Chem.*, 18, 1431 (1997).

[37] Y. Cai, S. Wu, R. Xu, *et al. Phys. Rev. B*, 73, 184104 (2006).

[38] A. M. Contreras-García, J Pendás and J. M. Recio. *J. Phys. Chem. B*, 112, 9787 (2008).

[39] C. J. Park, S. J. Lee, Y. J. Ko, *et al. Phys. Rev. B*, 59, 13501 (1999).

[40] M. S. Miao and W. R. L. Lambrecht. *Phys. Rev. Lett.*, 94, 225501 (2005).

[41] R. J. Hemley. *Annu. Rev. Phys. Chem.*, 51, 763 (2000).

[42] M. Eremets, R. J. Hemley, H. Mao, *et al. Nature*, 411, 170 (2001).

[43] S. Falconi and G. J. Ackland. *Phys. Rev. B*, 73, 184204 (2006).

[44] B. Militzer and R. J. Hemley. *Nature*, 443, 150 (2006).

[45] G. Weck, P. Loubeyre, and R. LeToullec. *Phys. Rev. Lett.*, 88, 035504 (2002).

[46] J. M. McMahon, M. A. Morales, C. Pierleoni, *et al. Rev. Mod. Phys.*, 84, 1607 (2012).

[47] N. W. Ashcroft. *Phys. Rev. Lett.*, 21, 1748 (1968).

[48] N. W. Ashcroft. *J. Phys.: Condens. Matter*, 12, 129 (2000).

[49] N. W. Ashcroft. *J. Phys. A: Math. Gen.*, 36, 6137 (2003).

[50] C. S. Zha, Z. Liu, and R. J. Hemley. *Phys. Rev. Lett.*, 108, 146402 (2012).

[51] C. J. Pickard and R. J. Needs. *Nat. Phys.*, 3, 473 (2007).

[52] V. Labet, P. Gonzalez-Morelos, R. Hoffmann, *et al. J. Chem. Phys.*, 136, 074501 (2012).

[53] V. Labet, R. Hoffmann, and N. W. Ashcroft. *J. Chem. Phys.*, 136, 074502 (2012).

[54] V. Labet, R. Hoffmann, and N. W. Ashcroft. *J. Chem. Phys.*, 136, 074503 (2012).

[55] V. Labet, R. Hoffmann, and N. W. Ashcroft. *J. Chem. Phys.*, 136, 074504 (2012).

[56] K. B. Wiberg. *Tetrahedron*, 24, 1083 (1968).

[57] F. A. La Porta, L. Gracia, J. Andrés, *et al. J. Am. Ceram. Soc.*, 97, 4011 (2014).

[58] K. A. Prior, C. Bradford, I. A. Davidson, *et al.* *J. Cryst. Growth*, 323, 114 (2011).

[59] Y. Ding, X. D. Wang, and Z. L. Wang. *Chem. Phys. Lett.*, 398, 32 (2004).

[60] L. Hou and F. Gao. *Mater. Lett.*, 65, 500 (2011).

[61] F. Huang and J. F. Banfield. *J. Am. Chem. Soc.*, 127, 4523 (2005).

[62] F. A. La Porta, M. M. Ferrer, Y. V. B. Santana, *et al.* *J. Alloys Compd.*, 555, 153 (2013).

[63] M. Durandurdu. *J. Phys. Chem. Solids*, 70, 645 (2009).

[64] F. A. La Porta, T. C. Ramalho, R. T. Santiago, *et al.* *J. Phys. Chem. A*, 115, 824 (2011).

[65] F. A. La Porta, R. T. Santiago, T. C. Ramalho, *et al.* *Int. J. Quantum Chem.*, 110, 2015 (2010).

[66] L. Gracia, A. Beltrán, and J. Andrés. *J. Phys. Chem. C*, 115, 7740 (2011).

[67] M. A. Alpuche-Aviles and Y. Y. Wu. *J. Am. Chem. Soc.*, 131, 3216 (2009).

[68] D. Segev and S. H. Wei. *Phys. Rev. B*, 71, 125129 (2005).

[69] X. Shen, J. Shen, S. J. You, *et al.* *J. Appl. Phys.*, 106, 113523 (2009).

[70] E. Y. Tonkov and E. G. Ponyatovsky. *Phase Transformations of Elements under High Pressure.* CRC Press, London (2005).

[71] B. Rousseau and N. W. Ashcroft. *Phys. Rev. Lett.*, 101, 046407 (2008).

[72] M. Marqués, G. J. Ackland, L. F. Lundegaard, *et al.* *Phys. Rev. Lett.*, 103, 115501 (2009).

[73] Y. Ma, A. R. Oganov, and Y. Xie. *Phys. Rev. B*, 78, 14102 (2008).

[74] A. Vegas, D. Santamaría-Pérez, M. Marqués, *et al.* *Acta Cryst. B Struct. Sci.*, 62, 220 (2006).

[75] E. Gregoryanz *et al.* *Phys. Rev. Lett.*, 94, 185502 (2005).

[76] E. Gregoryanz *et al.* *Science*, 320, 1054 (2008).

[77] M. Marqués, M. Santoro, C. L. Guillaume, *et al.* *Phys. Rev. B*, 83, 184106 (2011).

II

EXPERIMENTAL TECHNIQUES

CHAPTER 6

High-Pressure Generation and Pressure Scales

Valentín García Baonza
MALTA Consolider Team and Department of Physical Chemistry, Faculty of Chemistry, Complutense University of Madrid, Spain

Javier Sánchez-Benítez
MALTA Consolider Team and Department of Physical Chemistry, Faculty of Chemistry, Complutense University of Madrid, Spain

CONTENTS

6.1 INTRODUCTION

IF one asks for a definition of high pressure (HP), the answer will certainly depend on the scientific environment in which the question arises: physics, chemistry, materials science, geophysics, biology, or food technology, just to name a few areas in which the pressure variable plays an important role. Such uncertainty in the answer also appears if we consider other variables like temperature or the electric and magnetic fields, so the term "extreme conditions" is often used to refer to extreme values of all these variables within the context of the phenomena under investigation. Among our scientific community, there is a certain agreement in that high pressures refer to the study of those phenomena occurring above 1000 bar (kbar). Other terms like "ultra-high pressure" were coined to refer to phenomena involving pressures above 1000000 bar (Mbar). Of course, the above limits are somewhat arbitrary and the fundamental laws studied in other chapters apply uniformly regardless of the range of pressures considered. This means that we share methodology with many other scientific areas involving only tens or few hundreds of atmospheres. In other words, the same fundamental thermodynamics and kinetics models apply to all phenomena in which the pressure variable is considered.

In this chapter we briefly discuss the general problem of pressure generation and measurement because the details of a given device will vary with the sample characteristics and the properties or phenomena that we want to study under HP conditions. We have no room here for step-by-step descriptions of measurement protocols or safety measures, as required for large-volume and industrial devices (the operation and use of diamond anvil cells usually presents a negligible risk, perhaps only to our budgets).

Many specific examples of HP devices appear throughout this book and we shall sparingly refer to them. Our aim here cannot go beyond offering the reader an overview of the different types of devices and kinds of experiments that can be carried out under HP conditions. The advantages and limitations of each type of device are discussed sporadically and we shall try to provide general advice to enable the reader to decide which method would be most appropriate for a particular HP study or measurement. We again emphasize that the high-pressure term is itself relative, so we must keep in mind that it is not always possible to adapt HP devices for performing all our desired experiments. Perhaps the golden rule is that any HP apparatus is, at most, a multi-purpose device, but never an all-purpose device in terms of the properties that can be accessed and the kind of samples that we may really study.

The basis of the modern HP instrumentation was settled in the first decades of the 20th century, when Percy W. Bridgman (Nobel Prize in Physics, 1946) performed a huge number of experiments that covered various scientific fields [1]. Among other advances, Bridgman designed techniques for generating and measuring pressures up to 100000 atmospheres (10 GPa). He also applied all these techniques to the study of fundamental properties of matter: volumetric properties and the equation of state, thermal conductivity, viscosity, and electrical resistance. Several decades later, the successful synthesis of diamond by scientific teams at General Electric [2, 3] and ASEA (Swedish electrical company) transformed high-pressure studies into something interesting from a technological view, leading to the development and advancement of large-volume presses. The final milestone of high pressure on the scientific scene was the development in the late 1950s almost simultaneously at the National Bureau of Standards (NBS) and the University of Chicago of the first design of what is now known as the diamond anvil cell [4, 5] (DAC). It is worth noting that although the anvil design was known since the times of Bridgman anvils, this was the first time that diamond was used to generate HPs. These and other achievements are described in many books and monographs, but we recommend the story as narrated by R. Hazen [6].

Since then, HP technology has experienced a rapid development and is now a multidisciplinary field. For instance, the modern development of anvil cells has made it possible to introduce the pressure variable in different studies, thus promoting the progress of and interactions between different branches of science [7]. The exceptional hardness of diamond allows us to generate static pressures up to several million atmospheres on a sample, and its optical properties permit the in situ observation of a completely new range of phenomena. Thus, most of the available characterization techniques (structural, spectroscopic, electrical and magnetic, including using synchrotron sources and neutron radiation) have been successfully adapted to different DAC configurations to study microscopic samples, as described in Chapters 7, 8, and 9.

6.2 HIGH-PRESSURE GENERATION

Pressure, as a physical variable, is defined by the amount of force acting per unit area. This definition is of key importance for measuring the absolute pressure and for classifying the pressure scales that will be discussed later. However, the design and principle of operation of all equipment used to generate HPs can be rationalized better if we recognize that the product of pressure and volume has units of energy. Thus, the maximum pressure that can be generated in a given apparatus depends on the amount of energy that we are able to store and hold within the operation volume. This imposes some limits on the volume of the sample that can be compressed or processed, and indirectly limits the diagnostic techniques that can be coupled to our device for studying a sample. Unfortunately, these limitations apply even to pressure gauges, thus requiring secondary pressure scales to measure or sometimes just to estimate the pressure to which our sample is subjected. In our opinion, this is undoubtedly a fundamental problem and another source for intense discussion within the HP community. Hence the relevance of pressure metrology—a fact not recognized adequately by many researchers.

As pressure is defined as applied force over an area from an engineering view, the problem of creating pressure splits into two tasks: (i) generating the force, and (ii) defining the area over which it is applied, since different surface areas will result in different pressures for the same force. Due to the relatively limited commercial availability of pressure cells, you may have to develop your own HP instrumentation. You could adapt the cell to your particular sample requirements (state of aggregation, weak or strong response to light scattering, magnetic and electric fields,) sample environment (cryostats, heaters, goniometers, magnets,) and characterization technique (optical or x-ray sources, microscopes, spectrometers.)

The material used in the construction of a HP apparatus needs to be carefully considered according to the sample requirements and sample environment, and taking into account some key properties of the material such as: thermal (expansion and conductivity), electrical, magnetic, optical (transparency), and chemical (reactivity, resistance to corrosion). The most important are the mechanical properties. Considering that stress is a measure of the average amount of force exerted per unit area, the failure strength (tensile or compressive) is defined as stress applied to a part at the point when it yields. A quantity used to characterize materials is the elastic modulus (Young's modulus) which is the stress-to-strain ratio—the response of the material to stress in the elastic deformation regime. Although metals and metal alloys are normally chosen for construction of HP devices due to their excellent mechanical strength, good surface stability, and corrosion and oxidation resistance; other materials such as ceramics (e.g., zirconia or alumina) or composites (e.g., Zylon or Kevlar) can be chosen for some parts of the apparatus.

A scheme for developing HP instrumentation can be as follows. First, you must decide on the pressure limit to achieve and the sample volume to study to define the type of pressure cell to design. Then you need to select the materials for the cell by considering the key thermal, electrical, magnetic, optical, and chemical properties. The most important are the mechanical properties as noted earlier. After these decisions are made, you can design the cell and its components. At this step it is very useful to test your design using finite element analysis (FEA) to evaluate the reaction forces, stresses, and deformations of the cell under work. FEA is a numerical technique used for solving complex physics and engineering problems and is based on the finite element method (FEM). By breaking down a large complex problem into many smaller problems and connecting these through the boundary conditions using complex numerical algorithms, the FEM provides an approximate solution to the overall problem. Depending on the problem, this tool will simulate the stress distribution, deformation of the cell under load, temperature gradients, magnetic field distribution, and other factors. Finally you can deliver the drawings to the workshop for manufacturing the instrumentation.

The pressure-transmitting media also play a key role in most HP experiments, since they actually reduce the stress gradient within the sample chamber. The most relevant property of a pressure-transmitting medium is its

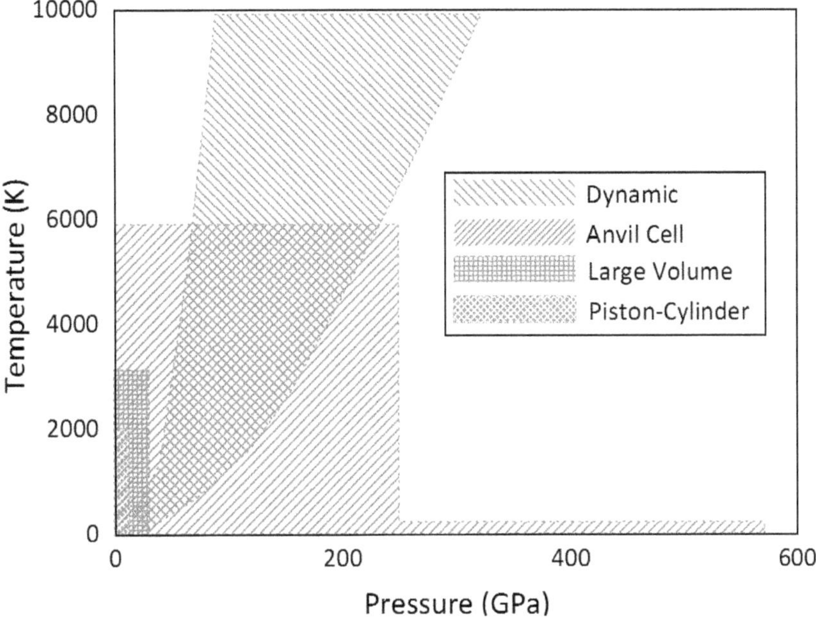

Figure 6.1 Typical pressure and temperature ranges accessible by various high-pressure techniques.

hydrostatic or pseudo-hydrostatic limit, but its stability and chemical inertness (unless we plan reactivity or synthetic studies) are also important. An additional desirable requirement is that the transmitting medium would provide no signals that affect the measured property or, at least, such signals would be easily identifiable and separable from those of the sample under study.

A initial classification of HP techniques can be established by dividing them into static and dynamic types. Static and dynamic experiments are complementary approaches, but the two techniques operate at different time scales and within different pressure-temperature ranges, so their results are sometimes difficult to compare directly (see Figure 6.1).

6.2.1 Static Pressure Techniques

The vast majority of techniques used in both academic laboratories and industrial facilities fall within the static category. Some general and schematic designs of some generic static-pressure devices are shown in Figure 6.2. With the exception of the DAC, we shall briefly discuss here some general aspects of the static techniques, and we refer the reader to specific monographs that treat each static device in more detail [7, 8].

The so-called piston-cylinder type HP cell system is one of the most obvious methods of pressure generation and provides good hydrostatic conditions for

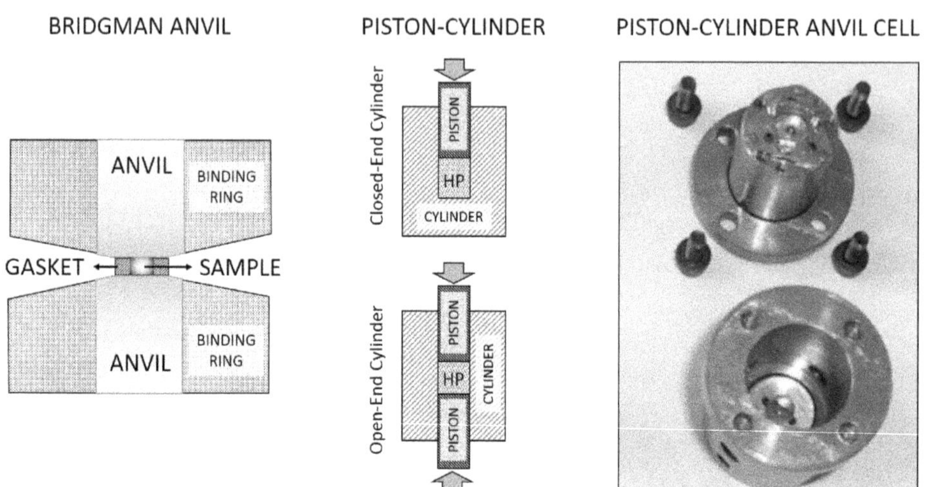

Figure 6.2 Designs for generating static pressures.

experiments. This system basically consists of a cylinder, which is the body of the cell, and one or two pistons that compress the transmitting medium and therefore the sample inside the cell along the cylinder axis. These cells underwent several modifications throughout the 20th century to allow them to perform electrical resistance, compressibility and optical studies in the infrared region [7]. These cells can achieve up to 3 GPa, depending on the construction material. Sealing elements, normally made of soft materials, must be used to avoid leakage. It was soon realized by Bridgman that if a piston has a conical shape, the pressure that can be achieved is higher, and thus the concept of anvil press emerged. The simplest design operates with two opposed anvils that compress a thin gasket that holds the sample. This design is sometimes known as the Bridgman-type anvil.

6.2.1.1 Diamond Anvil Cell

The diamond anvil cell, also known as the DAC, is perhaps the most popular and versatile device in high pressure research, as DACs provide a convenient way for pressure confinement and a good optical window for coupling several diagnostic techniques. For some practical applications, however, the DAC has the obvious limitation of the small size of the sample, since the diameter of the chamber rarely exceeds 200–300 microns, thus requiring the use of microscopic probes or techniques.

In 1950, Lawson and Tang used diamonds for the first time to generate pressure carrying out x-ray diffraction experiments at the University of Chicago. The pressure limit achieved was about 20 kbar (2 GPa), although with an improved design, 30 kbar could be reached. The current DAC designs are based

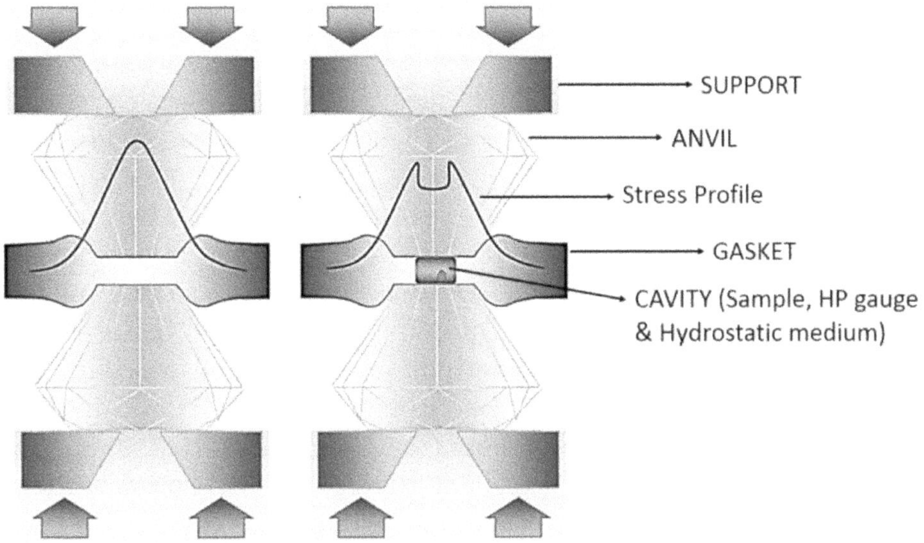

Figure 6.3 Stress profiles across the gasket in a generic anvil cell with (right) and without (left) a sample cavity. Apart from the sample, the cavity normally also hosts a pressure gauge and a hydrostatic or quasi-hydrostatic transmitting medium. Note the roughly constant pressure value obtained within the sample cavity.

on the first DAC constructed by the National Bureau of Standards by Weir, Lippincott, van Valkenburg and Bunting in 1959 [4]; since then a large number of DACs have been developed.

In general, a basic design of a DAC consists of a system of two opposed diamonds surrounding a sample. Diamonds act as pressure generators as well as optical windows. Depending on the diamond sizes, a pressure of 100 kbar could be easily reached by the original DAC. The first experiments were carried out with a solid sample directly located between diamonds. With this configuration, non-hydrostatic conditions were generated, along with a stress gradient from the center of the anvil to the edge; thus the measured pressure was in fact an average stress along the anvil. This problem was solved in 1964 by the introduction of a metallic gasket with a hole to host the sample, thus reducing the pressure gradient and increasing the stability of the sample. It also allowed HP experiments on fluids and solutions (under these conditions the stress profile inside the cavity is similar to the one shown in Figure 6.3). This figure illustrates a comparison between this stress profile and the one generated during the indentation process of the gasket (described later). This design presents several advantages: (i) it allows optical and spectroscopic measurements, (ii) an optical microscope can be used for observation and taking photographs, (iii) both liquids and solids can be studied, and (iv) microscopic amounts of sample are enough to carry out the experiments.

During the 1970s, another breakthrough was the discovery of pressure-transmitting media providing hydrostatic (or quasi-hydrostatic) conditions up to about 100 kbar. Since then, different media have been used, but one of the most commonly used is a mixture of methanol and ethanol (4:1 in volume). Although each of these components crystallizes at lower pressures, the mixture forms a glass allowing quasi-hydrostatic conditions above 100 kbar. In 1976, H. K. Mao and P. M. Bell [9] designed the ultra-high pressure DAC (also known as megabar DAC), making possible to reach pressures of several millions of atmospheres. These cells resemble the piston-cylinder design, and one diamond is mounted on a piston and the other one on the base of the cylinder. The vertical force is applied along the piston axis, moving it through the bore of the cylinder. Diamonds are mounted on tungsten carbide semicylinders (rockers) and the sample is located in a hard metal gasket, normally rhenium. The design of the slits allowing optic access to diamonds depends on the kind of experiment. As pressure is forced over an area to generate a pressure of 1 Mbar with a load of 1000 kg, the surface of the diamond culet should be a few hundred microns.

The use of diamond to generate high pressures using typical DAC designs led to progress in the study of HP phenomena, but it sometimes also presents drawbacks that could limit the success of an experiment. For instance, the maximum temperature inside the cell is around 700 K due to the graphitization of diamond. In addition, some spectroscopic experiments can be restricted by the relatively limited optical transparency of the diamonds (only allowing light transmission of wavelength above ca. 330 nm). In that case, alternative gems can be used as long as ultra-high pressures are not required. As most characterization techniques involve electromagnetic radiation, gems of certain high hardness and optical properties are chosen. Thus, the anvil selection depends on the mechanical properties and also on the proper optical transparency to ensure the sample characterization. One has to find a compromise between spectroscopic requirements and the demand of the HP cell design. As an example, the most used material to study the near UV, visible, and medium infrared regions is the synthetic sapphire, presenting transparency for the wavelength range from 144 nm to 6.5 microns. The maximum pressure achieved with sapphire cells is typically about 10 GPa [10], although some authors claim higher pressures. Another recently used gem is synthetic moissanite (6H-SiC) which allows routine experiments above 40 GPa [11].

6.2.1.2 Principles of Operation of Anvil Cell

The main obstacle of using anvil cells for a new researcher is that it is indeed a craft technique. An experienced user knows that sometimes the most difficult task in HP experiments is to get a suitable sample and, of course, the difficulty increases as an experiment requires higher pressures. In general, the maximum reachable pressure in any anvil cell depends on several factors, and all of them have to be taken into account in an experiment. Results depend on the hardness of the anvils, the size of the culets, the gasket material, the size of the hole in the gasket and its thickness, the studied sample, and other factors. In any case, independently of the desired maximum pressure, there are general common rules to follow for any HP experiment. Although we explain them below, a detailed account of the operation principles of anvil cells can be found elsewhere [12–15].

As can be seen in Figure 6.4, the physical principle on which anvil cells are based is very simple: the pressure is generated by the compression of a sample confined between two diamond culets and a practiced hole in the gasket. However, there exist different mechanisms to compress the anvils against each other. In the piston-cylinder configuration, which is among the most common, the pressure can be applied by the simple actions of screws, clamps, a membrane, or double-diaphragm. There also exist several alternative designs of the so-called Merrill–Bassett cell, in which the piston-cylinder tandem is replaced by rod guides. Nowadays, there are many commercially available cells designed for different kinds of experiments, although the procedure to carry out the experiments is very similar in all of them.

The selection and design of the anvils (modified brilliant, Drukker, or other cut or shape) are important steps in planning HP experiments, since the maximum reachable pressure is defined by the hardness of the anvil. Special attention has to be paid to possible defects in the anvil since they can be responsible for internal stress and cause the anvil to break under pressure. For correct operation, it is always necessary to check whether anvils are free of dislocations, surface scratches, etc. under an optical microscope (preferably using cross-polarizers). The selection of the anvil culet is defined by the desired maximum pressure, since the pressure is inversely proportional to the work surface. There are approximate equations to determine the ratio between the maximum reachable pressure and the size of the culet, but it is best to conslut an expert. For a fixed anvil culet, a beveled anvil will reduce the stress between anvil and gasket, allowing higher pressures (above 1 Mbar diamonds should be beveled or double-beveled).

Most of the cells have two supports (seats) for the anvils, which are made of tungsten carbide or hard alloys like Inconel. One of them may present a hemispherical shape to provide the degree of freedom required for the axial alignment, while the other one is cylindrical in shape, allowing the alignment in the x and y directions to center both anvils, although rotating wedges and other different alignment mechanisms have been proposed. The supports must

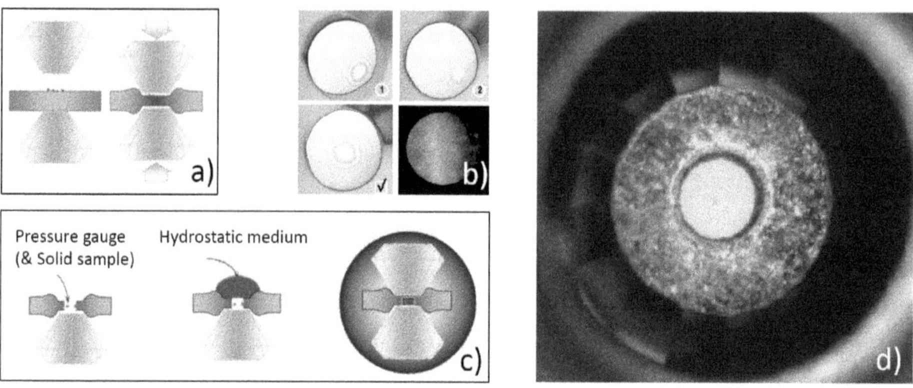

Figure 6.4 a) Gasket indentation procedure; some pressure markers can be spread in the gasket for monitoring. b) Interferometric alignment of the anvils; the first series of frames corresponds to sapphire anvils slightly curved to exaggerate the interference effect and the bottom-right frame shows a single interference fringe in a diamond anvil cell; the interference pattern indicates that the difference in parallelism between anvil culets is less than one micron. c) Charging pressure marker (and/or solid sample), loading of hydrostatic medium and anvil cell closing. d) Final view across a DAC charged with methanol-ethanol (4:1) and a small ruby chip in the center of the sample cavity.

have an opening allowing optical access to the sample and sometimes backing plates are used to protect the supports from indentation of the diamond table. The shape and angle of the opening limit the type of experiment to be performed, since it is not possible to substantially increase the optical angle without reducing the resistance of the cell. Current cell designs are based on changing the cleavage of the gems to allow a wide observation angle necessary for diffraction experiments. Fixing the anvils to each support must be centered to align the opening with the culet of the anvil.

To do so, an anvil support pair is mounted and coupled to a device that allows optical access with a microscope. Once the anvil is centered, it is fixed to its corresponding support with epoxy resin or other suitable adhesive for the temperature conditions of the experiment. It may be desirable to incorporate some gem-crimped systems to the cell to prevent the anvils from separating in successive operations. The supports or rockers are commonly adjusted into position by a set of precision screws located in the cell body. The process of adjusting screws and checking under the microscope is repeated until gem culets are fully aligned and concentric. Before starting any experiment, it is essential to check the alignment of the anvils. This is one of the most important steps for the success of an experiment with anvil cells, since it determines the stability of the cell under increasing pressure. This process has to be done under a microscope and it is considered completed when the observed interference rings are perfectly centered when aligning the anvils to each other, as described in Figure 6.4b. During this process, special attention has to be paid to the anvils to avoid damage by accidental impact.

Once the anvils are aligned, the next step is the preparation of the gasket, which is also one of the most delicate operations of cell preparation. The gasket confines the sample and provides additional stability to the cell. The selection of the gasket is very important since the material and its thickness determine the maximum reachable pressure in an experiment. One has to choose a hard material, not harder than the anvil, but not so soft that it might lead to an eventual contact between the anvils. By increasing pressure, the thickness of the gasket decreases and it hardens, possibly becoming harder than the gems, and this may cause an undesirable breaking of the anvils. Due to the stress profile on the gems during this operation (Figure 6.3), it is convenient to follow the pressure reached at the center of the cell during the compression of the gasket. The selection of a chemically inert gasket is desirable to avoid unwanted reactions with the sample.

The first step in the gasket preparation is the so-called indentation (or imprint), and it is shown in Figure 6.4a. After this operation, gasket thickness is reduced to the desirable height (typically below 50 microns) and an orientation is recommended to align the anvils during the following operations. During this process, the gasket is work-hardened, minimizing further deformation during the experiment. It is convenient to check whether the anvils are still aligned after indentation.

The next step for the cell preparation is to drill the hole that will host the sample at the center of the indentation mark. The size of the hole depends on the maximum pressure for the experiment, the anvil culet size, and the hydrostatic transmitting medium. An experimental rule to follow is that the diameter of the hole must be about one third the diameter of the anvil culet. After indentation, the gasket is carefully removed from the cell and is moved to the drilling machine. There are several ways to obtain a suitable hole in the gasket. The most common method suited for holes larger than 100 microns is simple mechanical drilling, but this process becomes increasingly difficult to implement with decreasing hole diameter. Laser drilling is another method for preparing holes but, of course, is dependent on the availability of a high-power laser with appropriate focusing optics. A third alternative is the electric discharge machine that allows drilling holes from 25 microns up to a few tenths of a millimeter [16]. All these devices have an xy base with micrometer accuracy under a monocular microscope to center the hole in the gasket indentation. It is essential that the gasket remains perpendicular to the drill or electrode during the drilling to ensure the centering of the hole in both anvils. It is common to find a hole centered on one side of the indentation but not in the other one due to non-perpendicular drilling. In this case, the process has to be repeated until the desirable result is obtained. On the other hand, it is essential to have the hole as centered as possible from both sides; otherwise unstable samples will migrate under load from the center to the edge of the culet, limiting the highest pressures of the experiment and often causing premature failure of the anvils.

Once the drilled gasket is back to the original cell position determined by carefully checking the orientation mark, the next step is the introduction of the sample and the pressure gauge, normally a ruby microchip (see below). Using many sensor pieces is not desirable because this can disturb the signal coming from the sample, while using just a few sensor pieces can impact searching for the microchips. Moreover, pressure gauge crystals have to be as close to the sample as possible and never at the edge of the hole where pressure can be different from the one to which the sample is subjected.

For experiments where true hydrostatic conditions are not required, the gasket hole is completely filled by the sample and the pressure gauge. In hydrostatic conditions, a pressure-transmitting medium is needed to ensure requirements are met. Typical pressure-transmitting media are: (i) liquids (alcohol, alcohol mixtures (methanol-ethanol 4:1), silicone oil), which are readily available but have relatively low hydrostaticity limits; (ii) soft solids (CsI, NaCl, KBr) which are good for laser heating, provide good insulation, and can be annealed at HP, but their hydrostatic limit is also low (around 7–8 GPa) and they are hygroscopic; (iii) powdered hard solids (MgO or alumina) that provide good thermal insulation, but are essentially non-hydrostatic; and (iv) condensed gases (He, Ne, Ar, nitrogen) that are the best pressure media for room and low temperature experiments, but are difficult to load; dedicated gas-loading or cryogenic installations coupled to the anvil cell are required.

In the specific case of the methanol-ethanol mixture, once the sample and pressure gauge are pre-loaded in the cell, a drop of this mixture is placed on the gasket with the help of a micropipette. The cell should be closed carefully to ensure that the sample and the pressure marker are well placed into the cavity.

6.2.2 Dynamic Pressure Techniques

The measurement of dynamic properties of materials requires specialized and often sophisticated equipment and techniques. In dynamic techniques, high-power projectiles (via explosives or with a gun) or pulsed lasers are used to generate shock waves in the specimen under study. This results in nearly steady waves in the material, achieving simultaneously extreme conditions of pressure and temperature in a very short time. A shock wave can be viewed as an adiabatic step increase in state and hydrodynamic variables that travels at supersonic velocity with respect to media ahead of the shock front. The shock front represents the transition region in which atoms are rapidly compressed with associated turbulent frictional heating and entropy production.

Shock waves produce a wide variety of physical and chemical transformations in materials. Shock wave studies involve material characterization using: stress-strain, elastic constants, equations of state (EOS) and dynamic fracture. Dynamic experiments have been crucial to the general advancement of HP science and they have a special place in the study of the Earth's interior (Chapter 15) because virtually complete pressure and temperature ranges can be achieved in the laboratory. Some compounds subjected to dynamic studies have interdisciplinary interest, like carbon and hydrocarbons. They are present in chondrites, but are also excellent systems for studying phase transformations (e.g., formation of diamond particles in explosive detonation products) or chemical effects (e.g., formation of organic molecules). Thus, comparisons of the compositions and morphologies of samples subjected to shock compression in the laboratory provide important clues to the processes in which some minerals and other compounds participate. These interdisciplinary aspects of the investigations in shock wave phenomena are especially noteworthy, as these studies have been pursued by scientists from a broad range of areas involving extreme pressures and temperatures. Close coordination between shock wave experimentalists with theoretical and computational researchers have provided guidance to formulate highly accurate predictions of material behavior under extreme conditions.

Dynamic shock wave techniques were mainly developed in large facilities of the US and the former USSR and are typically devoted to fundamental studies of matter under extreme conditions: equation of state (EOS), polymorphic transformations, amorphization, melting, metallization, fragmentation and spall fracture, explosive, and nuclear reactions.

Historically, most experiments were performed with a single shock wave, leading to an adiabatic process and achieving high temperatures but relatively low compressions. Later, the use of multiple shock waves (e.g., a reverberating shock wave in a compressible liquid contained between two anvils) provided a quasi-isentropic route to produce much higher compressions but relatively lower temperatures. The early single-stage devices (guns) provided measurements to approximately 20 GPa, while two-stage light-gas guns increased this range to over 500 GPa, involving temperatures up to 4000 K and times in the range of tens of nanoseconds. Current shock wave techniques allow access beyond 1 TPa and 10000 K.

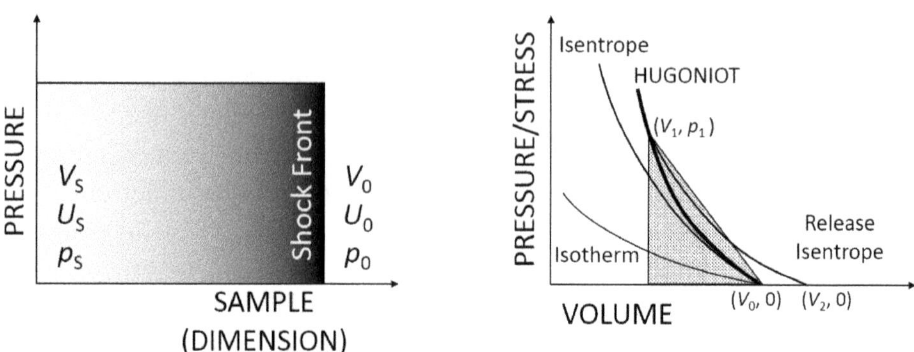

Figure 6.5 a) Schematic view of a steady shock wave passing through a material: p_S, U_S and V_S are state variables of shock pressure, internal energy on shock compression, and shock-compressed volume, respectively. Variables with zero subscript are initial values ahead of the shock front. b) Comparison of several curves in a pressure-volume diagram.

The formalism to interpret dynamic shock wave experiments is based on the Rankine–Hugoniot equation which expresses the conservation of mass, momentum, and energy across a shock wave, assuming that equilibrium exists on either side of a shock front. Historically, these experiments were intended to explain the thermodynamic changes produced at the discontinuity of the shock wave, and one may find different derivations and notations [17–19]. The most important output is the conservation of energy principle can be expressed as:

$$(U_S - U_0) = \frac{1}{2}(V_0 - V_S)(p_S + p_0), \tag{6.1}$$

where U is the energy, p is the pressure (or stress), V is the volume, and the subscripts S and 0 refer to the values of these properties on the target material during and ahead of the progress of a plane shock wave, respectively. Equation (6.1) can be interpreted graphically, since the internal energy increase across the shock is given by the integrated area (shaded) indicated in Figure 6.5. If the initial state of the material is $p_0 = 0$ and $U_0 = 0$, the energy input is then divided equally into increases in the internal and the kinetic energy of the material [18].

The states in the p-V plane obtained from experimental data using Equation (6.1) are called the Hugoniot curve (or EOS), dynamic adiabat, shock adiabat, or simply the Hugoniot, which is the locus of equilibrium states reached after shocking of a material. The Hugoniot curve has the same slope and curvature as the adiabat at the initial state of the material. Thus, for low amplitude shock waves the difference between the Hugoniot and the adiabat is quite small, but the difference increases as the strength of the shock increases, so the Hugoniot curve always lies above the corresponding adiabat.

Key information regarding the responses of materials at HP can be obtained from Hugoniot measurements under different initial densities or temperatures. However, a complete thermodynamic description of HP states requires assuming an EOS form to fit the available experimental data. A simple form which has been widely used is the Mie-Grüneisen EOS. Within this approximation the thermal energy of the system is assumed to be described by the sum of the energies of a set of independent harmonic oscillators, and the pressure can be expressed as:

$$p = p_c + \frac{1}{V}\sum_{i=1}^{3N} \frac{\gamma_j^G h\nu_i}{\exp\left(h\nu_i/k_B T\right) - 1}, \tag{6.2}$$

where p_c is often called the cold curve: $p_0 = -\left(\frac{\partial U_0}{\partial V}\right)_{T=0}$, with U_0 the cohesive energy at $T = 0$.

The frequencies of the vibrational modes (ν_j) scale with volume through the mode Grüneisen parameter γ_j^G are defined as:

$$\gamma_j^G = -\frac{V}{\nu_j}\left(\frac{\partial \nu_j}{\partial V}\right)_T = -\left(\frac{\partial \ln \nu_j}{\partial \ln V}\right)_T. \tag{6.3}$$

Different approximations to account for the volume dependence of the vibrations (Slater, Dugdale–MacDonald, or Zubarev) lead to a macroscopic form of the Grüneisen EOS that includes the thermodynamic Grüneisen parameter (γ_G), defined as:

$$\gamma^G = \frac{V}{C_v}\alpha_p \kappa_T, \tag{6.4}$$

where α_p, κ_T, and C_v are the isobaric thermal expansion, the isothermal compressibility and the isochoric heat capacity, respectively. The Grüneisen EOS can be written as:

$$p(V,T) = -\left(\frac{\partial U_{\text{coh}}}{\partial V}\right)_{T=0} + \frac{\gamma^G}{V} U_{\text{vib}}(T). \tag{6.5}$$

Since the volume dependence of γ^G is not usually available, assuming γ^G/V to be constant is a common approximation for applying Equation (6.5) with reasonable results.

The use of shock waves to determine the EOS of condensed matter under extreme conditions began with the classic papers of Walsh and Christian [20] and Al'tshuler and coworkers [21]. Since then, Hugoniot data has provided accurate EOS data for many materials. Several analytical and empirical relations have been developed for relating pressure, volume, and bulk modulus including those of Birch and Murnaghan (Chapter 1), which can be compared to the shock wave EOS. Some EOS are useful for developing accurate pressure standards for static experiments in the megabar range, where other pressure scales are extraordinarily difficult to implement (e.g., ruby luminescence).

More recently, laser-driven shocks are also used to study the dynamic behaviors of materials over small areas on short time scales [22]. In this manner, the available pressure regime has been raised to tens of TPa. Short laser pulses allow studies of the behavior of matter over ranges of extremely high-strain rates and tiny propagation distances because absorption of laser radiation by a target material takes place within a very thin layer. The rapid temperature increase in this layer and the associated temperature gradient induce ablation of the material plasma, which expands and drives a strong compression shock wave into the material. When the shock reaches the back surface of the target specimen, a tensile wave is reflected, and if the tensile stress becomes larger than the dynamic tensile strength, spall on the target material takes place. Therefore, fracture and fragmentation occur when a material is subjected to such extreme stress at such high rates. Under these conditions, the deformation of the target specimen is usually small, so analyses of the recovered specimen and possible ejected fragments are also feasible and reveal the physical or chemical changes in a sample. At such short-time scales, kinetic effects are extremely important, non-equilibrium states can be probed and, metastable states can be quenched for later study.

We finish this section with a reference to other dynamic techniques that are less known or atypical. First, a recently developed device called dynamic DAC (dDAC) can bridge the gap between static and dynamic experiments. Modest strain rates can drive systems out of chemical equilibrium and induce transformations to metastable local minimum energy configurations. Processes in this regime include phase transformations such as the kinetics of crystallization and growth in liquid-solid transitions [23]. Another example concerns shock wave generation in liquids, which is intimately related to cavitation phenomena and a subject of active investigation for its relevance to chemistry and bioscience applications. Cavitation can be distinguished between tension-induced (acoustic) and other methods (e.g., electrical sparks, focused laser). The interpretation of these phenomena again requires the equations for the conservation of mass, momentum, and energy for a compressible fluid. Since equations of state are needed to integrate the constitutive equations, the Tait equation (Chapter 10) suffices to solve the problem in most cases. If dissipation and viscous forces are neglected, the changes of state can be considered isentropic and application of the Rayleigh-Plesset equation is customary. An emerging technology related to these phenomena known as ultra-high pressure homogenization (UHPH), is used for the simultaneous sterility and stabilization of food (see Chapter 12), cosmetics, and pharmaceutical products (see

Chapter 13). Recent studies [24] show that this technology can reduce the particle sizes of fat globules and oil droplets to a greater extent than traditional homogenization, and that the pressures generated successfully inactivate microorganisms by rupturing cell walls through sudden pressure drops, shear stresses, cavitation, and turbulence.

6.3 PRESSURE METROLOGY AND PRESSURE SCALES

A major difficulty in HP science is providing a precise method to determine the pressure quickly, accurately, and reproducibly. Thus, pressure measurement and the development of pressure scales are intimately related to the stress conditions of HP experiments [25]. We now recall that the stress tensor has nine components: three (diagonal) normal stresses and six (off-diagonal) shear stresses. Under hydrostatic stress conditions, the shear stresses are null and the diagonal stresses are identical and equal to the pressure (isotropic). Under non-hydrostatic conditions, all the components may have non-zero values and this is what makes characterizing the stress distribution of the sample extremely difficult. What we may refer to as mean pressure is in fact an average of the three normal stress components that may be different, but we must keep in mind that the effect of the stress inhomogeneities can have a dramatic impact on many properties of the sample. In truly hydrostatic studies, calibrated piezoelectric transducers and Bourdon gauges are good choices for measuring the (hydrostatic) pressure in fluids and solutions below 1 GPa (Chapter 10). Death weight gauges are used as unique primary pressure scales for calibration of these systems only up to tens of MPa.

In the particular case of diamond anvil cells, the most common way to determine the pressure is based on the displacement observed on the fluorescence of ruby (Cr^{3+}-doped Al_2O_3) under increasing pressure [26]. An example of measured fluorescence spectra of ruby at room pressure is shown in Figure 6.6. The fundamentals of ruby luminescence are further discussed in Section 8.4.1 of this book. The two main luminescence bands (sometimes called peaks) are labeled as $R1$ and $R2$ and their pressure shift shows a linear dependence at low pressures. However, the slope changes at higher pressure and the dependence becomes sublinear. The main advantage of this secondary (not absolute) scale is that a (not so) simple measurement of the frequency shift of the $R1$ band gives an accurate estimate of the pressure shift using one of the several relationships proposed in the literature. Note that the ruby luminescence scale has been reviewed and reformulated several times over the past decades [27], and a good practice is to recalculate and update the old results to the most recent and accurate scale. Unfortunately, this is not always possible because the original spectra may not be available or the exact

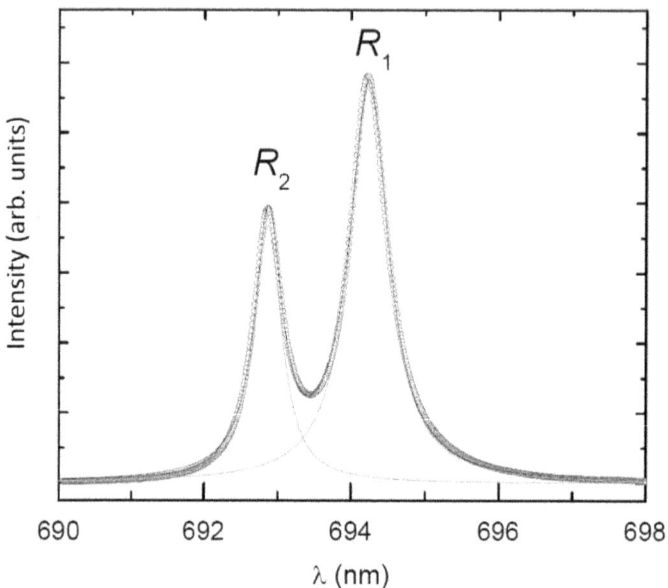

Figure 6.6 Ruby luminescence with a profile analysis based on two Lorentzian functions.

relationship is not clear, so care must be taken when comparing old data to a current study particularly at very high pressures.

At moderately HP (e.g., below 10 GPa) and room temperature, the following linear relationship holds approximately:

$$p(\mathrm{GPa}) = 2.74(\lambda_{R1} - \lambda_{R1}^{0}), \tag{6.6}$$

where the emission wavelength, λ, is expressed in nanometers. A good practice is to measure the luminescence spectra of the ruby marker at zero pressure to determine the exact value of λ_{R1}^{0}, since it may vary slightly in different rubies. Other expressions must be used at higher pressures and different temperature regimes. Normally, the error in determining the pressure of a single ruby crystal can be as high as ±0.05–0.10 GPa, but it also depends on the spectral resolution of the spectroscopic system used for the luminescence measurement and also on the numerical analysis of the spectra. In fact, while the λ_{R1} and λ_{R2} wavelengths are directly related to the mean pressure, their linewidths and the wavelength splitting $(\lambda_{R1} - \lambda_{R2})$ are sensitive to the stress state of the ruby specimen. Thus, the ruby fluorescence is also effective for evaluating non-hydrostatic conditions. We have to be careful, however, about the fact that $(\lambda_{R1} - \lambda_{R2})$ also depends on the orientation of the ruby, so the use of single crystals with known crystallographic orientations is preferred over ruby grains or balls. Of course, these refinements are to be considered mainly when the pressure range is increasingly large. At pressures above 100 GPa, the luminescence intensity of ruby becomes very weak and alternative scales are required.

There are several ways to analyze the spectral characteristics of ruby fluorescence. The most suitable involve a simultaneous fit of the $R1$ and $R2$ bands to Lorentzian or Voigt profiles. This procedure allows determination of the pressure and existing stress inhomogeneities. However, the fit of the whole ruby fluorescence spectrum can be very sensitive to the underlying spectral background and also to the range of wavelengths considered. Furthermore, the nominal spectral resolution of the spectroscopic system can be distorted by an incorrect signal analysis, so practice with the numerical treatment is advised. Peak-finding algorithms included in most data treatment software or visual tracking are not advised, except perhaps for a quick monitoring of the progress of an experiment. Many scientists have devoted great efforts to developing a high-precision pressure scale, but pressure may be underestimated by a poor signal analysis. Doped oxide garnets such as lanthanum lutetium gallium (LLGG), yttrium aluminum (YAG) and calcium gallium germanium garnet crystal (CGGG) are being investigated because of their importance in different technological applications. The combination of the luminescence properties of rare earth-doped bulk garnet crystals and their hardness, high transparency, and mechanical stability makes them useful as materials for HP gauges (see Section 8.4.2). The shielding of the $4f$ electrons produces very sharp absorption and emission lines in the UV-VIS-near-infrared (NIR) optical range that are less sensitive to the environment compared to those of the $3d$ electrons of Cr^{3+}.

Although ruby fluorescence can be easily measured, care must be taken to properly analize the spectra for obtaining the right pressure and do not misuse this practical scale by wrong interpretation of the spectra. Alternative pressure scales based on the Raman shift (Chapter 9) of diamond including the tip [28], c-BN or SiC have been proposed for pressure determination over a wide range of temperatures in DACs and other gem anvil cells.

If a material is used as a pressure gauge in diffraction experiments (e.g., NaCl, gold, or lead), it is important to evaluate the stress distribution acting upon the calibrant since the pressure will be deduced from a known thermodynamic EOS. Theoretical modelling is helpful in this respect, and the study of the behavior of gold in the Mbar range is a good example of how to proceed correctly.

If the HP device has no optical access to the sample chamber, a common, convenient, and sensitive secondary pressure sensor is the Manganin gauge, which has a known pressure dependence of electrical resistivity. It was first used by Bridgman in 1940, who found that the resistivity of the Manganin alloy increased significantly with increasing pressure, but was only slightly affected by temperature changes. This electrical gauge needs to be connected with feed-through wires in order to measure its resistivity and thus the pressure inside the cell. The pressure coefficient depends on the particular wire provided by the supplier, but it can be easily calibrated. The gauge will need little space; a 0.1 mm wire is normally wound in the form of a free-standing coil, without any frame and glue to avoid stresses, and a length shorter than 1 cm is sufficient for measurements of pressure to better than 0.01 GPa. This gauge is also a good indicator of the degree of hydrostaticity of a medium, and when the medium becomes viscous, the Manganin resistance slowly relaxes after a change in pressure.

Manganin gauges are also frequently used in shock wave studies for recording the stress or pressure history in the interior of a sample, although commercial gauges are often limited to 40–50 GPa [29]. The gauge is usually embedded in the sample, forming a plane zigzag strip, and electrically insulated from the sample material by thin polymer films or mica. A constant electrical current is passed through the gauge and to increase the precision of the measurements, a resistance bridge is usually added to the electrical system. Thus, when a shock pulse passes through the gauge plane, an increase in the measured voltage is recorded. Above several GPa, the change in the resistivity of the gauge is reversible and does not depend on whether dynamic compression occurs by single or multiple shocks [17]. As pressure is released, a slight hysteresis in the resistance at zero pressure is observed, which is attributed to the hardening of the Manganin gauge due to the shock compression.

6.4 SELECTED EXAMPLES OF PROPERTY MEASUREMENT

6.4.1 Static Pressure Cell for Measuring Magnetic Behavior under High Pressure

The magnetic behaviors of materials can be investigated through magnetization (M)—a fundamental physical magnitude which characterizes the responses of materials under an applied magnetic field. We can study the magnetic dependence of a sample on temperature, magnetic field and pressure to determine the origins of magnetic interactions (obtaining the exchange parameters), critical points for magnetic/structural transitions, and other factors. Magnetometers with precise control of temperature and magnetic field are now available. In this scenario, HP instrumentation for this equipment represents a powerful tool making it possible to investigate the magnetic behavior of a sample subjected to temperature and magnetic field and also under pressure. Commercial magnetometers are readily available and the most popular bases its excellent sensitivity on the so-called integrated superconducting quantum interference device (SQUID).

To design a pressure cell for direct current (DC) magnetic measurements

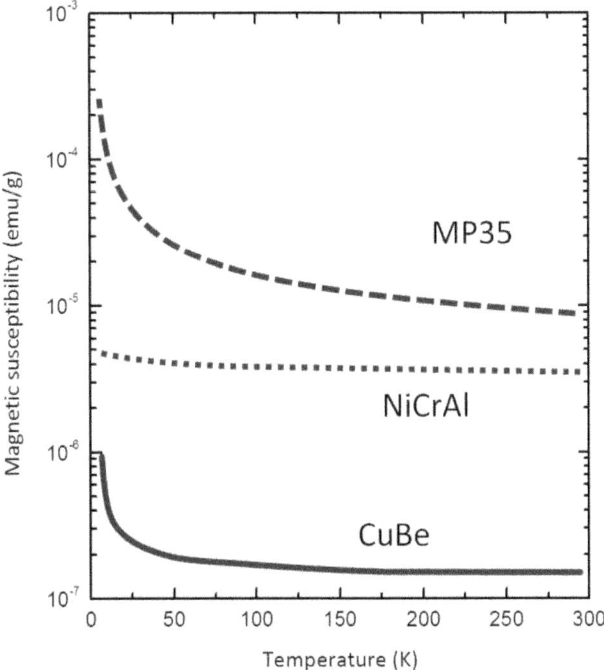

Figure 6.7 Temperature dependence of the magnetic susceptibility for some non-magnetic alloys.

on such equipment, the first challenge to overcome is the size limitation, as the sample chamber is only 9 mm in diameter. This means that the external diameter of the cell must be smaller than 9 mm to work when fitted on the instrument, which is much smaller than conventional HP devices. In this case, a piston-cylinder type cell would be the suitable choice [30].

To select the material for the cell, some technical considerations are necessary. As we will carry out magnetic measurements involving high magnetic fields and low temperature, the material must be chosen for stability under these extreme conditions. The construction material has to have strength appropriate to the desirable maximum working pressure of the cell. Ideally, the magnetization of this material must be as insensitive as possible to magnetic fields to produce the least magnetic background possible. There is no perfect material because lowering the magnetization leads to a decrease in yield strength. Among the commercially available materials taking into account the prices and the best balance among the previous constraints, one must choose the proper material for the HP cell. Figure 6.7 shows the magnetic susceptibilities of materials for this kind of pressure cell.

The yield strengths for the three non-magnetic materials in Figure 6.7 are: MP35N (Co-Ni-Cr-Mo) 2.3 GPa, NiCrAl 2.2 GPa and CuBe 1.5 GPa. There is another material (Ti_6Al_4V alloy) with very good mechanical properties but less sastisfactory magnetic ones. As can be seen in the plot, as yield strength increases, the materials become more magnetic. At this point we can construct several different piston-cylinder non-magnetic pressure cells. In all cases, the body cell can be designed as a double-wall cylinder with interference fit, allowing us to operate at higher pressure compared to the single-layered cylinder. Then, we can choose between two general options: (i) a double-layered body cell made of CuBe alloy for weakly magnetic samples (maximum pressure ca. 1.2 GPa) or (ii) a double-layered cell with outer cylinder made of CuBe and the inner one made of NiCrAl alloy for strongly magnetic samples that can achieve higher pressure than the previous one (maximum pressure 2 GPa, with higher cell background signal). Metal alloys normally can be hardened by heat treatment after machining the parts. The dimension of the cell has to be optimized to provide enough room for magnetic samples, but strong enough to operate at HP. The integration software of the magnetometer requires the pressure cell to be symmetric with respect to the sample.

There are several ways to calibrate the pressure in these HP cells. In the conventional design, the pressure inside the cell can be measured by manometers such as lead, tin, or indium, all of which have superconductive transitions of known pressure dependence. The transition temperatures of these manometers are in the range 4–7 K and due to their large response to the magnetic field below 4 K, they can disturb the measurement of the magnetic signal from the sample at that low temperature. To avoid the use of a superconductive pressure manometer, the deformation of the cell body with pressure can be used. When pressure is applied to the inner wall of the cylinder, it deforms linearly as a function of applied pressure. Therefore, if the pressure cell is well calibrated, the deformation can be used to determine the pressure inside the cell and the pressure manometer can be avoided. Alternatively, a Manganin pressure sensor provides an accurate way to measure the pressure in a piston-cylinder cell.

Since the pressure cell is symmetric in this case, the common way to apply pressure is by tightening the clamp screws on the two ends of the cell. Sometimes a hydraulic press is used to apply a load and control the applied pressure before it is locked inside with the end screws. The sample (around 10–15 mg) is normally pressed into a pellet or introduced on a PTFE capsule (smaller than the bore) inside the cell with a pressure-transmitting medium (Fluorinert FC75 or Daphne 7373 oil) [31] to ensure hydrostatic conditions. In order to properly seal the cell, disks or washers normally made of soft materials such as tin, indium and copper are used.

Once every part is carefully assembled and the cell is properly sealed, pressure can be applied and measurements can be carried out.

6.4.2 Transport Properties Measurement under Pressure Using Different Devices

High-pressure transport measurements provide very useful information about electronic and structural transformations that materials can suffer upon extreme conditions [1]. These measurements can also be used to study the metallization of non-metallic materials as well as superconductivity and phase transitions. With piston-cylinder cells, the pressure limit extends to 5 GPa, while with Bridgman cells a pressure of 13 GPa can be reached. Higher pressures to about 50 GPa can be achieved by using diamond anvil cells [15], with the advantage of optic access to the sample. Due to their difficulty, transport measurements with DACs have become experimental challenges. Four ohmic contacts are needed to use the van der Pauw technique [32] and measure only the resistivity from the studied sample, eliminating the signal coming from contacts and wires.

A piston-cylinder apparatus is used in this case to ensure hydrostatic conditions. As noted before, the way to generate pressure in these cells is with an external press that pushes the piston, compressing the transmitting media with the sample inside a cylinder. The electrical contacts are connected with feed-through wires to measure the resistivity of the sample and the resistivity of the Manganin pressure gauge. This system can reliably operate up to 3 GPa. Once the desirable pressure is reached, it can be blocked and the piston-cylinder cell can be removed from the external press. At this step, the pressure cell is ready to connect to a transport measurement device to carry out the measurements. As an example of this kind of cell, the Unipress LC-10 has been used for many years; it is made of CuBe alloy and can be used at low temperatures and under magnetic fields.

In a Bridgman-type cell, two opposed tungsten carbide anvils compress a pyrophyllite gasket. The sample is contained in the gasket and surrounded by a solid transmitting medium (sodium chloride, boron nitride, etc.). A pressure of 13 GPa can be reached and a relatively large volume of sample, about 5 mm^3, can be used. Pyrophyllite deforms plastically, avoiding anvil breaks, sealing the transmitting medium, and isolating the electrical contacts from the anvils. Regarding the electrical connections, Au, Ag, or Pt wires can be used and placed through the pyrophyllite gasket. These cells are used for resistivity measurements at low and high temperatures using an internal furnace or connection to an external measurement device with a controlled temperature system. The main disadvantage is that pressure cannot be measured *in situ* and must be calibrated as a function of the pressure from the external hydraulic press, using the resistivity of different calibrants such as Ba, Bi, and Gd. [33]

Transport measurements in DAC allow us to study samples at pressures as high as 200 GPa. The sample size for reaching that high pressure needs to

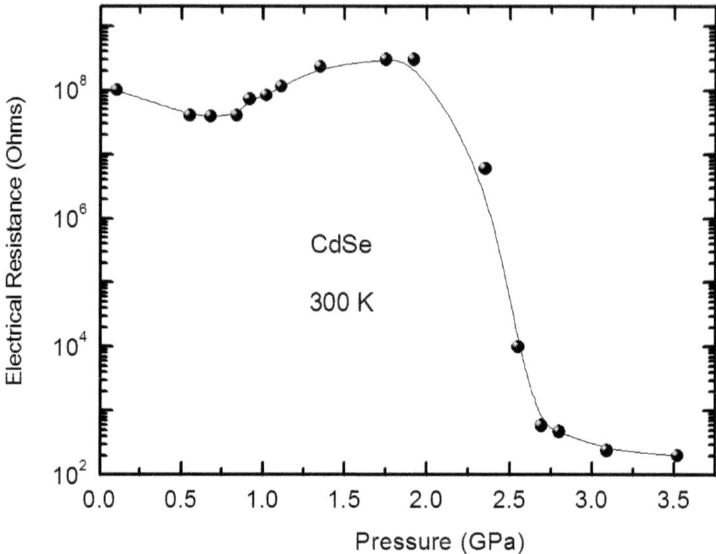

Figure 6.8 High pressure electrical resistance dependence of polycrystalline CdSe [34].

be very small—a maximum of 150 × 150 microns and 20 microns in height. This makes these experiments extremely challenging and normally several attempts may be required to get a successful result. Different designs have been developed for transport measurements on DACs for solid [35] and liquid [36] transmitting media. Two main problems to overcome are(i) diamond edges or hole edges that cut electrical wires and (ii) short circuits between wires or on the metal gasket. To minimize these problems, beveled anvils and very thin Au or Pt wires (5 microns in diameter) have to be used. The wires must be flattened to better reach the sample. To isolate the wires from the gasket, disks and layers of insulator materials such as MgO, Al_2O_3, or CsI are used. An advantage of using DACs is the optic access to the sample room which allows measuring the pressure *in situ* on a ruby luminescence scale. The wires coming from the cell have to be connected to an external measurement system. These cells are used for resistivity measurements at low temperature although some designs can also be used at high temperatures. Figure 6.8 shows a phase transition of polycrystalline CdSe measured by electrical resistance under pressure at room temperature. The drop in resistance at around 2.5 GPa is of six orders of magnitude and represents a drastic electrical property change due to the structural phase transition.

A recent development in the design of diamond anvils utilizes a tungsten thin-film probe patterned on the surface via the focused ion beam (FIB) deposition technique [37]. The FIB technique allows the deposition of a thin electrical wire (a few microns thick) that serves as a metal film on diamond culets. For the four-probe resistivity measurements on designed diamonds are made with four strips of tungsten of 5 micron width.

We can study the dependence of transport properties of materials on temperature, magnetic field and pressure and thus study electronic transitions, metallization, superconductivity, and phase transitions. Piston-cylinder, Bridgman and diamond anvil cells can be easily adapted to commercially-available measurement systems. Commercial systems allow a maximum magnetic field of ca. 9 Tesla combined with a temperature range of 2 to 400 K, which makes it a powerful tool in materials science. Some available commercial pressure cells are already adapted to these systems. In any case, a pressure cell fitting into the device can be adapted by a simple mechanical piece to the standard connector puck of the system.

6.5 SOME GENERAL REMARKS

In other chapters of this book, the importance of the pressure variable in the thermodynamic behavior of matter is demonstrated (Chapters 1 and 10). Also covered are pressure-induced phase transitions (Chapters 2, 15, and 16) and chemical (Chapter 5) transformations which also play a key role in the synthesis of novel materials assisted by pressure (Chapter 14). Theoretical understanding and advanced computational capabilities have made possible the simulation of HP phenomena with high levels of accuracy (Chapters 3 through 5). An effective partnership of simulation method and experimental skill is required for most HP investigations. A computer simulation should be considered an HP experiment in its own right; a computer functions in essence as a "silicon anvil cell". However, many comparisons of experimental and computational results are heavily biased by the real experimental conditions, and may differ from constraints imposed into a computer simulation. The current expertise from both experimental and computational sides give us confidence for reaching a consensus on the interpretation of most HP phenomena. As in many other scientific fields, experimenters and simulators must recognize the results of others simply as a guide for interpretation of their own results.

An important aspect of this book for HP practitioners is recognizing both the advantages and limitations of the experimental-computational partnership. Our understanding of HP phenomena is still limited by the lack of theoretical support despite the availability of excellent computational methods and resources. We still face an urgent need to develop new theoretical models that will aid our understanding of the chemical transformations, biological and geological processes, and other HP dynamics. This need should serve as an incentive to develop additional novel experimental techniques.

Some typical questions that arise during the interpretation of an HP experiment concern how the pressure is measured or estimated and whether deviations from hydrostatic conditions occurred during an experiment. Consequently, any comparison between a truly hydrostatic simulation or a theoretical model and a given experiment must be as honest as possible and performed with extreme care. Achieving truly hydrostatic conditions is not always easy during HP experiments and at sufficiently high pressures, certainly is an impossible task. Another general warning is that more often than desired the formation of metastable states related to non-hydrostatic conditions complicates the reproducibility of the experiments.

We should discuss the results in terms of a quantity closely related to pressure, namely stress. Of course, theory may easily deal with different stress distribution scenarios, but in most cases it is unlikely that the experimental information could allow us to assess the real stress distribution in a sample. In general, the estimation of non-hydrostatic components in many HP experiments is not at all straightforward. We can the analyze pseudo-hydrostatic or quasi-hydrostatic conditions in the hope that the existing deviations from hydrostaticity do not affect the conclusions of the study greatly. The reliability of some pressure scales is also often disputed due to the non-hydrostatic natures of the experiments, and such debate will very likely continue for a long time. In any case, most natural phenomena including those we are trying to replicate in our laboratories are non-hydrostatic, and their understanding is probably among the most luring and challenging aspects of HP research. In fact, it is increasingly common to perform non-hydrostatic experiments [38], which paradoxically returns us to the early days when the DAC was developed [4, 5] and before gaskets and hydrostatic media were introduced.

Finally, let us remember that the content of this chapter is general and introductory. An extensive collection of articles and monographs focus on characteristics and principles of operation of the HP devices discussed herein. We provide a non-exhaustive list of classical and general references that the interested reader should consider consulting. We also encourage the reader to check the specialized articles published in the *High Pressure Research* journal, where most of the experimental protocols are described in detail. The best reference is to ask an experienced researcher already working in a particular field and to attend specialized schools, meetings, conferences, and workshops related to high-pressure science and technology.

Bibliography

[1] P. W. Bridgman. *Collected experimental papers.* Harvard University Press, Cambridge (1964).

[2] F. P. Bundy, H. T. Hall, H. M. Strong, *et al. Nature*, 176, 51 (1955).

[3] H. P. Bovenkerk, F. P. Bundy, R. M. Chrenko, *et al. Nature*, 365, 19 (1993).

[4] C. E. Weir, E. R. Lippincott, A. Van Valkenburg, *et al. J. Res. Natl. Bur. Stand. A*, 63, 55 (1959).

[5] J. C. Jamieson, A. W. Lawson, and N. D. Nachtrieb. *Rev. Sci. Instrum.*, 30, 1016 (1959).

[6] R. M. Hazen. *The diamond makers.* Cambridge University Press, Cambridge (1999).

[7] M. I. Eremets. *High pressure experimental methods.* Oxford University Press, Oxford (1996).

[8] J. Loveday. *High-pressure physics.* CRC Press, Boca Raton (2012).

[9] H. K. Mao and P. M. Bell. *Science*, 191, 851 (1976).

[10] W. B. Daniels, M. J. Lipp, D. Strachan, *et al.* In A. K. Singh, ed., *Recent trends in high pressure research*, 809. IBH Publishing Co., Mumbai (1991).

[11] J.-A. Xu and H. K. Mao. *Science*, 290, 783 (2000).

[12] A. Jayaraman. *Rev. Mod. Phys.*, 55, 65 (1983).

[13] D. J. Dunstan and I. L. Spain. *J. Phys. E Sci. Instrum.*, 22, 913 (1989).

[14] I. L. Spain and D. J. Dunstan. *J. Phys. E Sci. Instrum.*, 22, 923 (1989).

[15] A. Jayaraman. *Rev. Sci. Instrum.*, 57, 1013 (1986).

[16] H. E. Lorenzana, M. Bennahmias, H. Radousky, *et al. Rev. Sci. Instrum.*, 65, 3540 (1994).

[17] G. I. Kanel, S. V. Razorenov, and V. E. Fortov. *Shock-wave phenomena and the properties of condensed matter.* Springer, New York (2004).

[18] W. M. Isbell. *Shock waves: measuring the dynamic response of materials.* World Scientific, Singapore (2005).

[19] W. J. Nellis. *Rep. Prog. Phys.*, 69, 1479 (2006).

[20] J. M. Walsh and R. H. Christian. *Phys. Rev.*, 97, 1544 (1955).

[21] L. V. Al'tshuler, K. K. Krupnikov, B. N. Ledenev, *et al.* *Soviet Phys. JETP*, 34, 606 (1958).

[22] M. Millot, N. Dubrovinskaia, A. Černok, *et al.* *Science*, 347, 418 (2015).

[23] G. W. Lee, W. J. Evans, and C.-S. Yoo. *Phys. Rev. B*, 74, 134112 (2006).

[24] A. Zamora, V. Ferragut, P. D. Jaramillo, *et al.* *J. Dairy Sci.*, 90, 13 (2007).

[25] K. Takemura. *High Pressure Res.*, 27, 465 (2007).

[26] R. A. Forman, G. J. Piermarini, J. D. Barnett, *et al.* *Science*, 176, 284 (1972).

[27] K. Syassen. *High Pressure Res.*, 28, 75 (2008).

[28] M. Hanfland and K. Syassen. *J. Appl. Phys.*, 57, 2752 (1985).

[29] Z. Rosenberg, A. Ginzberg, and E. Dekel. *Int. J. Impact Eng.*, 36, 1365 (2009).

[30] X. Wang and K. V. Kamenev. *Low Temp. Phys.*, 40, 735 (2014).

[31] S. Klotz, J. C. Chervin, P. Munsch, *et al.* *J. Phys. D Appl. Phys.*, 42, 075413 (2009).

[32] L. J. van der Pauw. *Philips Res. Rep.*, 13, 1 (1958).

[33] D. Errandonea, D. Martínez-García, A. Segura, *et al.* *High Pressure Res.*, 26, 513 (2006).

[34] A. Andrada-Chacón. Master thesis, Universidad Complutense de Madrid (2013).

[35] H. K. Mao and P. M. Bell. *Rev. Sci. Instrum.*, 52, 615 (1981).

[36] J. Gonzalez, J. M. Besson, and G. Weill. *Rev. Sci. Instrum.*, 57, 106 (1986).

[37] D. D. Jackson, C. Aracne-Ruddle, V. Malba, *et al.* *Rev. Sci. Instrum.*, 74, 2467 (2003).

[38] E. Del Corro, M. Taravillo, J. González, *et al.* *Carbon*, 49, 973 (2011).

CHAPTER 7

Structure Determination

Julio Pellicer-Porres
MALTA Consolider Team and Departamento de Fisica Aplicada-ICMUV, Universidad de Valencia, Valencia, Spain

Daniel Errandonea
MALTA Consolider Team and Departamento de Fisica Aplicada-ICMUV, Universidad de Valencia, Valencia, Spain

Alfredo Segura
MALTA Consolider Team and Departamento de Fisica Aplicada-ICMUV, Universidad de Valencia, Valencia, Spain

CONTENTS

7.1 INTRODUCTION

SINCE the first synchrotron radiation (SR) x-ray diffraction (XRD) experiment in a diamond anvil cell (DAC) in 1977 [1], the combination of high-pressure techniques with different SR spectroscopic methods has become one of the most fruitful experimental techniques. The growing presence of high pressure instruments in the third-generation synchrotrons in different countries suggests that these techniques will remain so in the coming years. The present chapter offers a brief introduction to the applications of some x-ray spectroscopic techniques to materials under extreme conditions, focusing on

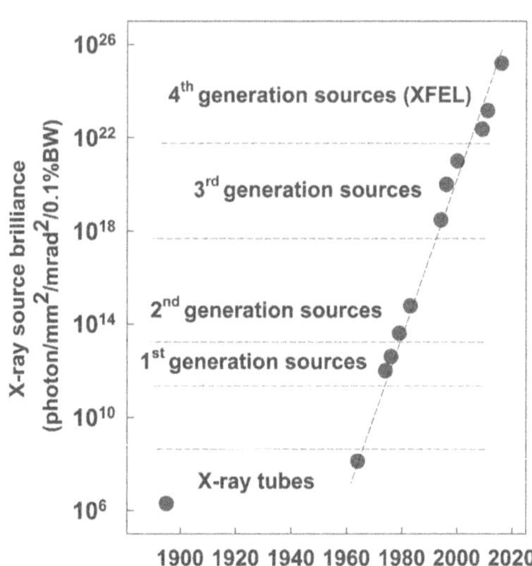

Figure 7.1 Historical evolution of the the brilliance of x-ray sources.

structure determination. We do not include examples involving the use of large volume presses (Paris–Edinburgh or multianvil). For these devices, that are increasingly used in large facilities, we refer to References [2, 3]. After a brief introduction on the generation of SR and the basic structure of an SR beamline, we will describe the spectroscopic techniques most commonly used in combination with DACs, then focus on two techniques used for structure determination under high pressure: x-ray diffraction (XRD) and x-ray absorption (XAS), illustrated with several examples in which these techniques have yielded significant results.

7.2 SYNCHROTRON RADIATION: GENERATION AND CHARACTERISTICS

7.2.1 Synchrotron Radiation versus X-Ray Tubes

Figure 7.1 shows the time evolution of x-ray sources spectral brilliance and illustrates the advantages afforded by the use of SR. While improvements in x-ray tubes (rotary anodes, refrigeration, etc.) resulted in barely a single order of magnitude increase in brilliance, the use of **first-generation** SR sources (particle accelerators using bending magnets) led to a four orders of magnitude increase in brilliance with respect to the best x-ray tubes. In turn, the introduction of the **second-generation** SR sources, i.e. accelerators designed to optimize SR emission, resulted in a further three orders of magnitude increase in brilliance. The **third-generation** sources, characterized by the

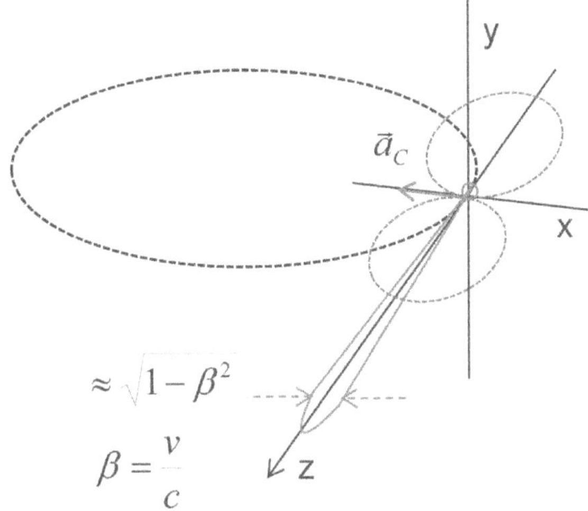

Figure 7.2 Relativistic narrowing of the dipole radiation pattern.

generalized use of insertion devices (undulators and wigglers), afforded a further six orders of magnitude increase in brilliance. Finally, in **fourth-generation** sources (x-ray free electron lasers or XFELs) a further increase by four orders of magnitude is expected. XFELs are pulsed sources and it is relevant to stress that this figure refers to the average brilliance, in contrast to the peak brilliance of a single x-ray pulse. In the case of the European XFEL (Hamburg), the peak brilliance is expected to be eight orders of magnitude larger than the average brilliance [4].

A third-generation SR source essentially consists of the following elements:

- An electron gun that generates a beam from a field emission cathode
- A linear accelerator that accelerates the electrons to a fraction of a GeV
- A booster or, properly speaking, synchrotron, that accelerates the electrons to maximum energy (6 to 8 GeV in the case of large sources such as ESRF or 2.5-3 GeV in the case of national sources like SOLEIL or ALBA)
- A storage ring in which bending magnets are also used as SR sources and insertion devices are installed in the straight sections

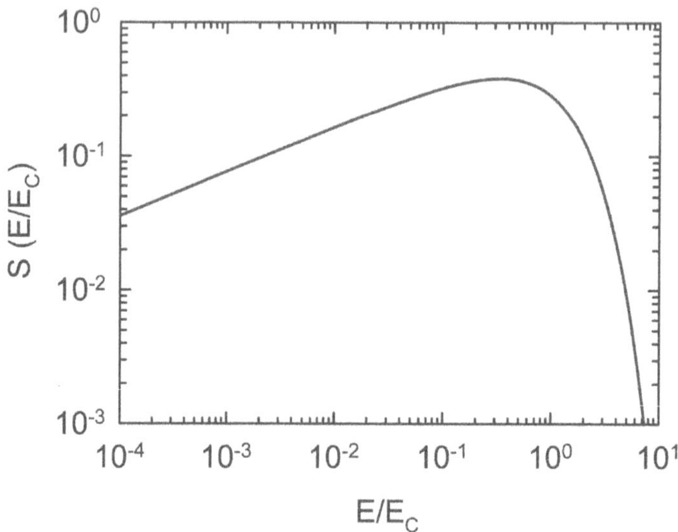

Figure 7.3 Universal curve function (S) of synchrotron radiation.

7.2.2 Nature of Synchrotron Radiation

Synchrotron radiation is emitted by a charged particle when accelerated by an electrical field (which increases its kinetic energy) or deviated by a magnetic field (centripetal acceleration). The latter situation is the mechanism used in SR sources. An accelerated charge behaves as an electric dipole, parallel to the electron centripetal acceleration, that emits a radiation pattern similar to that of a dipole antenna (arrow along the x-axis in Figure 7.2). For a relativistic electron in its own motion frame, the radiation pattern has cylindrical symmetry around the x-axis (light dashed lobes in Figure 7.2 correspond to constant-field lines in the z-x plane). As a result of the Lorentz transformation, in the laboratory motion frame the radiation pattern is strongly deformed (lobes drawn in continuous lines in Figure 7.2); consequently, SR emission is increasingly collimated in the direction of the electron movement as its velocity (v) gets closer to the speed of light (c). The SR beam angular aperture is proportional to $\sqrt{1-\beta^2}$ [5], where $\beta = \frac{v}{c}$. This is the explanation for the enormous increase in brilliance of SR sources with respect to x-ray tubes.

The first SR experiments were based on bending magnet beam emission. The emission spectrum is smooth and follows the so-called universal curve function (S) of synchrotron radiation (Figure 7.3).

The spectral density increases up to the critical energy E_c and then rapidly decreases. E_c depends on the energy of the electrons in the ring E and on the curvature radius of the beam R (or of the magnetic field B in the bending magnet) [5],

$$E_c(\text{keV}) = 2.2183\frac{E^3(\text{GeV}^3)}{R(m)} = 0.66503E^2(\text{GeV}^2)B(T). \tag{7.1}$$

The critical energy basically corresponds to the spectral width of a very short electromagnetic pulse (lasting as long as the SR of a single electron is observed through a slit placed in the direction of the SR line). In order to extend the range of energies towards hard x-rays, use can be made of so-called superbend magnets, which are shorter but generate a more intense magnetic field [6].

7.2.3 Insertion Devices

Insertion devices have been developed for applications requiring very intense x-ray photon flows. The simplest insertion device is the commonly named wavelength shifter that consists of three or five superconducting magnets with alternating direction magnetic fields. The central magnet is more intense, while the lateral magnets compensate the deviation of the central magnet so that the beam regains its straight-line trajectory. The wiggler is an extension of this device but with a larger number of magnets following a periodic configuration. The SR spectrum emitted by these systems is essentially identical to that of a bending magnet but with larger critical energy and a gain of about an order of magnitude in brilliance.

The undulator is an insertion device which, in the same way as the wiggler, uses alternating magnets that force the beam to trace a wave-like trajectory. The difference in this case is that the number of periods is much larger and the magnetic field intensity is much weaker, thereby resulting in very small beam deviations with respect to the straight-line trajectory. The electron behaves like a charge oscillating at a frequency determined by its velocity and the period of the undulator. In the electron motion frame, the emission wavelength would be $\lambda^* = L_p\sqrt{1-\beta^2}$ [5], where L_p is the period of the undulator. In the laboratory frame, the wavelength would undergo Doppler effect reduction to $\lambda_1 = \frac{L_p(1-\beta^2)}{2}$. This would be the wavelength of the first harmonic of the undulator. In the case of an undulator with a 2 cm period and electrons of 6 GeV, the wavelength of the first harmonic is of the order of 0.1 nm. Since the number of periods of the undulator is finite, widening of emission wavelength occurs. The situation is further complicated by increasing the magnetic field and forcing greater electron deviation. Undulation ceases being sinusoidal and transverse velocity becomes relativistic. All this gives rise to the emission of odd order harmonics of the fundamental emission frequency. Superconducting magnet undulators have recently been developed that make it possible to further increase the generated photon energies and the brilliance of the beam

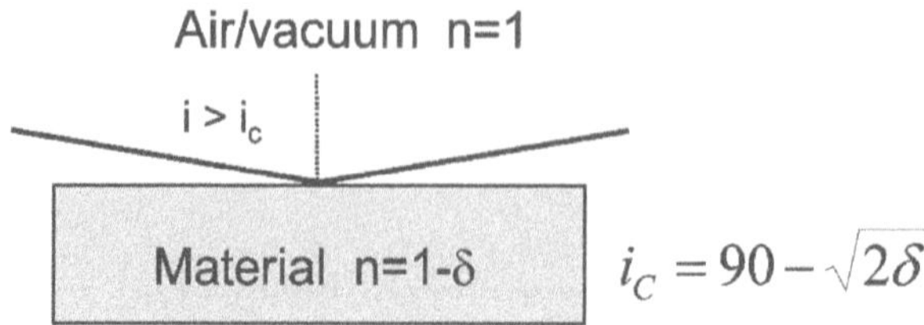

Figure 7.4 Total reflection of an x-ray beam on a material surface.

by more than one order of magnitude with respect to permanent magnet undulators [7].

7.2.3.1 X-Ray Optics

X-ray optical devices are based on different physical effects resulting from the interactions of radiation and matter. In particular, many reflexive and refractive devices are based on the fact that the refractive index of materials for high energies is $n = 1 - \delta$, where the parameter δ is proportional to the electronic density of the material.

X-ray mirrors. Since the refractive index of a material at x-ray energies is less than 1, there is a critical angle of incidence beyond which total reflection of an x-ray beam occurs on a well polished surface (Figure 7.4). Obviously, since δ is very small, the angle of incidence of the beam must be very close to 90° and very high quality surface polishing is required. In order to improve reflecting efficiency and wavelength band pass, a variety of multilayer mirrors have been developed by x-ray optics groups at different synchrotrons [8, 9].

Focalizing mirrors. A curved elliptical surface with a very small angle of incidence behaves as a focalizing element. The x-ray source is positioned at one of the foci of the ellipse and the mirror focuses the beam onto the other focus, obviously only in the plane of the ellipse. In order to achieve focalization at a point, one needs a second mirror with a surface perpendicular to that of the first mirror, forming the so-called Kirkpatrick–Baez configuration [10].

Diffractive lenses. A hollow sphere of radius R in a material behaves as a converging lens with a focal length $f = \frac{R}{\delta^2}$ [11].

Fresnel lenses. The quality achieved by nanolithography with particle or electron beams has allowed the design of Fresnel lenses (zone plates)

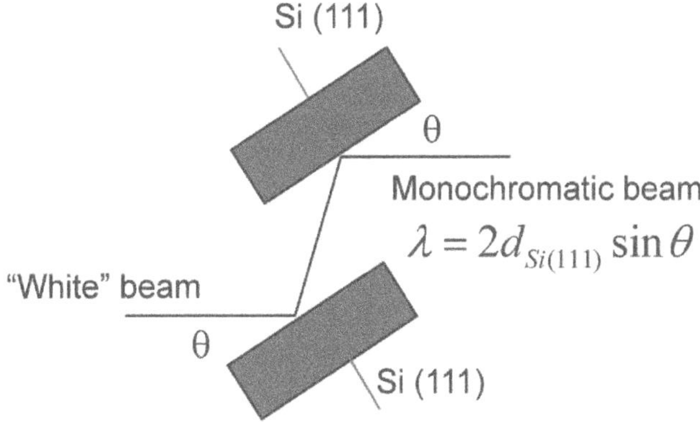

Figure 7.5 X-ray monochromator.

for x-rays based on the distribution of diffractive elements of nanometric size. A distribution comprising concentric circles of nanometric widths and distances, when adequately calculated, behaves as a converging lens [12].

Monochromators. In order to obtain a monochromatic x-ray beam, high crystalline quality silicon (Si) or diamond single crystals are used. Diamond crystals are very expensive and (111)-oriented Si crystals are therefore more widely used. The polychromatic beam falls upon the first crystal and the resulting monochromatic beam is then diffracted a second time by an additional crystal identical to the first one (Figure 7.5). Inelastic scattering experiments, which require very monochromatic beams (up to $\frac{\Delta E}{E} \sim 10^{-7}$), involve the use a second monochromator of up to four single crystals, with very precise temperature control.

7.2.4 General Structure of Synchrotron Radiation Beamline

Figure 7.6 shows the general structure of a SR beamline, in this case an x-ray diffraction line. For a detailed description of XRD beamline (ID27 at the ESRF) see Reference [13].

The source is an insertion device (ID). The beam generated by the ID first passes through the front end – the separation between the storage ring and the beamline fitted with a safety obturator that prevents beamline vacuum problems from affecting the storage ring. An attenuator limits the intensity of the beam that falls upon the monochromator. A filter eliminates the higher-order harmonics that might be diffracted by the monochromator. The Kirkpatrick–Baez mirrors [10] focalize the beam onto the sample (which may be included

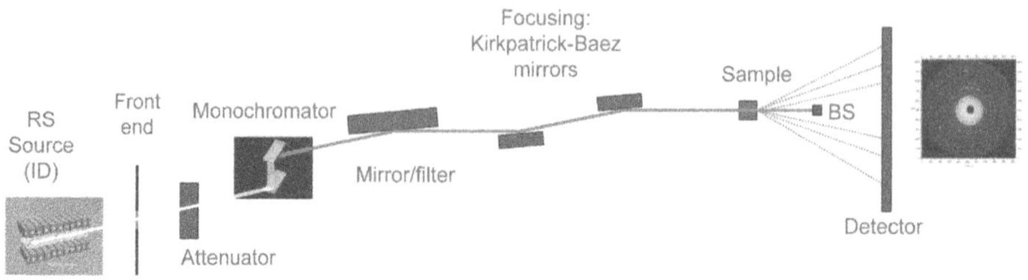

Figure 7.6 Schematic representation of a XRD synchrotron radiation line.

in a diamond anvil cell), and an absorbent beam stop (BS) blocks the direct beam transmitted by the sample. The diffraction spectrum is recorded using a bidimensional detector (CCD or image plate).

7.3 INTERACTION OF ELECTROMAGNETIC RADIATION WITH MATTER

Figure 7.7 shows some of the interaction processes between electromagnetic radiation and a material sample. An electromagnetic beam is targeted on the sample from the left. Part of the beam is absorbed, another part is transmitted and a further part is re emitted or scattered. Different spectroscopic techniques have been developed to extract physical information from these processes:

X-ray absorption (XAS). Part of the beam is absorbed by the sample, depending on its chemical composition, density and thickness. From the transmittance of the sample we can obtain the absorption spectrum, which necessarily includes the different x-ray absorption edges or thresholds (K, L, M, etc.) of the elements of the sample whose energy is within the explored range. The energy of these edges and the fine structure appearing for energies above each edge contain information on the composition and local order around the absorbing atoms. The XANES, EXAFS analyses and other tests are designed to obtain electronic and structural parameters from absorption spectra.

Elastic scattering. Defects in the crystals produce elastic scattering which can be used to study these defects, as well as other sources of disorder (ferroelectric domains, for example).

Inelastic scattering (IXS). The electrons excited by the x-rays in turn interact with the elementary excitations of the material (phonons, plasmons, etc.), absorbing or generating them, and re-emitting x-rays of

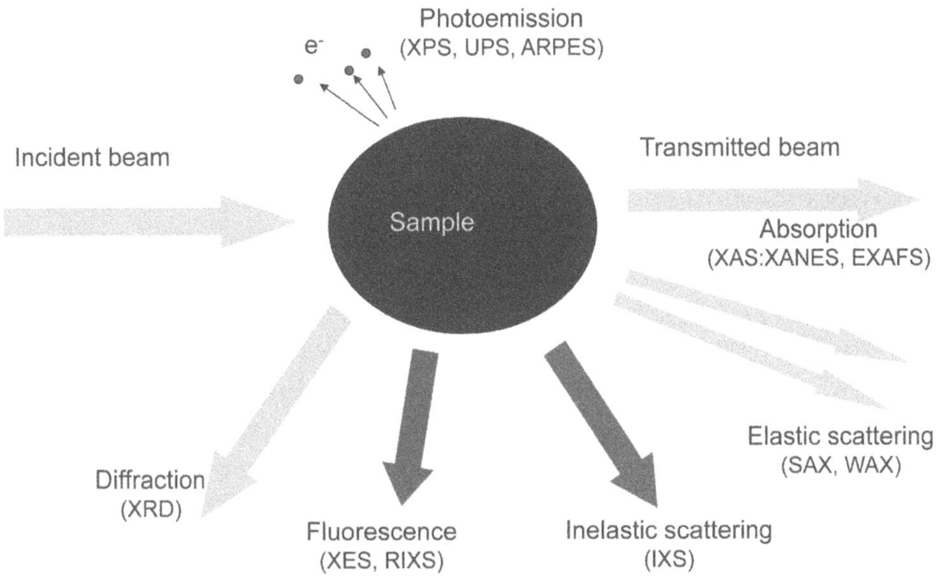

Figure 7.7 Physical processes in radiation-matter interaction.

different energies and wave-vectors. The analysis of the energy and direction of the scattered x-rays allows us to obtain the dispersion curves of the involved elementary excitations from energy and momentum conservation laws. For an introduction to IXS under high pressure, see Ref. [14]. Among other fields of application, this technique has provided very reliable data and deep insights onto many geophysical problems, as illustrated in Ref. [15].

Diffraction (XRD). In crystalline solids, constructive interference between the radiation scattered by each atom gives rise to very well defined peaks in certain directions. From the diffraction angle (the angle between the incident and diffracted radiation) and using Bragg's law, one obtains the characteristic crystal plane family sequence from which the material Bravais lattice and unit cell parameters are determined. Based on the intensities of the diffraction peaks and using different refining techniques, one also obtains the distribution of the atoms within the unit cell. Although known by different names, the WAX and SAX techniques (wide angle and small angle scattering) are essentially diffraction techniques applied to amorphous solids, disordered polymers or other materials.

Fluorescence. When the radiation photon energy exceeds an atomic absorption edge, electrons are excited to high energy levels, leaving a deep empty state (hole). Electrons from different higher energy atomic levels can transit to that empty state, emitting x-rays with well defined energies for each atom (K_α, K_β, etc.). The analysis of the energy and intensity of these emission lines allows us to study the composition of the material.

X-ray emission spectroscopy (XES). When fluorescence spectra are analysed using a high-resolution spectrometer, one can determine the electronic structure of the studied element through the change in energy and form of the fluorescence spectrum. Typical information obtained includes the oxidation state or spin configuration. A deeper insight on the electronic structure is provided by **resonant inelastic x-ray scattering (RIXS)**, a technique in which high resolution fluorescence spectra are recorded for excitation energies covering an interval across the absorption edge [16].

Photoemission (XPS, UPS). Electrons with binding energies below the exciting beam photon energy can be emitted to the vacuum from the zones of the solid close to the surface (typically less than twice the mean free path of the electrons, i.e., a few tenths of nanometers). The analysis of the kinetic energy spectrum of the photoemitted electrons and their angular distribution allows for the investigation of very fine details of electronic structure (band dispersion, Fermi surface, for example).

With the exception of spectroscopic techniques involving the use of ultra-high vacuum (such as photoemission), the rest of the spectroscopic procedures are compatible with the high-pressure techniques, particularly with diamond anvil cells.

7.4 PRESSURE CELLS FOR SYNCHROTRON RADIATION EXPERIMENTS

Diamond anvil cells (DACs) for SR experiments are basically identical to those used in optics experiments (see Section 8.2.3), with the limitation that diamond absorption is significant for energies below 10 keV (Figure 7.8). Diamond absorption can be an important limitation in XAS experiments with transition metals. The limitation is overcome either by using semi-perforated diamonds with a total effective thickness of less than 0.5 mm or using beryllium (Be) gaskets. In the latter case the beam passes through the Be gasket.

In diffraction experiments, much higher photon energies, on the order of 30 keV, are used and diamond absorption is far less relevant. The key parameter is the angular aperture of the cell that limits the maximum detectable Bragg angle. Usual membrane cell designs allow for measuring deviation angles up

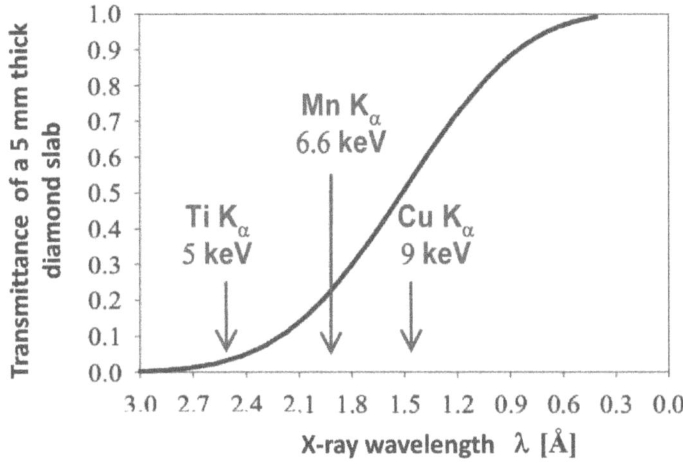

Figure 7.8 Transmittance of diamond anvils.

to 40°. The introduction of Boehler type diamond anvils has further increased the angular aperture [17]. Obviously, it is always possible to use higher-energy x-rays to reduce the Bragg angle.

An added advantage of the DACs is their compatibility with laser heating techniques. An optical microscopic assembly compatible with the SR system directs one or two high power infrared lasers (10 to 100 W) towards the cell and focalizes them onto the sample (or a mixture of the sample with a metal powder) [18]. Temperatures of several thousand Kelvins are reached at the focal point. The analysis of the thermal emission spectrum of the hot area is used to determine the sample temperature by comparison with calibrated black body-like thermal sources.

7.5 X-RAY DIFFRACTION UNDER EXTREME CONDITIONS

One can safely say that the most widely used DAC-compatible SR technique is x-ray diffraction (XRD). Powder and single crystal XRD experiments in DAC are routinely performed up to pressures beyond 1 Mbar [19, 20]. The relative simplicity of powder XRD setups (as compared to single crystal XRD diffractometers) has made the former the technique of choice in extreme conditions experiments. Details of a typical powder XRD beamline dedicated to high-pressure research can be found in a recent article which describes the MSPD beamline of ALBA [21]. Such facilities are used to study the structural properties of materials under extreme conditions and especially in geophysics for the determination of crystalline structures and equations of state (EOS) of the Earth's minerals in the conditions in which they are found in the crust, mantle or core of the planet. The SR beam can be focused to a diameter of 5 μm or below. In this way the pressure range can be extended to several megabars. The counterpart in comparison with single crystal XRD is the lowering of resolution that can complicate the identification of new crystalline structures.

X-ray diffraction under pressure is also used in material science to study the mechanical properties of materials and, when combined with *ab initio* calculations, predict their electronic properties. There have been great advances in this technique since 1990. The development of powder angle dispersive XRD (ADXRD), the use of bidimensional detectors and the development of high-pressure dedicated SR beamlines have led to the precise determination of many complex crystalline structures. However, it is now clear that in certain cases the structures discovered using powder diffraction are so complex that the drawbacks inherent to this technique have become limiting factors. Powder diffraction is unidimensional: peaks with the same spacing overlap. Only some of the crystals that constitute the powder loaded in a DAC diffract if the cell remains at rest. Furthermore, the orientation of the crystals is not necessarily random in a DAC (preferential orientations). Both circumstances can lead to erroneous assignation of the crystalline structure. We can counter this limitation by using detectors of greater resolution, larger samples, or rotating the DAC while it is exposed to the x-rays. However, the inherently unidimensional character of powder diffraction requires one to be extremely careful when indexing unknown crystalline structures. It is therefore always advisable to confirm the assigned structures by means of additional theoretical and experimental studies. In particular, combination of the powder and single crystal diffraction techniques has been shown to be extraordinarily powerful, allowing the resolution of structures as complex as the incommensurable structures of Ba, Bi, Sr, Te, etc. [22] We will discuss some examples of successful structure determination using x-ray diffraction under high pressures.

Our first example refers to the study of the mechanical properties and phase diagrams of tungstates [23], a family of materials widely used as scintillators and in other technical applications. Figure 7.9 shows the evolution of the diffraction spectra of $BaWO_4$ and $PbWO_4$ between ambient (room) pressure and 20 GPa. At ambient pressure both materials crystallize in a scheelite tetragonal phase composed of alternating chains of WO_4 tetrahedrons and BaO_8 dodecahedrons (PbO_8).

The displacement of the diffraction peaks towards larger diffraction angles reflects the material compression under pressure and can be used to determine the evolution of the lattice parameters and EOS. In this case we see that compression is anisotropic (Figure 7.10). From pressures of 7 GPa ($BaWO_4$) and 9 GPa ($PbWO_4$), signs of a phase shift that does not appear to imply dras-

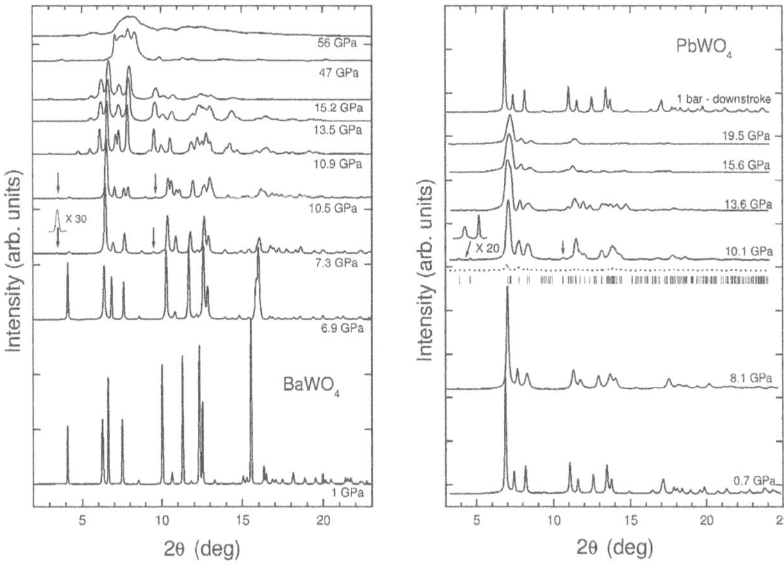

Figure 7.9 Evolution of the XRD diagrams of two tungstates under pressure. Adapted, with permission, from reference 23.

tic transformation of the spectrum are observed: disappearance of low-angle peaks and splitting of other peaks (Figure 7.9). In the experiments shown in the figure, the high-pressure phase is found to have a monoclinic fergusonite structure. No volume change in the transition is observed, which would be consistent with the occurrence of a second-order phase transition: above a certain pressure the tetragonal unit cell becomes distorted, giving rise to a monoclinic cell (Figure 7.10). At higher pressures a second phase transition is observed which, with the help of first principle calculations, has been assigned a much more compact monoclinic structure. Finally, above 50 GPa, the disappearance of the Bragg peaks and the appearance of a diffuse halo below $2\theta = 10°$ suggest amorphization of the studied compounds (Figure 7.9). More recently, it has been found that deviatoric stresses may influence the structural sequences in tungstates [24], transforming some of them directly from the scheelite structure to the dense monoclinic HP phase.

The second example, corresponding to the EOS of solid hydrogen, is particularly relevant for astrophysics (structure of Jovian type planets). It is probably one of the best examples of experiments that would be impossible to perform without SR, due to the low scattering factor of hydrogen (this factor being proportional to the square of the atomic number). It is also one of the most remarkable experiments performed in the 1990s. The group led by Loubeyre grew a hydrogen monocrystal in a liquid helium medium, loading the diamond cell with a He-H_2 mixture [25].

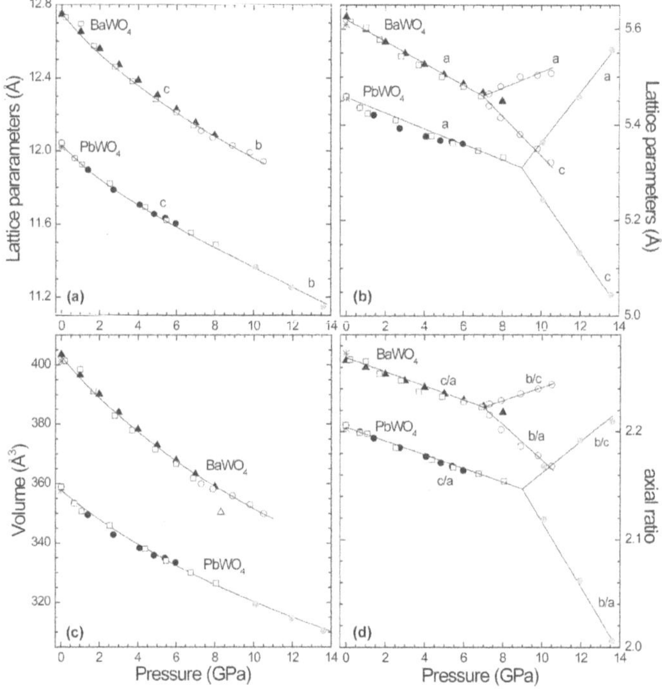

Figure 7.10 Lattice parameters and equation of state of two wolframates. Adapted, with permission, from reference 23.

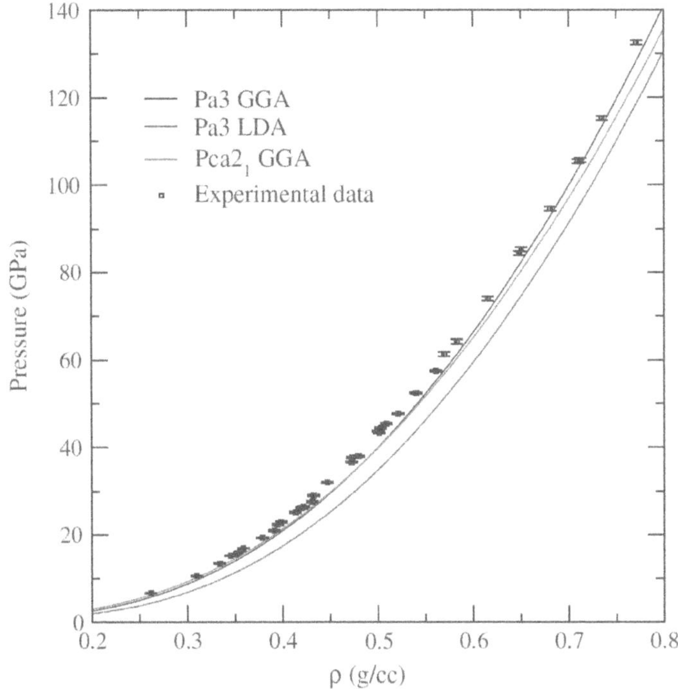

Figure 7.11 Equation of state of hydrogen. Adapted, with permission, from reference 26.

In order to avoid problems in identifying hydrogen diffraction peaks, because of their low intensity, x-ray diffraction was performed in the energy dispersive configuration, in which excitation is carried out with a polychromatic "white beam", and diffraction is measured at a fixed angle with a semiconductor detector measuring the photon energy at which the Bragg condition is met for that angle.

The use of collimators for the incident beam and fixed-angle diffraction allows for an unambiguous selection of the diffraction volume, with detection of the diffracted peaks originated in the sample and elimination of those corresponding to sample environment (gasket, diamonds, etc.).

Three diffraction peaks (each measured at a different angle) were followed under high pressure. Solid hydrogen in that range of pressures possesses a hexagonal structure; as a result, the three observed peaks were sufficient to determine the EOS with great precision. The experimental data as reported in Reference [26] is presented in Figure 7.11. The continuous lines in the figure represent *ab initio* calculations corresponding to two structures proposed in different domains of the hydrogen phase diagram.

Accurate determination of the hydrogen EOS is essential for understanding the structures of Jovian type planets. Knowing their mass and volume, the hydrogen EOS is crucial to determine which proportion of the mass of such planets is attributable to hydrogen.

The third example, corresponding to the melting curve of iron, lies at the heart of an important problem in geophysics: the structure of the Earth's nucleus and more specifically the frontier between its liquid and solid parts. The experimental determination of the melting curve of iron involves the use of pressures above 1 Mbar (100 GPa) and temperatures up to 4000 K, which are obtained by laser heating. The melting of Fe and other transition metals has been studied under such conditions [27–30]. The determination of a point on the curve is made by taking diffraction spectra along an isobar (increasing the temperature at constant pressure). Melting should be observed as a sharp change in the form of the diffraction pattern from that of a solid, characterized by well defined Bragg peaks, to that of a liquid, characterized by a diffuse scattering diffractogram composed of relatively wide bands, as can be seen for the case of Mo in Figure 7.12. In the figures the integrated XRD also shows the disappearance of Bragg peaks associated to Mo and the appearance of the difuse scattering. Pictures taken from the sample before and after melting are shown to illustrate the result of melting. A deeper discussion on melting determination can be found in the articles by Santamaría *et al.* [27] and Anzelini *et al.* [28]. The phase transition latent melting heat makes it very difficult to maintain temperature stability and obtain clear spectra around the melting temperature.

Figure 7.13 provides a summary of the experimental determinations of the melting curve of iron performed by different research groups [28, 30, 31]. The same studies were also used to determine the frontier between the cubic (γ) and hexagonal phases (ϵ) of iron. At the pressures and temperatures of the solid Earth core, iron is in a non-magnetic compact hexagonal phase (the ϵ phase). At pressures close to the solid-liquid core boundary, the experimental results indicate melting temperatures around 3600 K, which is considerably lower than the values predicted by *ab initio* calculations and also below the results of shockwave experiments. In contrast, more recent studies provide higher melting temperatures which agree better with calculations. The solid and dashed lines represent the results of different experiments. However, there are still issues to be clarified on the HP-HT phase diagram. In particular, discrepancies with shock experiments are not fully solved and temperature determination at Mbar pressures and thousands of kelvins encountered difficulties in obtaining fully reliable temperatures [32].

The last example corresponds to the complex structures found in simple metals such as Ba and Sr under high pressure (120 GPa). These elements have extremely simple structures at ambient conditions. However, under

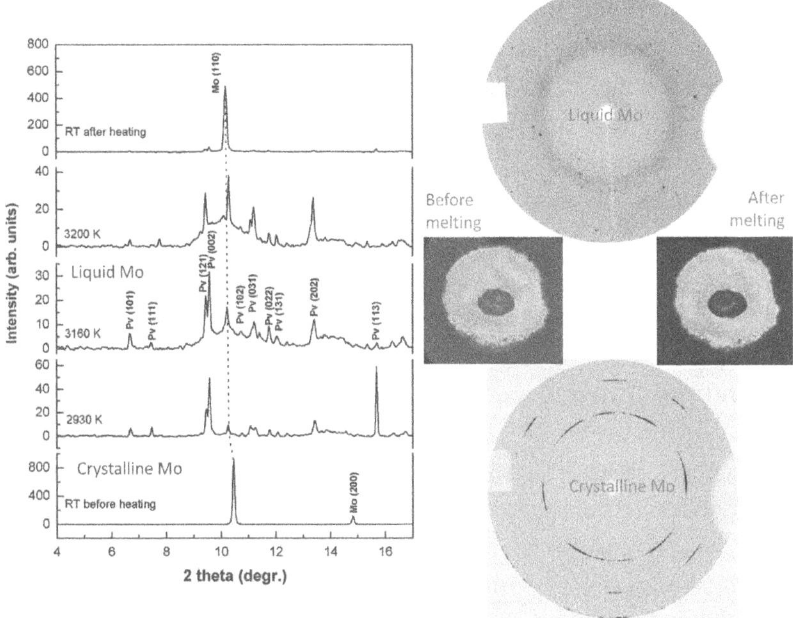

Figure 7.12 Plate images taken before and after melting in Mo. Integrated XRD patterns are also shown with pictures taken from the sample. Pv refers to the perovskite used as the pressure-transmitting medium.

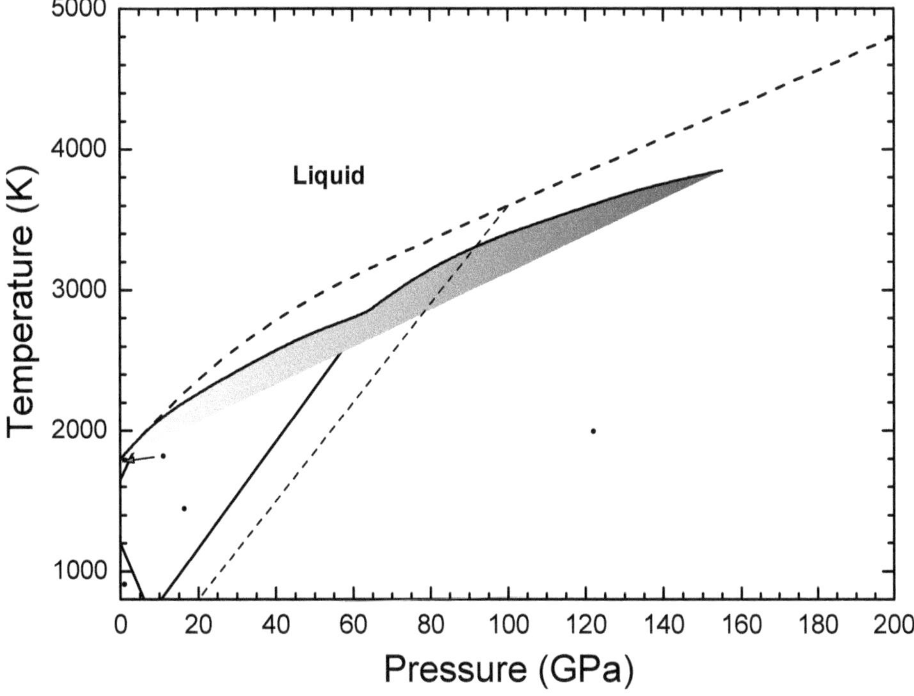

Figure 7.13 Pressure-temperature (P-T) phase diagram of iron.

pressure, changes induced in their crystal structures lead to extremely complex structures that defied elucidation for over 30 years. In the past decade, the combination of powder and monocrystal diffraction has made it possible to see that the configuration consists of the interspacing of two structures (a so-called "host-guest structure") [22] that has been referred to by the *Nature* journal as "the weirdest known atomic structure of any pure element". The new configuration of Ba consists of a tetragonal structure (host) that gives rise to the most intense reflections of the diffraction pattern. Within this configuration there are two unidimensional chain structures located along axis c that produce diffuse scattering (as a result of positional disorder) and a series of satellites around the principal reflections. One of the most remarkable characteristics of this structure is that the ratio between the c parameter of the host and guest structures is an irrational number. This means that it is impossible to find a distance for which the combined structure repeats periodically, hence leading to its description as an incommensurable structure. The same types of structures have also been found in other simple elements such as Sr, Bi, Sb, As, Se, Te, and S.

7.6 X-RAY ABSORPTION (XAS, XANES, EXAFS)

The x-ray absorption (XAS) spectrum of a material is determined by its electronic structure and contains different absorption edges due to transitions from the deep occupied shells to the higher energy empty states. For photon energies corresponding to the emission of an electron from an internal layer of an atom of a sample (absorption edge or threshold), we observe a sharp increase in the absorption coefficient. In a gas sample these edges appear over a monotonously decreasing background. Figure 7.14 shows the the x-ray absorption of krypton gas [33] where the Kr K edge occurring at 14326 eV can be observed. The absorption jump is associated with the excitation of a 1s electron.

In solids and liquids the edge is modified. To describe the features of x-ray spectra we usually distinguish two spectral regions. The energetic range from some electronvolts below the absorption edge up to about 50 eV above is referred to as the XANES (x-ray absorption near edge structure) range. It is related to multiple scattering processes and is sensitive to middle range orders (in most cases up to the third or fourth neighbor shell). In addition, the XANES region below the edge is called pre-edge. The peak which often appears around the absorption edge energy is called the white line (present at 14330 eV in solid Kr, see Figure 7.14). As the core electron is excited to an empty state, from a complementary view it can be thought that x-ray absorption probes the unoccupied part of the electronic structure of the system. With

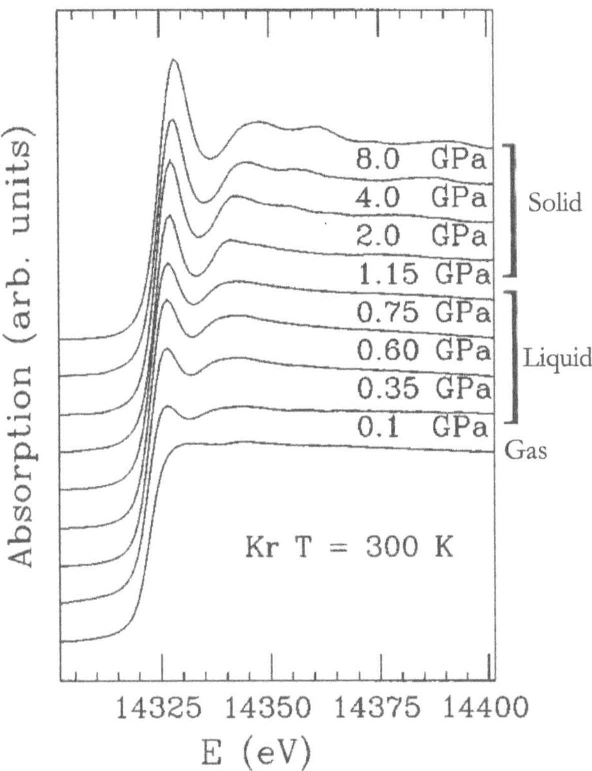

Figure 7.14 XAS spectra of krypton as pressure induces phase transitions from the gas to the liquid and solid phases. Spectra are taken at the Kr K edge. Adapted, with permission, from reference 33.

this is mind, we can relate the white line to $1s \rightarrow 4p$ transitions. In the gas there is no white line because final states are filled in krypton (noble gas). In the liquid phase the presence of neighbors around the absorbing atom begins to distort the $4p$ electronic states and a weak white line is observed. The effect is intensified in the solid phase.

The second spectral range has to do with oscillations that appear about 50 eV above the absorption edge and typically extend approximately up to 1000 eV. They are referred to as the EXAFS (extended x-ray absorption fine structure) range. In the EXAFS part of the spectrum, the kinetic energy of the photoelectron is significant when compared with the interaction with neighbouring atoms. As we are going to develop in a moment, the oscillations are originated by the interaction of the excited electron with the neighboring atoms and dependent upon the local order around the absorbing atom. Continuing the description of krypton spectra in Figure 7.14, EXAFS oscillations are absent in the monoatomic gas. The formation of a first neighbor shell in the liquid induces the apparition of EXAFS oscillations that are weak due to structural disorder. They become intense in the solid.

The usefulness of XAS spectroscopy is explained by the information it provides on the local environment of the absorbing atom. In this regard, XAS spectroscopy has the advantages of not requiring long range orders (crystalline and non-crystalline solids can be studied in the same way) and of allowing separate investigations of the surroundings of each atom.

According to quantum mechanics, x-ray absorption intensity is determined by the electric dipole matrix element between the initial and final states. In our case, the initial state is the electron at the deep level and the final state is the excited electron. The waves scattered by the neighboring electrons overlap the emergent spherical wave, producing interference – the effect of which depends on the phase shift between them and therefore on the wavelength of the photoelectron, determined by the exciting photon energy. Therefore, the amplitude of the final state and the value of the matrix element are modulated by the photon energy, thereby explaining the EXAFS oscillations. The phase shift is a function of the relative distance between the central atom and the scatterers. The scattered wave amplitude depends on the nature of the scattering atom. This is why information on the surroundings of the atom under study can be derived from the EXAFS spectrum. Theoretical calculation of the EXAFS spectrum shows the oscillations to be given by equation [34]:

$$\chi(k) = \frac{\mu - \mu_0}{\mu_0} = \sum_j \frac{N_j F_j}{k R_j^2} e^{-2k^2\sigma_j^2} e^{-\frac{2R_j}{\lambda}} \sin\left[2kR_j + \delta_j\right]. \tag{7.2}$$

The parameter j refers to the sum over the coordination spheres, each with N_j atoms at a distance R_j from the absorbing atom. k in turn is the wavevector

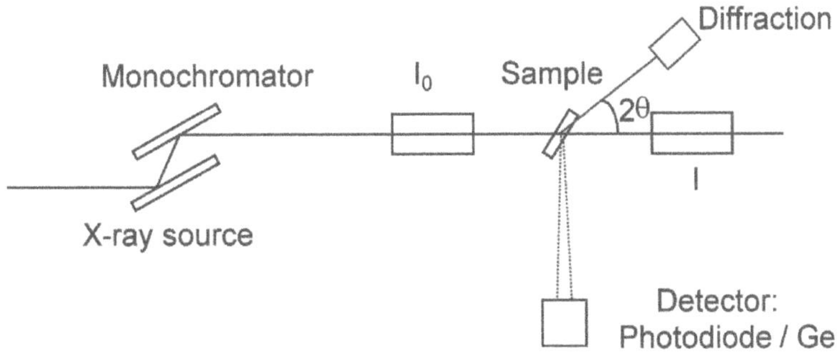

Figure 7.15 Schematic representation of a classical XAS beamline.

of the excited electron (calculated from its kinetic energy). The sinus function reflects the effect of interference between the emergent wave and the dispersed wave, where $2kR_i$ is the phase shift due to travel and return of the dispersed wave and $\delta_j(k)$ is the dephasing effect induced by the atomic potentials. The factor $e^{-\frac{2R_j}{\lambda}}$ is introduced phenomenologically and represents the probability that the photoelectron will travel to the backscattering atom and return without scattering and without the hole in the absorbing atom being filled. These two processes tend to destroy the coherence needed to obtain an EXAFS spectrum. λ represents the electron mean free path. The term $e^{-2k^2\sigma_j^2}$ is obtained after averaging the EXAFS equation in the presence of minor Gaussian type disorders or when taking into account the thermal vibrations within the harmonic approximation. σ_j is the standard deviation in the position of an atom in the layer j, and is referred to as the Debye–Waller pseudofactor. Finally, $F_j(k)$ is the backscattering amplitude, which is related to the scattering capacity of the atoms surrounding the absorbing atom.

The measurement of XAS spectra is carried out in the same way as the measurement of optical absorption spectra (see Chapter 8), based on the transmittance of a sample of known thickness. In the same way as in optical spectroscopy, the absorption spectrum can be measured by scanning techniques (classical XAS) or multichannel procedures (dispersive XAS).

In a classical XAS experiment (Figure 7.15), the incident beam is monochromatic and its intensity (I_0) is measured with an ionization chamber positioned before the sample. The transmitted intensity (I_T) is measured with a second ionization chamber. The absorption coefficient is obtained from the transmittance of the sample and its thickness, d:

$$\alpha = -\frac{1}{d}\ln\frac{I_t}{I_0}. \tag{7.3}$$

In the case of impurities or diluted elements, the difference between I_0 and

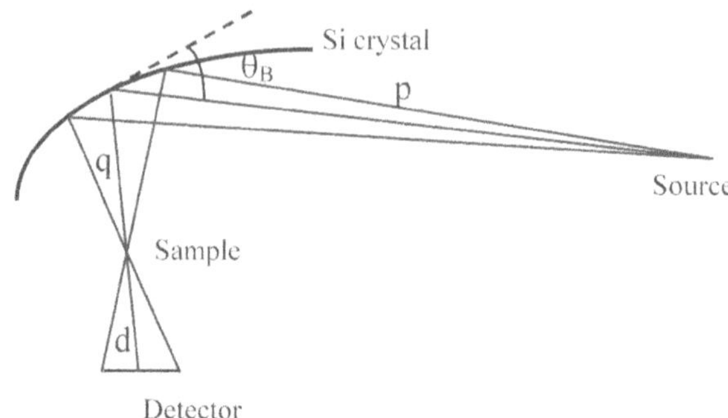

Figure 7.16 Schematic representation of a dispersive XAS beamline.

I_T is small – on the order of measurement noise. In this case the XAS spectrum is determined by using fluorescence detection (as in photoluminescence excitation optical spectroscopy). Instead of measuring the transmitted beam, the sample fluorescence spectrum is measured with a photodiode or dispersive energy detector, selecting the range corresponding to the emission lines of the diluted element. On scanning the energy of the incident beam, the fluorescence intensity directly measures the intensity absorbed by the sample, which is proportional to the absorption coefficient for small values of the latter.

In dispersive EXAFS (Figure 7.16) a white x-ray beam is diffracted by a curved crystal, normally made of silicon, that focalizes the diffracted beam onto the sample and at the same time selects a bandwidth around the absorption edge of the atom under study. The crystalline planes responsible for diffraction must define a portion of an ellipse with the source and sample as foci. In the simplest configuration, the diffraction planes are parallel to the bent crystal surface. The beam diffracted by each element of the ellipse will have the wavelength that complies with the Bragg condition for the corresponding angle of incidence. In this way beams of different energies passes through the focus in different directions and the incident and transmitted spectra can be obtained by means of a plane multichannel detector.

This configuration is particularly suitable for measurements in diamond cells due to the inherent focusing configuration and simultaneous acquisition of the whole spectrum. The small acquisition times are very useful to remove the glitches introduced in the absorption spectra by diffraction from the single crystal diamond anvils. Until recently, the common procedure used to avoid the glitches consisted of observing how they shifted on rotating the cell. For energies under 10 keV it is generally possible to obtain orientations for which no diamond diffraction peak is observed. At higher energies several orientations are needed to recompose the absorption spectrum. However, the apparition of nanocrystalline diamonds [35] solved the glitch problem (at least with the cells

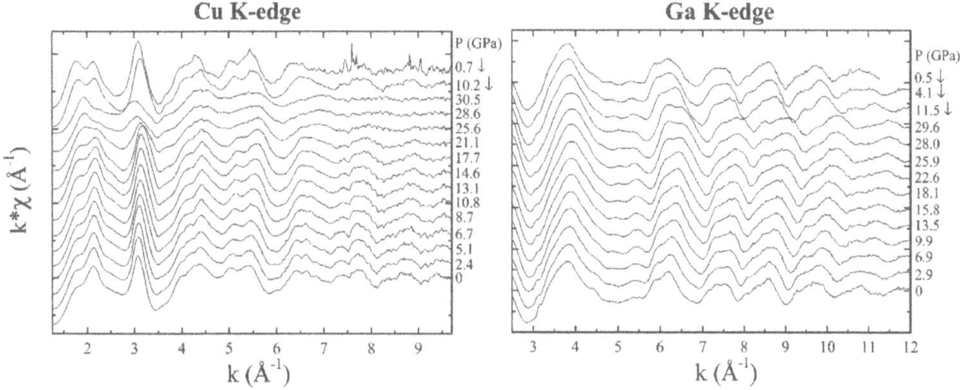

Figure 7.17 XANES-EXAFS spectra under pressure at the Cu K and Ga K edges. Adapted, with permission, from reference 37.

employed at moderate pressures). It has been possible, for example, to measure [36] high quality EXAFS data up to 13 Å^{-1} at the Se K edge (12658 eV). The eruption of nanocrystalline diamonds, as well as improvements in x-ray generation and conditioning, will probably change the way x-ray absorption under high pressure is measured, favoring the classical scanning technique.

Once the absorption spectrum has been obtained, the interpretation of the EXAFS zone follows the steps described below:

1. Determination and subtraction of the baseline to obtain the oscillations
2. Extraction of the function $\chi(k)$
3. Fourier transform of the function $\chi(k)$
4. Adjustment of the Fourier transform to obtain the interatomic distance, the number of first neighbors and the Debye–Waller pseudofactor, using the phases and amplitudes of the scattering atoms calculated numerically or extracted from reference compounds

The first example we will discuss is the study of $CuGaO_2$ delafossite under pressure [37]. This is a material with a rhombohedral structure, $R\bar{3}m$ spatial group, which can be described as a stacking of compact hexagonal atomic planes, whereby Ga is in octahedral six-fold coordination and Cu is in linear two-fold coordination. The experiment was performed in the dispersive XAS beamline of the DCI (LURE), with powder samples, up to 30 GPa and using silicone oil as transmitting medium.

Figure 7.17 shows how the XANES-EXAFS oscillations in the Cu and Ga K edges shift slightly towards high energies under pressure, due to reduction

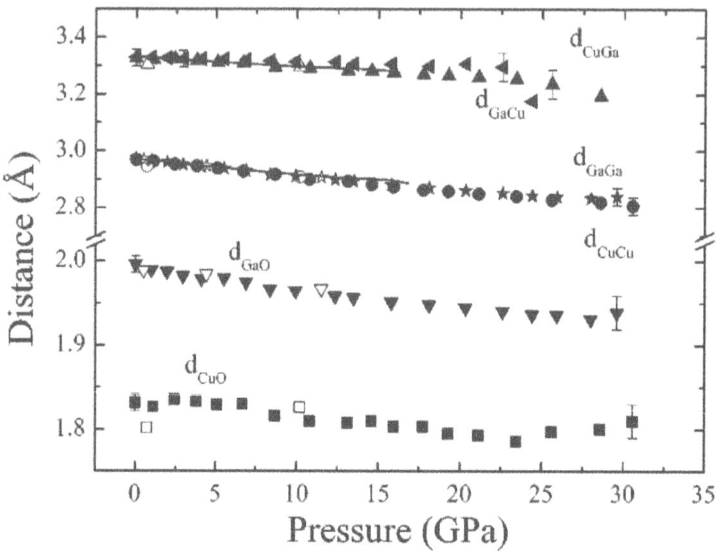

Figure 7.18 Interatomic distances obtained fitting the EXAFS oscillations of $CuGaO_2$ measured at both the Cu and Ga K edges. Adapted, with permission, from reference 37.

of the interatomic distances, with no change in shape until a structural phase transition occurs around 26 GPa.

Beyond this pressure we can see the changes in the spectrum of the Cu K edge, while no drastic changes are seen in the Ga edge. This result indicates that the phase transition basically affects the surroundings of Cu, the coordination of which increases, while that of Ga is maintained, the latter already being octahedral in the delafossite phase.

The quality of the spectra and the broad spectral range allow precise determination of the distances to first, second and third neighbors, as can be seen in Figure 7.18. The consistency of the results is confirmed by the coincidence of the distances to second (Cu-Cu, Ga-Ga) and third neighbors (Cu-Ga, Ga-Cu) obtained from spectral analysis at different thresholds. The high level of symmetry of the delafossite structure and the number of distances obtained allow us to determine evolution under pressure of the complete crystalline structure using EXAFS; this would be impossible with less symmetrical structures.

The second example deals with the evolution under pressure of the surroundings of a diluted atom, manganese in the dilute magnetic alloy $Zn_{0.95}Mn_{0.05}O$ [38]. The experiment, performed in the classical XAS configuration with fluorescence detection, studied the K edge of Mn in the wurzite and NaCl phases of the material. At the Mn K edge the absorption of

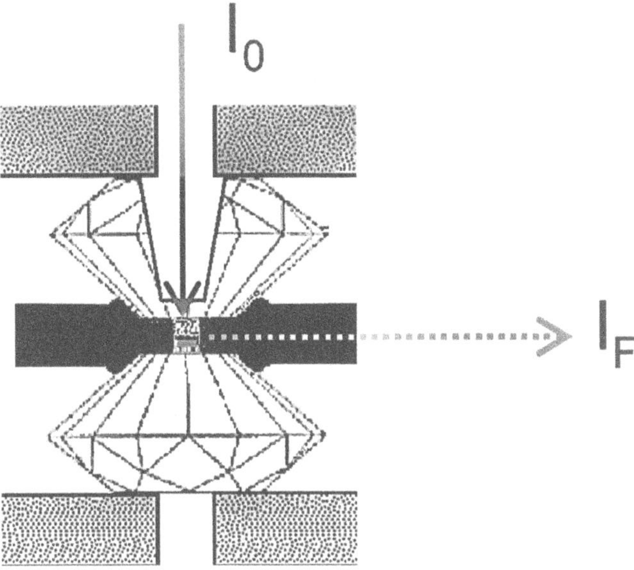

Figure 7.19 Diamond cell for XAS detected by fluorescence.

diamond is very high and a semiperforated diamond was used (Figure 7.19). Mn fluorescence was detected through the Be gasket [39].

Figure 7.20 (left) shows the pressure evolution of the XAS spectra and, extracted from them, the pressure dependence of the first and second neighbors distances. The wurzite-NaCl phase transition is visible from the modification of the XANES and EXAFS ranges, as well as the drastic reduction of the pre-peak located at 6542 eV. This absorption peak, typical of atoms with incomplete *d* shells, corresponds to a transition from the 1*s* level to hybridized states in which the empty states of the *d* shell participate. In a central-symmetrical structure, the inversion symmetry limits the hybridization possibilities and dipolar transitions between same-parity states (*s* and *d*) are forbidden (though only in the Brillouin zone center). As a result the pre-peak, while much less intense, is still seen in the NaCl phase.

Figure 7.20 (right) shows the pressure dependence of the first and second neighbor distances. The local compressibility modulus can be determined from the monotonous reduction of the distances under pressure. The phase transition involves an increase in the first neighbor distance and a decrease in the second neighbor distance. The increase in the first neighbor distance is a consequence of the increase in the coordination number from 4 to 6 in the NaCl phase. In contrast, and since the second coordination sphere basically exhibits the same configuration in both phases, the decrease in the second

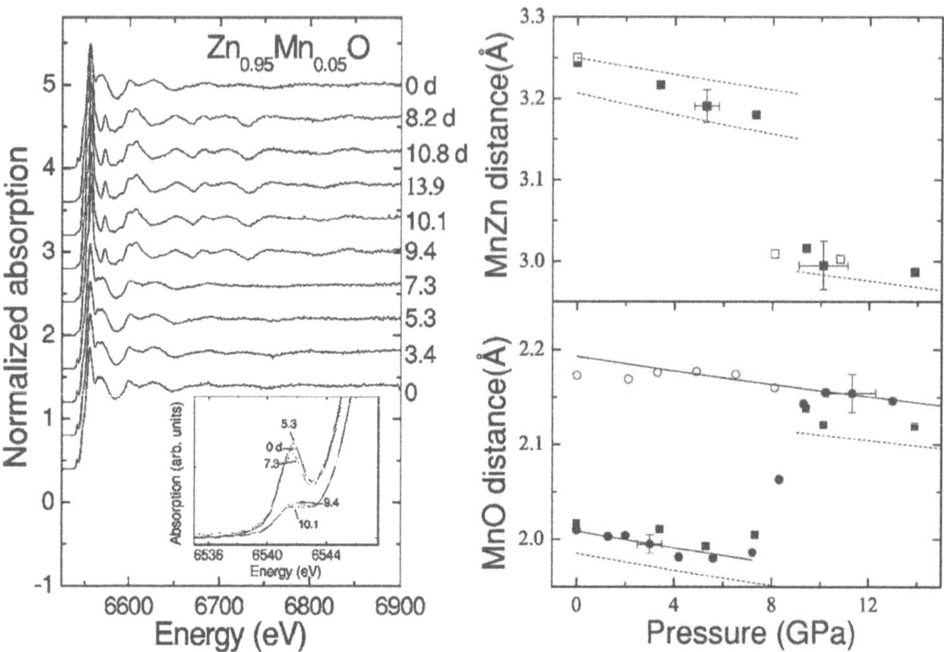

Figure 7.20 XAS spectra and interatomic distances under pressure in ZnMnO. Adapted, with permission, from reference 38.

neighbor distance basically reflects the density increase on transiting to the high pressure phase.

The last example has to do with titanate perovskites. These materials are technologically relevant. They yield rich phase diagrams that show strong competition between ferroelectric and antiferrodistorsive instabilities [40]. XRD depends on long-range order and is thus sensitive to the TiO_6 rotations and distortions characteristic of the antiferrodistorsive instabilities. However, XRD averages the atomic arrangements over a large number of unit cells and sometimes fails to provide clear information about the local structure. A high pressure XAS experiment clarifies the situation in $BaTiO_3$. At room pressure, $BaTiO_3$ is ferroelectric and crystallizes in a tetragonal structure. At 2 GPa a phase transition to a cubic structure is observed. Figure 7.21 shows the XANES spectra of $BaTiO_3$ at the Ti K edge. A large intensity of resonance B has been associated to the Ti off-center position. In Figure 7.21, resonance B decreases from ambient conditions to 10 GPa. The large intensity still observed after the phase transition indicates that Ti remains in off-center position in the cubic phase up to 10 GPa. In order to reconcile this fact with the cubic symmetry, the Ti off-center position must be disordered. XANES results have been correlated with the high pressure behavior of the diffuse reflectance observed

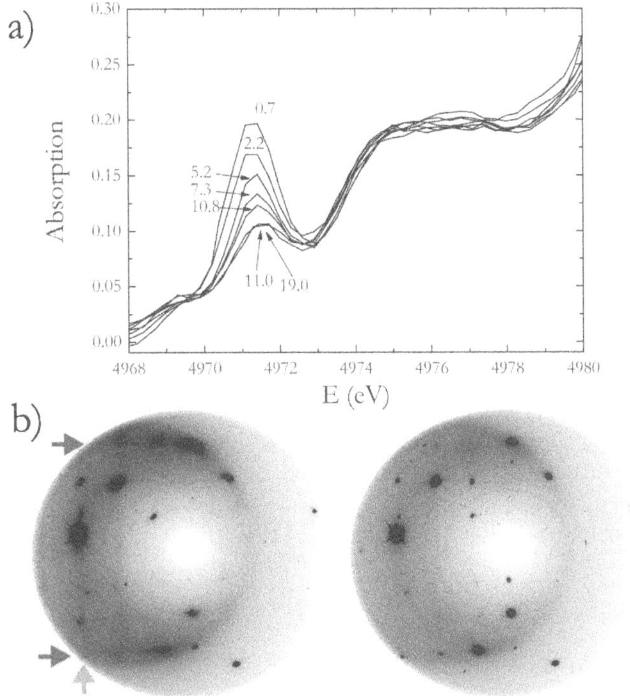

Figure 7.21 a) XANES spectra of $BaTiO_3$ at the Ti K edge. Numbers next to the spectra indicate pressure in GPa. (Figure courtesy of J. P. Itié) b) Selected XRD image plates. Adapted, with permission, from reference 41.

in XRD images in Figure 7.21. The behavior is related to the pseudocubic symmetry and disappears at 11 GPa [41].

Bibliography

[1] B. Buras, J. S. Olsen, L. Gerward, *et al.* *J. Appl. Crystallogr.*, 10, 431 (1977).

[2] S. Klotz, G. Hamel, and J. Frelat. *High Pressure Res.*, 24, 219 (2004).

[3] R. C. Liebermann. *High Pressure Res.*, 31, 493 (2011).

[4] European X-ray Free Electron Laser. Webpage (2014).

[5] H. Wiedemann. *Synchrotron radiation.* Springer, Berlin (2003).

[6] J. Zbanik, S. Wang, J. Chen, *et al.* *IEEE Trans. Appl. Supercond.*, 11, 2531 (2001).

[7] S. Casalbuoni, M. Hagelstein, B. Kostka, *et al.* *Phys. Rev. ST Accel. Beams*, 9, 010702 (2006).

[8] E. Ziegler, A. K. Freund, J. Susini, *et al.* *Opt. Eng.*, 29, 928 (1990).

[9] I. V. Kozhevnikov, I. N. Bukreeva, and E. Ziegler. *Nucl. Instrum. Methods Phys. Res. A*, 460, 424 (2001).

[10] P. Kirkpatrick and A. V. Baez. *J. Opt. Soc. Am.*, 38, 766 (1948).

[11] R. K. Smither, A. M. Khounsary, and S. Xu. *Proc. SPIE*, 3151, 150 (1997).

[12] A. Snigirev, V. Kohn, I. A. Snigireva, *et al.* *Nature*, 384, 49 (1996).

[13] M. Mezouar, W. A. Crichton, S. Bauchau, *et al.* *J. Synchrotron Radiat.*, 12, 659 (2005).

[14] F. A. Gorelli. *High-pressure physics.* CRC Press, Boca Raton (2012).

[15] G. Fiquet, J. Badro, F. Guyot, *et al.* *Phys. Earth Planet Inter.*, 143-144, 5 (2004).

[16] P. Glatzel and U. Bergmann. *Coord. Chem. Rev.*, 249, 65 (2005).

[17] R. Boehler and K. De Hantsetters. *High Pressure Res.*, 24, 391 (2004).

[18] E. Schultz, M. Mezouar, W. Crichton, *et al. High Pressure Res.*, 25, 71 (2005).

[19] G. W. Stinton, S. G. MacLeod, H. Cynn, *et al. Phys. Rev. B*, 90, 134105 (2014).

[20] M. Merlini and M. Hanfland. *High Pressure Res.*, 33, 511 (2013).

[21] F. Fauth, I. Peral, C. Popescu, *et al. Powder Diffraction*, 28, S360 (2013).

[22] M. I. McMahon, O. Degtyareva, and R. J. Nelmes. *Phys. Rev. Lett.*, 85, 4896 (2000).

[23] D. Errandonea, J. Pellicer-Porres, F. J. Manjón, *et al. Phys. Rev. B*, 73, 224103 (2006).

[24] O. Gomis, J. A. Sans, R. Lacomba-Perales, *et al. Phys. Rev. B*, 86, 054121 (2012).

[25] P. Loubeyre, R. LeToullec, D. Hausermann, *et al. Nature*, 383, 702 (1996).

[26] L. Caillabet, S. Mazevet, and P. Loubeyre. *Phys. Rev. B*, 83, 094101 (2011).

[27] D. Santamaría-Pérez, M. Ross, D. Errandonea, *et al. J. Chem. Phys.*, 130, 124509 (2009).

[28] S. Anzellini, A. Dewaele, M. Mezouar, *et al. Science*, 340, 464 (2013).

[29] D. Errandonea, M. Somayazulu, D. Häusermann, *et al. J. Phys. Condens. Matter*, 15, 7635 (2003).

[30] G. Shen, V. B. Prakapenka, M. L. Rivers, *et al. Phys. Rev. Lett.*, 92, 185701 (2004).

[31] D. E. R Boehler, D Santamaria-Pérez and M. Mezouar. *J. Phys. Conf. Series*, 121, 022018 (2008).

[32] A. Dewaele, M. Mezouar, N. Guignot, *et al. Phys. Rev. Lett.*, 104, 255701 (2010).

[33] A. Di Cicco, A. Filipponi, J. P. Itié, *et al. Phys. Rev. B*, 54, 9086 (1996).

[34] E. A. Stern. *Principles, applications, techniques of EXAFS, SEXAFS and XANES.* Wiley Interscience, New York (1988).

[35] N. Ishimatsu, K. Matsumoto, H. Maruyama, *et al. J. Synchrotron Radiat.*, 19, 768 (2012).

[36] M. Bendele, C. Marini, B. Joseph, *et al. Phys. Rev. B*, 88, 180506 (2013).

[37] J. Pellicer-Porres, A. Segura, Ch. Ferrer-Roca, *et al. Phys. Rev. B*, 69, 024109 (2004).

[38] J. Pellicer-Porres, A. Segura, J. F. Sánchez-Royo, *et al. Appl. Phys. Lett.*, 89, 231904 (2006).

[39] J.-P. Itié, A.-M. Flank, P. Lagarde, *et al. AIP Conf. Proc.*, 879, 1329 (2007).

[40] J. P. Itié, A. M. Flank, P. Lagarde, *et al. High-pressure crystallography: from fundamental phenomena to technological applications*. Springer, Berlin (2010).

[41] S. Ravy, J.-P. Itié, A. Polian, *et al. Phys. Rev. Lett.*, 99, 117601 (2007).

CHAPTER 8

Optical Spectroscopy

Fernando Rodríguez
MALTA Consolider Team and DCITIMAC, Facultad de Ciencias, Universidad de Cantabria, Santander, Spain

Rafael Valiente
MALTA Consolider Team and Departamento de Física Aplicada, Facultad de Ciencias, Universidad de Cantabria, Santander, Spain

Víctor Lavín
MALTA Consolider Team and Departamento de Física, Universidad de La Laguna, San Cristóbal de la Laguna, Spain

Ulises R. Rodríguez-Mendoza
MALTA Consolider Team and Departamento de Física, Universidad de La Laguna, San Cristóbal de la Laguna, Spain

CONTENTS

8.1 INTRODUCTION

OPTICAL spectroscopy allows us to explore the structure of matter through the radiation-matter interaction processes (absorption, inelastic light scattering, emission/excitation, etc.). Given that optical spectroscopy utilizes UV-Vis-NIR photons to probe matter, this technique is important for studying matter under extreme conditions of pressure and temperature in diamond anvil cells (DACs) or other alternative transparent anvils such as sapphire (SAC) and moissanite (MAC) [1–14]. The huge advance of high-pressure techniques came with the design of the diamond anvil cell (DAC) that displays multidisciplinary character. The transparency range of diamonds in the UV-NIR region (0.07 eV to 5 eV) makes possible the adaptation of traditional spectroscopic techniques used in material characterization laboratories. Hence optical absorption, luminescence, excitation, lifetime or Raman techniques (see Chapter 9) can be systematically applied to work with microsamples and long working distances [15–17]. It is possible to investigate the changes in the electronic properties induced by local or crystal modifications of a material under high pressures without changing the chemical composition. We can determine different pressure effects like energy level shifts, local modifications or structural phase transitions and their impact on electronic and vibrational properties, changes in color due to pressure (piezochromism), high spin to low spin transitions, or excited-state crossovers.

This advantage is not gained in other spectral domains of interest for studies of atomic, vibrational and electronic structure of materials such as x-rays ($\lambda \geq 1.5$ Å) and vacuum UV ($E > 5$ eV), in which the strong absorption of the diamond hinders the use of experimental techniques probing matter with photons in such spectral ranges. Figure 8.1 illustrates the light transmission of diamond in comparison with other materials. For comparative purposes, Figure 8.2 shows the optical transmission spectra of sapphires (corundum) [18–20]. Interestingly, the diamond absorption spectrum strongly depends on the impurity content (Figure 8.1), and this can seriously limit the spectral range of operation in a DAC. This is particularly important in the UV region near the diamond band gap ($E = 5.5$ eV; $\lambda \geq 225$ nm). Due to this, optical spectroscopy in UV with DACs deserves the use of ultralow fluorescence, Type IIa, diamond anvils; otherwise such spectral range is inaccessible.

Optical spectroscopy techniques like optical absorption and photoluminescence (including time-resolved emission/excitation spectroscopy) are very

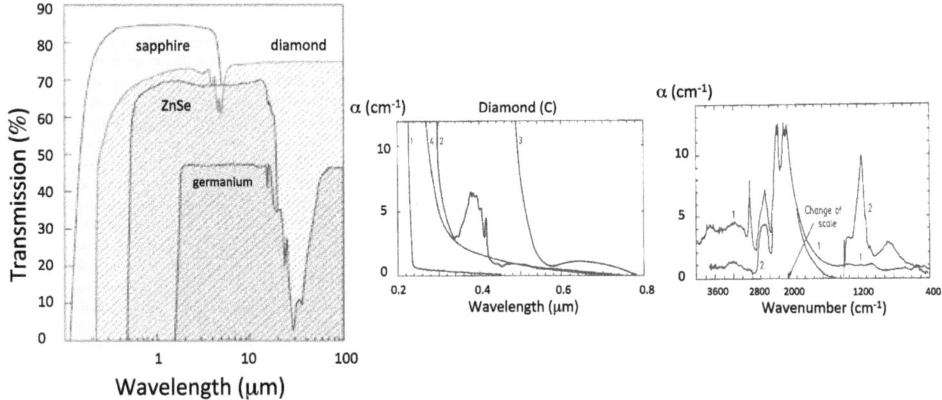

Figure 8.1 (Left) Transmission spectra of diamond, sapphire, zinc selenide, and germanium. (Right) Optical absorption spectra of diamond with different amounts of impurities in the UV-VIS and IR regions. Numbers 1-4 refer to the dominant impurities in the material: 1. Boron atoms; 2. N3 centers (nitrogen trimer); 3. nitrogen atoms; 4. brown diamond. Adapted, with permission, from Reference [16].

flexible and easy to adapt to DAC, and therefore well suited for high-pressure research. They can be applied to investigate different types of materials like semiconductors, insulators, and organic materials in different structural conformations such as single crystals, powders, nano-structures, etc., and a wide variety of physical phenomena (electronic levels, resonance, excited state crossovers, and spin transitions).

In general, high-pressure optical spectroscopy allows us to study the structure of matter as a function of volume (i.e, interatomic distances), explore materials properties along the P-T phase diagram, and unveil new phenomena associated with structural changes induced by pressure. An important aspect of high-pressure optical spectroscopy concerns the interpretation of the variations of optical parameters with temperature (thermal peak shifts, vibrational thermal shifts, etc.). In general, the variation of an optical parameter like the energy of an electronic or vibrational level E with temperature contains two different contributions related to 1) the variation of E due to changes of temperature at constant volume (explicit contribution), and 2) the variation of E induced by changes of volume due to thermal expansion (implicit contribution):

$$\begin{aligned}\left[\frac{\partial E}{\partial T}\right]_P &= \left[\frac{\partial E}{\partial T}\right]_V + \left[\frac{\partial E}{\partial V}\right]_T \left[\frac{\partial V}{\partial T}\right]_P \\ &= \left[\frac{\partial E}{\partial T}\right]_V + \left[\frac{\partial E}{\partial P}\right] \left[\frac{\partial P}{\partial V}\right] \left[\frac{\partial V}{\partial T}\right]_P \\ \left[\frac{\partial E}{\partial T}\right]_P &= \left[\frac{\partial E}{\partial T}\right]_V - B_0\alpha \left[\frac{\partial E}{\partial P}\right]_T, \end{aligned} \tag{8.1}$$

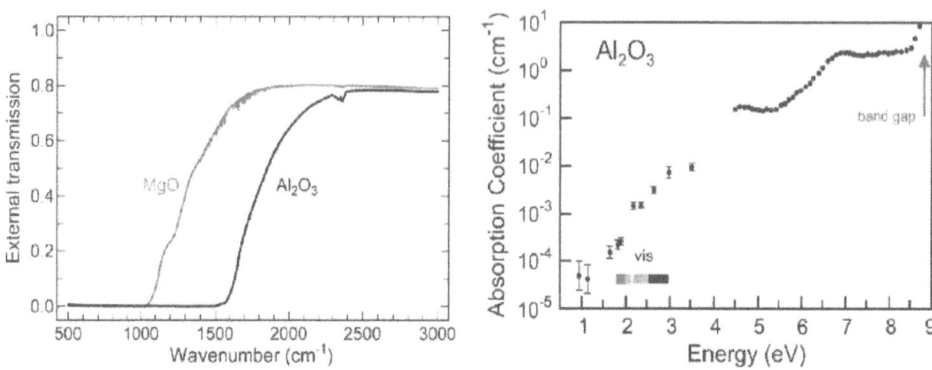

Figure 8.2 (Left) Transmission spectra of magnesium oxide (periclase) and aluminum oxide (corundum). (Right) Optical absorption spectrum of corundum in the gap spectral window (0–9 eV) [18–20]. Adapted, with permission, from Reference [27].

where B_0 and α are the bulk modulus and the thermal expansion coefficient, respectively (see Chapter 1 and Chapter 2 for complete descriptions of B_0 and α). The experimental thermal variation is difficult to interpret due to these two terms. Actually, it may induce confusion given that temperature variations have often been interpreted in terms of the explicit contribution (Boltzmann population effects at constant volume) or the implicit term (thermal expansion effects). However, both contributions may compete depending on the material and the parameter under investigation [21–25]. A direct way to discriminate these two contributions consists of measuring the implicit term by means of high-pressure or uniaxial stress techniques [25, 26].

Optical spectroscopy is beneficial for measuring physical properties, like phonons, transitions probabilities, and exchange coupling interactions, that are associated with the electronic ground state and also with excited states. It enables us, for example, to explore vibrational mode softening or hardening when passing from the ground state to excited states [28].

In this chapter we provide an overview of the different experimental setups adapted to different types of optical spectroscopy experiments (absorption, emission/excitation, time-resolved spectroscopy, lifetime, and others), along with selected examples of reference systems to illustrate the feasibility of optical spectroscopy to study matter at high pressure. Our focus is to explore the experiment viability and the difficulties and limitations imposed by the technique in each selected case. Finally, we show the suitability of optical spectroscopy for obtaining information on the local structures of pure or doped optically active (transition-metal and lanthanide) ion systems by means of structural correlations between the electronic and atomic structures, both

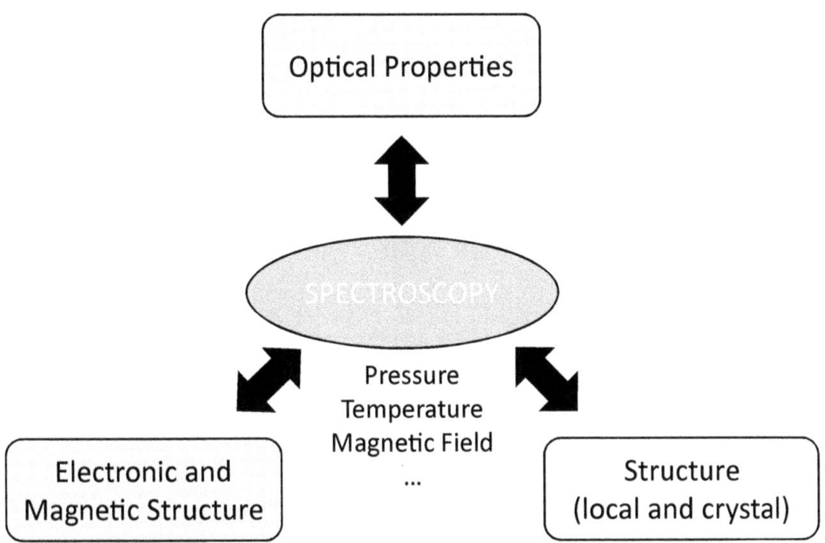

Figure 8.3 Scheme showing how high-pressure spectroscopy correlates local and crystal structures with electronic properties and structures.

modified by the application of external pressure. The basic idea is to establish the relationship between the structural changes of a system induced by pressure (interatomic distances and bond angles) with associated changes in energy levels (level energy, lifetime, bandwidth, and oscillator strength) determined by optical spectroscopy on insulators [1–4] and semiconductor materials [5] (see Figure 8.3).

8.2 OPTICAL ABSORPTION

8.2.1 Background and Basic Ideas

8.2.1.1 Macroscopic Parameters: Complex Refractive Index

The radiation-matter interaction processes can be macroscopically described through optical parameters like transmission T and reflectivity R which are related to the dielectric function of the material. Indeed, the absorption and refraction of light by a material can be described by the complex refractive index, $\tilde{n}$, defined by: $\tilde{n} = n + i\kappa$. In general, the real n and imaginary κ parts of $\tilde{n}$ depend on the wavelength. The real part is the same as the normal refractive index, while the imaginary part is directly related to the absorption coefficient of the material. These parameters are related through the equations:

$$\epsilon = \epsilon_1 + i\epsilon_2 = \tilde{n}^2 = (n + i\kappa)^2 \qquad \epsilon_1 = n^2 - \kappa^2 \qquad \epsilon_2 = 2n\kappa. \tag{8.2}$$

When a beam of light of intensity (I_0) incides on a sample along the x direction, a part of the intensity is reflected (RI_0), another part is absorbed while passing through the sample (I_{abs}) and finally, a part is transmitted (I). The equation governing the transmitted light through the sample (neglecting multiple reflections) is given by:

$$T_{exp}I_0 = (1-R)^2\,TI_0 = (1-R)^2\,I_0e^{-\alpha(\lambda)l}, \tag{8.3}$$

where the reflectivity R depends on both n and κ through the equation:

$$R = \left|\frac{\tilde{n}-1}{\tilde{n}+1}\right|^2 = \frac{(n-1)^2+\kappa^2}{(n+1)^2-\kappa^2}, \tag{8.4}$$

and the light transmission T is related to the absorption coefficient, $\alpha(\lambda)$, as follows. The fraction of absorbed intensity dI by a volume element dx is proportional to the incident intensity at the point $I(x)$. The constant of proportionality is by definition $\alpha(\lambda)$, which in general depends on wavelength: $dI = -\alpha(\lambda)\,Idx$. This equation is the differential form of the Lambert-Beer law:

$$I(\lambda) = I_0(\lambda)\,e^{-\alpha(\lambda)l}, \tag{8.5}$$

where l is the sample thickness. It must be taken into account that in an absorption experiment in a DAC, we get the intensities $I_0(\lambda)$ and $I(\lambda)$ representing the intensity passing through the pressure-transmitting medium (reference), and through the material in the DAC (sample), respectively. The experimental transmission $T_{exp} = {}^{I}/_{I_0}$ or the experimental absorbance $A_{exp} = \log({}^{I_0}/_{I}) = -\log(T_{exp})$ is related to the absorption coefficient $\alpha(\lambda)$ and reflectivity $R(\lambda)$ through the equation:

$$T_{exp}(\lambda) = [1-R(\lambda)]^2\,e^{-\alpha(\lambda)l}. \tag{8.6}$$

The precise measurement of the absorption coefficient requires the correction of $R(\lambda)$ unless the dispersion of the refractive index with wavelength is negligible ($R(\lambda) = \text{cte}$). It must be noted that relationship between absorbance and absorption coefficient is given by:

$$\begin{aligned} A &= \log\frac{I_0}{I} = \log e^{\alpha(\lambda)l} = \alpha(\lambda)\,l\log e = \frac{\alpha(\lambda)\,l}{2.3} \\ \alpha(\lambda) &= 2.3\frac{A(\lambda)}{l}. \end{aligned} \tag{8.7}$$

Other commonly used optical absorption parameters are the *molar extinction coefficient* ϵ defined (in units of cm^{-1} $Mole^{-1}$ L) as: $\epsilon(\lambda) = \frac{A(\lambda)}{M\times l}$, where M is the molar concentration of absorbing centres (in mole L^{-1}), the *absorption cross section* σ is defined (in units of cm^2) as: $\sigma(\lambda) = \frac{\alpha(\lambda)}{N} = \frac{1000\times\alpha(\lambda)}{M\times N_0}$,

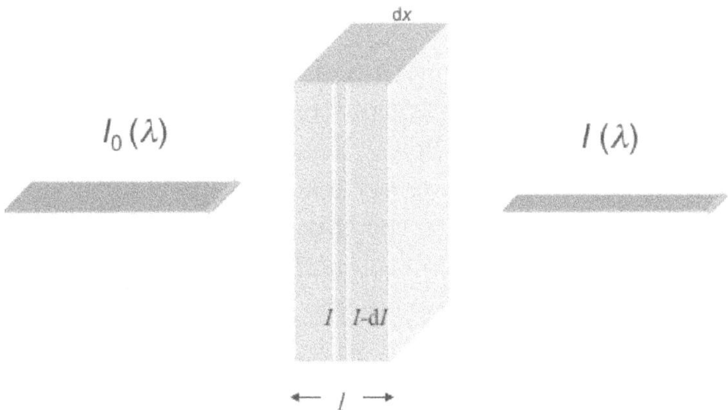

Figure 8.4 Light pathway through a parallelepipedic sample of thickness l. The infinitesimal volume of thickness dx absorbs an intensity, dI, that is proportional to the thickness and the incident intensity, I. The proportionality constant is defined as the absorption coefficient, $\alpha(\lambda)$, and depends on the wavelength.

and N and N_0 are the absorbing centres per cubic centimetre and Avogadro's number, respectively.

The relationship between the absorption coefficient and the imaginary part of the refractive index comes directly from the wave equation of a propagating electromagnetic field in a medium. The time-dependent electric field is given by $E(x,t) = E_0 e^{i(kx-\omega t)}$, where k is the wave vector of the light, ω is its angular frequency, and $|E_0|$ the amplitude at $x = 0$. In a non-absorbing medium of refractive index, n, the wavelength is reduced by a factor n with respect to vacuum: $k = \frac{2\pi}{\lambda} n = \frac{n\omega}{c}$; in an absorbing medium, this can be generalized taking the complex refractive index: $E(x,t) = E_0 e^{i\left(\frac{\tilde{n}\omega}{c}x-\omega t\right)} = E_0 e^{i\left(\frac{[n+i\kappa]\omega}{c}x-\omega t\right)} = E_0 e^{-\frac{\kappa\omega}{c}x} e^{i\left(\frac{n\omega}{c}x-\omega t\right)}$. The attenuation of light from $x = 0$ to l, is $I = |E(l,t)|^2 = |E_0|^2 e^{-2\frac{\kappa\omega}{c}l} \equiv I_0 e^{-\alpha l}$, and therefore $\alpha = \frac{2\kappa\omega}{c} = \frac{4\pi\kappa}{\lambda}$.

Optical absorption spectroscopy aims to measure the absorbed intensity as a function of the wavelength of light, and its interpretation takes into account the microscopic mechanisms governing the radiation-matter interactions and their relationship with the atomic, vibrational or electronic structures of the material. In order to relate the macroscopic parameters defined above with structural parameters of the system, we need to analyze microscopically the absorption and emission of photons by the matter in terms of the transition probability between the levels and the densities of states associated with such levels.

8.2.1.2 Spectroscopic Parameters: Microscopic Correlation

The relationship between the absorption coefficient and the microscopic mechanisms responsible for photon absorption (or emission) can be described in terms of Fermi's golden rule as follows. In a photon absorption process associated with an electronic transition between two discrete energy levels, the absorption cross section is related to the transition probability of one-photon absorption by the equation: $P_{a\to b} = \frac{2\pi}{\hbar}\left|\langle a\left|H_{int}\right|b\rangle\right|^2 \rho(E)$, where H_{int} is the radiation-matter interaction Hamiltonian and $\rho(E)$ is the radiation field mode density. The interaction Hamiltonian can be expanded in series of $\left(\frac{2\pi}{\lambda}\right)$ in the range of wavelengths of interest (UV-Vis-NIR), yielding different multipolar terms [1–4]. In the most common case of electric-dipole transitions $\left(H_{ED} = e\vec{r}\cdot\vec{E} = -\vec{\mu}_{ED}\cdot\vec{E}\right)$, the absorption coefficient α (or the absorption cross section σ) relates to the microscopic parameters of the model through the equation:

$$\frac{dI}{dx} = -\left[\left(\frac{\pi}{3\hbar^2}\right)\hbar\omega\left(\frac{\left|\langle a\left|\mu_{ED}\right|b\rangle\right|^2}{\epsilon_0 cn}\right)g(\omega)\,N\right]I, \tag{8.8}$$

where $g(\omega)$ is the spectral distribution of the transition (shape of the absorption/emission peak: $\int g(\omega)\,d\omega = 1$), ϵ_0 the vacuum permittivity, and n the refractive index of the medium. Therefore the optical absorption spectrum through the absorption coefficient provides information on the energy difference between the two states involved in the transition $\hbar\omega$ (energy conservation by one photon absorption) and the peak profile $g(\omega)$. Integrating the absorption cross section σ, which is equivalent to calculating the area under the peak in the absorption spectrum, we can obtain the electric-dipole matrix element: $\langle a\left|\mu_{ED}\right|b\rangle$. Generally, the oscillator strength is a widely used dimensionless parameter that gives us an idea of the transition probability between states. It relates to the electric-dipole matrix element of the transition through the expression $f = \left(\frac{2m\omega}{3\hbar e^2}\right)\left|\langle a\left|\mu_{ED}\right|b\rangle\right|^2$. The equation providing the transition rate between two states should be modified by the density of states associated with absorption bands in insulators or semiconductors or the phonon density of states if we consider two discrete electronic levels coupled to the crystal lattice. The refractive index of Equation (8.8) must be replaced by the factor ${}^{3n}/{(n^2+2)^2}$ within the local electric field approximation in a material medium $(n \neq 1)$.

Transition oscillator strength can take values of about 10^{-1} for internal atomic $s \to p$ transitions; 10^{-2}-10^{-3} for aromatic $\pi \to \pi^*$ or ligand-to-metal charge-transfer transitions; 10^{-4}-10^{-5} for $d-d$ crystal-field transitions in transition-metal ion systems; or 10^{-7} in spin- and parity-forbidden

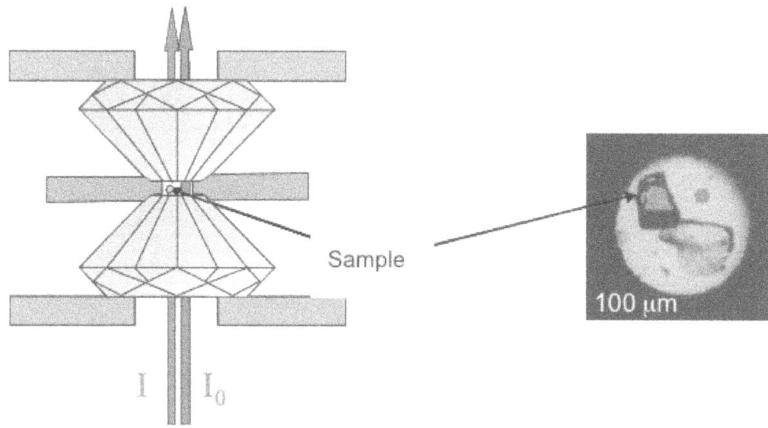

Figure 8.5 Scheme of a light beam passing through a DAC for measuring the sample (I) and reference (I_0) intensities and hence the determination of the absorbance (optical density) $A(\lambda)$. The image on the right side shows a single crystal of $CuMoO_4$ inside the cavity and the size of the focused beam in the sample, I (white) and near the sample I_0 (grey) taken as references. Note that the beam size must be smaller than the sample to ensure that the light beam fully passes through the sample (single crystal).

transitions (Laporte's rule) in lanthanides or transition-metal ions. Examples of these two cases are shown in Figures 8.11 and 8.16.

8.2.2 Optical Absorption Setups: Single Beam and Double Beam Mode Operation

Although the optical absorption (OA) technique under high-pressure conditions has enormous advantages in material characterization, some problems may arise in obtaining suitable spectra from the DAC. Due to the small size of the cavity, the optical devices and samples should be adapted when using a DAC. The hydrostatic cavity consists of a cylinder drilled in a pre-indented gasket (Figure 8.5) approximately 50 to 300 μm in diameter and 20 to 100 μm thicknesses, depending on the final pressure limit. Samples are typically about 100 μm in size and can be loaded as single crystals, polycrystalline, powders, or diluted samples. It is convenient to use plane-parallel plate samples of adequate thickness to get suitable absorption coefficient measurements. OA spectroscopy at high pressure requires the adaptation of spectrometers to DACs, i.e., micro-samples, and this constitutes the main experimental challenge in high-pressure spectroscopy as every experiment must be adapted to the specific requirements imposed by the sample (shape, size, reactivity, etc.).

To obtain precise measurements of the absorbance (or optical density) in microsamples, we must be aware of the light beam passing through the

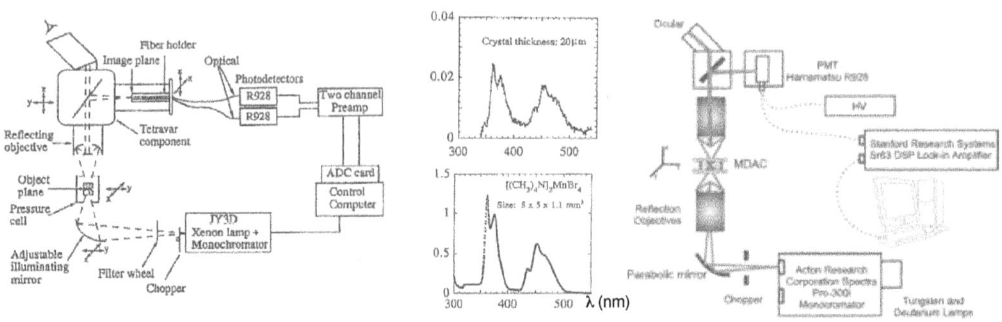

Figure 8.6 (Left) Double beam spectrophotometer for use with diamond (or sapphire) anvil cells. Note that I and I_0 are obtained with two identical detection systems to ensure a simultaneous measurement of the two intensities. The figures on the right compare the OA spectra of $[(CH_3)_4N]_2MnBr_4$ single crystals obtained with a conventional spectrophotometer using 8 mm size samples and a microscope setup using 80 μm samples. Setup taken from Reference [29]. (Right) Scheme of single beam spectrophotometer using synchronous detection. The microscope setup for focusing into the cavity is similar to that shown left for the double beam spectrophotometer and operates with membrane DACs. Adapted, with permission, from Reference [29].

sample (Figure 8.5), avoiding light diffusion effects. The light spot should be as small as possible in the explored wavelength range ($\sim$ 20 μm). This, together with high detection sensitivity required for a precise absorbance measurement, makes high pressure OA a complex non-standard technique if compared, for example, with photoluminescence (PL). In OA, the light intensity passing through the sample is usually lower than the incident intensity (an absorbance $A = 2$ means that I is 100 times lower than I_0). The measurements of I_0 and I can be difficult since the intensity may be contaminated by scattered light coming from multiple reflexions and dispersion by diamonds and the hydrostatic medium within the cavity. To compensate for these effects, it is necessary to measure the intensity of the corresponding reference to the beam of light that passes through the hydrostatic medium and diamonds from the sample (I_0) at each spectrum/pressure. Figure 8.5 shows a DAC containing a single crystal of $CuMoO_4$, a ruby as pressure gauge, and the sizes of the I and I_0 spots. In standard spectrophotometers, I and I_0 are normally measured in two different modes: single beam or double beam. In single beam mode, I and I_0 are not measured at the same time, but using two independent scans or independent data collection: first I and later I_0, or vice versa. In double beam mode, I and I_0 are measured simultaneously (within the period of the modulation beam time) in the same scan. This second method is more accurate because it corrects immediate intensity fluctuations of the light source and

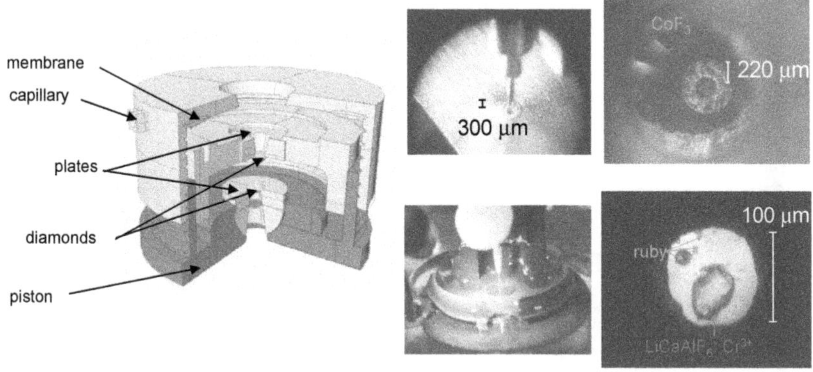

Figure 8.7 Representation of a membrane-type DAC (left). Photographs showing the perforation of a gasket to construct a pressure cavity: mechanical drilling (top) and electro-eroder machine (bottom). Sample loading with powder and single crystal, respectively [39].

instrumental misalignments (spectral drifts, system detection, and mechanical instability). However, double beam mode is not easy to implement in DACs due to size limitations [29–37]. A double-beam prototype setup adapted for DACs is shown in Figure 8.6 [29]. The setup is well suited for measuring weak absorbing samples (absorption peaks with $A < 0.01$) and hence is more adequate for diluted or doped systems. However single beam operation is better suited for high absorbing materials ($A > 1$) like direct transitions in molecular systems or semiconductors having high absorption coefficients ($\alpha > 10^5$ cm^{-1}).

Interestingly, OA in powders, liquids, and nanoparticle conglomerates may become difficult since the sample occupies the whole cavity, thus making the measurement of I_0 unfeasible. This difficulty can be overcome using large culet anvils (0.5 mm) and double drilled gaskets of 50 μm diameter each. One cavity is loaded with sample and the other one with a reference sample, providing a simple way of measuring I and I_0 [38].

8.2.3 Diamond Anvil Cell

Figure 8.7 shows a loaded DAC with two different samples: CoF_3 powder and $LiCaAlF_6$:Cr^{3+} single crystal with ruby balls serving as pressure gauge. The drawing corresponds to a membrane-type DAC. The optical experiments, both absorption and luminescence, are done with low-fluorescence type IIa 0.5 mm culet diamond anvils. The gasket pre-indentation and subsequent perforation (the hydrostatic cavity) are done by mechanical drilling or by electro-eroder.

As a pressure transmitting medium, a mixture of methanol-ethanol-water in the ratio 16:3:1, is commonly used at room temperature (RT) since it

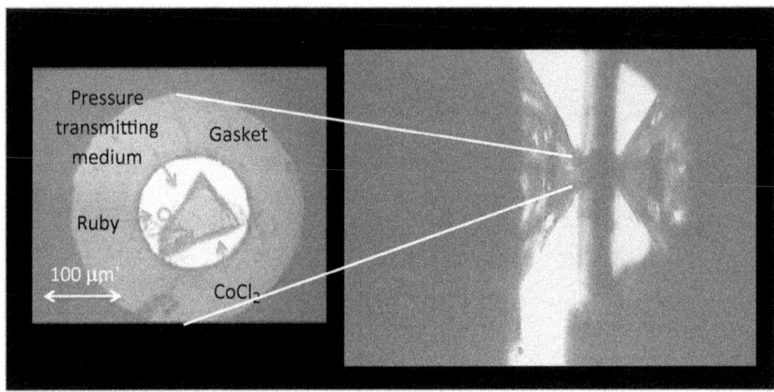

Figure 8.8 Drilled gasket forming the hydrostatic cavity of a DAC loaded with single crystal of $CoCl_2$ and ruby for pressure calibration. The diamond anvils compressing the gasket are shown on the right side. (Courtesy of J.A. Barreda-Argüeso, Ph.D. thesis, University of Cantabria, Santander, Spain)

provides ideal hydrostatic conditions below 10 GPa, although the ideal pressure-transmitting medium is highly pressurized (2–5 GPa) He gas. This medium provides the best hydrostatic conditions, even above the solidification pressure (12 GPa; 300 K) [40]. However in some special cases of highly hygroscopic, deliquescent or unstable materials, the use of alternative pressure-transmitting media like spectroscopic paraffin that provides good performance in the UV-Vis range. Also, a sample (Figure 8.7) can maintain adequate high-pressure conditions for optical spectroscopy experiments. In such cases, an analysis of the ruby emission peak profile (asymmetry and broadening) provides useful information about non-hydrostatic conditions [41]. DAC development permits application of external pressures in wide pressure ranges from 1 to 300 GPa, depending on the type of cell, Merrill–Basset or Mao–Bell [30], and the diamond anvil culet diameter, 0.7 mm or 0.1 mm, respectively.

Figure 8.8 shows a loaded DAC with single crystal of $CoCl_2$ and a 10 μm diameter ruby sphere for pressure calibration. In this case, the use of paraffin (or silicon oil) protects the deliquescent and hygroscopic $CoCl_2$ (blue phase below 2 GPa) from degradation into the hydrate cobalt chlorides (red color phases). In Figure 8.9 we show a single crystal of $KCoF_3$ compressed up to 1Mbar for exploring possible changes of electronic structure associated either with a spin crossover transition or metallization. The optical transparency of the crystal over the whole explored pressure range indicates that both the high-spin and insulator Co^{2+} states are stable in the explored pressure range.

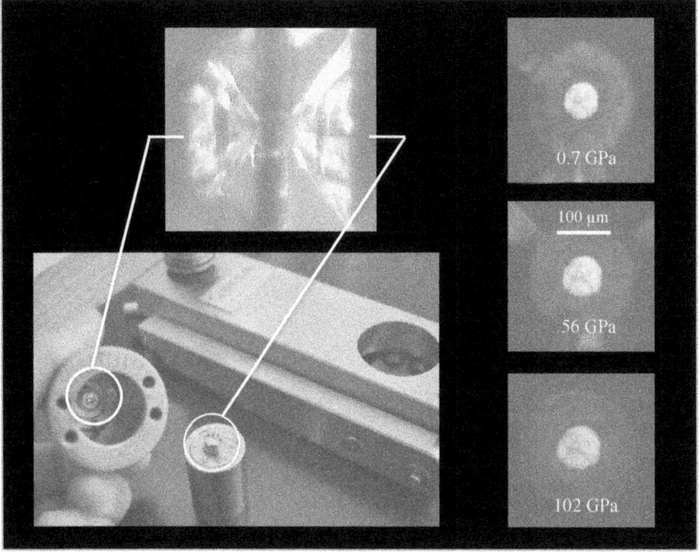

Figure 8.9 Mao–Bell-type DAC (LeverDAC-Mega DiaCell from Almax EasyLab) for operating in the Mbar range. Photographs show the diamond anvils and the hydrostatic cavity containing $KCoF_3$ (perovskite structure) at different pressures. (Courtesy of J.A. Barreda-Argüeso, Ph.D. thesis, Universidad de Cantabria, Santander, Spain)

8.2.4 Applications: Bandgap Absorption in Semiconductors and Spin Crossover Transition in Transition Metal Systems

The study of the band structures of semiconductors as a function of the interatomic distances is a widely investigated topic in high-pressure spectroscopy. The variation of the band gap associated with direct and indirect transitions constitutes a fundamental body of knowledge needed to understand the semiconductor band structure as well as reference systems for developing *ab initio* methods to calculate electronic bands from their structures (atomic positions and composition).

Figure 8.10 shows the variations of the band gap absorption threshold of ZnTe as a function of pressure. OA spectroscopy clearly shows that ZnTe is a direct gap semiconductor at ambient conditions as the increase of the absorption coefficient $\alpha(E)$ near the band gap follows the typical square-root dependence on $E = \hbar\omega$:

$$\alpha(\hbar\omega) = C\left|M_{VB\rightarrow CB}\right|^2 g(\omega) \propto (\hbar\omega - E_g)^{1/2}, \tag{8.9}$$

where C is a proportionality constant, $M_{VB\rightarrow CB}$ is the matrix element connecting the valence band and conduction band levels, and $g(\omega)$ is the joint

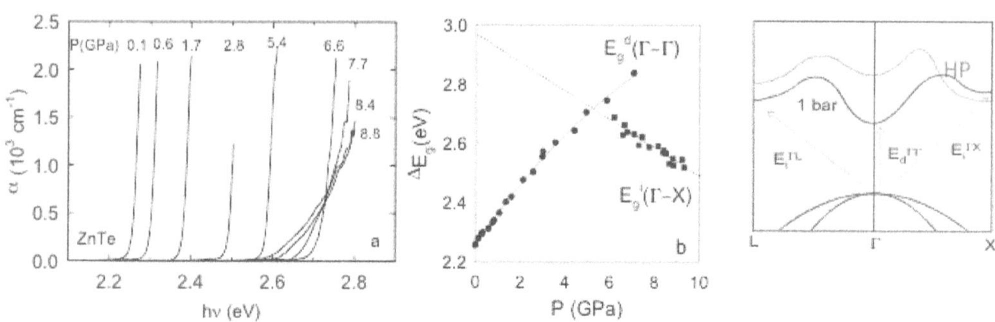

Figure 8.10 Pressure dependence of absorption threshold of the ZnTe semiconductor. The corresponding variation of the direct (2.24 eV) and indirect (2.95 eV) gap energy is shown. Energy dispersion curves $E(k)$ for the valence and conduction bands for II-V semicondutors and their evolution with pressure (HP) are presented [42]. (Courtesy of A. Segura, University of Valencia, Valencia, Spain)

hole-electron density of states, defined as:

$$\begin{aligned} g(\omega) &= 0 \quad \text{for} \quad \hbar\omega < E_g, \\ g(\omega) &= \frac{1}{2\pi^2}\left(\frac{2\mu}{\hbar^2}\right)^{3/2}(\hbar\omega - E_g)^{1/2} \quad \text{for} \quad \hbar\omega > E_g, \end{aligned} \tag{8.10}$$

where E_g is the gap energy and μ the effective electron-hole reduced mass [42].

This behaviour of the absorption coefficient contrasts with the indirect band gap transitions involving electron and hole states of different crystal momenta. Momentum conservation is preserved through electron-phonon coupling: $\hbar k'_{CB} = \hbar k_{VB} + \hbar k_{photon} \pm \hbar k_{phonon}$. Energy conservation of indirect transitions implies: $E'_{CB} = E_{VB} + \hbar\omega \pm \hbar\omega_{phonon}$. The plus-minus sign refers to the creation or annihilation of one phonon in the absorption process, respectively. Therefore the absorption coefficient for indirect transitions has a spectral dependence as:

$$\alpha^{ind}(\hbar\omega) \propto \left(\hbar\omega - E_g^i \pm \hbar\omega_{phonon}\right)^2, \tag{8.11}$$

where E_g^i is the indirect gap energy, and $\hbar\omega_{phonon}$ is the energy of the coupled phonon. Given that the probabilities of creation and annihilation of phonons are modulated by the Bose–Einstein statistics, $f(E) = \left(e^{\frac{E}{K_BT}} - 1\right)^{-1}$, the band gap absorption for indirect transitions depends on temperature; at low temperature only annihilation are possible.

The optical absorption threshold associated with the band gap in ZnTe indicates a direct gap behaviour with pressure in the 0 to 6.6 GPa range, the 2.25 eV direct $\Gamma - \Gamma$ gap shifting to higher energy with pressure at a rate of $\frac{\partial E_g^{\Gamma-\Gamma}}{\partial P} = 95$ meV/GPa. This behaviour abruptly changes from a direct $\Gamma - \Gamma$ gap to a pressure-induced redshifted $\Gamma - \mathrm{X}$ indirect gap at 6.6 GPa $\left(E_g^{ind}(\Gamma - \mathrm{X}) = 2.75\mathrm{eV}\right)$; the indirect gaps shifting at a rate: $\frac{\partial E_g^{ind}(\Gamma-\mathrm{X})}{\partial P} = -48$ meV/GPa. This behaviour of the band gap is a common characteristic of III-V and II-VI semiconductors. The direct gap at Γ increases sharply with pressure (about 100 meV/GPa), the indirect $\Gamma - \mathrm{L}$ also increases but less rapidly (about 50 meV/GPa), while the indirect $\Gamma - \mathrm{X}$ gap slightly decreases with pressure, as illustrated in Figure 8.10.

Another example of interest is the electronic structure of transition-metal systems (d^n electronic configuration) and its dependence on pressure. *d*-orbital electrons interact strongly with the surrounding ions or ligands (strong electron-ion coupling interaction), giving rise to electronic levels whose energy can be easily modified by the application of external pressure. Pressure-induced shifts can generate profound changes associated with the ground state or excited states yielding fascinating electronic (optical) phenomena. For example, the high spin (HS) to low spin (LS) transition in Fe^{2+} ($S = 2 \rightarrow 0$) cyanocompounds [13] or the ${}^4T_2 \leftrightarrow {}^2E$ excited-state crossover in Cr^{3+} [43] strongly affects the optical and magnetic properties in the former case, and a change in the photoluminescence from a broad band emission to a ruby-like narrow-line emission in the latter. In all these systems, the spectral range of interest includes typically UV (250-400 nm), visible (400–800 nm), and infrared (800–1700 nm) ranges.

To learn about pressure-induced spectral shifts in transition-metal ions, let us consider the application of an external pressure of 7 GPa in Mn^{3+} ($3d^4$) or Fe^{3+} ($3d^5$) on the basis of a sixfold octahedral symmetry (Figure 8.11(a)). Given that the crystal-field (CF) splitting between the O_h-split e_g and t_{2g} one-electron orbitals in fluorides is $10Dq = E(e_g) - E(t_{2g}) = 1.8 - 2.0$ eV at ambient conditions [43, 46], and the fluoride bulk modulus is $B_0 = 70$ GPa, the relative volume variation is approximately $\frac{\Delta V}{V} = -10$ %. For comparison purposes, this volume reduction can be compared with the variation of about $\frac{\Delta V}{V} = 1$ % due to thermal expansion from 0 K to 300 K or the variations obtained by uniaxial stress within the elastic limit of the material. The spectral variations or CF splitting variations induced by an external pressure of 7 GPa can be estimated on the assumption of a dependence of $10Dq$ with the metal-ligand bond distance (R) as $10Dq \propto R^{-5}$ [12, 47]:

$$\frac{\Delta 10Dq}{10Dq} = -\frac{5}{3}\frac{\Delta V}{V}. \tag{8.12}$$

Taking into account that $10Dq \simeq 2$ eV in TM fluorides and oxides (TM = Cr^{3+}, Mn^{3+}, and Fe^{3+}), the increase of $10Dq$ in the 0 to 7 GPa range would

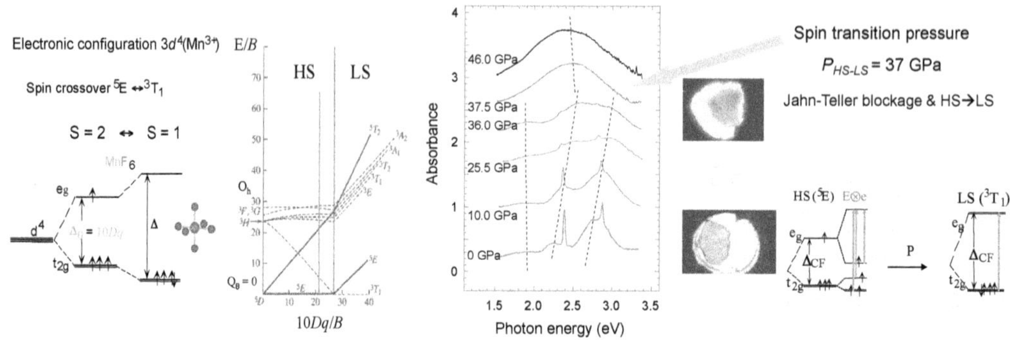

Figure 8.11 (a) Splitting of the one-electron $3d$ levels for Mn^{3+} ($3d^4$) in weak field (high-spin octahedral symmetry; $S = 2$) and strong field (low spin; $S = 1$). The Tanabe–Sugano diagram on the right indicates that the spin crossover transition $^5E \rightarrow {}^3T_1$ takes place at $10Dq/B = 27$. The vertical line indicates the electronic structure of Mn^{3+} in fluorides at ambient conditions [7, 8]. (b) OA spectrum variation of $CsMnF_4$ with pressure in the 0 to 46 GPa range. Note the effect of the spin transition at 37 GPa. Adapted, with permission, from References [44] and [45].

be approximately 0.4 eV (17 %), which obviously is a significant and readily observable spectroscopic change [12]. This estimate also illustrates the suitability of optical spectroscopy as a tool for obtaining information about the variations of structural parameters with pressure. In particular, if we look at the sensitivity of the optical spectra to changes in the bond length in a TM complex, the OA technique provides an accuracy of about 10^{-3} Å for $R = 2$ Å, if we are able to measure spectral variations of $10Dq = 5$ meV (40 cm^{-1}; $\Delta\lambda \simeq 2$ nm). Another point is the possibility of exploring the effects of structural phase transitions or spin crossover phenomena induced by pressure through the optical spectra. In the latter case, changes of magnetic moment (spin) due to electronic ground-state crossover have been profusely investigated through application of high pressure due to the relevance of the topic in physics, chemistry, biology, and geophysics (spin state of Fe^{2+} and Fe^{3+} at the Earth's lower mantle conditions) [1, 48–51].

Interestingly, the Jahn–Teller effect in $Mn^{3+}(3d^4)$ also provides a remarkable example of how this effect can affect the spin transition (interplay between spin state and $E \otimes e$ electron-ion coupling). Looking at the Tanabe–Sugano (TS) diagram for an octahedral MnF_6^{3-} complex (Figure 8.11(a)), the HS-LS transition [$^5E\,(S = 2) \rightarrow {}^3T_1\,(S = 1)$] would take place at $10Dq/B \simeq 27$. If we take $B \simeq 0.1$ eV (800 cm^{-1}) and $10Dq = 2$ eV (16000 cm^{-1}) for these complexes [12, 14, 28, 43, 46, 47], the relative volume variation required to achieve this transition ($10Dq_{HS \rightarrow LS} = 2.7$ eV) would be $\frac{\Delta V}{V} \simeq 17$ %, which would be attained by applying an external pressure of about $P_{HS \rightarrow LS} = 15$ GPa.

Therefore, Mn^{3+} would be an ideal candidate to observe this phenomenon in coordination compounds and its effects on material properties. However, Mn^{3+} in sixfold coordination, like the ferromagnetic $CsMnF_4$, is subjected to a strong Jahn–Teller effect as evidenced by the large tetragonal distortion exhibited by the MnF_6^{3-} octahedra in its antiferrodistortive layered perovskite structure [45]. Correlation studies by high-pressure techniques [12, 47] indicate that the strong Jahn–Teller distortion of Mn^{3+} stabilizes the HS ^5E ground state, preventing spin crossover to occur at the expected crossover pressure (Figure 8.11(b)). The characteristic three-band structured absorption spectrum of $CsMnF_4$ due to the Jahn–Teller effect is stable up to 37 GPa. At this pressure the spectrum abruptly changes to a single broadband type, characteristic of a LS $^3T_1$ ground state. The band energy, $10Dq = 2.5$ eV at 37 GPa, indicates that the spin transition occurs once the Jahn–Teller distortion is suppressed by pressure [45].

8.3 EMISSION AND EXCITATION SPECTROSCOPY

8.3.1 Fundamentals: Wavelength and Time Domains

Once an ion has been promoted to an upper excited state using a high-power light source (Xe lamp) or a laser with an excitation wavelength resonant with an absorption band, there is a competition between non-radiative and radiative deactivation mechanisms giving rise to the emission spectrum. The transition probability between excited and ground states is proportional to the oscillator strength. The higher the oscillator strength, the shorter the lifetime of the state. Therefore, when the non-radiative deactivation can be neglected the reciprocal lifetime is proportional to the transition oscillator strength.

In the luminescence spectrum, the emitted light intensity is measured as a function of the wavelength for a fixed excitation wavelength, whereas in the excitation spectrum, the emission intensity is monitored at a fixed emission wavelength upon scanning the excitation wavelength. Contrary to absorption, luminescence and excitation are both selective techniques. Typical light sources in the UV-Vis-NIR range are high power Xe lamps, tunable lasers like dye or Ti:sapphire types, and more recently tunable pulsed OPO laser or super-continuum light sources.

Time domain emission spectroscopy consists of the detection of the emitted intensity of light from a sample as a function of the time after excitation with a short pulse of light. It is possible to establish delay time after the pulse to separate emissions from different species with different lifetimes but in the same spectral region. The time-resolved spectrum can be obtained using chopped continuos wave excitation sources or pulsed light sources like Xe flash lamps coupled to a monochromator or tunable OPO laser with excitation pulse

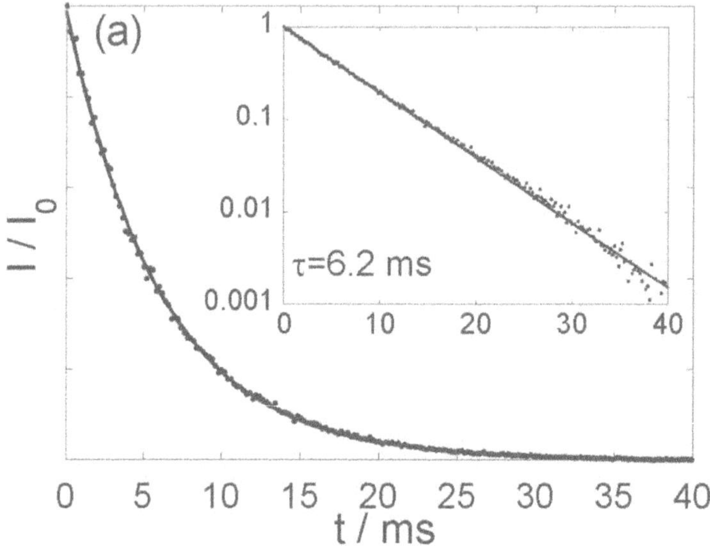

Figure 8.12 Temporal evolution of normalized Mn^{2+} $^4T_1 \rightarrow {}^6A_1$ emission intensity after pulsed excitation at Mn^{2+} excited state in $LaMgAl_{11}O_{19}$:Mn^{2+}, Yb^{3+} [52, 53]. Adapted, with permission, from Reference [53].

width in the 10 ns to 100 fs range with different repetition rates in the range 410 to 2400 nm. This range can be extended by appropriate optics up to UV. The light detection depends on the time resolution and required sensitivity. The most useful equipment includes intensified CCD (iCCD), streak cameras, micro-channel plate detectors, and electronic components with multichannel scaler, photon counter or time-correlated single-photon counting. The selection of the electronics and detection setup depends on the range of the fluorescence lifetime. Time resolution depends on the response time of the detection system (microchannel plate photomultiplier, photon avalanche detector, etc.). The best temporal resolution is obtained for a streak camera (*ca.* 1 ps). Micro-channel plate PMTs allow detection of decay times down to 10 ps. When the decay time is shorter than the system response time, signal de-convolution is necessary.

In order to obtain information about the radiative and non-radiative probabilities of the transition from an excited state, the temporal evolution of the luminescence intensity is recorded as a function of time. Performing the temporal evolution can be achieved in most spectroscopic laboratorie by collecting the dependence of the emission intensity as a function of time $I(t)$ at a fixed wavelength. In the simplest case, when the excitation and emission take place in the same center, the temporal evolution presents an exponential decay

$$I(t) = I_0 e^{-t/\tau}, \tag{8.13}$$

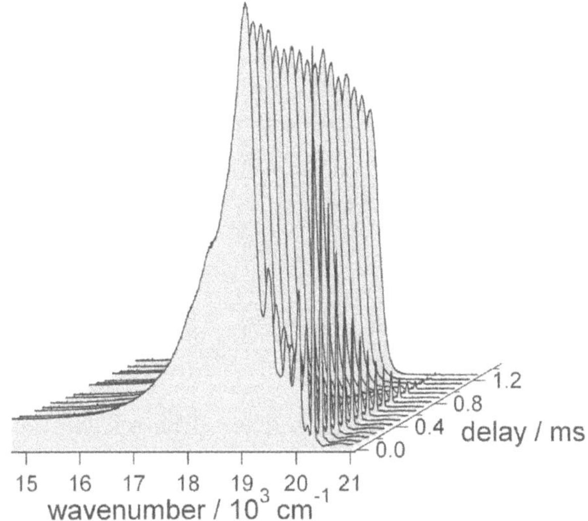

Figure 8.13 15 K upconversion time-resolved emission spectra of $LaMgAl_{11}O_{19}$:Mn^{2+}, Yb^{3+} after excitation at increasing delay intervals of 70 μs. While the Yb^{3+} pair emission has a fast decay, the Mn^{2+} luminescence presents a longer lifetime of about 6.2 ms [52, 53]. Adapted, with permission, from Reference [53].

where τ is the lifetime of the excited state (Figure 8.12).

Collecting the whole emission spectra at different delay times after the excitation pulse during different gate times (integration times) is called time-resolved luminescence (Figure 8.13). This technique is extremely useful to study energy transfer processes and to separate the contributions from different optically active centers activated within the sample if they have distinct luminescence dynamics.

8.3.2 Experimental Setups

Figure 8.14 shows the experimental scheme of the time-resolved emission and excitation spectroscopy experiment. The excitation source is a tunable (407 to 710 nm in the signal mode and 710 to 2400 nm in the idler mode) pulsed optical parametric oscillator (OPO) laser system pumped by the third harmonic of a Q-switched Nd:YAG. The pulse of 10 ns width and variable repetition rate are guided through an optic fiber to the sample and DAC. The emitted light is collected with a microscope objective and focused on a fiber bundle coupled with F/# matcher to a monochromator equipped with an intensified CCD (iCCD) camera. For fast emission lifetimes, the tunable OPO laser can be substituted by a supercontinuum laser source coupled to a monochromator

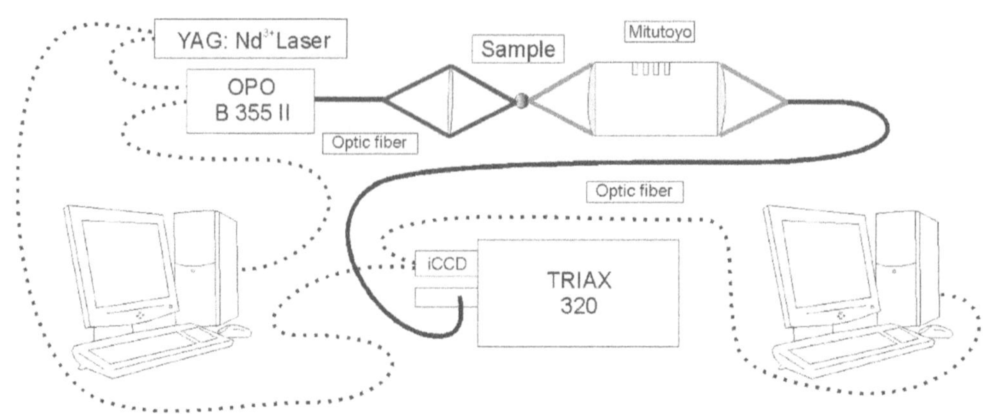

Figure 8.14 Experimental arrangement used for time-resolved emission and excitation spectroscopy experiments at high pressure. The DAC is placed in the sample position.

that allows us to tune the excitation from 370 to 2400 nm with femtosecond pulse width and repetition rate from 10 kHz to MHz.

For lifetime measurements, the 10 ns pulses from the OPO laser or modulated laser diode excitation can be used. An example of the experimental setup is shown in Figure 8.15. The emitted light is dispersed by a monochromator and the dispersed light detected with a photomultiplier and recorded with a multichannel scaler or oscilloscope. To increase the signal an appropriate load resistor can change the time constant of the RC circuit, but this time should be much shorter than the emission lifetime. The setup of Figure 8.14 can also be used to perform lifetime and time-resolved luminescence experiments. In high-pressure experiments the 90° configuration setup of Figure 8.15 should be modified to 180° or back scattering configuration due to the accessible windows of the DACs.

8.3.2.1 Correction of Spectra

Emission spectra must be corrected to allow different responses of the detection system to wavelength of radiation. In an excitation configuration it is also necessary to take into account the spectral distribution of the excitation source. This correction is important for obtaining suitable information on the peak positions and on the line or band shapes (profile analysis): $g(\omega)$. The absence of spectral correction can lead to errors in the peak position of several nanometers (of the order of the bandwidth). For emission spectra, the usual procedure consists of recording the spectrum of a black body at approximately 2500 °C. Similarly to a black body spectrum, a tungsten lamp provides adequate emission spectrum at the filament temperature for

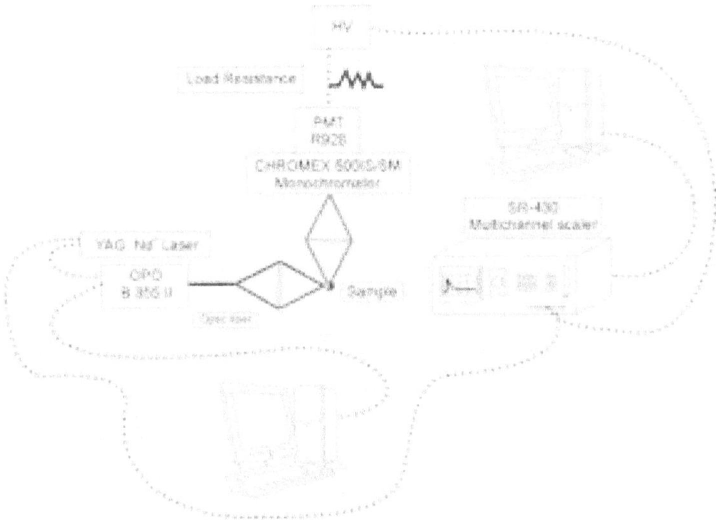

Figure 8.15 Example of an experimental setup for lifetime measurements at high pressures.

spectral corrections. After obtaining the black-body emission spectrum $I_B(\omega)$, the correction is carried out by normalizing the emission intensity to the number of photons emitted per unit of time and per unit of energy according to the energy $(E = (\hbar\omega))$ of the photons emitted by a black body at a given temperature:

$$J_E = \frac{2\pi\left(\hbar\omega\right)^2}{c^2 h^3}\frac{1}{e^{\frac{\hbar\omega}{k_B T}} - 1}. \tag{8.14}$$

The corrected spectrum is then given by the equation $I_{EM}(\omega) = J_E(\omega) \cdot S(\omega)/I_B(\omega)$ where the signal $S(\omega)$ corresponds to the uncorrected (raw) spectrum [1-4, 53].

The correction of the excitation spectrum is more intricate since it is necessary to know the spectral distribution of the excitation.

8.4 PRESSURE CALIBRATION: LUMINESCENCE PROBES

8.4.1 Ruby Gauge

In addition to the sample under study, the hydrostatic cell cavity has to accommodate a pressure sensor. As already introduced in Chapter 6, ruby is the most widely used sensor in high-pressure experiments. The pressure measurement is performed through the displacement of the $R1$ and $R2$ emission lines. This method is simple because ruby is luminescent at a wide range of temperatures up to 900 K [54] and maintains a dependence of the wavelength

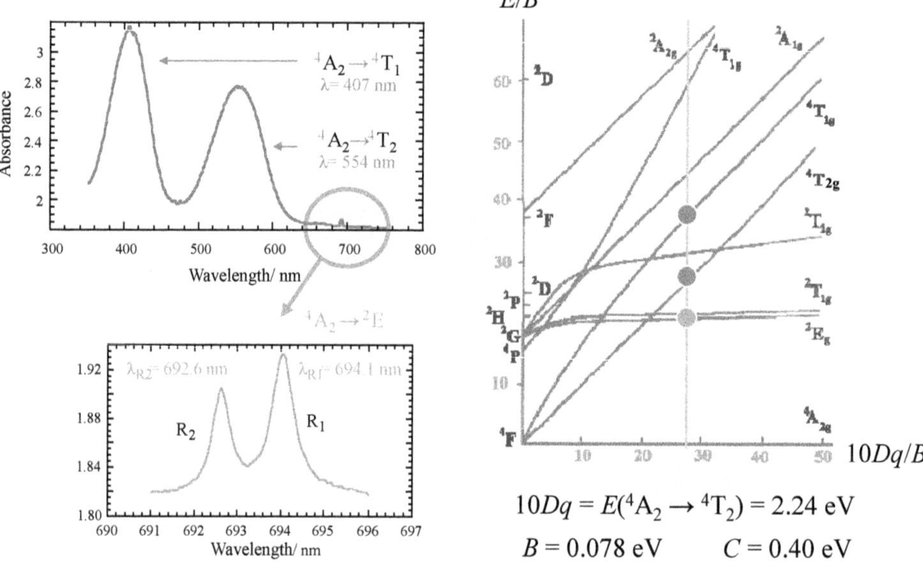

Figure 8.16 Optical absorption spectrum of ruby (Al_2O_3: Cr^{3+}) at ambient conditions showing a magnification of the $R1$ and $R2$ ($^4A_2 \rightarrow {}^2E$) absorption lines. The electronic structure of Cr^{3+} ($3d^3$) is shown through the corresponding Tanabe–Sugano diagram. The experimental transition energies from the OA spectrum are indicated by circles in the diagram at the corresponding point $10Dq/B = 28$. The spectroscopic parameters B and C and the crystal-field splitting $10Dq$ are included. (Courtesy of R. Martín-Rodríguez, graduate thesis, University of Cantabria, Santander, Spain, 2005)

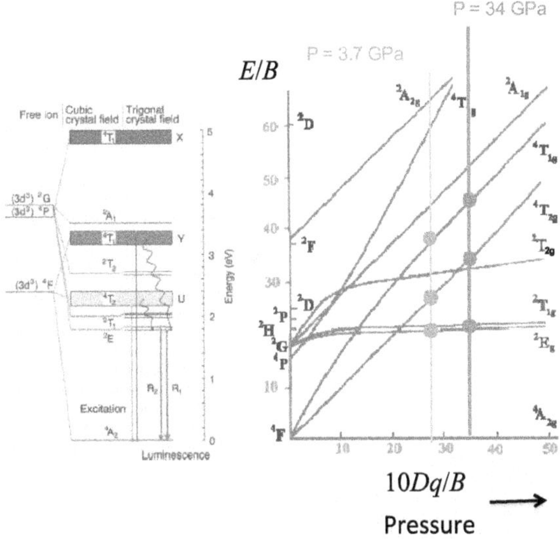

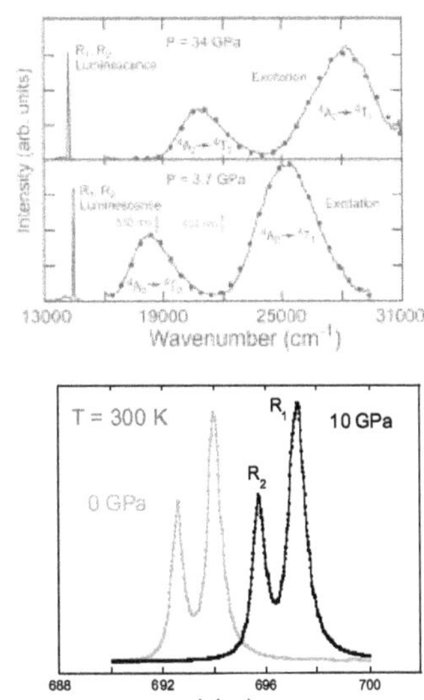

Figure 8.17 (Left) Energy level diagram of ruby (Al_2O_3: Cr^{3+}) at ambient pressure and Tanabe–Sugano diagram showing the evolution of the Cr^{3+} energy levels with pressure. (Right) Luminescence spectrum of ruby showing the $R1$ and $R2$ lines associated with ${}^4A_2 \rightarrow {}^2E$ transition under pulsed excitation at 550 nm at 0 and 10 GPa. The excitation and emission spectra of ruby at $P = 3.7$ and 34 GPa are also shown. Their corresponding transition energies are shown in the Tanabe–Sugano diagram. Adapted, with permission, from Reference [56].

of emission with the pressure that is practically linear in the range of 0 to 20 GPa [55]. Above this pressure, the scale of ruby ceases to be linear and extends to several Mbar. A review of the equations that relate the positions of the lines $R1$ and $R2$ and pressure is given elsewhere [55]. Ambient pressure lines are located at 692.8 nm ($R2$) and 694.3 nm ($R1$), and correspond to electronic transitions from the excited state doublet ${}^2E$ to the ground state ${}^4A_2$ of Cr^{3+} in Al_2O_3. Figure 8.16 shows the absorption spectrum of ruby in which two spin-allowed intense bands at 407 nm (${}^4A_2 \rightarrow {}^4T_1$) and 554 nm (${}^4A_2 \rightarrow {}^4T_2$) and two spin-flip-type weak lines at higher wavelengths (${}^4A_2 \rightarrow {}^2E$), give rise to the so-called emission lines $R1$ and $R2$. Figure 8.17 shows the energy levels of ruby with its emission spectrum [54].

Lines $R1$ and $R2$ move toward lower energy (higher wavelengths) with pressure as a result of the decrease of the Racah B parameter due to an increase of covalency of the Cr-O bond. Spectral positions depend on pressure and are calibrated in a wide pressure range. $R1$ and $R2$ lines at RT vary linearly with pressure in the 0–20 GPa range

$$\lambda_{R1} = 694.3 + 0.364P\,(\text{nm, GPa}),$$
$$\lambda_{R2} = 692.8 + 0.364P\,(\text{nm, GPa}), \tag{8.15}$$

where P is given in GPa and λ in nanometers. The pressure as a function of the luminescence shift $\Delta\lambda$ is given by $P(\text{GPa}) = 2.75\times\Delta\lambda$ (nm). As illustrated in Figure 8.17, the broad absorption bands $^4T_1$ and $^4T_2$ shift to shorter wavelengths (higher energies) with pressure following the trend of the Tanabe–Sugano diagram at a higher rate than the $R1$ and $R2$ lines. In fact, the transition energy of $^4T_1$ and $^4T_2$ essentially depends on $10Dq$ which increases with pressure following a volume dependence as $10Dq \propto V^{-5/3}$ [5, 12]. According to results of Figure 8.17, the $^4T_1$ and $^4T_2$ bands shift with pressure at a rate of −1.7 nm/GPa ($^4T_2$) and −1.4 nm/GPa ($^4T_1$) [56], which is about five times greater than the $R1$ and $R2$ shift rate (0.364 nm/GPa).

Above 20 GPa, the relationship between λ and P is no longer linear and other analytical forms of $P(\lambda)$ have been proposed to be valid up to 150 GPa [18, 57]:

$$P\,(\text{GPa}) = 1884\left(\frac{\lambda}{\lambda_0} - 1\right)\left[1 + 5.5\left(\frac{\lambda}{\lambda_0} - 1\right)\right], \tag{8.16}$$

with $\lambda_0 = 694.3$ nm ($R1$ line).

It must be pointed out that the inhomogeneous broadening of the ruby lines evidences the hydrostaticity loss of the pressure-transmitting medium at high pressure. Although the PL line broadening in ruby (or other PL probes) is due to different factors such as stress distribution, sample size, and orientation, the lack of hydrostaticity in the pressure chamber can be examined through the PL peak width. The more the peak broadening is, the more severe the non-hydrostatic conditions. An adequate characterization of the pressure distribution within the cavity for different transmitting media using a homogeneous distribution of 10 μm diameter ruby spheres is given elsewhere [58].

8.4.2 Pressure and Temperature Probes

Accurate measurements of the pressure and temperature in a hydrostatic chamber of a DAC requires calibrated standards. Thanks to the transparency of the anvils to visible light, it is common to know the working P and T through an *in-situ*, indirect measurement of the $P-T$ sensitive luminescence of optically active species (lanthanide ions, transition-metal ions, dye molecules or fluorescent proteins). The most common optical method employed for P

determination involves the pressure-induced shift of luminescence lines of previously calibrated optically active materials, although luminescence lifetime calibration has also been reported. Less standardized is the method to measure the exact T of the sample in the hydrostatic chamber [7]. Several methods of thermometry may be employed to determine *in situ* the T of the sample, such as transport [59], magnetic and optical measurements. In laser heating experiments in DACs, temperatures are determined by fitting the visible portion (600 to 800 nm) of the thermal radiation to the Planck radiation function [60]. Another optical method for T determination is based on the existence of two emitting levels of an optically active ion close enough in energy to be considered in quasi-thermal equilibrium. The fluorescence intensity ratio (FIR) of these two closely spaced energy levels ($E_1 - E_2 = E_G$) can be calibrated as a function of T and is proportional to their transition energies, emission cross-sections, and population distributions, which follow a Boltzmann-type distribution, i.e., $FIR \sim \exp\left({}^{E_G}/_{K_B T}\right)$, and can be easily measured with the same setup used for P calibration [61]. All these methods have limitations and are restricted to determined $P - T$ ranges.

The ideal optical pressure sensor for high-pressure techniques needs to meet some general requirements described by Barnett et al. [62]: (i) a single emitting line; (ii) no significant broadening or weakening and little or no surrounding background; (iii) a large shift with pressure ${}^{d\lambda}/_{dP}$; (iv) a small temperature-dependent line shift ${}^{d\lambda}/_{dT}$; (v) for high sensitivity and precision, its linewidth Γ should be small compared to the line shift; and, (vi) the host lattice should be highly stable at high P and T.

As shown in the previous section, the wavelength shift of the $R1$-$R2$ doublet lines ($^2E \rightarrow {}^4A_2$), emission from ruby (Al_2O_3:Cr^{3+}) is the most widely used standard optical P calibrant. Its high quantum efficiency and high P coefficient and linear behavior in an ample range of P make it an ideal probe for certain ranges of $P - T$. At low T's and in hydrostatic conditions, it can be used simultaneously as a $P - T$ sensor. However, the $R1$-$R2$ lines are very sensitive to non-hydrostatic stress and to T, in the sense that in both cases the lines broaden, and due to the small splitting (around 30 cm^{-1}), the overlapping of the two lines results in the formation of a broad asymmetrical band. As a consequence, the accuracy of the P-T measurement is then significantly reduced for high T. This fact, together with the increased photoluminescence quenching with T, limits the P measurement to T below 700 K [18]. In addition, the attainable P range is dependent on the hydrostaticity of the selected pressure-transmitting medium, which is difficult to maintain beyond 100 GPa. Due to this, other luminescent P sensors such as Cr^{3+} in alexandrite ($BeAl_2O_4$) [63, 64] have also been suggested, although they are less frequently used than ruby. In this sense, materials doped with lanthanides offer

Table 8.1 Characteristics of selected fluorescence pressure sensors at RT.

Sensor	Transition	$d\lambda/dP$ (nm GPa^{-1})	Linear pressure range (GPa)	Reference
Al_2O_3:Cr^{3+}	$^2E \rightarrow {}^4A_2$	0.36	20	[55, 62]
YAG:Sm^{3+}	Y_1	0.33	0-65	[76]
YAG:Sm^{3+}	Y_2	0.32	0-73	[76]
SrB_4O_7:Sm^{3+}	$^5D_0 \rightarrow {}^7F_0$	0.35	0-20	[69–72]
YAG:Nd^{3+}	$R2\left({}^4F_{3/2}\right) \rightarrow Z_5\left({}^4I_{9/2}\right)$	0.87	11	[77]
GSGG::Nd^{3+}	$R2\left({}^4F_{3/2}\right) \rightarrow Z_5\left({}^4I_{9/2}\right)$	0.99	12	[78]
GSGG::Nd^{3+}	$R2\left({}^4F_{3/2}\right) \rightarrow Z_5\left({}^4I_{9/2}\right)$	0.97	22	[79]
$YALO_3$:Nd^{3+}	$R2\left({}^4F_{3/2}\right) \rightarrow Z_2\left({}^4I_{9/2}\right)$	-0.13	< 20	[65, 80]

Note: The pressure shifts are approximate and depend on the bibliographic source.

better performances as fluorescence P sensors since they produce very sharp lines that show less sensitivity to T, although with less intensity compared to the Cr^{3+} R-lines of ruby. Therefore, the line shifts of f-f transitions offer an interesting possibility to establish high P-T gauges. Some interesting pressure sensors have been reported in the literature. $YAlO_3$:Nd^{3+} [62, 65], MFCl:Sm^{2+} (M=Ba, Sr) [66, 67], and $Y_3Al_5O_{12}$:Tm^{3+} (YAG) [68] are examples.

As commented above, an important point that must be taken into account is related to good mechanical and thermal stabilities. In this context, an important system used for high P sensor purposes is the SrB_4O_7:Sm^{2+} crystal, in which the P-induced shift of the $^5D_0 \rightarrow {}^7F_0$ intense fluorescence singlet line at 685.41 nm has been calibrated [69–74]. This line presents the same advantages described for ruby along with very small T dependence (-10^{-4} nm/K), slow increase of the bandwidth with P-T, and low sensitivity to non-hydrostatic stresses [69, 75]. The system has been calibrated at RT in hydrostatic conditions up to 48 GPa and at non-hydrostatic conditions up to 127 GPa [70, 71]. However, it has two major drawbacks concerning its behavior at high T: a decrease of the luminescence intensity above 500 K, which limit its use up to 900 K, and possible chemical reactions with some pressure-transmitting media.

For more than 50 years, garnets have been extensively used as host matrices for solid state lasers in which mechanical and thermal stabilities are crucial. In particular, lanthanides-doped YAG has played an important role in the laser industry. It has been observed that luminescence wavelengths of the most intense electronic transitions in this system are not influenced by T up to 1100 K therefore this system should be used for luminescent P sensors at high T. This has motivated the study of the YAG:Eu^{3+} as a candidate for P sensor up to 7 GPa at high T (1000 K) [81], testing the $^5D_0 \rightarrow {}^7F_1$ transition line. Other attempts with the same matrix have been made, particularly with

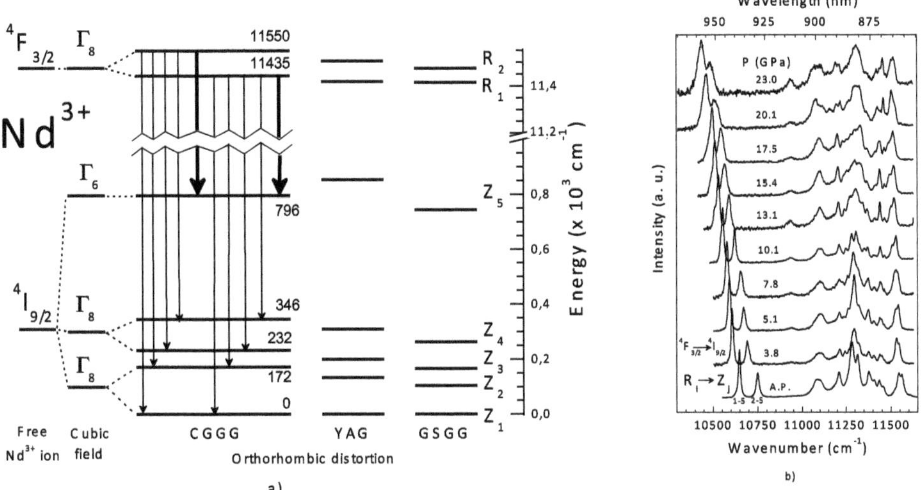

Figure 8.18 (Left) Partial energy level diagram of the Nd^{3+} ion in $Ca_3Ga_2Ge_3O_{12}$ (CGGG), $Gd_3Sc_2Ga_3O_{12}$ (GSGG), and $Y_3Al_5O_{12}$ (YAG) garnets at ambient pressure corresponding to the ${}^4F_{3/2}$ and ${}^4I_{9/2}$ Nd^{3+} multiplets and the breakdown of their degeneracy into three and five Stark levels in cubic and orthorhombic-distorted environments, respectively. (Right) Room temperature emission spectra of the ${}^4F_{3/2} \rightarrow {}^4I_{9/2}$ transition as a function of pressure.

YAG:Sm^{3+} [76, 77, 82–91]. It is known that garnet crystals belong to the cubic space group Ia-$3d$ and the lanthanide ions substitute mainly for the dodecahedral sites, with an orthorhombic D_2 local symmetry site. Since the luminescence properties of the lanthanide ions are ruled by the symmetry of their local sites in the host (see Section 8.5.2), the D_2 orthorhombic crystal-field interaction completely removes the degeneracy of the ${}^{2S+1}L_J$ multiplets. In the case of Sm^{3+}-doped YAG, the most intense emission lines correspond to the ${}^4G_{5/2} \rightarrow {}^6H_{7/2}$ transition, resulting in a triplet for the ${}^4G_{5/2}$ state and a quartet for the ${}^6H_{7/2}$ one. The transitions between the low-lying ${}^4G_{5/2}$ sublevel and the four ${}^6H_{7/2}$ sublevels are denoted as Y_1, Y_2, Y_3 and Y_4 in order of increasing energy. Y_1 is the most intense. Linear P dependences were found for Y_2 line up to 100 GPa and for Y_1, Y_3 and Y_4 up to 60 GPa, with slopes comparable to ruby and with T effects negligible up to 820 K (see Table 8.1) [91]. Other studies involving different Nd^{3+}-doped garnet crystals as possible high P sensors have been considered the near infrared ${}^4F_{3/2} \rightarrow {}^4I_{9/2}$ transition of the Nd^{3+} ions [92]. In this case, the D_2 orthorhombic crystal-field interaction split the ${}^4F_{3/2}$ multiplets in the $R1$ and $R2$ Stark levels and the ${}^4I_{9/2}$ ground state split in Z_N ($N = 1 - 5$) Stark levels [93, 94] (Figure 8.18).

The P-induced shifts of the Stark levels of the ${}^4F_{3/2} \rightarrow {}^4I_{9/2}$ transition of Nd^{3+} in gadolinium scandium gallium garnet (GSGG), gadolinium gallium

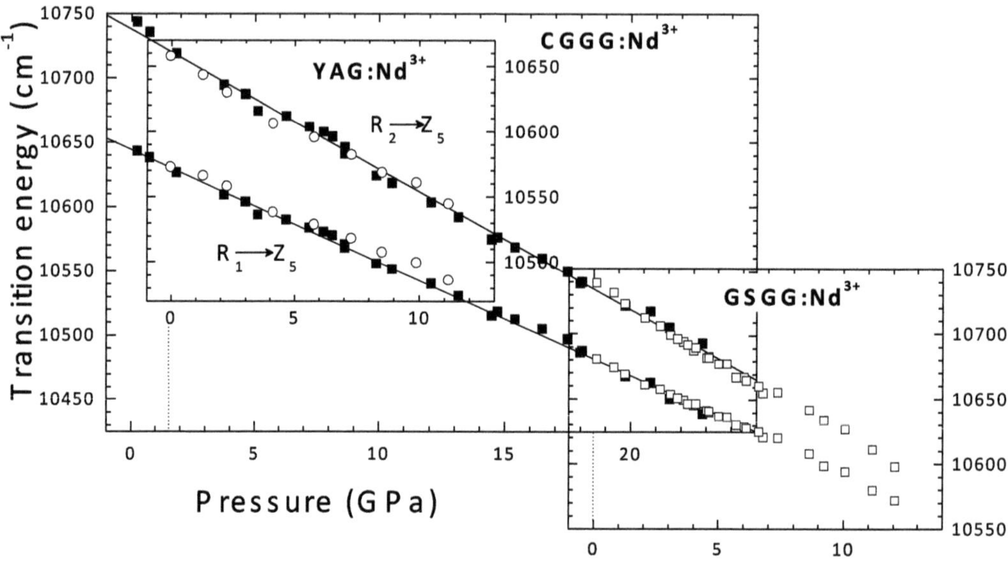

Figure 8.19 Pressure evolution of the $R1$,$R2$ $\rightarrow$ Z_5 peaks in Nd^{3+}-doped CGGG (■), GSGG (□), and YAG (○). All graphs use the same scale for the XY axes, but those for YAG and GSGG garnets are shifted to fit the same splitting as that of the CGGG garnet under study. Linear fits of the peak transition energies for the CGGG garnet are also included.

garnet (GGG), and YAG were obtained at very high P (< 80 GPa), focusing on the P evolution of the $R1, R2 \rightarrow Z_1$ peaks (Figure 8.19) [92]. More recently, the $R1, R2 \rightarrow Z_5$ emission lines in YAG [77], GSGG [78] and calcium gallium germanium garnet (CGGG) [79] have been analyzed, taking advantage of the high P sensitivity of the Z_5 line (see Section 8.5.3). Linear P coefficients were obtained for these lines, ranging from -7.5 to -11.3 cm^{-1}/GPa in YAG, GSGG and CGGG, up to 11 GPa, 12 GPa and 23 GPa, respectively (see Figure 8.19). Additionally, the GSGG:Nd^{3+} crystal has been tested as a possible T sensor considering the low T dependence of the fluorescence intensity ratio of the $R1, R2 \rightarrow Z_1$ transition [77]. Table 8.1 summarizes some characteristics of the pressure sensors described above.

8.5 APPLICATIONS

8.5.1 Excited–State Crossover in $LiCaAlF_6$:Cr^{3+}.

Cr^{3+}-doped $LiCaAlF_6$ presents a crystal-field (CF) strength that is very close to the $^2E-^4T_2$ excited-state crossover (ESCO) in the Tanabe–Sugano diagram for d^3 ions (Figure 8.20). These excited states exhibit different PL behaviours; the spin-allowed $^4T_2 \rightarrow {}^4A_2$ transition is strongly coupled to the lattice

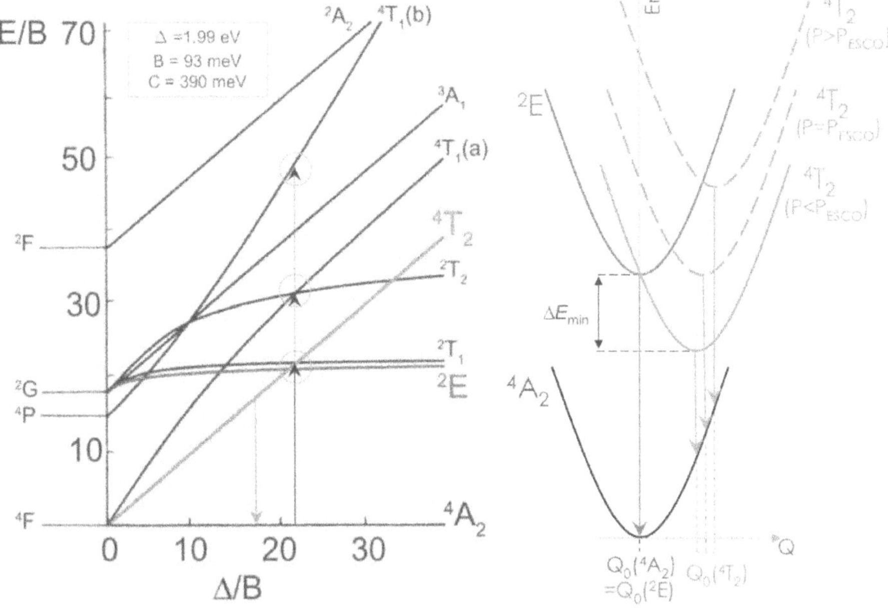

Figure 8.20 (Left) Tanabe–Sugano diagram of Cr^{3+} in O_h symmetry. (Right) Single-coordinate configurational diagram for ground and low energy-excited states with the ESCO requirements. Adapted, with permission, from Reference [43].

providing short-lived broadband PL, while the spin-forbidden $^2E \rightarrow {}^4A_2$ transition (spin-flip) is weakly coupled to the lattice, giving rise to long-lived narrow-line emissions. At ambient conditions, the PL-excited state of Cr^{3+} in fluorides lies below the ESCO, leading to a $^4T_2 \rightarrow {}^4A_2$ broadband emission. Pressure increases CF towards the ESCO, providing ideal conditions for drastic PL transformations at pressures well above the ESCO (Figure 8.20).

RT high pressure is required to place the emitting state $^4T_2$ broadband PL far above the ^2E emitting state narrow-line PL to obtain the most populated ^2E. The change from a broadband emission to a ruby-like emission is accompanied by a dramatic change of the corresponding lifetime (Figure 8.21) [43].

8.5.2 Crystal–Field and Energy Transfer in Lanthanides

Lanthanides are a technologicaly important series of optically active ions that have been extensively studied for applications in devices and materials such as lasers, optical amplifiers in fibers, displays, optical sensors, and non-linear optical materials. The optical properties of lanthanides in solids are directly related to their partially filled $4f$ shells and the inter-electronic interactions with the charges of the host ligands, all distributed in a particular local point

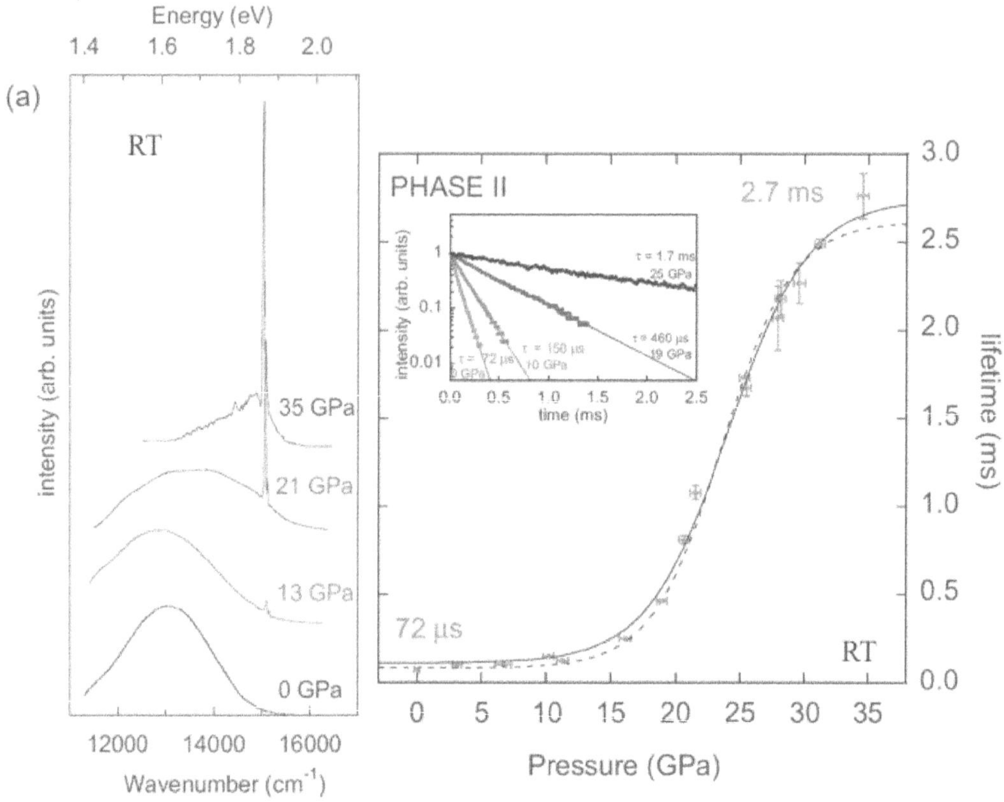

Figure 8.21 (a) RT time-resolved emission spectra of $LiCaAlF_6:Cr^{3+}$ in the 0 to 35 GPa range. (b) RT luminescence lifetime of $LiCaAlF_6:Cr^{3+}$ in the 0 to 35 GPa. The lines represent the fitting on the basis of two emitting excited states: $\tau\left({}^4T_2\right) = 72$ µs and $\tau\left({}^2E\right) = 2.7$ ms (inset) with O_h symmetry (solid line) and C_4 symmetry (dashed line). Adapted, with permission, from Reference [43].

symmetry, thus providing a detailed fingerprint of the surrounding coordinated ligands. Although their rich electronic structure is only weakly perturbed by the environment, their radiative transition probabilities are very sensitive to the host matrices. In the study of this interaction, the different ligands, coordination numbers, local distortions, and limited number of isostructural crystals restrict the empirical information that can be obtained. High-pressure techniques offer the possibility to vary the interatomic distances in a continuous way and study the effects of the lanthanide-ligand interactions in an energy level diagram and the quantum efficiencies of the emitting levels, following a systematic strategy that reinforces the predictive capabilities of the optical material engineering.

The lanthanide series, associated to the filling of the $4f$ inner shell covers the cerium ($Z = 58$) to lutetium ($Z = 71$) elements. The neutral lanthanides share the electronic configuration of xenon ($Z = 54$) plus two or three electrons in the $5d$ and $6s$ shells. Specifically, the f electrons begin to appear in the ground configurations of the optically active trivalent lanthanide ions [95].

As already mentioned, the optical properties and the energy level diagram of a trivalent lanthanide ion in a solid are ruled by the inter-electronic interactions between the electrons of its incompletely filled inner $4f$ shell and the charge of the host ligands, all distributed in a particular local point symmetry. The influence of the host lattice on the optical transitions within the $4f^n$ ($n = 1-14$) configuration is weak since, on one hand, the $4f$ orbital lies inside the ion and is shielded from the surroundings by the filled $5s^2$ and $5p^6$ orbitals and, on the other hand, the filling of the $4f$ orbital through the lanthanide series poorly screens the increasing nuclear charge, which strongly attracts the outer electrons and gives rise to a decrease in the dimensions of the lanthanide electron shells and in the radial integrals over the series (lanthanide contraction).

The complete Hamiltonian (free-lanthanide ion and CF interactions) that describes the lanthanide ions in solids uses a parametric method, in which a small group of phenomenological parameters reproduce the energy level diagram [96–100]

$$
\begin{aligned}
H = H_0 \; & + \sum_{k=2,4,6} F^k f_k + \zeta_{4f} A_{so} + \alpha L\left(L+1\right) + \beta G\left(G_2\right) + \gamma G\left(R_7\right) + \\
& + \sum_{i=2,3,4,6,7,8} T^i t_i + \sum_{h=0,2,4} M^h m_k \sum_{f=2,4,6} P^f p_f \\
& + \sum_{\substack{-k \le q \le k \\ k=2,4,6}} B_q^k \sum_i C_q^{(k)}\left(i\right),
\end{aligned}
\tag{8.17}
$$

where the main ones are the inter-electronic repulsion $F^k f_k$, the spin-orbit $\zeta_{4f} S_{so}$ interactions, and the CF $B_q^k \cdot C_q^{(k)}$ interaction, whose magnitude is small compared to the spin-orbit interaction (weak interactions field limit) and is intimately related to the symmetry of the local CF acting on the optically active lanthanide ions. The $C_q^{(k)}$ represent angular one-electron tensor operators that can be evaluated, whereas the radial CF coefficients B_q^k are fitting parameters and the values of k and q are determined by the local point symmetry of the lanthanide site. In addition, an average, rotational invariant CF strength parameter S has been proposed for comparison among different materials [101]

$$
S = \left[\frac{1}{3}\sum_k \frac{1}{2k+1}\left(\left|B_0^k\right|^2 + s\sum_{q\le k,\, q>0}\left|B_q^k\right|^2\right)\right]^{1/2}. \tag{8.18}
$$

Apart from the slight red shift of the ${}^{2S+1}L_J$ multiplets, the main modifications introduced by the CF interaction are the appearance of the Stark levels, arising from the removal of the degeneracy of the ${}^{2S+1}L_J$ multiplets [20–24] and the electric-dipole transitions between multiplets in the optical (UV-Vis-NIR) energy range, forbidden in the free ion [102, 103].

When a lanthanide ion is embedded in a material (crystalline, glass, or liquid), the free-ion ${}^{2S+1}L_J$ energy levels characterized by the total angular momentum J are affected in two ways: the energy levels are, again shifted to lower energies and the $2J+1$-fold degeneracy is partially or totally lifted, giving rise to CF, or Stark levels, labeled with the irreducible representations of the particular point symmetry of the lanthanide site (double groups notation). The absorption and emission of photons give rise to optical spectra that consist of narrow peaks if the ions occupy the same site in an ordered structure or broad bands if there is a distribution of local sites with slightly different bond distances and angles typical of amorphous structures like glasses. Those peaks or bands are associated to electronic transitions between states of the $4f^n$ $(n = 1 - 14)$ ground configurations of the lanthanide series. The intensities of these bands show oscillator strengths of around 10^{-7} to 10^{-5}, keeping in mind that the optical absorption transitions are strongly forbidden by the parity selection rule, and have been successfully characterized by the Judd–Ofelt theory [102–105].

Once an absorption process has taken place, there are two ways of returning to the ground state: radiative and non-radiative processes. The radiative or luminescence process involves a de-excitation of the lanthanide ions from the excited state back to a low-lying excited state or to the ground state by the emission of photons of equal or less energy than those absorbed by the lanthanide ions. Moreover, the energy of the emission must be, as in the absorption case, resonant with the energy of a transition between two energy levels.

In principle, all the excited states of the lanthanide ions can emit photons. However, luminescences from many of these energy levels are only observed at low temperature or in host lattices with low-phonon energies, since the lanthanide ion de-excitation can be completed through radiationless processes, i.e., the energy absorbed by the optically active ions is not emitted as radiation of photons but is dissipated as heat in the solid. Two different processes may account for this radiationless de-excitation. First, an interaction between the lanthanide ions and the vibrating lattice allows the optically active ion to lose its energy and return to the ground state, i.e., the multiphonon relaxation. Second, if the distances between optically active ions or between lanthanides and luminescence traps in a solid are short enough and the emission and absorption bands overlap in energy, a very efficient transfer of the energy from the excited lanthanide ions called donors to other non-excited lanthanide ions or luminescence traps called acceptors takes place. While the former interaction depends only on the solid chosen as host, the latter can be controlled with the concentrations and types of optically active ions used in the initial composition of the solid.

Basically, the energy transfer probability depends on the spectral overlap of the donor's emission and the acceptor's absorption and on the interatomic interaction and distance between the donor and the acceptor ions, according to Fermi's Golden Rule [106–110]:

$$W_{DA} = \frac{2\pi}{\hbar^2} \left|\langle DA^* \left| H_{DA} \right| D^* A\rangle\right|^2 \int g_D(v)\, g_A(v)\, dv. \tag{8.19}$$

The theoretical treatment of the energy transfer processes at a microscopic level starts from obtaining the probability of transfer between a donor-acceptor pair in order to subsequently apply a statistical treatment that allows the connection of theory with the experimental results, such as the temporal evolution of the luminescence of the donor. In this way, the probability of donor-acceptor transfer for multipolar coulombic interactions can be expressed by [109]

$$W_{DA} = \frac{C_{DA}^{(6)}}{R^6} + \frac{C_{DA}^{(8)}}{R^8} + \frac{C_{DA}^{(10)}}{R^{10}} + \cdots, \tag{8.20}$$

where $C_{DA}^{(S)}$ is the energy transfer parameter and $S = 6, 8$ or 10, depending on whether the dominant mechanism of the interaction is dipole-dipole, dipole-quadrupole or quadrupole-quadrupole, respectively. Assuming a statistical distribution of optically active ions, energy transfer is negligible if the concentration of lanthanide ions in the solid is less than 0.1 mol%, while it becomes increasingly important when the concentration of acceptor ions increases to a few mol%.

Applying high pressure to a material with optically active lanthanide ions provides an opportunity to systematically vary all these processes, through the continuous change of its ligand-to-lanthanide bond distances and angles when the volume decreases, and may lead to a new understanding of its optical properties [111, 112].

8.5.3 Upconversion Processes

The photon upconversion (UC) is based on the generation of high-energy photons, usually in the visible or UV, after the absorption of two or more lower energy photons in the IR or NIR region [113]. The phenomenon was discovered and theoretically explained by F. Auzel [114] and Ovsyankin and Feofilov [115] in 1966. Unlike second-harmonic generation, UC luminescence

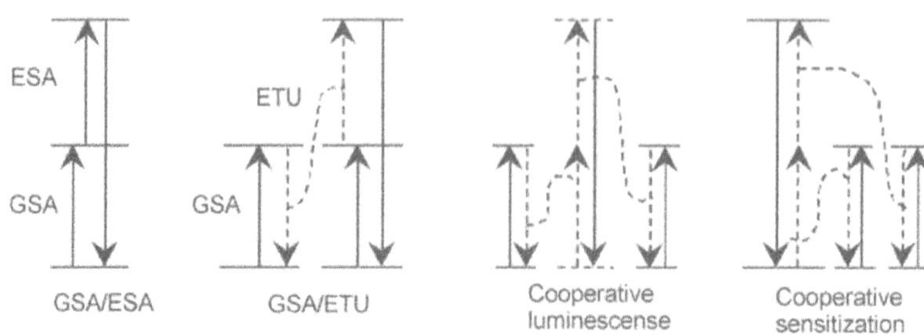

Figure 8.22 Most relevant UC luminescence mechanisms. GSA/ESA, GSA/ETU, cooperative luminescence, and cooperative sensitization.

does not require coherent radiation. Although most of the work has focused on lanthanide ion-doped crystals and glasses, transition-metal ion and *TM-RE* mixed systems are suitable to present UC luminescence. The most relevant UC mechanisms are shown in Figure 8.22. For simplicity, this figure shows the energy level schemes where the energy differences between the intermediate and upper excited states are equal. The excitation energy corresponds exactly to this energy difference. At least two metastable excited states are required to exhibit upconversion. f-f excited states in lanthanides are often metastable due to the reduction of the electron-phonon coupling by the well shielded f electrons. Multiple metastable excited states are much less common in *TM* ions where multiphonon relaxation processes dominate. As a rule of thumb for lanthanide ions, spectroscopists use the so-called *gap law* that establishes that the multiphonon relaxation rate constant (k_{nr}) between two states $|1>$ and $|2>$ with energies E_1 and E_2 decreases exponentially with the energy gap between these two states:

$$k_{nr} \propto e^{-\beta p} = \frac{1}{\tau_{nr}}, \tag{8.21}$$

where $p = \frac{\Delta E}{\hbar\omega_{max}}$ is the reduced energy gap and represents the number of maximum-frequency phonons of energy $\hbar\omega_{max}$ required to bridge the electronic energy gap $\Delta E = E_2 - E_1$ between the states and $\beta = \ln\left(\frac{p}{S}\right) - 1$. S is the Huang–Rhys parameter and represents the intensity of the electron-phonon coupling. For f-f transition $S \ll 1$ (weak coupling), and for f-f transition in lanthanides, multiphonon relaxation is usually competitive up to ($p \leq 6$). For energy smaller than $5\hbar\omega_{max}$ the nonradiative relaxation dominates over the radiative relaxation. This empirical evidence involves the maximum phonon energy of the host lattice.

Few examples of UC luminescence involving transition metal ions at high pressure can be found in the literature [116–118]. At low temperature,

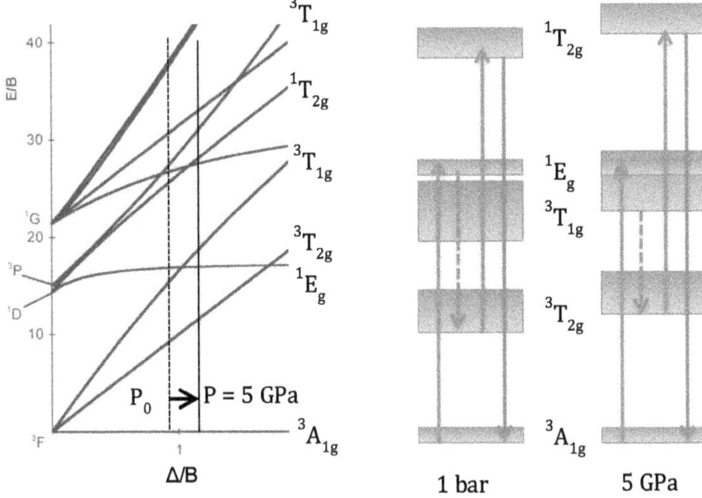

Figure 8.23 (Left) Tanabe–Sugano diagram for Ni^{2+} (d^8 electronic configuration). The vertical lines represent the ligand-field strength for the host lattice at ambient pressure and 5 GPa. (Right) Energy level scheme illustrating pressure effects on the UC process. The solid arrows represent the GSA, ESA and luminescence, and the dashed arrows indicate the non-radiative relaxation. More details can be found in Reference [116].

excitation of Ni^{2+} around 810 nm can lead to visible UC luminescence at around 610 nm. Excitation into the ^1E_g state leads to non-radiative relaxation to a reservoir excited state $^2T_{2g}$ and a second photon of the same energy can be absorbed, leading to UC luminescence from the $^1T_{2g}$ upper excited state (Figure 8.23), corresponding to the ground-state absorption/excited-state absorption (GSA/ESA) upconversion mechanism. Upon pressure increases, the ligand field strength increases and the emission band moves to higher energies; at 5 GPa the band has moved from 610 nm to 585 nm. Thus, the UC emission undergoes a color change from red to yellow in the 1 bar to 5 GPa pressure range. The pressure-dependent UC excitation spectra revealed a slight red shift of the relevant UC $^3T_{2g} \rightarrow {}^1T_{2g}$ excited state (ESA) transition, which was ascribed to a reduction of the Racah B and C parameters caused by an increase of covalency at high pressures.

Figure 8.22 (second drawing) shows the ground-state absorption/energy-transfer upconversion (GSA/ETU) process. Two interacting ions are excited via GSA to an intermediate level, then a non-radiative energy transfer process takes place to generate one ion in its ground state and the other one in an upper excited state. Then the emission takes place from this state with an energy that is almost twice the excitation energy level. The efficiency of this

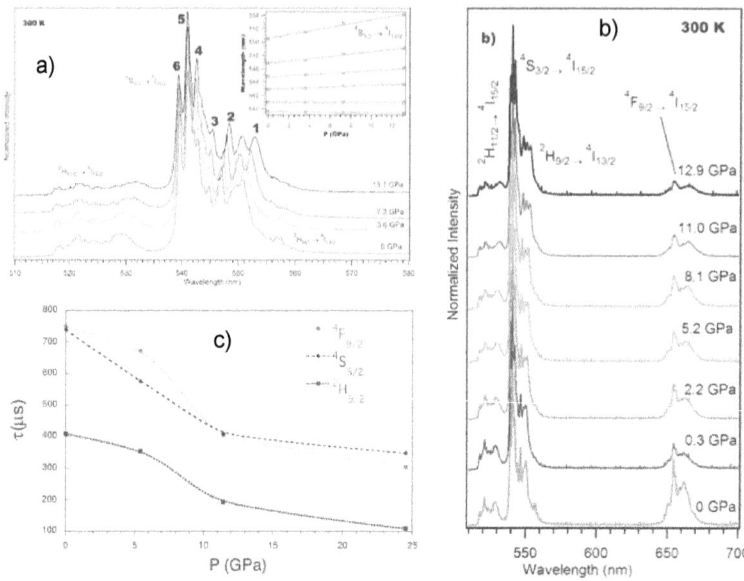

Figure 8.24 Pressure dependence of (a) the crystal-field splitting of the $^2H_{11/2} \rightarrow {}^4I_{15/2}$ and $^4S_{3/2} \rightarrow {}^4I_{15/2}$, (b) the intensity ratio between $^2H_{11/2}+ {}^4S_{3/2} \rightarrow {}^4I_{15/2}$ and $^4F_{9/2} \rightarrow {}^4I_{15/2}$ and (c) the lifetime of $^2H_{11/2}$, $^4S_{3/3}$ and $^4F_{9/2}$ multiplets, corresponding to the assigned Er^{3+} of the UC emission in $NaYF_4{:}Er^{3+}$, Yb^{3+} [119].

mechanism depends on the distance R of the two interacting ions as well as on the spectral overlap between the emission of the donor and the absorption of the acceptor modifying the experimental lifetime.

These two factors can be modified by hydrostatic pressure techniques. A paradigmatic example of this mechanism is the NIR-to-VIS UC emission in Er^{3+}, Yb^{3+} co-doped materials. In $NaYF_4{:}Er^{3+}$, Yb^{3+} the high-pressure experiments demonstrate (Figure 8.24) that the emission lifetime, the intensity ratio between different emission multiplets, and the CF splitting of the Er^{3+} are changed by hydrostatic pressure [119].

An outstanding pressure effect has been observed in $NaCl{:}Ti^{2+}$ [117]. At low temperature and ambient pressure, it shows two broad emission bands corresponding to the transition from the $^3T_{2g}$ and $^3T_{1g}$ excited states to the $^3T_{1g}$ ground state. With two metastable excited states, this system fulfills the basic requirements for UC processes. Therefore excitation at about 9400 cm^{-1} into $^3T_{2g}$ leads to emission at 5500 cm^{-1} and 12850 cm^{-1} from $^3T_{2g}$ and $^3T_{1g}$. The temporal evolution of the UC luminescence intensity corresponds to two consecutive absorption sites, i.e., GSA and ESA. Upon increasing pressure the $^3T_{2g}$ and $^3T_{1g}$ states shift to higher energies with a similar rate as expected from the same slope of these states in the Tanabe–Sugano diagram (Figure

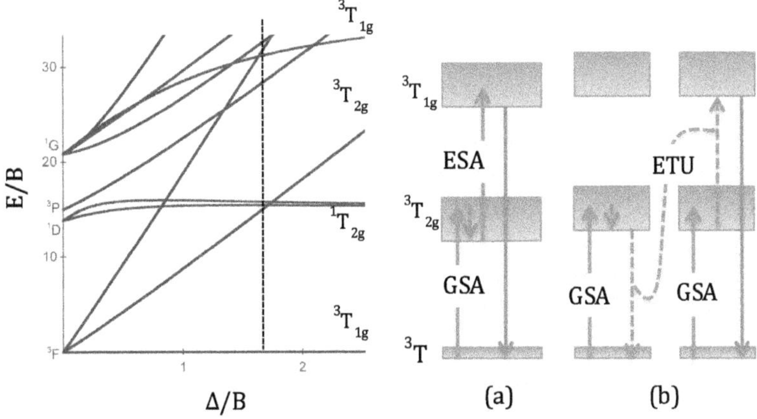

Figure 8.25 (Left) Tanabe–Sugano energy level diagram for Ti^{2+} (d^2 electronic configuration). The vertical dashed lines represent the ligand-field strength in NaCl at ambient pressure. (Right) Energy level schemes for NaCl:Ti^{2+} at (a) ambient pressure and (b) 3.5 GPa. Solid arrows represent the GSA, ESA and luminescence processes. The short dashed arrow corresponds to the multiphonon relaxation and the long dashed arrows to non-radiative ETU process. More details can be found in [116, 117].

8.25). At high pressure, the temporal evolution of the UC intensity after short pulses changes from a pure decay to a rise preceding the decay, which is a fingerprint for an UC mechanism involving non-radiative energy transfer. This corresponds to an ETU process. Therefore, pressure induces a change from GSA/ESA to a GSA/ETU mechanism [117], as shown schematically in Figure 8.25.

Figure 8.22 (third pannel) depicts the cooperative luminescence mechanism. Typically, two identical closely interacting ions combine their IR or NIR excitation energy creating one visible photon of exactly twice the excitation energy. The emitting level is an excited level of the ion pair. Therefore, contrary to the general thinking, the emission takes place from a real metastable state. In fact the lifetime of the emitting state is exactly half of the intermediate level [120]. This is the mechanism responsible for the blue-greenish emission of Yb^{3+} pairs in Yb^{3+}-doped materials [121]. According to multipolar interaction or exchange interaction, this mechanism is pressure dependent as a result of the shortening of the ion-ion distance upon increasing pressure.

The cooperative sensitization (Figure 8.22(fourth pannel)) can be understood as a simultaneous energy transfer between two identical ions excited in the IR, NIR or VIS to a third ion, which has no energy level resonant at this energy. The UC emission occurs from the upper excited state at almost twice the excitation energy. As a three-ion process, the interaction and the distance between ions are extremely important and suitable for modification in high-pressure experiments. Examples of this type of mechanism are the UCs in Yb^{3+}-Tb^{3+} and Yb^{3+}-Eu^{3+} [122–124].

8.6 CONCLUSIONS

High-pressure optical spectroscopy is a useful technique to investigate electronic structures of Transition-metal and Lanthanide-related compounds as well as semiconductors, and to establish correlations between material structures and optical processes through volume variation (compression) while maintaining the chemical compositions. This is fundamental to explain temperature variations of electronic (vibrational) levels as it enables us to separate the purely thermal shifts (at constant volume) from those due to thermal expansion. Furthermore, optical spectroscopy is a useful complement for structural studies. In some pressure experiments, it can provide local structure information, which at present is barely achieved by x-ray absorption (very diluted systems) or diffraction experiments (dynamical or disordered effects). Finally, pressure-induced phase transformations and energy level tuning enable us to find new photoluminescence phenomena not expected at ambient pressure conditions [125].

Bibliography

[1] W. Fowler. *Physics of color centers.* Academic Press, New York (1968).

[2] B. Henderson and G. Imbusch. *Optical spectroscopy of inorganic solids.* Monographs on the physics and chemistry of materials. Clarendon Press, Oxford (1989).

[3] A. Lever. *J. Chem. Educ.*, 45, 711 (1968).

[4] G. Blasse, K. Bleijenberg, and R. Powell. *Luminescence and energy transfer.* Luminescence and energy transfer series. Springer–Verlag, New York (1980).

[5] H. G. Drickamer and C. W. Frank. *Annu. Rev. Phys. Chem.*, 23, 39 (1972).

[6] J. Ferraro. *Vibrational spectroscopy at high external pressures: The diamond anvil cell.* Elsevier, New York (2012).

[7] W. Holzapfel and N. Isaacs. *High pressure techniques in chemistry and physics: A practical approach.* Practical approach in chemistry series. Oxford University Press, Oxford (1997).

[8] E. Tonkov. *High pressure phase transformations: A handbook.* Taylor & Francis, Boca Raton (1992).

[9] P. Bridgman. *The physics of high pressure.* G. Bell, London (1949).

[10] W. Sherman and A. Stadtmuller. *Experimental techniques in high-pressure research.* John Wiley & Sons Incorporated, New York (1987).

[11] I. Eremets. *High pressure experimental methods.* Oxford University Press, Oxford (1996).

[12] F. Rodríguez. In A. Katrusiak and P. McMillan, eds., *High-pressure crystallography*, Volume 140 of NATO Science Series, 341–352. Springer, Heidelberg (2004).

[13] P. Gütlich and H. A. Goodwin. In *Spin crossover in transition metal compounds I*, 1–47. Springer, Berlin (2004).

[14] F. Rodríguez. In E. Boldyreva and P. Dera, eds., *High-pressure crystallography*, NATO Science for Peace and Security Series B, 215–229. Springer, Heidelberg (2010).

[15] C. Clark, P. Dean, and P. Harris. *Proc. R. Soc. London, Ser. A*, 277, 312 (1964).

[16] A. Neves, M. Nazaré, I. E. Group, *et al. Properties, Growth and Applications of Diamond.* Institution of Electrical Engineers, New York (2001).

[17] A. Collins and A. Williams. *J. Phys. C Solid State Phys.*, 4, 1789 (1971).

[18] K. Syassen. *High Pressure Res.*, 28, 75 (2008).

[19] Z. Liu, J. Xu, H. P. Scott, *et al. Rev. Sci. Instrum.*, 75, 5026 (2004).

[20] R. Graham and W. Brooks. *J. Phys. Chem. Solids*, 32, 2311 (1971).

[21] S. Darwish and M. S. Seehra. *Phys. Rev. B*, 37, 3422 (1988).

[22] B. Di Bartolo. *Optical interactions in solids.* World Scientific, London (2010).

[23] M. M. de Lucas, F. Rodríguez, and M. Moreno. *J. Phys. Condens. Matter*, 7, 7535 (1995).

[24] F. Rodríguez, M. Moreno, J. Dance, *et al. Solid State Commun.*, 69, 67 (1989).

[25] F. Rodríguez, G. Davies, and E. Lightowlers. *Phys. Rev. B*, 62, 6180 (2000).

[26] G. Davies. *Phys. Rep.*, 176, 83 (1989).

[27] M. Innocenzi, R. Swimm, M. Bass, *et al. J. Appl. Phys.*, 67, 7542 (1990).

[28] F. Rodríguez, P. Núñez, and M. M. de Lucas. *J. Solid State Chem.*, 110, 370 (1994).

[29] B. A. Moral and F. Rodríguez. *Rev. Sci. Instrum.*, 66, 5178 (1995).

[30] A. Jayaraman. *Rev. Mod. Phys.*, 55, 65 (1983).

[31] J. Barnett, S. Block, and G. Piermarini. *Rev. Sci. Instrum.*, 44, 1 (1973).

[32] J. R. Ferraro and L. J. Basile. *Appl. Spectrosc.*, 28, 505 (1974).

[33] G. Piermarini and S. Block. *Rev. Sci. Instrum.*, 46, 973 (1975).

[34] B. Welber, M. Cardona, C. Kim, *et al.* *Phys. Rev. B*, 12, 5729 (1975).

[35] B. Welber. *Rev. Sci. Instrum.*, 47, 183 (1976).

[36] K. Syassen and R. Sonnenschein. *Rev. Sci. Instrum.*, 53, 644 (1982).

[37] B. Weinstein, R. Zallen, M. Slade, *et al.* *Phys. Rev. B*, 25, 781 (1982).

[38] A. Jaffe, Y. Lin, W. L. Mao, *et al.* *J. Am. Chem. Soc.*, 137, 1673 (2015).

[39] M. N. Sanz-Ortiz. *Estudio espectroscópico de cruzamiento de niveles en óxidos y fluoruros de Cr^{3+}, Ni^{3+} y Co^{3+} a alta presión: Influencia en las transiciones de espín y efecto Jahn-Teller.* Ph.D. thesis, Universidad de Cantabria, Santander, Spain (2010).

[40] H. Mao, R. Hemley, Y. Wu, *et al.* *Phys. Rev. Lett.*, 60, 2649 (1988).

[41] S. Klotz, J. Chervin, P. Munsch, *et al.* *J. Phys. D: Appl. Phys.*, 42, 075413 (2009).

[42] A. Segura, J. Sans, D. Errandonea, *et al.* *Appl. Phys. Lett.*, 88, 011910 (2006).

[43] M. N. Sanz-Ortiz, F. Rodríguez, I. Hernández, *et al.* *Phys. Rev. B*, 81, 045114 (2010).

[44] F. Aguado, F. Rodríguez, and P. Núñez. *Phys. Rev. B*, 76, 094417 (2007).

[45] R. Martín-Rodríguez. *Síntesis, caracterización estructural y estudio espectroscópico de materiales nanocristalinos y microcristalinos.* Ph.D. thesis, Universidad de Cantabria, Santander, Spain (2011).

[46] M. N. Sanz-Ortiz and F. Rodríguez. *J. Chem. Phys.*, 131, 124512 (2009).

[47] F. Aguado, F. Rodríguez, and P. Núñez. *Phys. Rev. B*, 67, 205101 (2003).

[48] M. Pasternak, G. K. Rozenberg, G. Y. Machavariani, *et al.* *Phys. Rev. Lett.*, 82, 4663 (1999).

[49] J. Badro, G. Fiquet, V. V. Struzhkin, *et al.* *Phys. Rev. Lett.*, 89, 205504 (2002).

[50] J.-F. Lin, V. V. Struzhkin, S. D. Jacobsen, *et al.* *Nature*, 436, 377 (2005).

[51] A. F. Goncharov, V. V. Struzhkin, and S. D. Jacobsen. *Science*, 312, 1205 (2006).

[52] R. Martín-Rodríguez, R. Valiente, and M. Bettinelli. *Appl. Phys. Lett.*, 95, 091913 (2009).

[53] R. Martín-Rodríguez, R. Valiente, F. Rodríguez, *et al.* *Phys. Rev. B*, 82, 075117 (2010).

[54] H. Seat and J. Sharp. *IEEE Trans. Instrum. Meas.*, 53, 140 (2004).

[55] G. Piermarini and S. Block. *Rev. Sci. Instrum.*, 46, 973 (1975).

[56] S. J. Duclos, Y. K. Vohra, and A. L. Ruoff. *Phys. Rev. B*, 41, 5372 (1990).

[57] A. D. Chijioke, W. Nellis, A. Soldatov, *et al.* *J. Appl. Phys.*, 98, 114905 (2005).

[58] J. Chervin, B. Canny, and M. Mancinelli. *Intl. J. of High Pressure Res.*, 21, 305 (2001).

[59] A. M. J. Schaeffer and S. Deemyad. *Rev. Sci. Instrum.*, 84, 095108 (2013).

[60] A. J. Campbell. *Rev. Sci. Instrum.*, 79, 015108 (2008).

[61] S. F. León-Luis, U. R. Rodríguez-Mendoza, E. Lalla, *et al.* *Sens. Actuators B*, 158, 208 (2011).

[62] J. Barnett, S. Block, and G. Piermarini. *Rev. Sci. Instrum.*, 44, 1 (1973).

[63] A. Jahren, M. Kruger, and R. Jeanloz. *J. Appl. Phys.*, 71, 1579 (1992).

[64] T. Kottke and F. Williams. *Phys. Rev. B*, 28, 1923 (1983).

[65] H. Hua and Y. K. Vohra. *Appl. Phys. Lett.*, 71, 2602 (1997).

[66] B. Lorenz, Y. R. Shen, and W. B. Holzapfel. *High Pressure Res.*, 12, 91 (1994).

[67] Y. R. Shen and W. B. Holzapfel. *Phys. Rev. B*, 51, 15752 (1995).

[68] P. Wamsley and K. Bray. *J. Lumin.*, 60-61, 188 (1994).

[69] F. Datchi, A. Dewaele, P. Loubeyre, *et al.* *High Pressure Res.*, 27, 447 (2007).

[70] A. Lacam and C. Chateau. *J. Appl. Phys.*, 66, 366 (1989).

[71] Q. Jing, Q. Wu, L. Liu, *et al.* *J. Appl. Phys.*, 113, (2013).

[72] J. M. Leger, C. Chateau, and A. Lacam. *J. Appl. Phys.*, 68, 2351 (1990).

[73] V. V. Urosevic, Z. M. Jaksic, L. D. Zekovic, *et al.* *High Pressure Res.*, 9, 251 (1992).

[74] S. V. Rashchenko, A. Y. Likhacheva, and T. B. Bekker. *High Pressure Res.*, 33, 720 (2013).

[75] F. Datchi, R. LeToullec, and P. Loubeyre. *J. Appl. Phys.*, 81, 3333 (1997).

[76] J. Liu and Y. K. Vohra. *Appl. Phys. Lett.*, 64, 3386 (1994).

[77] S. Kobyakov, A. Kaminska, A. Suchocki, *et al.* *Appl. Phys. Lett.*, 88, (2006).

[78] S. F. León-Luis, J. E. Muñoz-Santiuste, V. Lavín, *et al.* *Opt. Express*, 20, 10393 (2012).

[79] U. R. Rodríguez-Mendoza, S. F. León-Luis, J. E. Muñoz Santiuste, *et al.* *J. Appl. Phys.*, 113, (2013).

[80] S. Garcia-Revilla, F. Rodríguez, R. Valiente, *et al.* *J. Phys. Condens. Matter*, 14, 447 (2002).

[81] H. Arashi and M. Ishigame. *Jpn. J. Appl. Phys.*, 21, 1647 (1982).

[82] N. J. Hess and G. J. Exarhos. *High Pressure Res.*, 2, 57 (1989).

[83] N. J. Hess and D. Schiferl. *J. Appl. Phys.*, 68, 1953 (1990).

[84] N. J. Hess and D. Schiferl. *J. Appl. Phys.*, 71, 2082 (1992).

[85] J. Liu and Y. K. Vohra. *Solid State Commun.*, 88, 417 (1993).

[86] Q. Bi, J. M. Brown, and Y. Sato-Sorensen. *J. Appl. Phys.*, 68, 5357 (1990).

[87] J. Liu and Y. K. Vohra. *J. Appl. Phys.*, 79, 7978 (1996).

[88] Y. Zhao, W. Barvosa-Carter, S. D. Theiss, *et al.* *J. Appl. Phys.*, 84, 4049 (1998).

[89] C. Sanchez-Valle, I. Daniel, B. Reynard, *et al.* *J. Appl. Phys.*, 92, 4349 (2002).

[90] Q. Wei, N. Dubrovinskaia, and L. Dubrovinsky. *J. Appl. Phys.*, 110, (2011).

[91] A. F. Goncharov, J. M. Zaug, J. C. Crowhurst, *et al.* *J. Appl. Phys.*, 97, (2005).

[92] H. Hua, S. Mirov, and Y. K. Vohra. *Phys. Rev. B*, 54, 6200 (1996).

[93] J. B. Gruber, M. E. Hills, C. A. Morrison, *et al.* *Phys. Rev. B*, 37, 8564 (1988).

[94] U. Hömmerich and K. L. Bray. *Phys. Rev. B*, 51, 12133 (1995).

[95] R. Pappalardo. In *Luminescence of inorganic solids*, 175–234. Springer, Heidelberg (1978).

[96] B. G. Wybourne. In *Spectroscopic properties of rare earths*. Interscience, Wiley, New York (1965).

[97] S. Hüfner. In *Optical spectra of transparent rare earth compounds*. Academic Press, New York (1978).

[98] C. A. Morrison and R. P. Leavitt. Volume 5 of Handbook on the physics and chemistry of rare earths, 461 – 692. Elsevier, Amsterdam (1982).

[99] D. Garcia and M. Faucher. Volume 21 of Handbook on the physics and chemistry of rare earths, 263 – 304. Elsevier, Amsterdam (1995).

[100] C. Görller-Walrand and K. Binnemans. Volume 23 of Handbook on the physics and chemistry of rare earths, 121–283. Elsevier, Amsterdam (1996).

[101] N. C. Chang, J. B. Gruber, R. P. Leavitt, *et al.* *J. Chem. Phys.*, 76, 3877 (1982).

[102] B. R. Judd. *Phys. Rev.*, 127, 750 (1962).

[103] G. S. Ofelt. *J. Chem. Phys.*, 37, 511 (1962).

[104] R. Peacock. In *Rare Earths*, Volume 22 of Structure and Bonding series, 83–122. Springer, Heidelberg (1975).

[105] C. Görller-Walrand and K. Binnemans. Volume 25 of Handbook on the physics and chemistry of rare earths, 101 – 264. Elsevier, Amsterdam (1998).

[106] T. Forster. *Z. Naturf. A*, 4, 321 (1949).

[107] D. L. Dexter. *J. Chem. Phys.*, 21, 836 (1953).

[108] G. Blasse and B. Grabmaier. In *Luminescent materials*, 1–232. Springer, Berlin (1994).

[109] G. Blasse and B. Grabmaier. In *Energy transfer processes in condensed matter*, Volume 114, 1–232. Springer, New York (1984).

[110] I. Hernández, F. Rodríguez, and H. D. Hochheimer. *Phys. Rev. Lett.*, 99, 027403 (2007).

[111] K. Bray. In H. Yersin, ed., *Transition metal and rare earth compounds*, Volume 213 of Topics in Current Chemistry series, 1–94. Springer, Berlin (2001).

[112] T. Tröster. Volume 33 of Handbook on the physics and chemistry of rare earths, 515 – 589. Elsevier, Amsterdam (2003).

[113] F. Auzel. *Chem. Rev.*, 104, 139 (2004).

[114] F. Auzel. *C. R. Acad. Sci.*, 262, 1016 (1966).

[115] V. Ovsyankin and P. P. Feofilov. *JEPT. Lett.*, 3, 322 (1966).

[116] O. S. Wenger, R. Valiente, and H. U. Güdel. *High Pressure Res.*, 22, 57 (2002).

[117] O. S. Wenger, G. M. Salley, R. Valiente, *et al. Phys. Rev. B*, 65, 212108 (2002).

[118] O. S. Wenger, G. M. Salley, and H. U. Güdel. *J. Phys. Chem. B*, 106, 10082 (2002).

[119] C. Renero-Lecuna, R. Martín-Rodríguez, R. Valiente, *et al. Chem. Mater.*, 23, 3442 (2011).

[120] M. P. Hehlen and H. U. Güdel. *J. Chem. Phys.*, 98, 1768 (1993).

[121] E. Nakazawa and S. Shionoya. *Phys. Rev. Lett.*, 25, 1710 (1970).

[122] G. Salley, R. Valiente, and H. Guedel. *J. Lumin.*, 94-95, 305 (2001).

[123] F. W. Ostermayer and L. G. Van Uitert. *Phys. Rev. B*, 1, 4208 (1970).

[124] R. Martín-Rodríguez, R. Valiente, S. Polizzi, *et al.* *J. Phys. Chem. C*, 113, 12195 (2009).

[125] A. M. Srivastava, C. Renero-Lecuna, D. Santamaría-Pérez, *et al.* *J. Lumin.*, 146, 27 (2014).

CHAPTER 9

Vibrational Spectroscopy

Jesús González

MALTA Consolider Team and Grupo de Altas Presiones y Espectroscopía, CITIMAC, Facultad de Ciencias Universidad de Cantabria, Santander, Spain

Francisco Javier Manjón

MALTA Consolider Team and Grupo de Investigación de Materiales en Condiciones Extremas, EXTREMAT, Instituto de Diseño para la Fabricación y Producción Automatizada, Universitat Politècnica de València, Valencia, Spain

CONTENTS

9.1 RAMAN EFFECT

9.1.1 Introduction

THE inelastic scattering of light was initially predicted [1] in 1923; however, it was not until 1928 that Raman performed the first experiment [2] and confirmed the predictions of Smekal and thus obtained the Nobel Prize in Physics in 1930. Unlike Rayleigh scattering, which corresponds to an elastic light scattering (conservation of energy) [3], Raman scattering is an inelastic process due to various elementary excitations (quasi-excitations) where energy

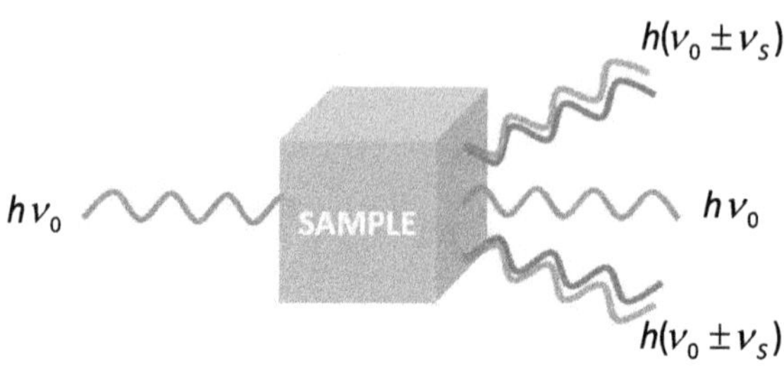

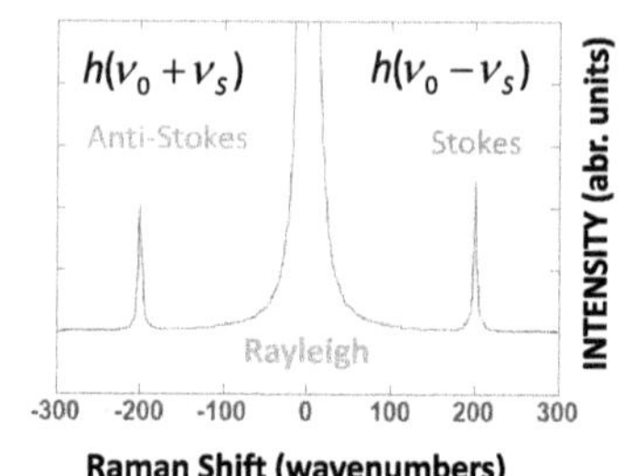

Figure 9.1 Mechanism of the Raman effect.

can be lost or gained in the scattering process. Excitations responsible for Raman scattering can be for example the internal vibrational modes in a molecule (vibrons), phonons in an ordered crystal structure (lattice vibrations) or magnons in systems with magnetic order transitions.

In conventional Raman spectroscopy, a sample is irradiated by an intense laser beam (see Figure 9.1) with initial frequency (ν_0), usually in the UV-visible region, and the scattered light consist of two types: (i) *Rayleigh scattering*, is strong and has the same frequency as the incident beam (ν_0), (ii) *Raman scattering*, is very weak (10^{-5} to 10^{-8} of the incident beam) and has the frequencies $\nu_0 \pm \nu_m$, where ν_m is the vibrational frequency of the excitation (vibrons, phonons, magnons).The $\nu_0 - \nu_m$ and $\nu_0 + \nu_m$ lines are called the *Stokes* and *anti-Stokes* lines, respectively. In Raman spectroscopy, one measures the vibrational frequency (ν_m) as a shift from the incident beam frequency (ν_0). Note that in the following we will discuss of a frequency (ν) or an angular frequency ($\omega = \nu/2\pi$) to refer to the frequency of a vibrational mode.

Figure 9.2 shows the three basic processes which occur in a molecule with a vibration of frequency (ν_m) when it is excited with a laser light source of frequency (ν_0). At room temperature, most molecules are present in the lowest energy vibrational level (m state). Under laser excitation, virtual states are created when the laser interacts with the electrons and causes polarization. The energy of these virtual states (which are not real states of the molecule) is determined by the frequency of the light source used (ν_0). The Rayleigh process is the most intense since most photons scatter this way. It does not involve any energy change and consequently the light returns to the same energy state. The Raman scattering process involves a change of energy in the scattered light due to the interaction with the molecule. Absorption of energy by the molecule leading to the promotion of the molecule from the ground vibrational state m to a higher energy excited vibrational level (n state) results in a process called Stokes scattering in which energy is transferred from the light to the molecule. However, due to thermal energy, some molecules may be present in an excited

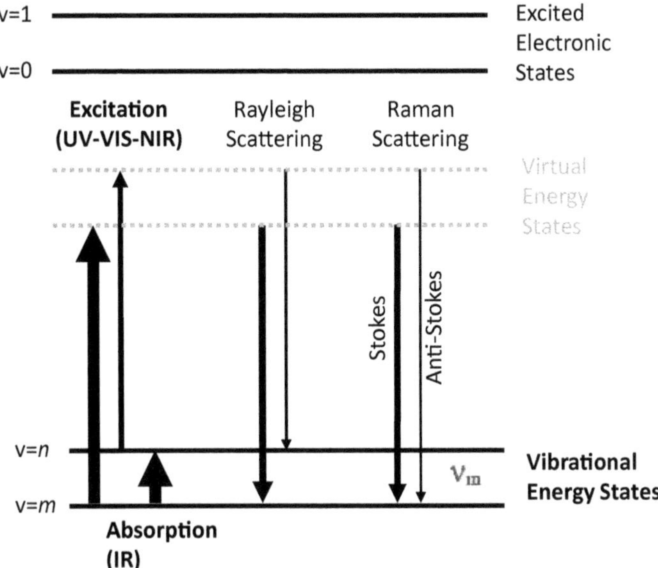

Figure 9.2 Rayleigh and Raman scattering processes. The lowest energy vibrational state m is shown at the foot with states of increasing energy above it. Both the exciting photon (upward arrows) and the scattered photon (downward arrows) in Stokes and anti-Stokes processes have much larger energies than the energy of a vibration (arrow in IR process).

state such as the excited state n in Figure 9.2. Scattering from these states to the ground state m is called anti-Stokes scattering and involves transfer of energy from the molecule to the scattered photon. The relative intensities of the two processes depend on the populations of the various states of the molecule. In a molecule in thermal equilibrium, the population of molecules at the ground state m is much larger than at the excited state n (Maxwell–Boltzmann distribution law). Thus, the Stokes lines are stronger in intensity than the anti-Stokes lines under normal conditions. Since both give the same information, it is customary to measure only the Stokes side of a spectrum.

Classically [4, 5], the Raman scattering process can be explained by the time-dependent polarizability of the molecules. When a molecule is subjected to the electric field of electromagnetic radiation $\mathbf{E} = \mathbf{E}_0 \cos \omega t$ the dipolar moment of the molecule $\mathbf{p}$ is given by:

$$\mathbf{p} = \boldsymbol{\mu} + \alpha \mathbf{E}, \tag{9.1}$$

where $\boldsymbol{\mu}$ is the permanent dipolar moment and $\alpha \mathbf{E}$ is the induced dipolar moment. The polarizability α is represented by a second-order tensor with components α_{ij} which depend on the symmetry of the molecule. The

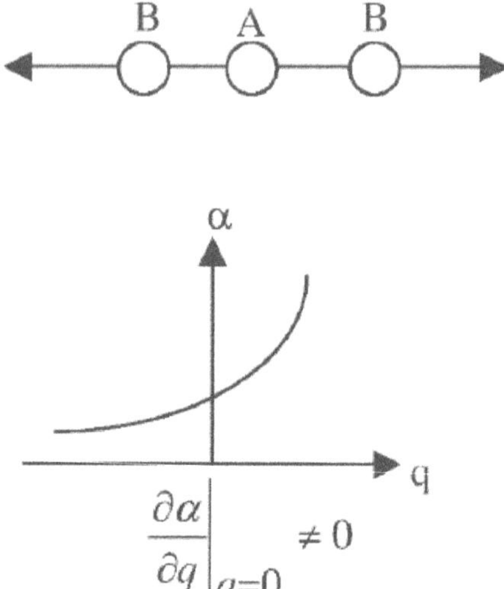

Figure 9.3 Dependence of $\partial\alpha/\partial q$ on the normal coordinate of vibration of an AB_2 type molecule.

polarizability can be expressed in terms of the normal coordinates $q_n(t)$ of a vibrating molecule which, for small vibrations can be approximated as $q_n(t) = q_{n0}\cos(\omega_n t)$, being q_{n0} the amplitude and ω_n the vibration frequency of the normal mode n. In this way, the total dipolar moment can be given by:

$$\mathbf{p} = \boldsymbol{\mu}_0 + \alpha(0)\,\mathbf{E}_0\cos(\omega t) + \sum_{n=1}^{Q}\left(\frac{\partial \boldsymbol{\mu}}{\partial q_n}\right)_0 q_{n0}\cos\omega_n t$$
$$+\frac{1}{2}\mathbf{E}_0\sum_{n=0}^{Q}\left(\frac{\partial \alpha}{\partial q_n}\right)_0 \left[\cos(\omega+\omega_n)t + \cos(\omega-\omega_n)t\right], \qquad (9.2)$$

where we have made a power series expansion of q_n, μ and α and taken into account only terms to first order of approximation. The second term in Equation (9.2) describes the Rayleigh scattering, the third term corresponds to the optical absorption in the infrared spectrum, and the fourth term to the Raman scattering. Therefore, if $|\partial\alpha/\partial q|_{q=0}$ is zero the molecular vibration is not "Raman active". Namely, to be Raman-active, the rate of change of polarizability α with the vibration must not be zero (see Figure 9.3).

From a quantum mechanical view, the inelastic scattering of an incident photon by the phonons (lattice vibrations) in a crystal is defined as an interaction given by a third order process, which includes the electron-radiation (photon) interaction represented by the Hamiltonian (H_{ER}) and the

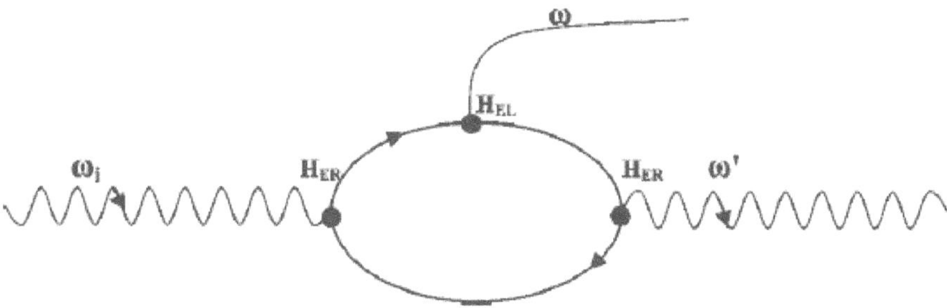

Figure 9.4 Scheme of the Raman scattering process. The interaction vertices are represented by black dots.

electron-lattice (phonon) interaction with the Hamiltonian (H_{EL}). Figure 9.4 illustrates the Raman scattering in terms of these elementary excitations [6, 7].

A photon of frequency ω_i and wave vector $\mathbf{K_i}$, cannot interact with the lattice directly, but rather interacts with an electron by the electron-radiation (photon) interaction creating in the process an electron-hole pair. The process creates (or destroys) a phonon of frequency ω and wave vector $\mathbf{K}$ via electron-lattice interaction and finally the electron recombines with the hole through the electron-radiation interaction. The vertices of interaction (see Figure 9.4) indicate that the process is third order: second-order is the electron-radiation interaction and first-order is the electron-lattice interaction. If the frequency (wave vector) of the scattered photon is ω_s ($\mathbf{K_s}$), the conservation laws of energy (momentum) can be expressed as:

$$\begin{aligned} \omega_i &= \omega_s \pm \omega, \\ \mathbf{K_i} &= \mathbf{K_s} + \mathbf{K}, \\ E_i &= \hbar\omega_i = h\nu_i = hc/\lambda_i = hcw_i. \end{aligned} \tag{9.3}$$

If ω_s is lower (higher) than the incident photon frequency ω_i, the dispersion is called Stokes (anti-Stokes). The frequency change of the incident photon ω_i-ω_s, is called the Raman shift. In this way, the Raman shift of a given vibration measures directly its frequency ω, which is independent of the excitation wavelength (energy) used. Instead of the energy or frequency, the Raman spectra displays usually the wavenumber of the vibration (given in cm^{-1}). The relationship between energy, frequency, and wavenumber for a vibration i is also included in Equation (9.3), where E_i is the energy, λ_i is the wavelength, w_i the wavenumber, and h and c are the Planck's constant and the speed of light, respectively.

The intensity of the Raman scattered light is proportional to the number of molecules or atoms illuminated by the excitation light. The Stokes and anti-Stokes intensities are proportional to the number of molecules in the lower and higher states, respectively. At a given temperature, the fraction of molecules in one vibrational state with respect to the other (N_{as}/N_a) is given by the Boltzmann distribution $N_{as}/N_s = (g_{as}/g_s)\exp(-\Delta E/k_BT)$ where N_{as} and N_s are, respectively, the number of molecules in the higher (anti-Stokes) and lower (Stokes) state, and g_{as} and g_s are, respectively, the degeneracies of these states. ΔE is the difference in energy between these two states and k_B is the Boltzmann's constant and T the temperature in Kelvin.

9.2 INSTRUMENTATION OF RAMAN MICROSCOPY

Raman scattering studies involve the excitation of vibrations in matter with light and the collection of the Raman signal scattered by the sample using an optical spectrometer. Figure 9.5 shows the instrumentation in a typical system of Raman microscopy where light from the sample is collected in the entrance slit of an optical spectrometer [8]. Raman scattering measurements are not easy to perform due to the small cross-section of the Raman scattering process (10^{-6} to 10^{-12} of the intensity of the incident light) and the small frequencies of vibrational modes, yield Raman signals close in frequency to the excitation light. To obtain a good signal-to-noise ratio in Raman scattering measurements, the excitation of Raman signal must be provided by a high intensity and highly monochromatic light source so lasers working in the UV-Vis-NIR range are the common choices for the excitation source. In particular, the small linewidth of laser lines in gas lasers is preferred to those of solid state lasers for Raman scattering studies. However, the recent improvement of solid state lasers in the last decade and their small price, low-cost maintenance, and ease of use compared to gas lasers has made them common choices in the most recent Raman scattering systems which have gone from the laboratory to industry and the field.

The spread of modern solid state lasers for Raman scattering studies has been aided by the improvement of interference (or dicroic) filters acting as band pass filters. This type of optical filter reflects one or more spectral bands and transmits others, while maintaining a nearly zero coefficient of absorption for all wavelengths of interest. An interference filter is located after the laser (see Figure 9.5) to filter undesired laser lines present in the emitted light of most types of lasers. This means that they are used to make laser light even more monochromatic by letting the laser light in a narrow energy range centered in the desired laser line and only a few cm^{-1} wide pass. The most recent interference filters allow laser light be monochromatized from 1 to 3 cm^{-1}.

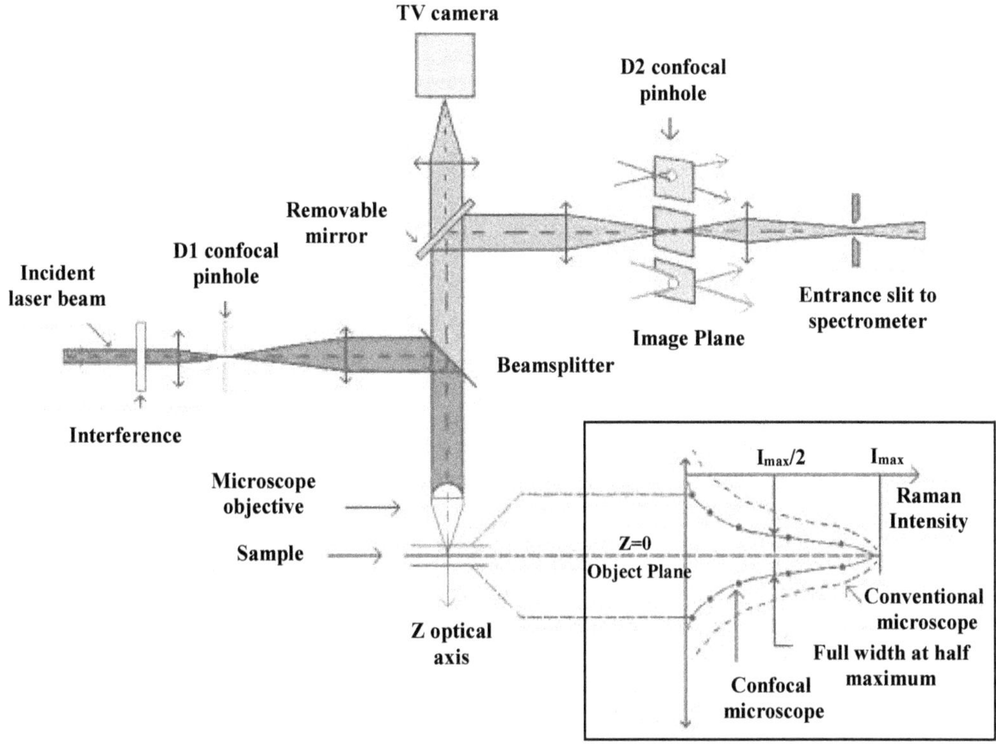

Figure 9.5 Laser focusing, sample viewing and scattered light collection geometry usually employed in Raman microscopy. Inset shows the signal collected from the focal plane and from outside the focal plane in the entrance slit of the spectrometer in this confocal microscopic system.

To excite Raman scattering in a compressed sample, usually of micron size and inside a diamond anvil cell (DAC), laser light must be focused onto the sample through the diamond windows. For that purpose, metallurgical microscope objectives with large or ultralarge working distances (10 to 30 cm) are commonly used (see Figure 9.5).

In order to avoid optical aberrations plan apochromatic objectives are used in Raman microscopes. Color aberrations are not important for Raman scattering since the useful spectral region is relatively narrow; however, in resonance Raman spectroscopy, the comparison of data from different excitation wavelengths must be performed with caution. Therefore, achromatic or apochromatic objectives are of help if one uses refractive objectives; otherwise, a reflective microscope objective like the Cassegrain–Schwarzschild objective with a perfect achromaticity can be used. Note that in this case the geometrical configuration shown in Figure 9.5 must be slightly modified. On the other hand, it is important to correct spherical aberrations so plan (or planar) objectives are mostly used instead of semi-plan or simple achromatic objectives. Plan objectives are much more expensive than the other two but allow larger working distances, which are mandatory when using DACs.

An important aspect of an optical system is the spatial resolution which can be evaluated in terms of the lateral (x-y direction) resolution and the depth (z-direction) resolution. The spatial resolution of a conventional wide field and a confocal microscope is influenced by the numerical aperture (NA) of the objective and the excitation wavelength λ. The numerical aperture is given by the simple expression: numerical aperture (NA) $= n \times \sin(\theta)$. In the numerical aperture equation, n represents the refractive index of the medium between the objective front lens and the specimen, and θ is the one-half angular aperture of the objective. The numerical aperture of a microscope objective is a measure of its ability to gather light and resolve fine specimen detail at a fixed object distance. Image-forming light waves pass through the specimen and enter the objective in an inverted cone. A longitudinal slice of this cone of light reveals the angular aperture, a value that is determined by the focal length of the objective. In practice, it is difficult to achieve numerical aperture values above 0.95 with dry objectives. Microscope objectives are now available that allow imaging in alternative media such as water (refractive index = 1.33), glycerin (refractive index = 1.47), and immersion oil (refractive index = 1.51). The numerical aperture of an objective is also dependent to a certain degree upon the amount of correction for optical aberration [6].

Furthermore, the spatial resolution is ultimately determined by the diffraction limit of the radiation. It is defined by the distance between the central maximum and the first minimum of the diffraction pattern, which is given by $r = 0.61\lambda/NA$, where λ is the wavelength of the light and NA the numerical aperture of the objective. In the focal plane (x-y direction), two objects are completely resolved if they are separated by $2r$ and barely resolved if they are separated by r (Rayleigh criterion of resolution) [9]. As an example, if we use an objective of $NA = 0.95$ and a laser of $\lambda = 514.5$ nm, features at a distance $r = 340$ nm can be distinguished and completely resolved at a distance of 680 nm. In this way, a confocal Raman microscope can reach 250 to 300 nm of lateral resolution. On the other hand, the depth resolution of a confocal microscope is proportional to $\lambda/(NA)^2$. Thus, a confocal Raman imaging microscope can achieve a depth resolution of 1 μm or less.

In order to collect the Raman signal, the light containing the Raman signal emitted by the sample exits the DAC through the diamond windows so a microscope objective is also used to collect and guide the light towards an spectrometer. Most Raman microscopes use the same microscope objective both to excite and collect the Raman signal as shown in Figure 9.5. This optical configuration for Raman scattering measurements is called *backscattering* geometry and it limits the selection rules that can be used to study Raman scattering of compressed samples.

The light collected by the microscope objective is composed of the incident laser light reflected by the diamond windows and the sample, the elastic scattered light from the sample and the diamond window (Rayleigh signal mainly with the same wavelength of the laser), and the inelastic scattered light from the sample and the window (Raman signal with slightly different wavelengths from the incident laser light). Since intensity of the Raman signal is very small compared to the reflected and the Rayleigh scattered light, both signals must be discriminated before being detected at the spectrometer. Otherwise, the Raman signal would not be seen on the top of the reflected and Rayleigh signal. The discrimination of the Raman signal can be performed by different means. The first and most versatile way to filter the Raman signal is the use of a double or a triple monochromator where the first or two first monochromators act as a filter and the last stage acts as a spectrometer. The main advantage of this configuration is that it is capable of rejecting most of the reflected and Rayleigh signal and it allows Raman scattering measurements in both Stokes and anti-Stokes configuration a few cm^{-1} away from the laser line. The main drawback of this configuration is the very low throughput (collected Raman signal intensity) of these configurations compared to detection performed using only a single spectrometer.

An alternative way of filtering Raman signal is to use a filter to block a narrow range of wavelengths close to the laser line while allowing the pass of the Raman signal. The first Raman filters were introduced in the 1990s and known as notch filters, band-stop or band-rejection filters. They are designed to transmit most wavelengths with little intensity loss while attenuating light within a specific wavelength range (the stop band) to a very low level. They allow observation of Stokes (long wavelengths) and anti-Stokes (short wavelengths) scattered light while blocking the laser line within 100 cm^{-1} from the laser line. Notch filters allow angle tuning to measure Raman signals closer to the laser line. The main drawback of notch filters is that they are very sensitive and degrade with time and are relatively expensive compared to other filters. Edge filters appeared in the market in the early 2000s and most modern Raman micro spectrometers make use of edge filters. They have several advantages compared to notch filters since they allow measurements at closer wavelengths to the laser line and they are cheaper and more durable than notch filters. Most recent edge filters allow measurements as low as 30 cm^{-1} from the laser line. The main disadvantage is that edge filters allow the measurement of either the Stokes or the anti-Stokes signal but not both with the same filter. Besides, they allow only a very small angle tunning compared to notch filters. The last development in Raman signal filtering is the use of volume Bragg grating (VBG). VBG consists of a phase volume diffractive grating that allows measurements of Raman modes as low as 5 cm^{-1} from the laser line with a transmittance of 95 %. The main inconvenience of these filters is that they are relatively expensive compared to edge filters and require a very good filtering of the laser line, resulting in a laser linewidth similar to those of gas lasers. In fact, volume Bragg gratings are used inside laser diodes to stabilize the wavelength of the laser line so a combination of volume Bragg gratings can be used both as filters of laser lines and of Rayleigh signal. The use of notch, edge, and VBG filters to block reflected laser and Rayleigh signals have allowed the use of single spectrometers with much higher throughput than double and triple spectrometers for Raman scattering measurements. The main drawback of these filters is that they are designed for specific laser wavelengths so they are in general not suitable for performing detailed resonant Raman scattering studies where the laser wavelength is usually tuned at will. The position of the Raman filter is not shown in Figure 9.5 and it would be ideally located immediately before or after the removable mirror (the one that allows optical inspection of the sample with the camera), i.e., before the lens that focuses light into the image plane of the confocal system.

Once the Raman signal is discriminated from the elastically scattered signal, it can be sent to the entrance slit of the spectrometer to register the Raman spectrum; however, the Raman signal collected in a conventional microscope setup would contain the signal coming from, (i) the sample (at the focal plane) and, (ii) the surrounding medium and the diamond windows of the DAC (out of the focal plane). To discriminate all these Raman signals and collect only the Raman signal coming from the sample, a confocal microscopic setup is used. The aim of confocal microscopy is to collect only light coming from the focal plane. To this purpose, light must be focused first on a pinhole (D1 in Figure 9.5) whose image is focused on the sample on a small spot (point illumination) at the focal plane and then light coming from the sample must be focused on a small pinhole (D2 in Figure 9.5) in the conjugate or image plane of the detection beam path. Due to the point illumination and detection, information only from a single point is determined at a time. The advantages of confocal microscopy over conventional wide-field microscopy are: (i) a higher resolution in the axial direction (giving the opportunity to collect serial optical sections from different focal planes to generate depth profiles), (ii) an improved resolution in the lateral plane, and (iii) a reduced background signal. The pinhole size must be a compromise between resolution and signal throughput. The signal should be as high as possible, while the spatial resolution should be as good as possible. Note that a small pinhole strongly increases depth and lateral resolution at the expense of a smaller signal reaching the detector. This is the reason for many systems to work with confocal diaphragms of variable size to adjust pinhole size to experimental conditions. The confocal system consists of two pinhole spatial filters (D1 and D2 in Figure 9.5) that allow focussing and spatially filter the laser beam (D1) and the Raman signal (D2). With this technique a significant improvement in both the contrast and the spatial resolution of the system along the z axis is obtained so that a signal coming from the focal plane ($z = 0$) is maximized compared to signals coming from planes with $z \neq 0$ as shown in the bottom part of Figure 9.5. Despite this improvement, a Raman signal coming from diamond windows is always observed together with Raman signal coming from the sample because of the strong Raman scattering cross section of diamond. Fortunately, the Raman signal of diamond consists of a single band near 1330 cm^{-1} which is much stronger than the Raman signals of most samples and it can be also used to calibrate pressure at very high pressures when the photoluminescence of ruby decreases considerably.

Finally, the Raman light is refocused in the entrance slit of the spectrometer. The spectrometer is used to decompose light into different wavelengths and send them to the detector to monitor the intensity of each wavelength. The most delicate part in the spectrometer for Raman measurements is the detector since the detection of the weak Raman signal needs a photodetector with a high signal-to-noise ratio. Since its discovery in the 1970s, the charged coupled device (CCD) of Si is the most widely used photodetector in the UV-VIS range. Photo-diode arrays (PDA) of Ge and InGaAs are mostly used in the NIR range. These detectors are usually cooled down to liquid nitrogen temperature to maximize the signal-to-noise ratio; however, in recent decades, the improvement of CCDs with high gain and low darkness background and the use of notch and edge filters in combination with single monochromators as spectrometers has increased the signal-to-noise ratio and many Raman microspectrometers now have CCDs cooled by Peltier effect only from -40 to -100 °C avoiding the use of liquid nitrogen and making possible the proliferation of compact Raman spectrometers that can be used outside the laboratory.

9.3 RAMAN CROSS SECTION

9.3.1 Cross Section

The scattering cross section is defined as:

$$\frac{d\sigma}{d\Omega} = \frac{Radiated\, power\, inside\, d\Omega}{Incident\, power\, per\, unit\, area} = \frac{PR}{PI}, \tag{9.4}$$

where the incident power per unit area, is equal to:

$$PI = \frac{N\hbar\omega c}{n} \tag{9.5}$$

where N is the number of incident photons per volume unit and c/n is the velocity of light. The radiated power within a solid angle element dW is given by

$$\begin{aligned} PR &= (Number\, of\, incident\, photons)\hbar\omega' \frac{dw}{dt}(d\Omega) \\ &= (NV)\hbar\omega' \frac{dw}{dt}(d\Omega), \end{aligned} \tag{9.6}$$

where $\frac{dw}{dt}(d\Omega)$ is the probability of transition per unit of time of a photon from one state $(\omega, \mathbf{k}, \exp(\alpha))$ to another state with a polarization $\exp(\alpha)$ and wave vector $\mathbf{k}'$ within the solid angle element $d\Omega$ plus a phonon with a frequency Ω_{ph} and wave vector $\mathbf{q}$. Then, substituting Equations (9.5) and (9.6) in Equation (9.4), we obtain

$$\begin{aligned} \frac{d\sigma}{d\Omega} &= \frac{(NV)\,\hbar\omega' \frac{dw}{dt}\,(d\Omega)}{\frac{N\hbar\omega c}{n}} \\ \frac{d\sigma}{d\Omega} &= \left(\frac{Vn}{c}\right)\left(\frac{\omega'}{\omega}\right)\frac{dw}{dt}\,(d\Omega)\,. \end{aligned} \tag{9.7}$$

Some authors define a differential cross section in a simplest form, namely:

$$\begin{aligned} \frac{d\sigma}{d\Omega} &= \frac{flow\, of\, particles\, dispersed\, per\, unit\, of\, time\, within\, d\Omega}{current\, density\, of\, incident\, particles\ \times d\Omega} \\ \frac{d\sigma}{d\Omega} &= \frac{(NV)\,\frac{dw}{dt}\,(d\Omega)}{\frac{Nc}{n}} \\ \frac{d\sigma}{d\Omega} &= \left(\frac{Vn}{c}\right)\frac{dw}{dt}\,(d\Omega)\,. \end{aligned} \tag{9.8}$$

9.3.2 Theory of Time-Dependent Perturbation

The theory of time dependent perturbation was developed by Dirac [10]. An atomic wave function can be defined as:

$$\psi = \sum_k c_k\,(t)\,u_k\,(\mathbf{r})\,e^{-iEt/\hbar}, \tag{9.9}$$

where $u_k\,(\mathbf{r})$ is the self-function with energy E_k satisfying,

$$H_0 u_k\,(\mathbf{r}) = E_k u_k\,(\mathbf{r})\,, \tag{9.10}$$

in the absence of a time-dependent perturbation. In other words, without perturbation, the Schrödinger equation in the presence of a potential $H_I(t)$ is given by the next expression where the Hamiltonian $H_I(t)$ of the perturbation is very small compared to the Hamiltonian without perturbation,

$$\begin{aligned}
&(H_0 + H_I)\psi = i\hbar\frac{\partial\psi}{\partial t}\\
&H_0\psi + H_I\psi = i\hbar\frac{\partial}{\partial t}\left(\sum_k c_k(t)\,u_k(\mathbf{r})\,e^{-iE_kt/\hbar}\right)\\
&H_0\sum_k c_k(t)\,u_k(\mathbf{r})\,e^{-iE_kt/\hbar} + H_I\sum_k c_k(t)\,u_k(\mathbf{r})\,e^{-iE_kt/\hbar} =\\
&i\hbar\sum_k \dot{c}_k(t)\,u_k(\mathbf{r})\,e^{-iE_kt/\hbar} - E_kc_k(t)\,u_k(\mathbf{r})\,e^{-iE_kt/\hbar}.
\end{aligned} \tag{9.11}$$

We are considering that the Hamiltonian H_0 contains both the kinetic energies of electrons and the Coulomb interactions between electrons and the nucleus whereas $H_I(t)$ is the Hamiltonian of the interaction of atomic electrons with the radiation. In this way, using Equation (9.6), we have

$$H_I\sum_k c_k(t)\,u_k(\mathbf{r})\,e^{-iE_kt/\hbar} = i\hbar\sum_k \dot{c}_k(t)\,u_k(\mathbf{r})\,e^{-iE_kt/\hbar}. \tag{9.12}$$

Multiplying the above equation by $u_m^*(\mathbf{r})$ and integrating over all the space, we obtain the following differential equation:

$$\dot{c}_k(t) = \frac{1}{i\hbar}\sum_k \langle m\,|H_I(t)|\,k\rangle\, e^{-i[E_m-E_k]t/\hbar}c_k(t). \tag{9.13}$$

This differential equation gives rise to the integral equation

$$c_k(t) = \frac{1}{i\hbar}\sum_{k=1}^{\infty}\int_0^t dt'\,\langle m\,|H_I(t')|\,k\rangle\, e^{-i[E_m-E_k]t'/\hbar}c_k(t'). \tag{9.14}$$

For an arbitrary k, the above equation simplifies to

$$c_k(t) = \frac{1}{i\hbar}\int_0^t dt'\,\langle m\,|H_I(t')|\,k\rangle\, e^{-i[E_m-E_k]t'/\hbar}c_k(t'). \tag{9.15}$$

The formal solution of the above integral equation is given by

$$c_k(t) = \sum_{n=1}^{\infty} c_k^n(t), \tag{9.16}$$

with

$$
\begin{aligned}
c_k(t) &= (i\hbar)^{-n}\int_0^t dt_n H(t_n)\int_0^t dt_{n-1}H(t_{n-1})\cdots\int_0^t dt_1 H(t_1)\\
&\quad t \geq t_n \geq t_{n-1}\cdots \geq t_1 \geq 0.
\end{aligned}
\tag{9.17}
$$

The term c_k^0 represents the order zero approximation; in our case this term is equal to zero, and c_k^n the nth order. We are interested in the first and second order approximations, which correspond, respectively, to

$$
c_k^{(1)}(t) = (i\hbar)^{-1}\int_0^t dt_1 \langle m|H_I(t_1)|k\rangle e^{-i[E_m-E_k]t_1/\hbar}
\tag{9.18}
$$

$$
\begin{aligned}
c_k^{(2)}(t) &= (i\hbar)^{-2}\int_0^t dt_2 \langle m|H_I(t_2)|k\rangle e^{-i[E_m-E_k]t_2/\hbar}\\
&\quad \int_0^{t_2} dt_1 \langle m|H_I(t_1)|k\rangle e^{-i[E_m-E_k]t_1/\hbar}\\
&= (i\hbar)^{-2}\int_0^t dt_2\int_0^{t_2} dt_1 \langle m|H_I(t_2)|k\rangle e^{-i[E_m-E_k]t_2/\hbar}\\
&\quad \langle m|H_I(t_1)|k\rangle e^{-i[E_m-E_k]t_1/\hbar}.
\end{aligned}
\tag{9.19}
$$

Thus, the approximation of the time-dependent perturbation to the second order is given by

$$
c_k^n(t) \approx c_k^{(1)}(t) + c_k^{(2)}(t)\,.
\tag{9.20}
$$

9.3.3 Transition Probability

For a photon emitted within a solid angle element $d\Omega$, the number of allowed states in an energy range $(\hbar\omega', \hbar(\omega' + d\omega'))$ can be written as $\rho_{\hbar\omega, d\Omega} d(\hbar\omega')$, where

$$
\begin{aligned}
\rho_{\hbar\omega_s, d\Omega} &= \frac{V}{(2\pi)^3}\frac{d^3|\mathbf{k}'|}{d(\hbar\omega')},\\
&= \frac{V}{(2\pi)^3}\frac{dS_{\mathbf{k}'}d|\mathbf{k}'|}{d(\hbar\omega')}\\
&= \frac{V|\mathbf{k}'|^2}{(2\pi)^3}\frac{d|\mathbf{k}'|\,d\Omega}{d(\hbar\omega')},\\
&= \frac{V(\omega')^2}{(2\pi)^3}\left(\frac{n'}{c}\right)^3\frac{d\Omega}{\hbar}.
\end{aligned}
\tag{9.21}
$$

Thus, the transition probability per unit time within the solid angle element $d\Omega$, is given by

$$\frac{dw}{dt}(d\Omega) = \int \left(\left| c_k^{(1)}(t)\, c_k^{(2)}(t) \right|^2 / t \right) \rho_{\hbar\omega', d\Omega} dE. \tag{9.22}$$

9.3.4 Cross Section of Kramers–Heisenberg

In the derivation of the above formula we considered the scattering of photons by atomic electrons [6, 7]. Before dispersion, the atom is in state A and the incident photon is characterized by $(\mathbf{k}, \exp(\alpha))$. After dispersion, the atom is in state B and the incident photon is characterized by $(\mathbf{k}', \exp(\alpha'))$. For simplicity, we will consider an atom with one electron and we neglect the spin magnetic moment interaction. In terms of current density and charge densities of electrons, the electromagnetic interaction of one vector potential with electrons is given by

$$\begin{aligned} H_I &= H_A + H_{AA}, \\ H_A &= -\frac{1}{c}\int \mathbf{j} \cdot \mathbf{A}(\mathbf{r})\, \mathbf{d^3 r}, \\ H_{AA} &= -\frac{e^2}{2mc^2}\int \rho(r)\, A^2(r)\, d^3 r. \end{aligned} \tag{9.23}$$

The corresponding quantum mechanical expressions for transverse photons, become

$$H_A = \sum_i \frac{e}{2mc}\left(\mathbf{p}_i \cdot \mathbf{A}(\mathbf{x}, \mathbf{t}) + \mathbf{A} \cdot (\mathbf{x}, \mathbf{t}) \cdot \mathbf{p}_i\right), \tag{9.24}$$

$$H_{AA} = \sum_i \frac{e^2}{2mc}\mathbf{A}(\mathbf{x}, \mathbf{t}) \cdot \mathbf{A}(\mathbf{x}, \mathbf{t}), \tag{9.25}$$

where the summation is over all atomic electrons involved in the interaction with:

$$\begin{aligned} \mathbf{A}(\mathbf{x_i}, \mathbf{t}) = \frac{1}{\sqrt{V}} \sum_{\mathbf{k}} \sum_{\alpha} c \sqrt{\frac{\hbar}{2\omega}} & \left(a_{\mathbf{k},\alpha}(t)\, \epsilon^{(\alpha)} e^{i\mathbf{k}\cdot\mathbf{x} - \omega \mathbf{t}} + \right. \\ & \left. a'_{\mathbf{k},\alpha}(t)\, \epsilon^{(\alpha)} e^{-i\mathbf{k}\cdot\mathbf{x} + \omega t} \right). \end{aligned} \tag{9.26}$$

The expression $\mathbf{A}(\mathbf{x_i}, \mathbf{t})$ is known as the operator acting on a photon state or photons in many states in $\mathbf{x_i}$, where x_i is related to the ith coordinate of the

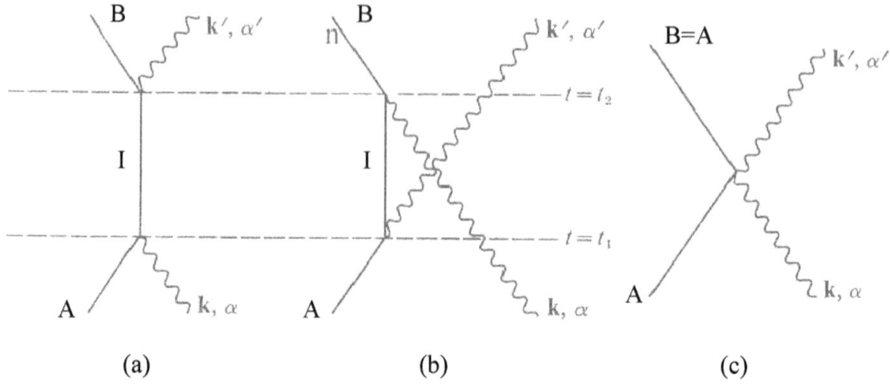

Figure 9.6 Space-time diagrams for the scattering of radiation.

electron and the operators a and a' represent the creation and annihilation operators, respectively. Since we are considering a one-electron atom, the sum in Equations (9.18) and (9.19) on i disappears.

Figure 9.6 shows the space-time diagrams (Feynman diagrams) representing the interaction of Hamiltonian H_A (left and center), and H_{AA} (right). The solid line represents the atom and the wavy line represents a photon. The time is assumed to move upward. For a type 1 process, represented by Figure 9.6(left), the atomic state A first absorbs the incident photon at t_1 and then passes to the state I, then at t_2 the atomic state I emits the scattered photon and switches to state B. For a Type 2 process represented by Figure 9.6(center), state A first emits the scattered photon at t_1 and then passes to the state I, then at t_2 the I state absorbs the incident photon (which has not been annihilated) and is then transformed in state B. Hence, using the theory of time-dependent perturbation will expand to the second order the Hamiltonian H_A and to first order the Hamiltonian H_{AA}. Therefore, from Equations (9.19) and (9.22) the Hamiltonian matrix element of the H_{AA}, will be:

$$\begin{aligned}
\left\langle B;\mathbf{k},\epsilon^{(\alpha)}\left|\mathbf{H_{AA}}\right|\mathbf{A};\mathbf{k},\epsilon^{(\alpha)}\right\rangle &= \left\langle B\left|\frac{e^2}{2mc^2}\mathbf{A}\left(\mathbf{x_i},\mathbf{t}\right)\cdot\mathbf{A}\left(\mathbf{x_i},\mathbf{t}\right)\right|A\right\rangle \\
&= \left\langle B\left|\frac{e^2}{2mc^2}\left(a_{\mathbf{k},\alpha}a'_{\mathbf{k},\alpha}+a'_{\mathbf{k},\alpha}a_{\mathbf{k},\alpha}\right)\frac{c^2\hbar}{2V\sqrt{\omega\omega'}}\epsilon^{(\alpha)}\cdot\epsilon^{(\alpha)}e^{[i(\mathbf{k}-\mathbf{k}')\cdot\mathbf{x}_i(\omega-\omega')t]}\right|A\right\rangle \\
&= \frac{e^2}{2mc^2}\frac{c^2\hbar}{2V\sqrt{\omega\omega'}}2\epsilon^{(\alpha)}\cdot\epsilon^{(\alpha')}e^{[i(\omega-\omega')t]}\left\langle A|B\right\rangle,
\end{aligned} \tag{9.27}$$

where we have replaced $\exp i\mathbf{k}\cdot\mathbf{x}$ and $\exp i\mathbf{k}'\cdot\mathbf{x}'$ by one, since in the approximation of the largest wavelength the atomic electron can be assumed to be located at the origin ($x = 0$). We should also note that we have considered the cross terms, i.e., $a_{\mathbf{k,a}}a_{\mathbf{k',a'}}$ and $a_{\mathbf{k',a'}}a_{\mathbf{k,a}}$ which describe the creation and annihilation of a photon, and vice versa, typical in dispersion processes. Then, the amplitudes of first order transitions, are obtained by substituting Equation (9.24) into Equation (9.7),

$$\begin{aligned} c_k^{(1)}(t) &= (i\hbar)^{-1} \frac{e^2}{2mc^2} \frac{c^2\hbar}{2V\sqrt{\omega\omega'}} 2\delta_{AB}\epsilon^{(\alpha)} \cdot \epsilon^{(\alpha')} \\ &\int_0^t dt_1 e^{[-i(\omega-\omega')]} e^{-i(E_B-E_A)t_1/\hbar} \\ &= (i\hbar)^{-1} \frac{e^2}{2mc^2} \frac{c^2\hbar}{2V\sqrt{\omega\omega'}} 2\delta_{AB}\epsilon^{(\alpha)} \cdot \epsilon^{(\alpha')} \\ &\int_0^t dt_1 e^{[i(\hbar\omega'-\hbar\omega+E_B-E_A)t_1/\hbar]}. \end{aligned} \tag{9.28}$$

Similarly, the amplitude of the second order transition is obtained by combining Equations (9.18) and (9.22), in the dipolar approximation,

$$\begin{aligned} c_k^{(2)}(t) &= (i\hbar)^{-2} \frac{c^2\hbar}{2V\sqrt{\omega\omega'}} \left(-\frac{e}{mc}\right)^2 \int_0^t dt_2 \int_0^{t_2} dt_1 \\ &\sum_I \left\langle B \left| \mathbf{p} \cdot \epsilon^{(\alpha')} \right| I \right\rangle e^{[i(E_B-E_I+\hbar\omega')t_2/\hbar]} \times \\ &\left\langle I \left| \mathbf{p} \cdot \epsilon^{(\alpha')} \right| A \right\rangle e^{[i(E_I-E_A+\hbar\omega)t_1/\hbar]} \\ &+ \sum_I \left\langle B \left| \mathbf{p} \cdot \epsilon^{(\alpha)} \right| I \right\rangle e^{[i(E_B-E_I+\hbar\omega)t_2/\hbar]} \times \\ &\left\langle I \left| \mathbf{p} \cdot \epsilon^{(\alpha)} \right| A \right\rangle e^{[i(E_I-E_A+\hbar\omega')t_1/\hbar]}. \end{aligned} \tag{9.29}$$

Integrating in t_1 and ignoring the terms that arise when evaluated at $t = 0$, we have

$$\begin{aligned} c_k^{(2)}(t) &= -(i\hbar)^{-1} \frac{c^2\hbar}{2V\sqrt{\omega\omega'}} \left(\frac{e}{mc}\right)^2 \\ &\sum_I \left(\frac{\left(\mathbf{p} \cdot \epsilon^{(\alpha')}\right)_{BI} \left(\mathbf{p} \cdot \epsilon^{(\alpha)}\right)_{IA}}{E_I - E_A + \hbar\omega} + \frac{\left(\mathbf{p} \cdot \epsilon^{(\alpha)}\right)_{BI} \left(\mathbf{p} \cdot \epsilon^{(\alpha')}\right)_{IA}}{E_I - E_A + \hbar\omega'} \right) \\ &= \int_0^t dt_1 e^{[i(\hbar\omega'-\hbar\omega+E_B-E_A)t_1/\hbar]}, \end{aligned} \tag{9.30}$$

where we have simplified the nomenclature of Equation (9.10), and using Equations (9.25) and (9.26), the transition probability per unit of solid angle is given by

$$
\begin{aligned}
\frac{dw}{dt}(d\Omega) &= \int \left(\left| c_k^{(1)}(t) + c_k^{(2)}(t) \right| / t \right) \rho_{\hbar\omega', d\Omega} d(\hbar\omega') \\
&= \frac{2\pi}{\hbar} \left(\frac{c^2\hbar}{2V\sqrt{\omega\omega'}} \right)^2 \left(\frac{e^2}{mc^2} \right)^2 \frac{V(\omega')^2}{(2\pi)^3 \hbar} \left(\frac{n'}{c} \right)^3 d\Omega \left| \delta_{AB} \epsilon^{(\alpha)} \cdot \epsilon^{(\alpha')} - \right. \\
&\left. \frac{1}{m} \sum_I \left(\frac{\left(\mathbf{p} \cdot \epsilon^{(\alpha')}\right)_{BI} \left(\mathbf{p} \cdot \epsilon^{(\alpha)}\right)_{IA}}{E_I - E_A + \hbar\omega} + \frac{\left(\mathbf{p} \cdot \epsilon^{(\alpha)}\right)_{BI} \left(\mathbf{p} \cdot \epsilon^{(\alpha')}\right)_{IA}}{E_I - E_A + \hbar\omega} \right) \right|^2
\end{aligned} \tag{9.31}
$$

where we have:

$$
\begin{aligned}
\left| \int_0^t dt_1 e^{[i(\hbar\omega' - \hbar\omega + E_B - E_A)t_1/\hbar]} \right|^2 &= \left(\int_0^t dt_1 e^{[i(\hbar\omega' - \hbar\omega + E_B - E_A)t_1/\hbar]} \right) \times \\
&\quad \left(\int_0^t dt_1 e^{[-i(\hbar\omega' - \hbar\omega + E_B - E_A)t_1/\hbar]} \right) \\
&= \int_0^t dt_1 \\
&= t\delta\left(\hbar\omega' - \hbar\omega + E_B - E_A\right).
\end{aligned} \tag{9.32}
$$

The conservation of energy $\hbar\omega' - \hbar\omega + E_B - E_A = 0$ is represented in the delta of Dirac.

Finally, combining Equations (9.5) and (9.27), we obtain the differential cross section of Kramers–Heisenberg,

$$
\begin{aligned}
\frac{d\sigma}{d\Omega} &= r_0^2 \left(\frac{\omega'}{\omega} \right) n (n')^3 \times \left| \delta_{AB} \epsilon^{(\alpha)} \cdot \epsilon^{(\alpha)} \right. \\
&- \left. \frac{1}{m} \sum_I \left(\frac{\left(\mathbf{p} \cdot \epsilon^{(\alpha')}\right)_{BI} \left(\mathbf{p} \cdot \epsilon^{(\epsilon)}\right)_{IA}}{E_I - E_A + \hbar\omega} + \frac{\left(\mathbf{p} \cdot \epsilon^{(\alpha)}\right)_{BI} \left(\mathbf{p} \cdot \epsilon^{(\epsilon')}\right)_{IA}}{E_I - E_B + \hbar\omega} \right) \right|^2,
\end{aligned} \tag{9.33}
$$

where r_0 is the classical electron radius, whose value is

$$
r_0 = \frac{e^2}{4mc^2} = \frac{1}{137} \frac{\hbar}{mc} = 2,82 \times 10^{-13} \text{cm}. \tag{9.34}
$$

9.3.5 Raman Cross Section

The cross section of the Raman effect can be obtained from the cross section of Kramers-Heisenberg (Equation (9.27)) taking into account that in the Raman effect the radiation scattering is inelastic; i.e., $\omega' \neq \omega$ and $A \neq B$. Thus using Equation (9.4) instead of Equation (9.5) we have

$$\frac{d\sigma}{d\Omega} = r_0^2 \left(\frac{\omega'}{\omega}\right)^2 \frac{n}{m} (n')^3 \left| \sum_I \left(\frac{\left(\mathbf{p} \cdot \epsilon^{(\alpha')}\right)_{BI} \left(\mathbf{p} \cdot \epsilon^{(\epsilon)}\right)_{IA}}{E_I - E_A + \hbar\omega} + \frac{\left(\mathbf{p} \cdot \epsilon^{(\alpha)}\right)_{BI} \left(\mathbf{p} \cdot \epsilon^{(\epsilon')}\right)_{IA}}{E_I - E_B + \hbar\omega} \right) \right|^2, \tag{9.35}$$

or

$$\frac{d\sigma}{d\Omega} = r_0^2 \left(\frac{\omega'}{\omega}\right)^2 \frac{n}{m} (n')^3 \left| \sum_I \left(\frac{\left(\mathbf{p} \cdot \epsilon^{(\alpha')}\right)_{BI} \left(\mathbf{p} \cdot \epsilon^{(\epsilon)}\right)_{IA}}{E_I - E_A + \hbar\omega} + \frac{\left(\mathbf{p} \cdot \epsilon^{(\alpha)}\right)_{BI} \left(\mathbf{p} \cdot \epsilon^{(\epsilon')}\right)_{IA}}{E_I - E_B + \hbar\omega'} \right) \right|^2. \tag{9.36}$$

The mechanism for the inelastic scattering of light can be seen as the modulation of the dielectric susceptibility by some elementary excitations of a solid. From Equation (9.36) we obtain the expression [11]:

$$\frac{d^2\sigma}{d\Omega d\omega_s} = vV \frac{\omega_s^4}{c^4} \left|\widehat{\epsilon_s} \cdot \chi' \cdot \widehat{\epsilon_l}\right|^2 \langle UU' \rangle_\omega, \tag{9.37}$$

for the differential cross section of scattering into the solid angle and frequency increments $d\Omega$ at Ω and $d\omega_s$ to ω_s, respectively. Here χ' is the higher order susceptibility tensor (often called Raman tensor) appropriate to the elementary excitation of amplitude U, $\widehat{\epsilon_l}$ and $\widehat{\epsilon_s}$ are the unit polarization vectors of the incident and scattered radiation, respectively, v is the volume of interaction and V is the volume of the sample. The notation $\langle UU' \rangle_\omega$ denotes the power spectrum of $|U|^2$.

Equation (9.37) is very useful because it separates the differential cross section in a form factor $\langle UU' \rangle_\omega$ describing the frequency spectrum of the excitation under study and a strength factor $\left|\widehat{\epsilon_s} \cdot \chi' \cdot \widehat{\epsilon_l}\right|$ containing relevant interactions of light with the elementary excitations through other intermediate excitations of the solid. To be more specific, for a single Stokes process phonon we have:

$$\langle UU' \rangle_\omega = \frac{\hbar}{2N\omega_i} (n_i + 1)\, g_i(\omega), \tag{9.38}$$

where N is the number of oscillators in the solid, $n_i = \left[\exp\left(h\omega_i/KT\right) - 1\right]^{-1}$

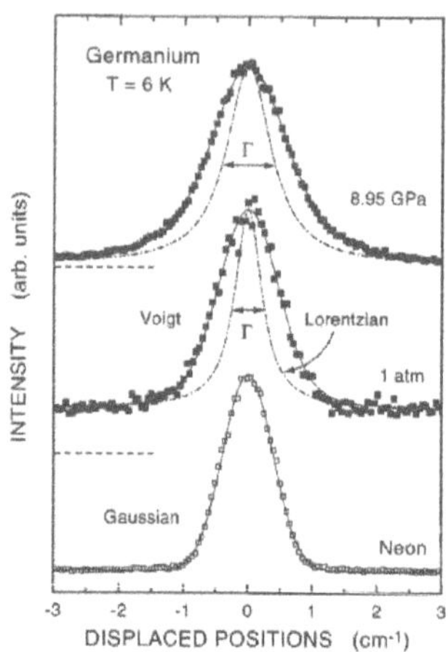

Figure 9.7 Germanium (Ge) Raman spectra at 6K at two different pressures. At the bottom we see the Gaussian response function of the neon lamp with a FWHM = 0.9 cm^{-1}, which represents the resolution of the spectrometer. At 1 atm the dots represent the experimental Raman peak and the dashed line corresponds to the Lorentzian of the phonon (TO, LO) after applying the Voigt function, where FWHM = 0.5 cm^{-1}. We can clearly see the additional width introduced by the resolution of the spectrometer. Adapted, with permission, from Reference 13.

is the Bose thermal population factor and $g_i(\omega)$ is the lineshape function represented by a Lorentzian:

$$g_i(\omega) = \frac{\frac{\Gamma_i}{2\pi}}{(\omega_i - \omega)^2 + \left(\frac{\Gamma_i}{2}\right)^2}, \tag{9.39}$$

where Γ_i represents the linewidth at half maximum (FWHM) of the Lorentzian function and ω_i the frequency at the peak maximum. The Γ_i value can be used to evaluate the phonon lifetime τ_i, $\Gamma_i \sim 1/\tau_i$ through the Heisenberg time-energy uncertainty relation $\Delta_i \mathrm{E}/\mathrm{h} = 1/\tau_i$, where ΔE_i is the FWHM of the Lorentzian Raman peak in units of cm^{-1} and $h = 5.3 \times 10^{12}$ cm^{-1}s [12].

In an ideal harmonic crystal, the lineshape is expected to be infinitesimally narrow, but experimental peaks of real materials exhibit an intrinsic linewidth. The presence of various decay channels shortens the phonon lifetimes by anharmonic processes involving multiple phonon recombinations which conserve both energy and crystal momentum. However, impurities and defects disturb the translation symmetry of the harmonic crystal. Therefore, they modify the Raman linewidth by elastic scattering processes that also contribute to the phonon lifetime shortening scenario. Finally, the natural isotopic dispersion yields a phonon frequency dispersion that induces an inhomogeneous broadening of the observed Raman modes. Assuming a Boltzmann equation approach, one can take separately into account impurities and anharmonic effects on τ as $(1/\tau) = (1/\tau_{imp}) + (1/\tau_{anharmonic})$.

The study of the linewidth in a Raman spectrum should be made using the Voigt profile to take into account the finite resolution of the spectrometer since the experimental lineshape is usually a convolution of the Lorentzian shape of the Raman mode with the Gaussian response function of the spectrometer. The following expression describes the Voigt profile [14]:

$$I\left(\tilde{v}\right) = \frac{a_0 a_3}{2\pi\sqrt{\pi}a_2^2}\int_{\infty}^{-\infty} \frac{e^{-t^2}}{\frac{a_3^2}{2a_2^3} + \left(\frac{\tilde{v}-a_1}{\sqrt{2}a_2} - t\right)^2}, \tag{9.40}$$

where a_0 is the area, a_1 is the frequency at the experimental peak maximum, a_2 is the Gaussian width (FWHM/2), and a_3 is the Lorentzian width (FWHM/2). An example of a fit performed with this profile is shown in Figure 9.7 [1].

For one-phonon scattering in semiconductors, a typical two-band term contributing to $|\widehat{\epsilon_s} \cdot \chi' \cdot \widehat{\epsilon_l}|$ has the form [11]:

$$\widehat{\epsilon_s} \cdot \chi' \cdot \widehat{\epsilon_l} \sim \frac{\langle v\,|\widehat{\epsilon_s} \cdot \mathbf{p}|\, c\rangle \left\langle c \left| H_{ep}^{(1)} \right| c \right\rangle \langle c\,|\mathbf{p} \cdot \widehat{\epsilon_l}|\, v\rangle}{(\omega_g + \omega_i - \omega_l)(\omega_g - \omega_l)}. \tag{9.41}$$

Equation (9.41) represents the process diagrammed in Figure 9.8. Here $\mathbf{p}$ is the momentum of the electron, $H_{ep}^{(1)}$ is the Hamiltonian of electron-phonon interaction and $|v\rangle$, $|c\rangle$ and ω_g are defined in Figure 9.8.

9.4 EXAMPLES OF EFFECT OF PRESSURE ON RAMAN SCATTERING CROSS SECTION

When analyzing Equations (9.37), (9.38) and (9.41), one can observe a Raman scattering cross section, that determines the shape of the Raman phonons experimentally measured and indicated by four factors which can vary with pressure: frequency (ω_i), linewidth (Γ_i), lineshape, defined in $g_i(\omega)$, and the intensity of the Raman phonon, mainly determined by the $|\widehat{\epsilon_s} \cdot \chi' \cdot \widehat{\epsilon_l}|$ factor. In the following, we will show how pressure influences all these factors.

9.4.1 Raman Frequency Shift Induced by Pressure

An increase of pressure over a material leads to a decrease of volume and consequently to a reduction of interatomic and intermolecular distances. In general, since phonon frequencies are proportional to the square root of the interatomic force constants that depend inversely on interatomic bond distances, the most evident effect of pressure on phonons is the shift of phonon

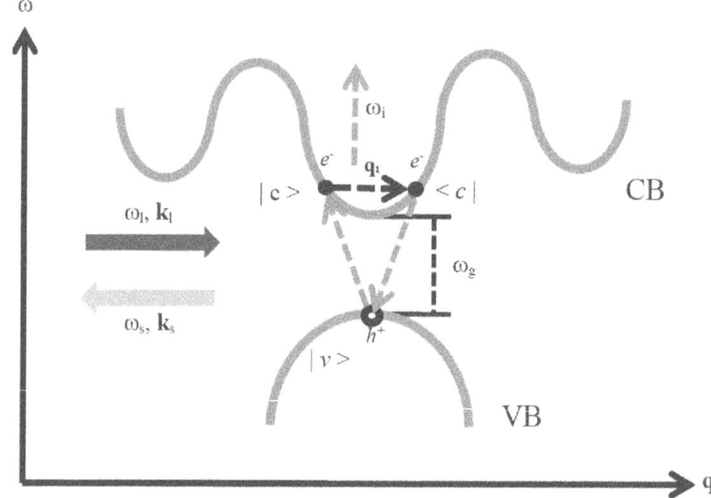

Figure 9.8 A two-band process in which incident light ω_l, K_l excites an electron from the valence state $|v\rangle$ to the conduction state $|c\rangle$. The electron scatters to a different q-state $\langle c|$ emitting a phonon ω_i, q_i, and then recombines with the hole in $|v\rangle$ emitting scattered radiation ω_s, K_s. The intermediate states $|c\rangle$ and $\langle c|$ should be summed over a complete set; ω_g is the direct bandgap a $q \approx 0$.

frequencies. An increase of pressure usually results in an increase of phonon frequencies due to the increase of the interatomic force constants. Figure 9.9 shows the shift of phonon frequencies towards higher wavenumbers with increasing pressure in the scheelite phase of YVO_4 [15]. There are some phonons whose frequencies decrease (soften) on increasing pressure. They are known as "soft" modes and show anomalous behavior. In a soft mode, the decrease in frequency with increasing pressure is mainly due to the decrease of the related interatomic force constant despite the decrease of the interatomic distances. A soft mode of T (B_g) symmetry is observed in the scheelite phase of YVO_4 in Figure 9.9. The existence of soft modes is related to the dynamic instability of the crystalline lattice. This instability can induce a collapse of the crystalline structure with increasing pressure and consequently to a change of the crystalline structure of the material (thus leading to a solid-solid phase transition) or to the destruction of the long-range order in the crystalline structure (thus leading to amorphization or disproportionation). In this respect, soft mode behavior monitoring helps to reveal the amorphization, disproportionation, or phase transition mechanisms occurring in compressed materials. In particular, soft phonons are frequent in second order phase transitions whose mechanisms can be elucidated by applying Landau theory. In this way, the soft mode of

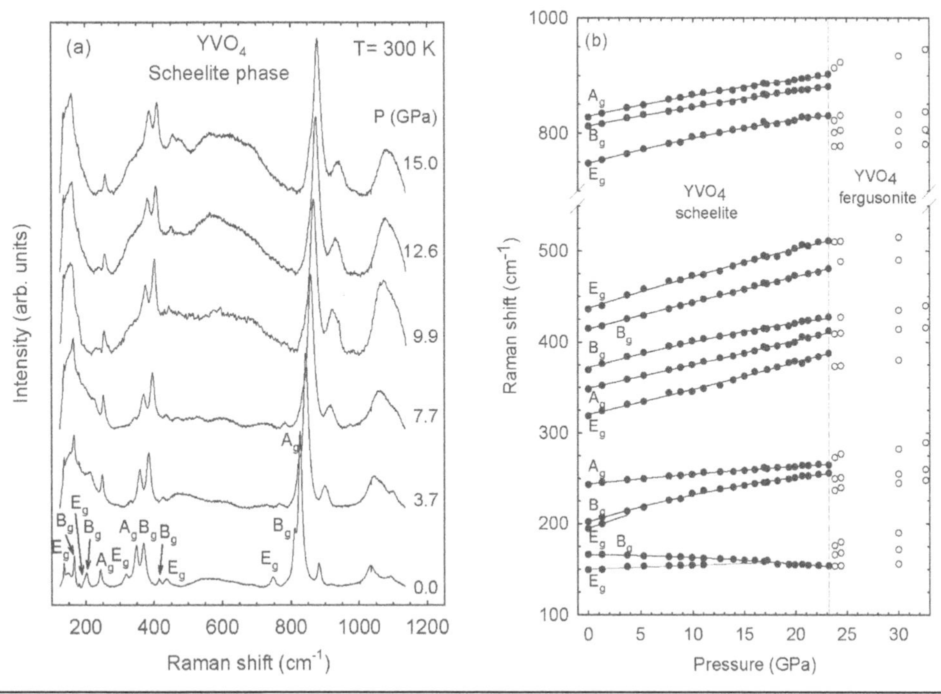

Figure 9.9 (a) Raman spectra of the high-pressure scheelite phase of YVO_4 up to 15 GPa. (b) Pressure dependence of the first order Raman mode frequencies in scheelite-type and M-fergusonite-type phases. Note the coincidence of frequencies of the lowest Raman modes of the two phases near 23 GPa in agreement with the second order nature of this phase transition. Adapted, with permission, from Reference 15.

Figure 9.9 is related to the second order phase transition between scheelite and fergusonite structures in YVO_4 taking place above 23 GPa [15, 16].

In general, the pressure dependence of lattice parameters and volume of a material in a given structure is not linear due to the hardening (increase of bulk modulus) of materials with the increase of pressure. Similarly, the pressure dependence of Raman mode frequencies is not linear. The reason is that the Raman mode frequencies scale mainly with the interatomic distances, which decrease in a non-linear way due to the increase of the bulk modulus of the material [17]. In general, the pressure dependence of the phonon frequencies is sublinear, just as the pressure dependence of lattice parameters and volume; i.e., at large enough pressures a variation in pressure, ΔP, leads to a smaller change in phonon frequency, volume and lattice parameters than that caused by the same variation ΔP at low pressures. Figure 9.10 shows the

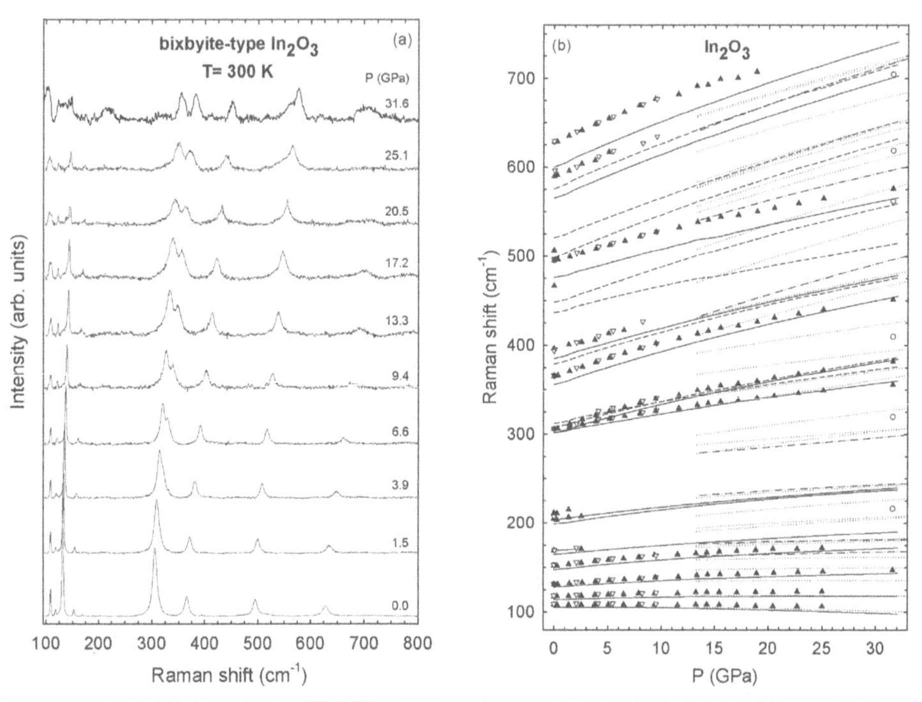

Figure 9.10 (a) Raman spectra of cubic In_2O_3 at different pressures up to 31.6 GPa. (b) Pressure dependence of the first order Raman mode frequencies in cubic In_2O_3. Comparison of experimental data (symbols) with lines (theory) shows that the Raman modes up to 31.6 GPa correspond to the cubic phase (solid lines). Cubic modes not observed are depicted with dashed lines. Theoretical Raman modes of rhombohedral and orthorhombic phases are plotted as dashed dotted and dotted lines, respectively. Adapted, with permission, from Reference 18.

sublinear pressure dependence of both the experimental and theoretical first order Raman mode frequencies in the cubic phase (bixbyite) of In_2O_3 [18].

9.4.2 Changes in Phonon Linewidth and Lineshape Induced by Pressure

Pressure can also affect the linewidth of Raman modes, experimentally characterized by the full width at half maximum (FWHM). Phonon linewidth is related to the imaginary parts of certain self-energies, usually that due to the anharmonic coupling of the Raman phonon to two phonons and that induced by impurities or isotope mass fluctuations [20]. The effect of pressure on the phonon linewidths induced by impurities and by isotope mass fluctuations has not been fully investigated; however, the effect of pressure on the variation of the linewidth of a Raman phonon due to anharmonic coupling of the Raman

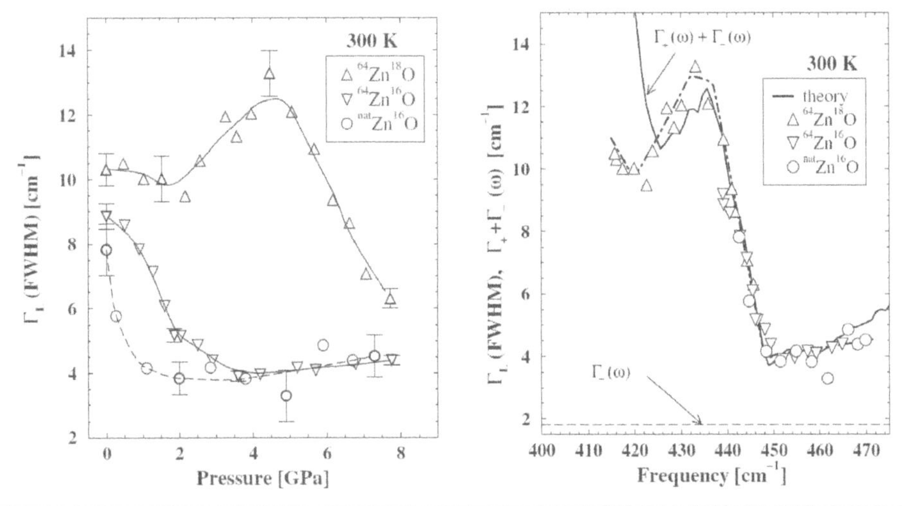

Figure 9.11 (a) Effect of pressure on the linewidth (FWHM) of the E_2^{high} mode in three ZnO samples with different isotopic compositions. Lines are guides to the eye. (b) Experimental points renormalized according to the isotopic composition and comparison with the two-phonon density of states (sum + difference). Adapted, with permission, from Reference 19.

phonon of frequency ω_0 to two phonons of frequencies ω_1 and ω_2 has been investigated. The decay of one phonon into two phonons depends on the two-phonon density of states (DOS). Usually, a high-energy phonon with frequency ω_0 can decay into two phonons of lower frequency provided that $\omega_0 = \omega_1 + \omega_2$; therefore, the linewidth of the mode with frequency ω_0 will depend on the two-phonon sum DOS $\Gamma_+(\omega)$. On the other hand, a low-frequency phonon with frequency ω_0 can decay into two phonons provided that $\omega_0 = \omega_1 - \omega_2$; thus, its linewidth will depend on the two-phonon difference DOS $\Gamma_-(\omega)$. In this way, a first-order Raman phonon has a small (large) linewidth if its frequency ω_0 matches with a low (high) two-phonon density of states. Therefore, pressure can affect the Raman mode linewidth because it induces a shift of the frequencies of all vibrational modes so that the phonon frequency of a Raman mode can sweep the maxima and minima of the two-phonon DOS as pressure changes. The strongest changes of linewidth with pressure and with isotopic composition have been recently observed for zinc oxide (ZnO). Figure 9.11 shows the change in the E_2^{high} linewidth with increasing pressure for different samples of isotopically modified zinc oxide (ZnO) [19]. It can be observed that in some samples the phonon linewidth is large (small) and it decreases (increases) with pressure (see Figure 9.11(a)). In general, the linewidth

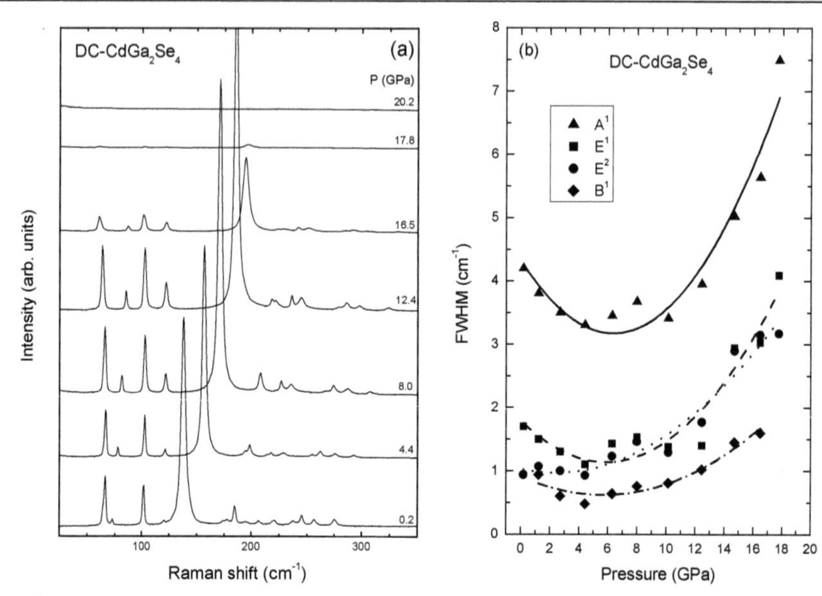

Figure 9.12 (a) Raman spectra of defect chalcopyrite $CdGa_2Se_4$ up to 20.2 GPa. (b) Pressure dependence of the FWHM of the most intense Raman modes. A clear increase of the linewidth of all modes is observed above 10 GPa. Adapted, with permission, from Reference 21.

change is correlated with the shift of the first order E_2^{high} phonon frequency with respect to the two-phonon DOS (see Figure 9.11(b)). In other words, the phonon linewidth as a function of pressure depends on the sweep of the phonon frequency over the two-phonon DOS (sum + difference).

It is noteworthy that the pressure dependence of the linewidth of Raman modes can also be affected by the loss of hydrostatic conditions of the pressure-transmitting medium or by the instability of the crystalline lattice close to a phase transition. In this last case, one can observe in some cases the appearance of precursor defects of the phase transition. In general, one can observe in all these two cases an increase of the linewidth of all Raman modes because both processes lead to a non-homogeneous compression of the sample which results in a Gaussian pressure distribution around a given (average) pressure. This non-homogeneous pressure distribution results in a large phonon linewidth caused by the convolution of the different overlapping Raman modes coming from different parts of the sample (of submicron size) excited by the laser beam. Figure 9.12 exemplifies the overall increase of the Raman mode linewidths in $CdGa_2Se_4$ above 10 GPa mainly due to the formation of precursor defects of a phase transition taking place above 18 GPa [21].

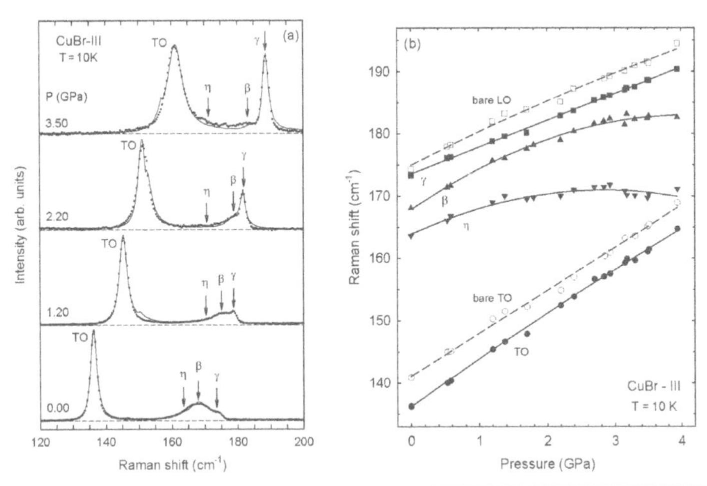

Figure 9.13 (a) Low-temperature Raman spectra of zincblende $^{63}Cu^{81}Br$ measured at different pressures up to 4 GPa (solid symbols). Solid lines correspond to fits of a Fermi resonance model. (b) Pressure dependence of the Raman mode frequencies (solid symbols) and the bare (harmonic) *TO* and *LO* frequencies obtained by fitting a Fermi resonance model (open symbols). Adapted, with permission, from Reference 22.

Pressure can also change phonon lineshapes. Two main causes can be considered: (i) Fermi resonance and (ii) Davydov splitting . In a Fermi resonance a phonon has a complex anharmonic lineshape because its frequency ω_0 is in resonance with a very sharp high density of two-phonon density of states (sum, $\Gamma_+(\omega)$, or difference, $\Gamma_-(\omega)$) due to the coincidence of ω_0 with $\omega_1 + \omega_2$ or $\omega_1 - \omega_2$, respectively [23]. This sharp feature is also known as a van Hove singularity. Since pressure is able to shift a phonon frequency (ω_0) at a different rate than the sum ($\omega_1 + \omega_2$) or difference ($\omega_1 - \omega_2$) of a two-phonon density of states, pressure is able to induce (or remove) a Fermi resonance in a material and consequently induce a lineshape change. Figure 9.13 shows an example of the evolution of a Fermi resonance in zincblende-type CuBr at low temperatures [22]. At low pressures, the *LO* phonon is decomposed into three components due to the resonance of its frequency with a high two-phonon density of states corresponding to *TA*+*LA* and *TA*+*TO* combinations at the *K* point of the Brillouin zone. On increasing pressure, the strong positive shift of the *LO* frequency compared to the frequency of the van Hove singularities leads to the decrease of the Fermi resonance (at 3.5 GPa the *LO* phonon almost has a normal Lorentzian lineshape). Conversely, the *TO* phonon, whose lineshape is rather normal at low pressure, broadens considerably on increasing

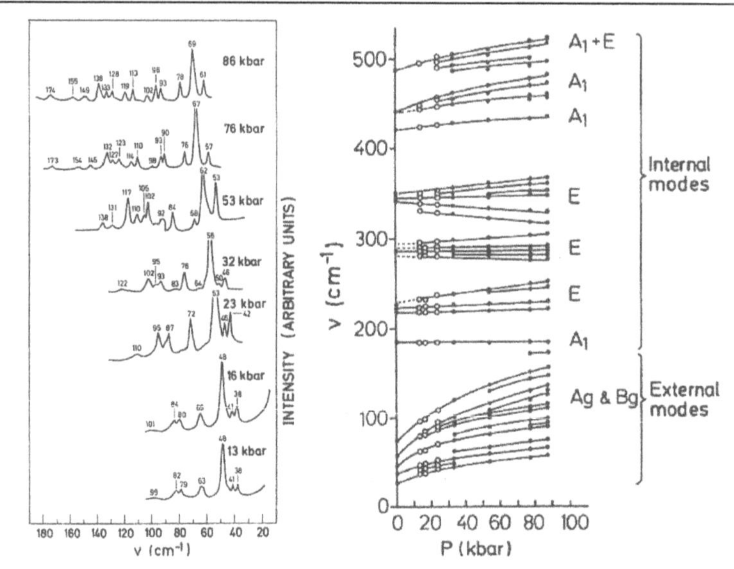

Figure 9.14 (a) Raman spectra of P_4S_3 measured at different pressures up to 8.6 GPa. (b) Pressure dependence of the Raman mode frequencies (symbols) in P_4S_3. Lines are guides to the eyes. Several splittings are noted beyond 2 to 3 GPa. Adapted, with permission, from Reference 25.

pressure as its frequency approaches the van Hove singularities corresponding to the TA+LA combinations at the L and W points of the Brillouin zone due to the larger pressure coefficient of the TO mode than those of the singularities.

On the other hand, Davydov splitting, also known as factor group splitting or correlation field splitting, is another phenomenon that can drastically change the lineshapes of phonon modes. Davydov splitting consists of the splitting of phonon modes in molecular crystals due to the existence of more than one formula unit or molecule per primitive unit cell [24]. Split modes can be very close to each other, leading to the observation of broad features unless one has extremely good resolution. Pressure can help separate the splitting between the different modes due to the different pressure coefficients of Raman frequencies so that they can be observed in a separate way. Figure 9.14 shows an example of Davydov splitting in the Raman spectrum of P_4S_3 [25].

9.4.3 Variations in Intensity of Raman Modes Produced by Pressure

Finally, pressure can also affect the intensity of first order Raman phonons, which mainly depends on the $|\hat{\epsilon}_s \cdot \chi' \cdot \hat{\epsilon}_l|$ factor containing the Hamiltonian matrix elements of electron-photon and electron-phonon interactions responsible

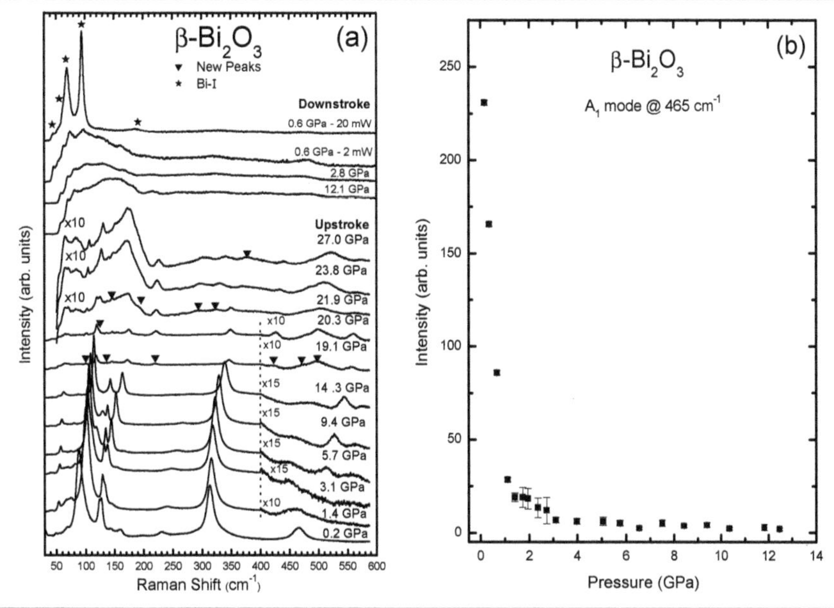

Figure 9.15 (a) Raman spectra of β-Bi_2O_3 measured at different pressures up to 14 GPa. (b) Pressure dependence of the intensity of the A_1 mode at 465 cm^{-1}. Adapted, with permission, from reference 26.

for the Raman effect. The value of the $|\hat{\epsilon_s} \cdot \chi' \cdot \hat{\epsilon_l}|$ factor also depends on the polarization of the incident light $\hat{\epsilon_l}$ and the scattered light $\hat{\epsilon_s}$ which are related by the selection rules affecting the above interactions. In this respect, the most evident effect of pressure on the change of the $|\hat{\epsilon_s} \cdot \chi' \cdot \hat{\epsilon_l}|$ factor occurs when a phase transition between different structures or a pressure-induced amorphization happens. In such cases, the whole Raman spectrum is completely modified due to the different selection rules applying to the original and final structures. Figure 9.12 shows a different Raman spectrum below and above 18 GPa due to the first order phase transition between the defect chalcopyirite phase and the disordered rock salt phase in $CdGa_2Se_4$. In the defect chalcopyrite phase (up to 18 GPa) Raman active modes could be observed, whereas in the disoredered rock salt phase (above 18 GPa) the Raman spectrum is flat because Raman activity of first order phonons is completely forbidden in the rock salt phase. On the other hand, an example of pressure-induced amorphization is observed in Figure 9.15, which shows the Raman scattering of β-Bi_2O_3 as a function of pressure [26]. A broadening of the whole Raman spectrum is observed in β-Bi_2O_3 above 20 GPa due to the loss of long-range order (translation symmetry) typical of amorphous materials. The original Raman spectrum is not recovered on decreasing pressure and only laser heating

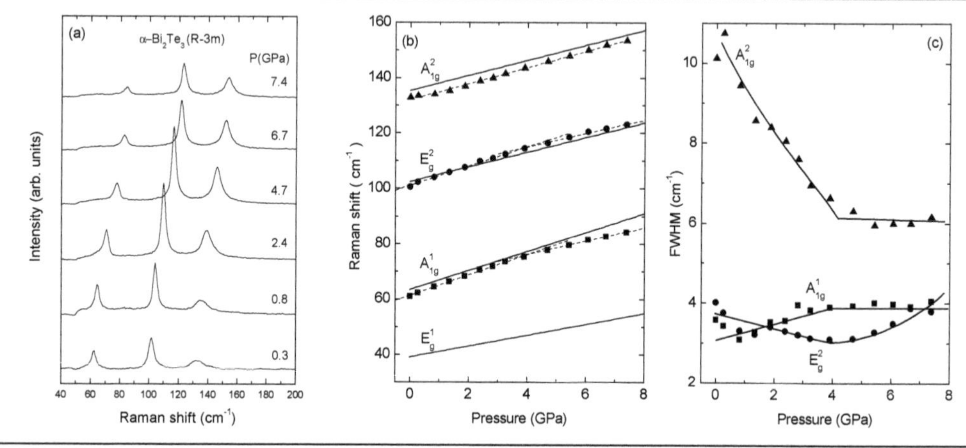

Figure 9.16 (a) Raman spectra of rhombohedral Bi_2Te_3 measured at different pressures up to 8 GPa. Pressure dependence of the frequency (b) and linewidth (c) of Raman modes in Bi_2Te_3. In (b) experimental data (symbols) are compared to theoretical data (solid lines) while dashed lines are guides which indicate the change in the pressure coefficient of some Raman-active mode frequencies near 4 GPa due to the presence of an ETT. A change in Raman mode linewidths is also observed near 4 GPa. In (c), solid lines are guides to the eyes. Adapted, with permission, from Reference 27.

near ambient pressure makes the sample recover a crystalline structure with well defined Raman peaks.

A especial case of pressure-induced phase transition which can affect the Raman spectrum in a rather subtle way is the pressure-induced electronic topological phase transition. An electronic topological transition (ETT) is a structural transition of order $2^{1/2}$ according to Ehrenfest, i.e., it is an isostructural transition caused by a change in the topology of the Fermi surface. The change in the topology is a consequence of the opening or closing of pockets in the Fermi surfaces of electrons and such a change can be induced by pressure. Figure 9.16 shows the pressure dependence of the frequency and FWHM of Raman modes in Bi_2Te_3 [27]. A change in the pressure coefficient of the frequency and linewidth of Raman modes is observed near 4 GPa due to the onset of a pressure-induced ETT.

Apart from the big changes in the Raman spectrum caused by pressure-induced phase transitions and pressure-induced amorphization, pressure can also affect the intensity of Raman modes in a given structure provided that the high-order susceptibility tensor χ' of the material changes as pressure varies.

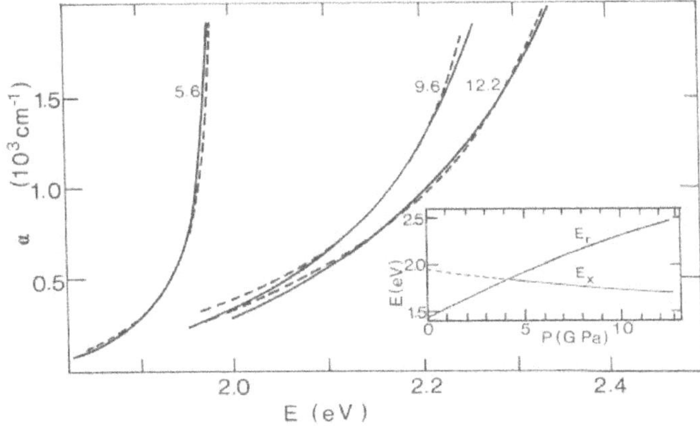

Figure 9.17 Absorption coefficient vs photon energy of GaAs at 300 K for three different pressures (in GPa). Solid lines, experiment. Inset, variation with pressure of the direct (E_Γ) and indirect (E_X) energy gaps, obtained from the fit of the absorption edge. The energy of the gaps varies with pressure as $E_\Gamma(\text{eV}) = 1.427 + 11.5 \times 10^{-2}P - 24.5 \times 10^{-4}P^2$, and $E_X(\text{eV}) = 1.950\text{-}2.6 \times 10^{-2}P + 4.7 \times 10^{-4}P^2$. Adapted, with permission, from Reference 29.

The change in the susceptibility with pressure can be due to the variation of different parameters, such as the number of scattering centers, the energy gap, and the phonon frequency [28]. Figure 9.15 shows the Raman scattering of β-Bi_2O_3 as a function of pressure. A decrease of the intensity of the A_1 mode located at 465 cm^{-1} is observed (see Figure 9.15(b)), which is ascribed to a change in the lattice susceptibility of the material as it approaches a second order isostructural phase transition around 2 GPa [26]. Note that at pressures close to a phase transition, the Raman modes of the initial phase decrease in intensity while the modes of the final phase increase in intensity. This effect is due to the decrease in the number of scattering centers of the initial phase and the increase in the number of scattering centers of the final phase as the phase transition pressure is overcome. This explanation is also responsible for the decrease of the intensity of the Raman modes in defect chalcopyrite $CdGa_2Se_4$ near 18 GPa (see Figure 9.12) because of the decrease in the number of scattering centers of the defect chalcopyrite phase and the increase in the number of scattering centers of the Raman-inactive rock salt phase.

A especial case of change of Raman intensity is observed when the laser energy, $\hbar\omega_l$, coincides with the bandgap energy, $\hbar\omega_g$. In such case, Equation (9.41) leads to a zero in the denominator, thus leading to an extraordinary increase of the Raman mode intensity. This case is known as *resonant Raman*

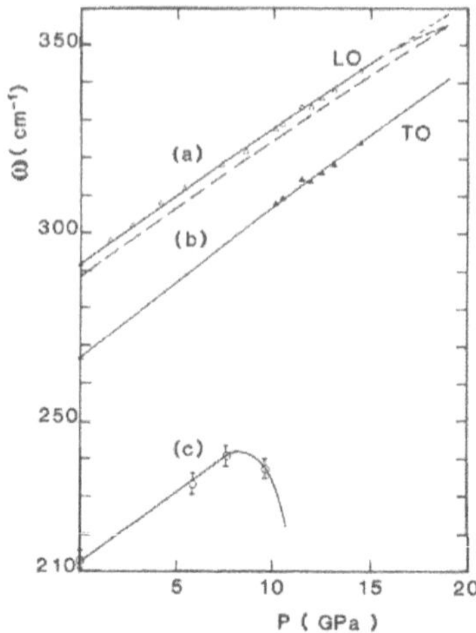

Figure 9.18 Pressure dependence of the wavenumber of the observed (a) *LO* (open triangles, solid line) and (b) *TO* (closed triangles, solid line) Raman modes in [100]-oriented cubic GaAs at increasing pressure. Dashed line is shifted down by -3 cm^{-1} with respect to the solid line of the *LO* mode to reproduce the disorder-induced shift observed in this mode on the downstroke as measured at ambient pressure. Adapted, with permission, from Reference 29.

scattering to differentiate it from the normal case, known as *non-resonant Raman scattering* which is evidenced in the previous examples. Pressure allows the tuning of the bandgap energy of a material to make it resonant or non-resonant with laser energy, thus leading to an increase or decrease of the Raman intensity, as an example, we will show the resonant Raman effect induced by pressure in an oriented single crystal of gallium arsenide in the [100] direction. This semiconductor belongs to the III-V family with the cubic zincblende structure (GaAs-I) and has a direct energy gap at the Γ point of the Brillouin zone of 1.427 eV at 300 K. Under hydrostatic pressure (at about 18 GPa), GaAs-I undergoes a first order phase transition to an orthorhombic structure (GaAs-II) [29]. Figure 9.17 shows the variation with pressure of the direct energy gap (E_Γ) in GaAs-I. A direct-to-indirect bandgap crossover is induced by pressure (around 4 GPa) in GaAs-I so that it becomes an indirect gap semiconductor (with bandgap E_X), which shows an electronic band structure similar to that of silicon (Si) [29].

The change of the frequencies of the *TO* and *LO* modes in GaAs-I as a function of pressure is shown in Figure 9.18 [29]. Since the *TO* mode is forbidden in this geometry, it becomes more easily observable by resonance near 10 GPa when the laser energy (2.409 eV) becomes comparable to that

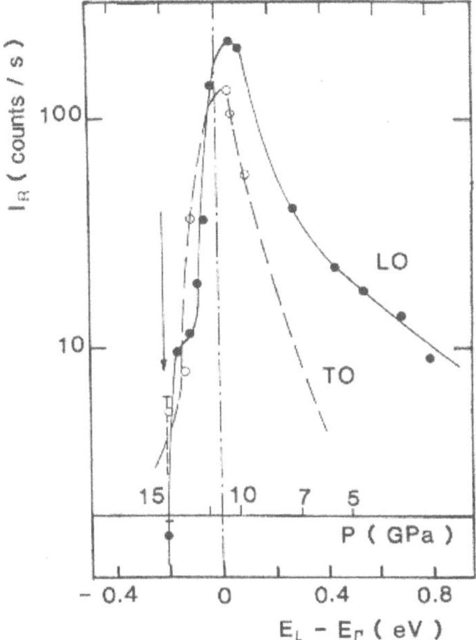

Figure 9.19 Raman intensity of the LO (closed circles) and TO (open circles) modes on the upstroke, at constant incident laser power. Lower abscissa scale represents the energy difference between the laser light (E_L = 2.409eV) and the direct gap energy (E_Γ). E_Γ is related to pressure (upper abscissa scale) as shown in the inset of Figure 9.17. The sharp decrease in intensity (arrow) above 16 GPa indicates the onset of the increasing opacity of the crystal due to the phase transition. Adapted, with permission, from Reference 29.

of the direct gap (see Figure 9.19). The lower part of the diagram of the LO frequency versus pressure below 10 GPa where pressure is homogeneous in the cell is taken as the internal pressure gauge for the sample. The pressure dependence of both modes has been fitted with

$$\nu_{LO} = 291.6 + 3.8P - 0.013P^2,$$
$$\nu_{TO} = 266.4 + 4.22P - 0.015P^2, \tag{9.42}$$

with ν in cm^{-1} and P in GPa.

The intensity of both LO and TO modes has a resonance with a maximum between E_Γ and $E_\Gamma + h\omega_{phonon}$, as expected (ingoing and outgoing resonances) [29]. A noteworthy feature is that the ratio of LO to TO intensities is not symmetrical with respect to the resonance as it should be within the small (25 cm^{-1}) difference in energy of the two modes. This result means that in the region of 15 GPa, the crystal is already perturbed enough to allow selection rules to be broken through local structural defects and/or crystallite misorientation. Further, the steepness of the decrease in intensity above 14 GPa is to be related to diffusion on the incident and outgoing light as was shown in transmission. Finally, the sharp decrease around 16 GPa marked by an arrow in Figure 9.19 just locates the pressure where the crystal irreversibly becomes opaque at the phase transition towards the orthorhombic phase.

Bibliography

[1] A. Smekal. *Naturwissenschaften*, 11, 873 (1923).

[2] C. V. Raman and K. S. Krishnan. *Nature*, 121, 501 (1928).

[3] J. W. Strutt. In *Scientific Papers*, Volume 4, 397–405. Cambridge University Press, Cambridge (2009).

[4] W. H. Weber and R. Merlin. *Raman scattering in materials science*, Volume 42. Springer, Berlin (2000).

[5] I. R. Lewis and H. Edwards. *Handbook of Raman spectroscopy: from the research laboratory to the process line.* CRC Press, Boca Raton (2001).

[6] R. P. Feynman. *QED: The strange theory of light and matter.* Princeton University Press, Princeton (2006).

[7] D. Kaiser. *Am. Sci.*, 93, 156 (2005).

[8] G. Turrell and J. Corset. *Raman microscopy: developments and applications.* Elsevier, London (1996).

[9] R. Salzer and H. W. Siesler. *Infrared and Raman spectroscopic imaging.* Wiley Online Library, Weinheim (2009).

[10] P. W. Langhoff, S. T. Epstein, and M. Karplus. *Rev. Mod. Phys.*, 44, 602 (1972).

[11] B. A. Weinstein and G. J. Piermarini. *Phys. Rev. B*, 12, 1172 (1975).

[12] M. Millot, R. Tena-Zaera, V. Munoz-Sanjose, *et al. Appl. Phys. Lett.*, 96, 152103 (2010).

[13] C. Ulrich, E. Anastassakis, K. Syassen, *et al. Phys. Rev. Lett.*, 78, 1283 (1997).

[14] B. Di Bartolo. *Optical interactions in solids.* John Wiley & Sons, Inc., New York (1968).

[15] F. J. Manjón, P. Rodríguez-Hernández, A. Muñoz, *et al.* *Phys. Rev. B*, 81, 075202 (2010).

[16] D. Errandonea and F. J. Manjón. *Mater. Res. Bull.*, 44, 807 (2009).

[17] M. Cardona. *J. Phys. Coll.*, 45, C8 (1984).

[18] B. García-Domene, H. M. Ortiz, O. Gomis, *et al.* *J. Appl. Phys.*, 112, 123511 (2012).

[19] J. Serrano, F. J. Manjón, A. H. Romero, *et al.* *Phys. Rev. Lett.*, 90, 055510 (2003).

[20] M. Cardona. *Phys. Status Solidi B*, 241, 3128 (2004).

[21] O. Gomis, R. Vilaplana, F. Manjón, *et al.* *J. Appl. Phys.*, 111, 013518 (2012).

[22] F. J. Manjón, J. Serrano, I. Loa, *et al.* *Phys. Rev. B*, 64, 064301 (2001).

[23] M. Krauzman, R. M. Pick, H. Poulet, *et al.* *Phys. Rev. Lett.*, 33, 528 (1974).

[24] A. S. Davydov. *Theory of molecular excitons.* McGraw-Hill, New York (1962).

[25] T. Chattopadhyay, C. Carlone, A. Jayaraman, *et al.* *Phys. Rev. B*, 23, 2471 (1981).

[26] A. Pereira, J. Sans, R. Vilaplana, *et al.* *J. Phys. Chem. C*, 118, 23189 (2014).

[27] R. Vilaplana, O. Gomis, F. Manjón, *et al.* *Phys. Rev. B*, 84, 104112 (2011).

[28] C. Trallero-Giner, K. Kunc, and K. Syassen. *Phys. Rev. B*, 73, 205202 (2006).

[29] J. M. Besson, J. P. Itié, A. Polian, *et al.* *Phys. Rev. B*, 44, 4214 (1991).

CHAPTER 10

Physical Properties of Fluid Media

Bérengère Guignon

MALTA Consolider Team, Departamento de Química Física I - Universidad Complutense de Madrid, and Departamento de Procesos—Instituto de Ciencia y Tecnología de Alimentos y Nutrición (ICTAN-CSIC), Madrid, Spain

Valentín García Baonza

MALTA Consolider Team and Departamento de Química Física I, Universidad Complutense de Madrid, Madrid, Spain

CONTENTS

10.1 INTRODUCTION

THIS chapter is devoted to materials in the fluid or liquid states and it is focused on the study of their physical properties at pressures below 1 GPa. This pressure level still marks the limit for most practical applications requiring an accurate knowledge of fluid properties and large-scale industrial processes. These areas of study span from simple liquids to very complex fluid systems related to living organisms and microbiology (Chapter 11), food technology (Chapter 12), cosmetic and pharmaceutical preparations (Chapter 13), geophysics and geochemistry (Chapter 15), and astrobiology (Chapter 16).

The knowledge of physical properties of fluids is therefore vital for understanding interdisciplinary phenomena and also required for performing and checking simulations by molecular dynamics or computational fluid dynamics. The potential benefits range from the optimization of industrial processes to a better comprehension of biological phenomena and geological events.

The objective of this chapter is to provide an overview of theoretical and experimental aspects of the determination of fluid physical properties under pressure. In Section 10.2, the difficulties inherent to the predictions of fluid properties will be explained and the approaches adopted to offer a generalized understanding will be presented. Section 10.3 will be dedicated to selected experimental methods. We will emphasize non-optical techniques, since structural and optical techniques have been addressed in Chapters 7 to 9. Techniques for the determination of volumetric, acoustic, and thermal properties will be described. The chapter closes with selected examples to illustrate how the study of such properties can lead to a better understanding of the behavior of fluids under pressure, promote progress of high-pressure science and technology, and enhance the development of novel industrial processes.

10.2 MODELS FOR PREDICTION OF FLUID PHYSICAL PROPERTIES

10.2.1 General Background

Historically, the study of fluid properties under pressure dates back to 1762, when physicist John Canton demonstrated for the first time that liquids were in fact compressible [1]. Since then, many high-pressure devices for physical property measurements have been developed, with special mention to those built by P. W. Bridgman during the first half of the 20th century [2]. However, in spite of these advances, few commercial systems for measuring physical properties are available and home-built devices are found in most laboratories.

The experimental studies of fluids under pressure have shown that, in comparison to normal gases and crystals, the properties of normal liquids and fluids are difficult to predict. There are evidences that some properties show a more or less general behavior, but numerous exceptions to the rules appear; the prototypical example is liquid water which accumulates the largest concentration of anomalies known (Section 10.4). The lack of a systematic behavior for liquids limits our knowledge in other areas of high-pressure research in which they are involved, for example, melting phenomena, high-pressure amorphization, glassy phases and liquid-liquid phase transitions, to name a few. Our knowledge of the high-pressure behavior of fluids is mostly derived from experiment, which explains why scientists have developed a variety of complicated devices and performed laborious measurements to obtain accurate material property data.

The primary fluid properties studied were volumetric. Volumetric properties are characteristic of the matter which defines the space it occupies in relation with the mass of material m and how this space varies with the pressure and temperature conditions: density ρ and its reciprocal the specific volume v and the mechanical coefficients: thermal expansivity or thermal expansion coefficient α_p, isothermal compressibility coefficient κ_T, piezothermal or thermal-pressure coefficient γ_v, and the bulk modulus B. These equations recall the basic thermodynamic relations between these properties:

$$\rho = \frac{m}{V} \quad \text{and} \quad v = \frac{1}{\rho}, \tag{10.1}$$

$$\alpha_p = \frac{1}{v}\left(\frac{\partial v}{\partial T}\right)_p, \tag{10.2}$$

$$\kappa_T = -\frac{1}{v}\left(\frac{\partial v}{\partial p}\right)_T, \tag{10.3}$$

$$\gamma_v = \left(\frac{\partial p}{\partial T}\right)_v = \frac{\alpha_p}{\kappa_T}, \tag{10.4}$$

$$B = -v\left(\frac{dp}{dv}\right). \tag{10.5}$$

As shown in these equations, the experimental determination of these properties essentially derives from mass, volume, pressure, and temperature measurements. After a series of volumetric measurements, one typically obtains triplets of (p, v, T) data that completely define the thermodynamic state of a system. Similar to the solid state (Chapter 1 and 2), the most common method to represent the thermodynamic properties of a fluid is to choose an analytical function (*e.g.*, $v = f(p, T)$) or equation of state (EOS) (*e.g.*, $g(p, v, T) = 0$) able to represent or more precisely to fit the available pressure-volume-temperature coordinates (pvT surface), thus allowing the calculation

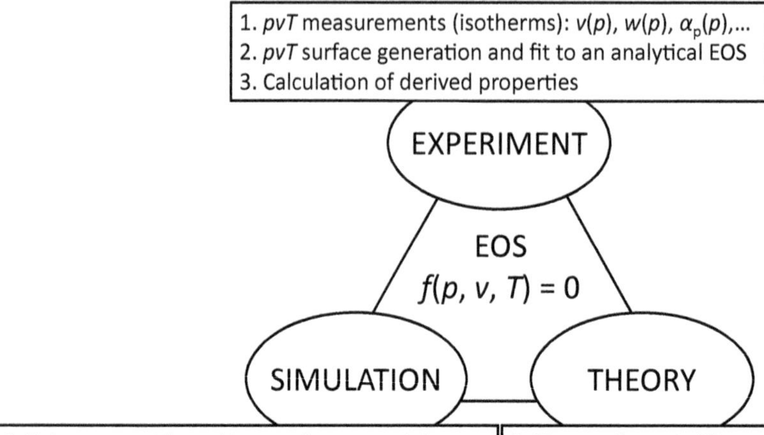
1. pvT measurements (isotherms): v(p), w(p), α_p(p),...
2. pvT surface generation and fit to an analytical EOS
3. Calculation of derived properties
EXPERIMENT
EOS
f(p, v, T) = 0
SIMULATION
THEORY
1. Select ensemble and interaction potential
2. Run Molecular Dynamics, Monte Carlo,...
3. Comparison/calculation of Derived Properties
1. Thermodynamic/Phenomenological Models
2. GvdW, Scaled Particle, Perturbation Theory,...
3. Comparison/calculation of Derived Properties

Figure 10.1 Workflow for a complete study of the EOS of a fluid.

of any desired derived property using standard thermodynamic relationships (Equations (10.2) to (10.5) and others). The selection of a proper EOS is mostly heuristic since there is no definitive theory or unique model. However, some general trends have already been defined from the available data that serve as guides for the development of a unified theory of fluid behavior under pressure. In Figure 10.1, the different steps constituting the complete study of the EOS of a fluid are summarized.

10.2.2 Observed Behaviors of Fluids under Pressure

The most intuitive behavior of fluids under pressure is the decrease of specific volume upon compression. The isothermal compressibility also follows this trend. However, distinct patterns can be observed when considering different temperature regimes (Figure 10.2). Indeed, several properties of fluids show barely general behavior as a function of pressure. For instance, the bulk modulus (i.e., the inverse of compressibility, often referred to as compression modulus in the literature of fluids) is also almost linear with increasing pressure [3]. This behavior was recognized by Tait in 1888 [4] for liquid water and has served as a basis for developing empirical EOS, which are essentially equivalent to the Murnaghan equation in solids (Chapter 1). Another observation is the linear relationship between the temperature-density coordinates whose compressibility factor ($Z = pV/RT$) is unity. This fact seems to be little known, despite discovery nearly a century ago. Huang and O'Connell [5]

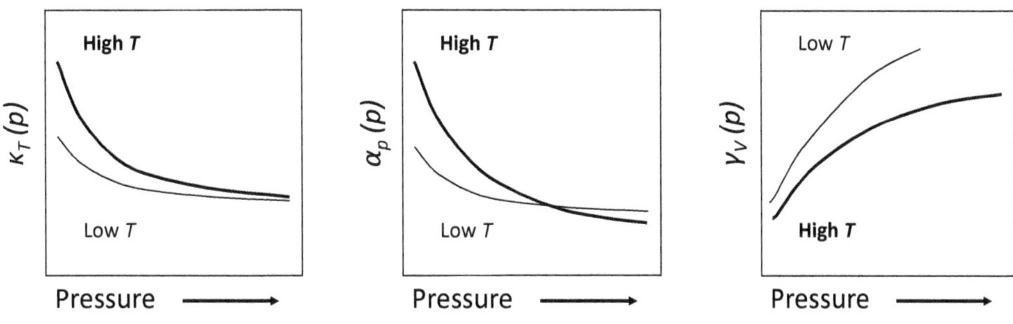

Figure 10.2 Pressure dependence of selected volumetric properties of normal fluids under different temperature regimes.

examined the high-pressure behavior of 250 fluids, and they concluded that, for each substance, there exists a density at which the reduced isothermal compression modulus ($B_{RED} = 1/(\rho\kappa_T RT)$) is independent of temperature. This feature has also served to reveal further regularities [6]. The thermal-pressure coefficient varies slowly with volume, so a plot of pressure versus temperature along isochoric paths is essentially a straight line, even in the vicinity of the critical point.

Finally, the pressure dependence of the thermal expansivity or thermal expansion coefficient has also been of interest since the isotherms of $\alpha_p(p)$ exhibit characteristic intersections at high pressures. From the view of other thermodynamic properties, this implies that $c_p(p)$ exhibits a shallow minimum at high-pressure, but this feature has also been considered a constraint for analytical EOS [7, 8]. Some of these regularities have been successfully interpreted within the framework of statistical thermodynamics and the general conclusion is that at high densities the quality of an EOS model depends mainly on the analytical form of the repulsive part of the interaction potential, while modifying the attractive part provides little improvement - a conclusion that has led to criticisms of the well-known cubic EOS derived from the van der Waals equation [9].

A successful approach to modeling the high-pressure behavior of fluids is based on power laws referred to the so-called spinodal curve [10], that essentially represents the limit of mechanical stability of a given phase (here, liquid). Although the spinodal curve can never be reached in practice, its existence is supported by a large number of experiments and its location is determined by extrapolation of data measured in thermodynamically stable states; the curve thus obtained is called the pseudospinodal curve, and spinodal and pseudospinodal terms are used interchangeably in the literature. An expansion of the Helmholtz potential shows that when approaching the spinodal curve, the mechanical coefficients, the viscosity and the diffusion coefficient, follow power laws along isothermal paths [11]. From a practical view, the question is: how large is the range over which these power laws hold? Experience reveals that the exponents remain almost constant over both metastable and stable regions of a liquid and several phenomenological models to represent $\kappa_T(p)$ and $\alpha_p(p)$ have been applied to the study of many liquids [12], including the development of a universal EOS model [13]. This approach was also employed to give a unified description of the thermodynamic properties of superheated, stretched, supercooled, and ordinary water under the so-called stability-limit conjecture [14].

10.2.3 Equation of State

Although the theoretical understanding of the liquid state is continuously evolving and improving, most phenomena occurring at high-pressure have not received theoretical support. In particular, a fundamental problem in high-pressure research is still to find a general EOS form valid for all types of fluids. However, the field has advanced over recent decades with the impressive progress resulting from by computer simulation, and now it is possible to combine theoretical, experimental and simulation aspects to develop increasingly accurate and realistic EOS models.

From the fundamental view, most theoretical EOS models have been derived to correlate the observed high-pressure behavior of fluids with physically meaningful parameters. The simplest and best example is the van der Waals (VdW) equation:

$$\left(p + \frac{n^2 a}{V^2}\right)(V - nb) = nRT, \tag{10.6}$$

that makes explicit for the first time the concept of intermolecular potential parameters (although not considered in the first derivation) through the covolume b and the attractive parameter a.

The VdW equation led to the development of many other forms: the well-known family of cubic EOS and, in combination with the statistical thermodynamics models, to the so-called generalized van der Waals (GVdW) EOS [15]. Many GVdW EOS have been adapted for different classes of fluids depending on the interactions that dominate the system. Pure fluids can be generically classified according to the strengths and types of intermolecular interactions. Simple fluids are systems having spherical or highly symmetrical molecules that can be represented by dispersion interactions (*e.g.*, noble gases or methane). The rest may be considered complex fluids: those composed of non-spherical or large molecules with many internal degrees of freedom (*e.g.*, nitrogen, hydrocarbons), associated liquids (*e.g.*, water and alcohols), highly polar liquids (*e.g.*, ionic liquids, alkaline halides), and liquid metals, which are dominated by long-range Coulomb forces. Most of the models developed to date are focused on establishing a principle of corresponding states for similar fluids in terms of the characteristic parameters of the EOS - an idea already introduced by van der Waals in terms of the critical parameters.

If we further consider fluid mixtures, even those obtained by mixing of simple fluids, the complexity of their phase diagrams and thermodynamic behavior increases dramatically; think, for instance, of mixtures involved in food technology, geological and biological fluids, and molten minerals. The enhanced complexity found in the studies of the high-pressure behavior of fluid mixtures gave rise to another extensive area of research around the EOS of fluids, namely the development of parameter mixing rules for developing extended principles of corresponding states.

The previous models of EOS are very important for our understanding of the fluid state, but their form is usually far from convenient to correlate experimental results for practical purposes. In other words, they are often too stiff or too complex to fit the thermodynamic properties of fluids. Thus, other forms are commonly employed to correlate experimental pVT data, which are usually available as tabulated (p, V) pairs along several selected isotherms. For this reason, there exists some confusion about the terminology employed in the field, since most expressions should refer to as isothermal EOS. A thermodynamically consistent EOS should conform the relationship $f(p, V, T) = 0$, so additional functions must be selected to account for the temperature dependence of the characteristic fitting parameters of a given isothermal EOS.

Among the most simple isothermal EOS is the linear secant modulus [3, 16] but the most successful is the Tait equation that we shall use here as an example and may be written as:

$$\kappa_T = -\frac{1}{V}\left(\frac{\partial V}{\partial p}\right)_T = \frac{1}{V}\frac{C_2}{C_1 + p}, \tag{10.7}$$

or in the integrated form:

$$V = V(p_0) - C_2 \log\frac{C_1 + p}{C_1 + p_0}, \tag{10.8}$$

where the parameters C_1 and C_2 are functions of temperature that are independent of pressure.

The pressure parameter C_1 has received several theoretical interpretations, but most of them involve identifying the quantity $(p + C_1)$ as the difference between the thermal and attractive pressures of a liquid. Leyendekkers [17] compares C_1 for water to the network stress derived from the application of the scaled particle theory, and Baonza *et al.* [18] identified C_1 as the negative value of the divergence pressure along the spinodal curve. This latter interpretation has the advantage that theoretical expressions for the temperature dependence of the spinodal curve are available [19], thus facilitating the derivation of a complete pVT EOS.

Of course, there exist several modified forms of the Tait equation that have been proposed in pV correlations that will not be considered here for the sake of brevity. Other isothermal equations that are frequently cited in the literature are those borrowed from the study of the solid state, namely the Murnaghan, Birch–Murnaghan, Vinet (see Chapter 1), and the spinodal equation discussed in the previous section.

Finally, let us emphasize that a common problem derived from pV correlations to a given isothermal EOS is that the parameters for the same substance may show considerable disagreement among different authors. This problem also occurs when one compares different values of the bulk modulus at atmospheric pressure B and its pressure derivative B' in solids (see Chapter 1). Sometimes the problem arises simply because the parameters are obtained from data covering different pressure ranges. This is in fact a numerical problem related to data analysis which is often also biased by the form of the EOS selected. Manipulation of numerical data analysis is undoubtedly one of the major problems when determining reliable values of the thermodynamic properties of fluids (and solids), and a subject of continuous controversy through the years. Another undesirable by-product of poor numerical analysis is the loss of experimental information, since many raw experimental pV results have been lost through decades as the original tabulated results were replaced by a poor fit procedure at the moment of publication.

10.3 EXPERIMENTAL DETERMINATION OF FLUID PHYSICAL PROPERTIES

In this section, after recalling the basic measuring principle, several experimental techniques will be briefly described. We will discuss their main characteristics (pressure range, versatility, relative accuracy), together with their advantages and limitations. Most high-pressure devices fall into the large-volume category and deal with specific properties. Anvil cell devices (including the Paris–Edinburgh type) coupled to structural or optical techniques are used to study fluid media as well but, as already mentioned in Chapter 6, they usually lack the required accuracy to study the thermodynamic properties of fluids, so they will not be considered here.

10.3.1 Volumetric Properties

At atmospheric pressure, the simplest device to measure the density of a liquid is a pycnometer: a given volume of liquid placed in a calibrated volumetric flask is weighed. Other more or less complex methods have been developed according to the characteristics of the liquids (volatile, highly viscous, corrosive, etc.) and the specific purpose of the measurement (non-invasive, on-line in the production chain, high cost of material, accuracy, etc.). All these methods usually involve accurate temperature control and regulation, rendering relatively easy the determination of thermal expansivity at atmospheric pressure from Equation (10.2).

Regarding experimental determinations at high-pressure, both direct and indirect methods exist. In the case of **direct methods**, the measurement starts with the determination of the volume of sample contained in the device at atmospheric pressure. Then the volume decrease due to compression is measured at the target pressure under constant temperature (isothermal devices). The devices used for direct volumetric measurements at high-pressure have already been reviewed by several authors [20–24]. Only a general description is provided here.

The devices used for direct measurement of volume under pressure have two main parts: the sample container with a defined volume (cell) and the volumometer (*i.e.*, device for volume change measurement). The sample container is usually cylindrical or bellows-like. The volumometer commonly consists of a liquid column in a capillary tube, a solid piston in a cylindrical cell (or reservoir), or a piece attached to the smooth part of a bellows. The position x of the piston or bellows extremity can be monitored by direct visual reading, a micrometer screw, a resistance wire, capacitance measurement, or a differential transformer. The way the cell and the volumometer are designed and arranged has led to classification as variable-volume devices and isochoric devices [22, 25]. In variable-volume devices, the mass of sample inside the cell is constant during all the measurements (Figure 10.3). A floating piston in a cylindrical cell or a bellows cell enables sample volume decrease upon compression. The volume change suffered by the sample is thus computed from the change in the position of the piston or bellows extremity. The section S (or effective section in the case of bellows) is considered to be approximately constant with pressure. Therefore, the sample volume change ΔV is calculated from the piston or bellows extremity position change Δx:

$$\Delta V(p) = S\Delta x(p). \tag{10.9}$$

The sample volume at a given pressure is then:

$$V(p) = V(p_0) - \Delta V(p), \tag{10.10}$$

where $V(p_0)$ is the sample volume at atmospheric pressure p_0. This initial sample volume can be calculated from the known sample density at atmospheric

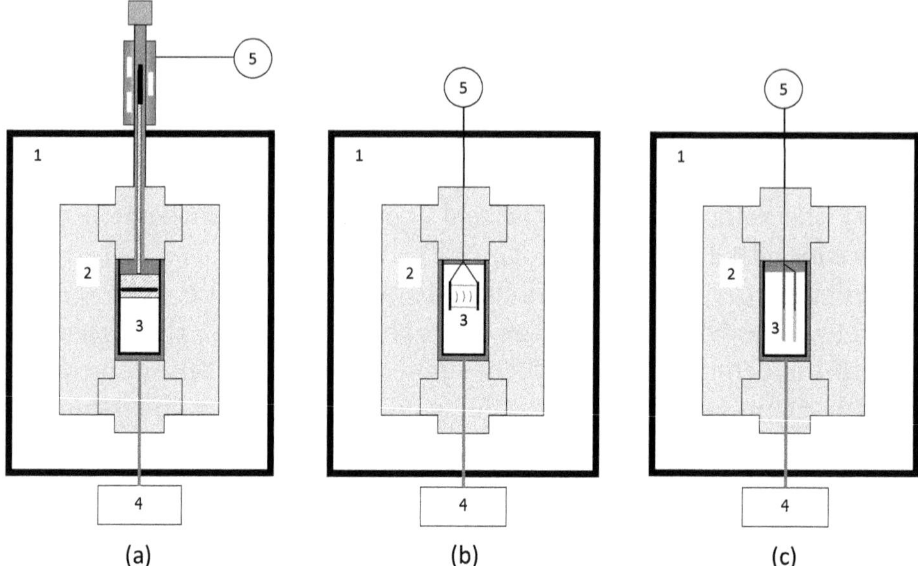

Figure 10.3 Typical apparatus used for direct physical property measurements in fluids at high-pressure: (a) variable-volume device, (b) pulse-echo ultrasonic cell, (c) dual-needle probe. 1: thermostatic bath, 2: high pressure vessel, 3: sample, 4: compression pump, 5: linear variable differential transformer in (a), pulse generator and oscilloscope in (b), heat source in (c).

pressure and from the mass of sample introduced in the device. In isochoric devices, the cell has a constant volume; the mass of sample contained in the cell varies. A capillary connection between the cell and a reservoir enables the entry (or exit) of a fluid sample in the cell during compression (or pressure release). The amount of fluid injected (or ejected) is determined from the fluid level in the reservoir (volumometer) or by direct weighing. The fluid in the reservoir is kept at a reference temperature (T_{ref}) different from the temperature of measurement. The sample density at this reference temperature must have previously be well characterized as a function of pressure, which enables us to calculate sample mass variation in the constant-volume cell. Hence, the sample mass in the cell at a given pressure is:

$$m(p) = \rho(p, T_{ref})\Delta V(p). \tag{10.11}$$

Each kind of device presents advantages and drawbacks. For volumometers using a liquid column, the solidification of this liquid under high-pressure, low temperature, conditions and its solubility in the liquid sample set some limits. For a solid piston volumometer, the piston seal is the problematical part: leaks may occur and its deformation under pressure contributes to volume variation in a small but significant way at the beginning of the compression (pressures below 10 to 20 MPa) [16]. For bellows, the distortion of the solid metal holder in the radial direction has to be corrected. The selection of the most suitable device depends on the temperature and pressure ranges of interest, the required accuracy, and the characteristics of the sample.

The most intuitive and practical procedure for density (specific volume) determination is to perform isothermal measurements. For this purpose, the high-pressure vessel is held at a given temperature (for example, by immersion in a bath). When the target temperature is reached, the pressure, temperature, and volume are recorded. Then the pressure is increased step by step at each point of measurement. During compression from one point to another, adiabatic heat is generated in both the sample and the compression fluid. Depending on the pressure increment, compression fluid, and sample characteristics, the temperature can rise from barely 0.01 K per MPa in aqueous samples to more than 0.10 K per MPa in oil or alcohol samples [26]. Between each step, it is necessary to wait for heat dissipation before recording the measurement. It is seldom possible to measure the temperature directly inside the sample during volume measurements so an estimation of the temperature rise with pressure increment and the time necessary to recover the target temperature is desirable (this can be done in a separate experiment or by calculation if data are available for this). Temperature control is indeed essential to limit the uncertainty in the isothermal compressibility determination. It is important to keep the pressure as constant as possible while measuring the volume at different temperatures with the aim of deriving the thermal expansion coefficient with the best accuracy possible. Other relevant factors to consider when measuring density (and other volumetric properties) under pressure are the compression of the device and the presence of air in the sample. Under the effect of pressure, the device dimensions change: for the early unjacketed devices, this deformation was anisotropic but current devices are designed with the sample holder completely surrounded by the compression fluid (hydrostatic conditions) so the deformation is considered isotropic. This leads to an over- or under-estimation of the sample volume if corrections are not handled properly. Such corrections can be obtained from calculations based on material properties and deformation or directly evaluated from a specific experiment with no sample (whenever possible), or indirectly by using a well-characterized standard. Gases are roughly 100 to 10000 times more compressible than liquids at pressures below 10 MPa. Therefore, even small volumes of air involve relatively large volume variations at low pressures in comparison to the volume variation of a liquid sample. At pressures above (10 to 20) MPa, such air contribution to sample volume drop becomes negligible and the sample volume starts to decrease linearly with the pressure increase. Different options exist to deal with this issue: sample degassing before loading, filling of the sample in the high-pressure vessel under vacuum, or linear interpolation between measurements at atmospheric pressure and measurements at pressures above 20 MPa. Over the rest of the explored pressure domain, an outlier in the set of collected data usually indicates sample leak or contamination by the compression fluid or a phase change during measurement. After correction and analysis of the raw data, the next step, the selection of a suitable EOS, can begin (Figure 10.1). After numerical fitting, the isothermal compressibility, thermal expansion coefficient, and other thermodynamic properties can be derived using Equations (10.2) to (10.5).

In the case of **indirect methods**, acoustic and calorimetric measurements are used in combination with thermodynamic equations to relate the measured magnitude with the volumetric property to be determined. These methods will be described in later sections.

Another indirect method which is commonly employed is the vibrating-tube technique, which is considered as the most commercially successful system for density determination up to 100 MPa. This device is composed by a U-shape tube filled with the fluid to be analyzed. It is connected to a volumetric pump to generate pressure, and the temperature is electronically controlled (thermostatic bath). The tube is excited to a vibration of constant amplitude and the oscillation period of the filled tube varies with the density of the liquid inside the tube. The density and the square of oscillation period are related by a very simple law:

$$\rho = A\tau^2 + B, \tag{10.12}$$

and several calibration procedures for A and B have been proposed [27, 28].

A rather different but interesting indirect way to obtain the EOS of fluids is the so-called expansion method, which provides an easy way to assess the mechanical coefficients with reasonable accuracy (*ca.* 2 % to 3 %) [29]. The intent is to determine the amount of fluid contained in a sample vessel of known volume under fixed conditions of pressure and temperature by expanding the sample to the gas phase into a much larger vessel of known volume, which is kept at constant temperature. The measurement of the pressure reached in the expansion vessel allows us to calculate the mass contained in the sample vessel with high accuracy using the virial equation for the gas. This method has been used to measure the pVT surfaces (*i.e.*, the complete EOS) of liquids and compressed gases up to several kbars and temperatures well above the critical one. In one modification the variation of the amount of liquid contained in the sample vessel was obtained along selected isotherms by performing successive partial expansions of the compressed fluid until the vapor pressure of the liquid at the selected temperature was reached. The volume of the liquid at coexistence was determined in a single final expansion step similar to that described for the original expansion method; this is known as the differential expansion method [30]. An additional modification involved measuring a reference isobar, usually room pressure, along which the variation with temperature of the amount of substance contained in the sample vessel is also obtained by differential expansions [31]. The final expansion of each isotherm is thus not required, and a single reference density is required to generate the whole pVT surface of the fluid. Although this method is very accurate, it is not well suited for studying fluids of low vapor pressure at the temperature of the expansion system.

Using the methods described above, many different fluids could have been characterized as a result of the works of several research groups. Among these researchers, Amagat and Bridgman were the first to publish data on fluid volumetric properties. Today databases contain a great variety of fluids whose properties are known over relatively wide ranges of pressure and temperature. However, with the development of high-pressure technology and new industrial fluids, there is a constant need for studying fluid volumetric properties. This is the case of ionic liquids: the density of two trialkylimidazolium-based ionic liquids was recently measured at temperatures from (278 to 398) K and pressures up to 120 MPa [32]. This is also the case of biofuels whose density was studied up to 200 MPa [33]. Other examples will be given in Section 10.4.

10.3.2 Acoustic Properties

Acoustic properties characterize the way that acoustic waves propagate in matter. For practical reasons, acoustic waves with a frequency above 20 kHz, that is, ultrasonic waves are applied for experimental study of acoustic properties. Ultrasounds propagate in fluids as longitudinal mechanical waves. In the case of viscous fluids, transversal components (shear waves) may develop and should also be taken into account. The velocity of the longitudinal wave propagation and its attenuation are the most useful acoustic properties to describe fluid responses to local deformations caused by such waves. The speed of sound w is simply defined as:

$$w = \frac{d}{t}, \tag{10.13}$$

where t is the time (time-of-flight) during which the wave travels between two points separated by the distance d.

The wave loses energy upon its travel through the matter and this is characterized by the linear attenuation coefficient of the wave propagation, α:

$$\alpha = \frac{1}{d} \ln \left(\frac{A_1}{A_2} \right), \tag{10.14}$$

where A_1 and A_2 are the amplitudes of the wave at points 1 and 2, respectively. Both points are separated in time by the wave travelling time corresponding to distance d.

In addition to these properties, the adiabatic (isentropic) compressibility κ_s is also of great interest. It reflects the ease by which the fluid particles locally move upon the passage of the mechanical wave. This property is related with the density and the speed of sound by the Newton–Laplace equation:

$$\kappa_s = \frac{1}{\rho w^2}. \tag{10.15}$$

For example, κ_s was calculated from density and speed-of-sound measurements for glycerol between (283 and 373) K at pressures up to 100 MPa [34].

At atmospheric and high-pressure, the principle for acoustic property determination remains the same. The experimental setups used at atmospheric pressure are modified to suit high-pressure constraints. There are several measuring techniques for acoustic property determination according to the arrangement of the piezoelectric sensor(s): pulse-echo [35–38], double reflector pulse-echo [39], transmission [40, 41], and multi-reflection [42]. Brillouin spectroscopy is also often employed as an indirect method to obtain the speed of sound. All these techniques were adapted for measurements under high-pressure by different researchers. Feed-throughs for cables are made in the high-pressure vessel wall or plug to connect piezoelectric sensors, while dedicated anvil cells are used in Brillouin spectroscopy.

The technical adaptation of direct ultrasound measurement to the high-pressure domain is relatively straightforward. Piezoelectric ceramics working at high-pressure are commercially available (*e.g.*, those manufactured by Ferroperm, Denmark). The main difficulty consists in ensuring a good sealing of electrical cable connections between the atmospheric and high-pressure parts of the setup: leaks and cable ruptures at this point are frequent. Sample mixing under pressure may or may not be necessary and would introduce a higher complexity of the experimental setup. In the case of attenuation coefficient measurements, it is interesting to be able to vary the distance between transducers directly under pressure. These technical complications can be overcome using magnetic stirring [21] and a rotating head, respectively [43]. If some gas is present in the measurement zone, inconsistencies appear between the low- and high-pressure measurements. This is due to the scattering effect of gas at low pressure which become negligible as soon as pressure is increased because the gas dissolves in the liquid phase and no longer interferes. This phenomenon has to be taken into account in the calibration procedure to choose the proper reference value(s) [44]. Filling under vacuum should be mandatory for accurate measurements below 100 MPa [39]. Signal treatment can be done in the same manner as at atmospheric pressure: frequency analysis, correlation function, etc. When the frame holding the piezoelectric sensors is located inside the high-pressure vessel, the distance used in the calculation of the speed of sound should decrease with pressure. This decrease can be calculated from frame material properties (deformation) or from experiments with a standard (*e.g.*, pure water). Under hydrostatic conditions, assuming isotropic deformation, the frame shrinkage in the direction of interest ϵ is obtained from:

$$\epsilon(p) = p\left(\frac{1-2\mu}{E}\right) d(p_0), \tag{10.16}$$

where μ is the Poisson coefficient and E is Young's modulus of the frame material. Thus, the distance set by the frame decreases from $d = d(p_0)$ at atmospheric pressure to $d(p) = d(p_0) - \epsilon(p)$ at pressure p.

Compared to direct methods, the method developed from Brillouin's theory works over a range of higher frequencies (1 to 10 GHz), even with small quantities of sample, and without any precise mechanical system. An incident laser beam is scattered by local fluctuations of the medium density and refractive index. The analysis of the inelastic scattering spectrum allows us to determine the sound wavelength and the speed of sound results from the product of this wavelength and the frequency. The knowledge of viscosity (shear and bulk viscosities) and isobaric and isochoric specific heats is a prerequisite.

The speed of sound increases with pressure as a result of matter compactness which favors ultrasounds propagation (Figure 10.4). Relative uncertainties of the speed of sound under pressure are typically between (0.01 and 0.5) % depending on the pressure level.
Examples of acoustic measurements under high-pressure will be given in Section 10.4.

The main value of acoustic measurements is that the speed-of-sound calculation is useful to indirectly determine density, thermal expansion coefficient and specific heat. The numerical method developed by Davis and Gordon [45] starts from the following relationships:

$$\left(\frac{\partial\rho}{\partial p}\right)_T = \left(\frac{\partial\rho}{\partial p}\right)_S + \frac{T\alpha_p^2}{c_p} = \frac{1}{u^2} + \frac{T\alpha_p^2}{c_p}, \tag{10.17}$$

$$\left(\frac{\partial c_p}{\partial p}\right)_T = -T\left(\frac{\partial^2 V}{\partial T^2}\right)_p = -TV\left[\left(\frac{\partial\alpha_p}{\partial T}\right)_p + \alpha_p^2\right]. \tag{10.18}$$

In their original method, Davis and Gordon used the following additional thermodynamic constraint:

$$\left(\frac{\partial\alpha_p}{\partial p}\right)_T = -\left(\frac{\partial\kappa_T}{\partial T}\right)_p, \tag{10.19}$$

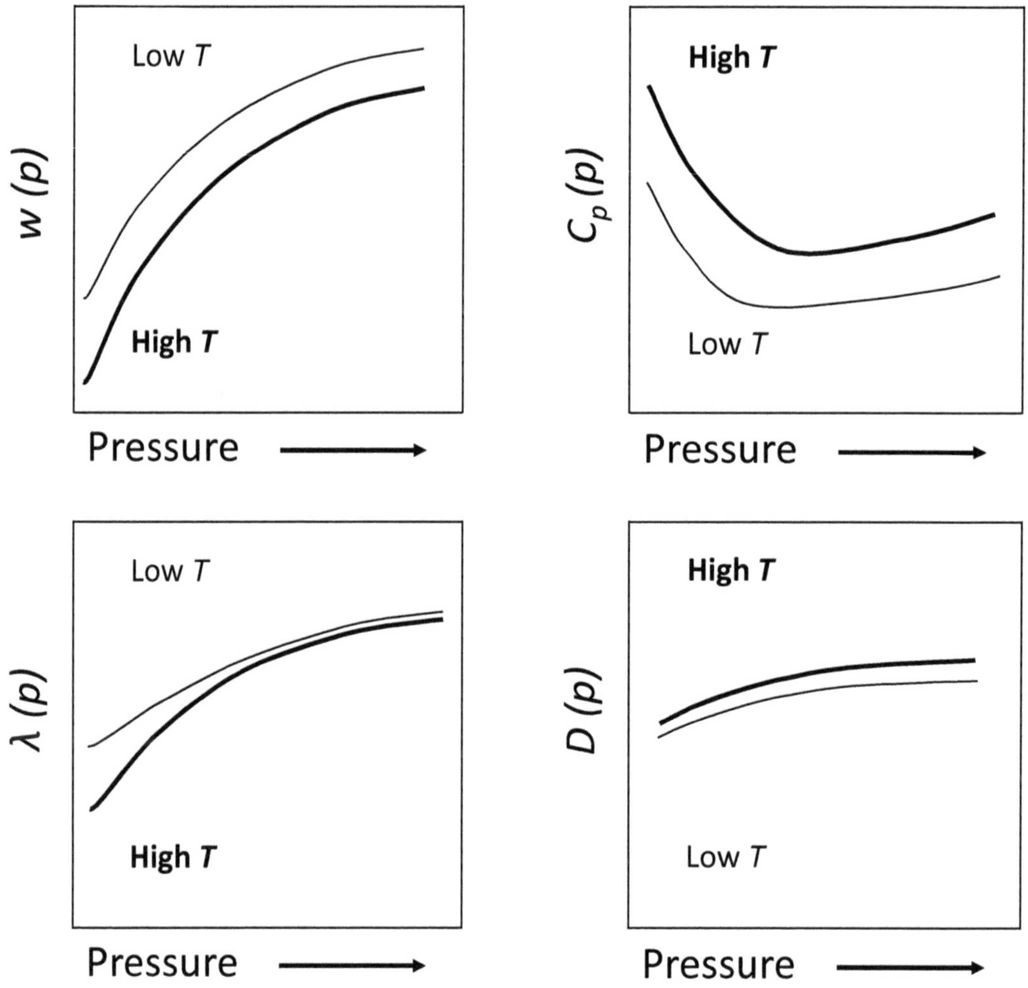

Figure 10.4 Pressure dependence of selected acoustic and thermal properties of normal fluids.

while Sun *et al.* [46] developed an alternative numerical procedure by introducing a polynomial fit to the density along an isobar of reference. Other authors use empirical EOS models [47] to fit the whole set of data to perform the numerical integration of the complete set of first order differential equations. A drawback of this method is that all these calculations require the knowledge of the variation of the density and c_p with temperature at atmospheric pressure. Recently, Lago *et al.* proposed the recursive equation method to obtain the uncertainty of the calculated ρ and c_p values [48] and avoid the need for c_p input from experiments [49].

10.3.3 Thermal Properties

The thermal properties addressed in this section are the isobaric specific heat c_p, the thermal conductivity λ, and the thermal diffusion coefficient D.

10.3.3.1 Isobaric Specific Heat

The isobaric specific heat is defined as the heat quantity needed to increase the temperature of one kilogram material by one Kelvin.

$$c_p = \frac{1}{m}\left(\frac{\partial Q}{\partial T}\right)_p . \tag{10.20}$$

At atmospheric pressure and at high pressure, the most popular way to obtain specific heat is by direct measurement through differential scanning calorimetry (DSC). Although thermal analysis (TA) is also possible at high pressure, it lacks accuracy [50]. According to these techniques, a small mass of the sample to analyze is cooled or heated at a given rate. Simultaneously, a standard is submitted to the same thermal cycle. In DSC, the heat flow necessary to keep the sample at the same temperature as the standard is electronically controlled and monitored. The analysis of the temperature and of the heat flow recorded over time provides the specific heat. In the high-pressure domain, the first differential calorimeters were developed by Schneider and co-workers [50–52]. Current commercial providers are Setaram, Netzsch, and Mettler Toledo, which make such devices relatively accessible to a wide range of disciplines. Heat capacity has been studied at pressures below 100 MPa with such DSC equipment [53, 54]. To reach higher pressures, other authors still build their own calorimeters [55, 56]. However, such devices are usually limited to pressures below 250 MPa. For c_p measurements at higher pressures, indirect methods are often applied. As mentioned in previous sections, the specific heat can be calculated using thermodynamical equations and computational techniques starting from specific volume or speed-of-sound data (Davis–Gordon's and other recursive methods). This method was widely used to derive c_p and other thermophysical properties of fluids over different ranges of pressure and temperature: water (251–293 K and $p < 350$ MPa) [35], alcohols (273–368 K and $p < 280$ MPa) [46, 57], n-hexane (293–373 K and $p < 100$ MPa) [58], and tomato paste [59] (273–323 K and $p < 350$ MPa). The influence of pressure on the c_p of liquids has long been questioned with no definitive conclusion about a general trend [60]. With the accumulation of data and the improvement of data quality, it could be proposed that: (1) c_p isotherm behavior depends on the pressure range and (2) c_p likely passes through a minimum whose location is substance-dependent [61] (Figure 10.4).

Calorimetric techniques also allow us to determine various thermodynamic properties. An interesting one concerns the heat of compression which measures the thermal effects due to a sudden change in the pressure under isentropic conditions. Thus the following thermodynamic relations hold:

$$\left(\frac{\partial T}{\partial p}\right)_S = \left(\frac{\partial v}{\partial S}\right)_p = \frac{vT\alpha_p}{c_p} = \frac{\kappa_T - \kappa_S}{\alpha_p}. \tag{10.21}$$

Under isothermal conditions, the variation in pressure causes a change in entropy, which can be related to α_p using the relationship:

$$\left(\frac{\partial S}{\partial p}\right)_T = -\left(\frac{\partial V}{\partial T}\right)_p = -V\alpha_p. \tag{10.22}$$

Using this method, α_p can be determined with an accuracy similar to that obtained in the measurement of the heat of compression (*ca.* 1 % to 2 %) and it has been applied to many liquids of different natures [62–64]. It should be noted that to obtain the density from this kind of measurement, a reference density isotherm as a function of pressure must be known, which in fact implies coupling of a direct volumetric technique to obtain a whole set of thermodynamic properties or the complete EOS of a fluid.

10.3.3.2 Thermal Conductivity

The thermal conductivity of a fluid characterizes its heat propagation when it is at rest. It is defined in homogeneous and isotropic materials from Fourier's law by:

$$\lambda(T) = -\frac{\vec{Q}}{\overrightarrow{\mathrm{grad}T}}, \tag{10.23}$$

where $\vec{Q}$ is the heat flow (heat input).

The determination of thermal conductivity under high-pressure conditions is carried out in the same way as at atmospheric pressure. The basic experimental setup used to determine thermal conductivity consists of a heat source, a sample holder, several thermocouples, and a data acquisition system. Once again, the main difficulty for adapting an experimental device to the high-pressure domain is the feed-through: good tightness and the resistance of electric cables to many compression and decompression cycles are required. The procedure for calibration and measurements at high pressure is the same as at atmospheric pressure. It depends on the measurement method and cell geometry. Two methods are traditionally distinguished: the steady state and transient methods. The steady state method considers two parallel and infinite surfaces between which the liquid sample is confined. These surfaces are held at two different constant temperatures by waiting for thermal equilibration (steady state) after the heat source has been switched on and delivers a given power. The temperature gradient is then perpendicular to the isothermal surfaces and constant inside the fluid sample. The cells are two coaxial cylinders or parallel plates. Compared to solid materials, the measurement of heat conductivity in fluids is challenging since heat transfers by convection and radiation must be made negligible. For this purpose, the design of the cell should contemplate: (1) the smallest gap possible between surfaces (0.2 mm to 0.7 mm), (2) the best parallelism and smoothness of the surfaces, and (3) the selection of a material with a high thermal conductivity and a low emissivity (for this purpose, sample holders are usually made of copper or silver).

Thermal conductivity is determined from temperature measurements, heat flow value, cell constant (obtained by calibration), and corrections specific to the setup configuration. In addition to the corrections usually applied at atmospheric pressure, the change in cell constant with pressure has to be taken into account.

A lot of fluids have been characterized with this method at pressures up to 1,500 MPa and at a wide range of temperatures. Abdulagatov and co-workers characterized several aqueous solutions in that way [65, 66]. This method is still in use but the transient method is nowadays more popular. The transient method relies on the hypothesis that convection in a fluid is negligible over a short period of time just after the emission of an intense and local heat pulse. For this purpose, a hot-wire apparatus is used. It consists of a thin vertical cable (wire) placed in a shield and immersed in the fluid to be analyzed (Figure 10.3). The wire is initially held at the equilibrium temperature of the fluid and then suddenly heated. The change in temperature at a given radial distance from the heat source depends on the heat dissipation in the surrounding fluid. After calculation, the thermal conductivity is deduced from the slope C of the line obtained when plotting the temperature change at this point as a function of time on a logarithmic scale:

$$\lambda = \frac{Q}{4\pi C}. \tag{10.24}$$

The design parameters should be chosen so that the wire can be considered a linear heat source of infinite length. The ratio of probe dimensions (length/diameter) should be above 30 to 100 to ensure that the conductive heat transfer is mainly axial. Viscous samples are more appropriate than fluid samples with low viscosity to prevent convection effects and avoid correction factors. The method is relatively simple to carry out even under pressure. Its weak point remains the fragility of the hot wire. Compression and heating cycles often lead to wire rupture. Nevertheless, measurements are possible at pressures up to 700 MPa and at a wide range of temperatures. In this way, Werner *et al.* [67], Ramaswamy *et al.* [68], and Nguyen *et al.* [69] determined the thermal conductivity of a variety of fluid foods. Measurements in avocado purée evidenced that the thermal conductivity remained almost constant at temperatures from (298 to 348) K while the thermal diffusion coefficient increased by about 50 % after compression to 700 MPa. The general trend of λ with pressure is shown in Figure 10.4.

10.3.3.3 Thermal Diffusion Coefficient

Thermal diffusivity D defines how quickly heat is transmitted through a material. It can be calculated from the ratio of the specific heat capacity to the thermal conductivity as follows:

$$D = \frac{\lambda}{\rho c_p}. \tag{10.25}$$

To our knowledge, dynamic light scattering [70], transient grating spectroscopy [71, 72] and the dual-needle technique [69, 73, 74] have been the principal methods used for obtaining thermal diffusivity of fluids under high pressure.

The dual-needle technique can be utilized for pure fluids and also for relatively complex fluids taking into account that the accuracy is usually worse in this last case but still between (2 and 7) %. It was found suitable for thermal diffusivity determinations in foods [69, 73, 74]. The dual-needle probe can be regarded as an extension of the hot-wire device employed for thermal conductivity measurement. The probe is provided with a second needle containing a thermocouple placed at a known distance r in parallel to the line heat source. After the emission of a heat pulse of duration t_{pulse}, the temperature is recorded at the distance r from the hot wire. This temperature passes through a maximum after the time Δt elapsed from heat pulse emission. The higher the thermal diffusivity of the medium, the shorter Δt is. The thermal diffusivity is calculated as a function of these characteristic times and of the distance r as follows:

$$D = \frac{r^2}{4}\frac{\Delta t^{-1} - (\Delta t - t_{\text{pulse}})^{-1}}{\ln \Delta t - \ln(\Delta t - t_{\text{pulse}})}. \tag{10.26}$$

For example, the thermal diffusivity was measured in samples of sucrose solution, protein solution, honey, oil, tomato paste and cheese with this method. In agreement with the tendency illustrated in Figure 10.4, it was found that D tends to increase slightly with pressure over the range of condition studied (0.1–600 MPa at 298 K) following a polynomial law [69].

The determination of the thermal diffusivity by dynamic light scattering is based on the theory of microscopic fluctuations. Microscopic fluctuations of temperature are caused by fluctuations in scattered light intensity. After an analysis of the fluctuation phenomena occurring at microscopic scale, D can be deduced from the characteristic decay time (by correlation function) or broadening of the unshifted frequency component (Rayleigh scattering) in the frequency spectrum [75]. The fluids to be analyzed have to be pure, transparent, and in thermal equilibrium. The method was applied, for instance, to study refrigerant fluids near the critical region. The sample was placed in a cell with quartz windows able to withstand pressures of 15 MPa at moderate temperatures. The uncertainty of the measurements was estimated to be less than 5 %.

Transient grating spectroscopy has been coupled to diamond anvil cells to derive the thermal diffusivity of water and other systems of interest in Earth and planetary science (*e.g.*, aqueous solution of sodium sulfate, methanol, or fluid oxygen) [71, 72]. The sample is excited by laser pulses which trigger an acoustic disturbance. The interference gives rise to a spatially periodic distribution of temperature and consequently of refraction index, which is monitored. This grating decays exponentially by thermal diffusion with the rate constant, τ:

$$\tau = \frac{4\pi^2 D}{l^2}, \tag{10.27}$$

with l representing wavelength of the periodic distribution of intensity. Thus the thermal diffusivity of water could be derived with a typical accuracy of 2 % [71].

In addition to these methods, experiments were designed to get determine thermal diffusivity in an indirect way through the resolution of the inverse problem from temperature measurements [76]. This approach consists of modeling the system under study, including the fluid sample of interest. The numerical simulation of heat transfers during the experiment requires thermal diffusivity as an input. An approximate value is initially introduced in the model and the temperature distribution resulting from the computation is compared to the experimental temperature profile. After that, a new value is tested to decrease the differences and this operation is repeated until a satisfactory agreement is found between simulation and experimental results. This procedure was followed to find the thermal diffusivity behavior with temperature under pressures of 400 MPa and 500 MPa for mashed potato and olive oil [77] and for pork meat paste and tomato puree [78]. The uncertainty in the estimated thermal diffusivity was claimed to be about (4 to 5) %.

10.3.4 Measurement Uncertainty

To analyze the results for a given physical property, a budget of uncertainties has to be established following the guidelines for the evaluation and expression of uncertainty in measurement results. Such guidelines can be obtained from the National Institute of Standards and Technology (NIST) and freely downloaded from the Internet [79, 80].

For indirect methods, the law of propagation of uncertainty must be applied and this explains why such methods are usually less accurate than the direct ones. Moreover, in addition to the measured property, the temperature and pressure parameters are also simultaneously measured. The uncertainties in pressure and temperature measurements also contribute to the uncertainty in the measured property. These uncertainty contributions have to be combined in quadrature with that of the property measurement to obtain the combined standard uncertainty u_c in the measured property. Considering the case of density, it gives:

$$u_c(\rho) = \sqrt{(u(\rho_T))^2 + (u(\rho_p))^2 + (u(\rho))^2}. \tag{10.28}$$

It is clear from this formula that high-pressure measurements will always tend to have higher uncertainties than atmospheric pressure measurements. The uncertainties in temperature and pressure measurements should be the lowest possible to achieve the best accuracy for the evaluated property. As for volumetric measurements, the uncertainty in ultrasonic and thermal measurements is higher at high pressure than at atmospheric pressure due to the contribution of the uncertainty in pressure measurement. Moreover, the uncertainty of pressure gauge usually increases with pressure.

10.4 SIGNIFICANCE OF STUDY OF FLUID PROPERTIES AT HIGH PRESSURE: SOME EXAMPLES

10.4.1 Revealing Structural Changes in Water

Water is very likely the most studied liquid under pressure due to its importance in science and technology, including several applications covered by this book (Chapters 11 to 13, 15 and 16). In addition, the study of the phase diagram of water and ices (Chapter 16) has been a recurrent topic in high-pressure science and technology since the pioneering work of Bridgman, but we shall focus here only on the liquid phase.

Tabulated values for the thermodynamic properties of water under pressure are accessible from the IAPWS (International Association for the Properties of Water and Steam). In 1995, this organization approved a new formulation for general and scientific use, and an archival paper describing this formulation was published [81].

It is widely known that the thermodynamic behavior of water is marked by thermodynamic anomalies that are still the subjects of scientific debate. *Anomaly* is to be understood as a certain property that displays a behavior different from that expected for a normal liquid. Of course, the existence of anomalies has historically been linked to changes of hydrogen bond configurations modulated by variations in the density. It should be stressed that studies of high pressures have been key in the development and validation of interpretative models of liquid water and will probably play a similar role in the discoveries of additional peculiar behaviors of liquid water. For further details, see the excellent summary on water structure and science updated and maintained by Prof. Martin Chaplin on a dedicated website (www1.lsbu.ac.uk/water/water_structure_science.html).

It is difficult to summarize the complex behavior exhibited by water under high pressure. However, it is clear that changes of the structural and physical properties (viscosity, self-diffusion coefficient, or compressibility) take place above *ca.* 200 MPa and densities around 1.1 kg·dm^{-3}. The explanation for all these changes appears to be related to an increase in interpenetration of hydrogen bonded networks (interstitial water molecules) in the liquid. Such structural changes are related to the so-called low density water (LDW) to high density water (HDW) transition, which has been a hot topic for the high-pressure community in the last decade. Our aim here is to discuss whether such subtle changes, at least changes in density, can be detected by a direct measurement of a thermodynamic property. Our target is the compressibility, either isothermal or isentropic, the latter deduced from the measurement of the speed of sound.

Although several hints about changes in liquid water structure with pressure were embedded in the early two-state models of water [82], perhaps the first attempt to detect this transition from purely thermodynamic measurements is the study of Mirwald [83]. This author performed a series of isothermal compressibility measurements between (273.15 and 353.15) K and pressures up to 1.6 GPa using a piston-cylinder technique. Besides the expected decrease in κ_T with increasing pressure, two slope changes were detected in the vicinity of (0.3 to 0.4 and 0.8) GPa, thus suggesting two abrupt changes in the density within the liquid phase stability limits. In a recent study of Koga *et al.* [84], using raw data of the speed of sound [85] and the specific volume [86] from the literature, the second derivative $\partial^2\kappa_T/\partial p^2$ was numerically evaluated, without using any fitting function. Their results showed a weak but clear step anomaly in $\partial^2\kappa_T/\partial p^2$ which they argued was due to a gradual crossover in the molecular organization of water that extends between *ca.* (50 and 200) MPa at room temperature. A detailed comparison of these two studies shows an agreement in the changes observed at room temperature, but some disagreement in their temperature dependence, particularly at high temperatures. The results of Koga *et al.* [84] show a clear decrease of the LDW-HDW pressure transition with increasing temperature, an observation which is in agreement with molecular dynamics simulations [87], Brillouin scattering (*i.e.*, speed of sound) experiments [88], ultrafast spectroscopy results [89], and a recent work using speed-of-sound measurements [42].

10.4.2 Understanding Amphiphilic Molecule Behaviors in Water under Pressure

The study of amphiphilic molecule aggregation in water also evidenced the effect of pressure on water structure. Micelle behavior under pressure was investigated from both the speed-of-sound and the attenuation coefficient measurements: the first parameter enabled to define the shift in the critical micellar concentration with pressure and the second one to reveal the formation of gel-like phases [90, 91]. The coincidence in pressure ranges for the LDW-to-HDW transition and for gel-like phase formation reveals the fundamental role of water in the molecular assembly of amphiphiles under pressure. These measurements demonstrate that beyond the intrinsic value of the property the study of phenomena under pressure can be addressed in a complementary way to the usual optical measurements.

10.4.3 Predicting Properties of Compression Fluid Mixtures and Complex Liquid Food Systems

As illustrated in the case of water and noted at the beginning of this chapter, the behaviors of even simple fluids are hard to predict accurately. Experimental measurements of their physical properties *in situ* under pressure are thus useful to reveal the behavior of a given fluid based on pressure and temperature. The same occurs in fluid mixtures and complex fluid systems. The results of predicting the properties of fluid mixture produced from pure fluids may be far from the real values due to interactions between components and impacts of pressure. Mixing rules are often employed for specific volume approximations in the high-pressure domain:

$$v_{mix}(p) = w_1 v_1(p) + w_2 v_2(p) \quad \text{and} \quad v_{mix}(p) = \frac{v_{mix}(p_0)}{v_1(p_0)} v_1(p) \qquad (10.29)$$

where w_1 and w_2 are the mass fractions of each fluid in the mixture and v_1 and v_2 their respective specific volumes at pressure p. The 1 is supposed to be the main component in the second expression. After comparison of the calculated values to the experimental ones, it was found that the specific volumes of binary mixtures of compression fluids could be approximated within 3 % over the whole range of pressures studied (to 350 MPa) [86]. However, the approximation was even better at the highest pressures (within 0.5 %), evidencing that the excess volume not taken into account in Equation (10.29) tends to decrease with pressure increase. In the case of κ_p and α_p, the approximation is less straightforward because of the derivatives.

Regarding aqueous systems, the displacement of water volumetric property anomalies toward lower temperatures is observed: κ_T minimum at around 319 K for water is shifted to 301 K in the case of orange juice with an estimated concentration in solutes of 12 % and to 297 K for 40 % [92]. The presence of solutes disturbs the water network by shifting the synergistic pressure and temperature effect. These data are of interest for the development of food high-pressure food processing [93].

10.4.4 Developing High Pressure Sensors for Industrial Applications

At atmospheric pressure, speed-of-sound measurements have been employed to characterize a wide range of phenomena. Under high pressure, this same approach is still possible. In this way, the crystallization of fatty acids, triglycerides and other oil components was evidenced at pressures up to 600 MPa from the sudden discontinuity in the speed of sound upon compression. The kinetics of the phase change could be monitored *in situ* [40]. In this way, the monitoring of other phenomena during high-pressure processing is of interest for process control. The speed of sound is often the most useful acoustic parameter but the attenuation or absorption coefficient may be more suitable depending on the situation: for supercritical fluid characterization, the absorption coefficient was found to be more relevant than the speed of sound to accurately define the critical point of carbon dioxide [43].

The main difficulty for the instrumentation of high-pressure vessels remains the thickness of walls and pressure leaks but ultrasounds may serve as a response to this challenge. Ultrasounds have given rise to numerous non-invasive sensors. Such sensors are particularly relevant in the context of high-pressure technology developments. The study of acoustic properties of fluids under pressure at laboratory scale constitutes a good basis for further advances.

10.5 OUTLOOK

We have demonstrated that the knowledge of the physical properties of pure fluids and mixtures is key in many scientific and technological areas of the high-pressure scene. In addition to the volumetric, acoustic, and thermal properties, there are a lot of other properties of interest in the field of high-pressure studies on fluids such as viscosity [94], pH [95, 96], and electrical conductivity [97]. The available technology is now sufficiently advanced for assessing a number of fluid properties in the few GPa range. Thereby, it is possible to combine moderate pressure experiments with structural (radial distribution function) and transport experiments performed in large facilities using anvil cell devices. This fact, together with the extraordinary capabilities of computer simulations (from molecular dynamics to coarse-grain models), allows us to face the challenge of understanding very complex fluids like those involved in biological and geological phenomena under extreme conditions.

Of course, many open questions still remain for liquid water and aqueous solutions and mixtures, and will be the subjects of many studies in the coming years. But, in our opinion, the next step concerns the understanding of chemical aspects involving both aqueous and non-aqueous solutions. Our focus in must be the study of complex biological and geological processes and other chemical transformations related to molecular fluids (*e.g.*, carbon dioxide trapping). After all, any advances in the study of the kinetics at high-pressure require strong data on both thermodynamic and transport properties, like those discussed in this chapter.

Bibliography

[1] W. A. Steele and W. Webb. *Compressibility of liquids*, 145–162. Academic Press, London (1963).

[2] P. W. Bridgman. *Collected experimental papers.* Harvard University Press, Cambridge (1964).

[3] A. T. J. Hayward. *Br. J. Appl. Phys.*, 18, 965 (1967).

[4] P. G. Tait. *Report on some of the physical properties of fresh water and sea water*, Physics and Chemisry II, 1–76. Neill Edinburgh (1989).

[5] Y. H. Huang and J. P. O'Connell. *Fluid Phase Equilib.*, 37, 75 (1987).

[6] M. Taravillo, M. Cáceres, J. Núñez, *et al.* *J. Phys. Condens. Matter*, 18, 10213 (2006).

[7] V. G. Baonza, M. Alonso, and J. Delgado. *J. Phys. Chem.*, 97, 10813 (1993).

[8] U. K. Deiters and S. L. Randzio. *Fluid Phase Equilib.*, 103, 199 (1995).

[9] R. Privat, F. Gaillochet, and J. N. Jaubert. *Fluid Phase Equilib.*, 327, 45 (2012).

[10] V. G. Baonza, M. C. Alonso, and J. N. Delgado. *J. Phys. Chem.*, 98, 4955 (1994).

[11] H. Kanno and C. A. Angell. *J. Chem. Phys.*, 70, 4008 (1979).

[12] P. Pruzan. *J. Phys. Lett.*, 45, 273 (1984).

[13] M. Taravillo, V. G. Baonza, M. Cáceres, *et al.* *J. Phys. Chem.*, 99, 8856 (1995).

[14] R. J. Speedy. *J. Phys. Chem.*, 86, 982 (1982).

[15] J. V. Sengers, R. F. Kayser, C. J. Peters, *et al.* *Equations of state for fluids and fluid mixtures.* Elsevier, Amsterdam (2000).

[16] A. T. J. Hayward. *J. Phys. D Appl. Phys.*, 4, 938 (1971).

[17] J. V. Leyendekkers. *J. Phys. Chem.*, 97, 1220 (1993).

[18] V. G. Baonza, M. Caceres, and J. Núñez. *Phys. Rev. B*, 51, 28 (1995).

[19] V. P. Skripov. *Metastable liquids.* Wiley, New York (1973).

[20] E. Whalley. In B. L. Neindre and B. Vodar, eds., *Experimental thermodynamics*, 421–500. Butterworth-Heinemann, London (1975).

[21] F. J. Millero. *ACS Symp. Ser.*, 133, 581 (1980).

[22] V. Tekáč, I. Cibulka, and R. Holub. *Fluid Phase Equilib.*, 19, 33 (1985).

[23] J. C. Holste, K. R. Hall, P. T. Eubank, *et al. Fluid Phase Equilib.*, 29, 161 (1986).

[24] W. Wagner and R. Kleinrahm. *Metrologia*, 41, S24 (2004).

[25] A. M. F. Palavra, M. A. Tavares Cardoso, J. A. P. Coelho, *et al. Chem. Eng. Technol.*, 30, 689 (2007).

[26] S. Buzrul, H. Alpas, A. Largeteau, *et al. J. Food Eng.*, 85, 466 (2008).

[27] Y. A. Sanmamed, A. Dopazo-Paz, D. González-Salgado, *et al. J. Chem. Thermodyn.*, 41, 1060 (2009).

[28] I. M. S. Lampreia and C. A. N. de Castro. *J. Chem. Thermodyn.*, 43, 537 (2011).

[29] W. B. Streett and L. A. K. Staveley. *J. Chem. Phys.*, 55, 2495 (1971).

[30] J. C. G. Calado and L. A. K. Staveley. *J. Chem. Soc. Faraday Trans.*, 67, 289 (1971).

[31] V. G. Baonza, M. Cáceres, and J. Nuñez. *J. Chem. Educ.*, 73, 690 (1996).

[32] F. M. Gaciño, T. Regueira, M. J. P. Comuñas, *et al. J. Chem. Thermodyn.*, 81, 124 (2015).

[33] M. Habrioux, S. V. D. Freitas, J. A. P. Coutinho, *et al. J. Chem. Eng. Data*, 58, 3392 (2013).

[34] H. Khelladi, F. Plantier, J. L. Daridon, *et al. Ultrason.* (2009).

[35] J. P. Petitet, R. Tufeu, and B. Le Neindre. *Int. J. Thermophys.*, 4, 35 (1983).

[36] W. D. Wilson. *J. Acoust. Soc. Am.*, 31, 1067 (1959).

[37] A. Zak, M. Dzida, M. Zorebski, *et al. Rev. Sci. Instrum.*, 71, 1756 (2000).

[38] M. Dzida, M. Chorazewski, M. Zorebski, *et al. J. Phys.*, 137, 203 (2006).

[39] S. Lago and P. A. Giuliano Albo. *J. Chem. Thermodyn.*, 41, 506 (2009).

[40] P. Kiełczyński, M. Szalewski, A. Balcerzak, *et al. LWT Food Sci. Technol.*, 57, 253 (2014).

[41] B. Lagourette, J. L. Daridon, J. F. Gaubert, *et al. J. Chem. Thermodyn.*, 26, 1051 (1994).

[42] E. Hidalgo Baltasar, M. Taravillo, V. G. Baonza, *et al. J. Chem. Eng. Data*, 56, 4800 (2011).

[43] L. Zevnik, M. Babič, and J. Levec. *J. Supercrit. Fluids*, 36, 245 (2006).

[44] E. Hidalgo Baltasar, M. Taravillo, V. G. Baonza, *et al. Mater. Sci. Eng.*, 42 (2012).

[45] L. A. Davis and R. B. Gordon. *J. Chem. Phys.*, 46, 2650 (1967).

[46] T. F. Sun, C. A. Ten Seldam, P. J. Kortbeek, *et al. Phys. Chem. Liq.*, 18, 107 (1988).

[47] M. Jr. *Revs. Modern. Phys.*, 41, 316 (1969).

[48] S. Lago and P. A. Giuliano Albo. *J. Chem. Thermodyn.*, 40, 1558 (2008).

[49] S. Lago and P. A. Giuliano Albo. *J. Chem. Thermodyn.*, 58, 422 (2013).

[50] M. Kamphausen and G. M. Schneider. *Thermochim. Acta*, 22, 371 (1978).

[51] R. Sandrock, U. Wenzel, H. Arntz, *et al. Thermochim. Acta*, 49, 23 (1981).

[52] C. Schmidt, M. Rittmeier-Kettner, H. Becker, *et al. Thermochim. Acta*, 238, 321 (1994).

[53] D. Bessières, H. Saint-Guirons, and J. L. Daridon. *J. Therm. Anal. Calorim.*, 62, 621 (2000).

[54] D. Bessières and F. Plantier. *J. Therm. Anal. Calorim.*, 89, 81 (2007).

[55] S. Zhu, S. Bulut, A. L. E. Bail, *et al. J. Food Process Eng.*, 27, 359 (2004).

[56] S. Zhu, A. Le Bail, and H. S. Ramaswamy. *J. Food Eng.*, 75, 215 (2006).

[57] D. González-Salgado, J. Troncoso, F. Plantier, *et al.* *J. Chem. Thermodyn.*, 38, 893 (2006).

[58] J. L. Daridon, B. Lagourette, and J. P. E. Grolier. *Int. J. Thermophys.*, 19, 145 (1998).

[59] C. Aparicio, B. Guignon, L. Otero, *et al.* *J. Food Eng.*, 104, 341 (2011).

[60] V. A. Drebushchak. *J. Therm. Anal. Calorim.*, 95, 313 (2009).

[61] J. Troncoso, P. Navia, L. Romaní, *et al.* *J. Chem. Phys.*, 134, 094502 (2011).

[62] L. Ter Minassian, K. Bouzar, and C. Alba. *J. Phys. Chem.*, 92, 487 (1988).

[63] S. Randzio, J. P. E. Grolier, and J. R. Quint. *Fluid Phase Equilib.*, 110, 341 (1995).

[64] P. Pruzan. *J. Chem. Thermodyn.*, 23, 247 (1991).

[65] I. M. Abdulagatov, L. A. Akhmedova-Azizova, and N. D. Azizov. *J. Chem. Eng. Data*, 49, 688 (2004).

[66] L. A. Akhmedova-Azizova and I. M. Abdulagatov. *J. Solution Chem.*, 43, 421 (2014).

[67] M. Wernery, A. Baars, C. Eder, *et al.* *J. Chem. Eng. Data*, 53, 1444 (2008).

[68] R. Ramaswamy, V. M. Balasubramaniam, and S. K. Sastry. *J. Food Eng.*, 83, 444 (2007).

[69] L. T. Nguyen, V. M. Balasubramaniam, and S. K. Sastry. *Int. J. Food Prop.*, 15, 169 (2012).

[70] M. Pitschmann and J. Straub. *Int. J. Thermophys.*, 23, 877 (2002).

[71] E. H. Abramson, J. M. Brown, and L. J. Slutsky. *J. Chem. Phys.*, 115, 10461 (2001).

[72] E. H. Abramson, J. M. Brown, and L. J. Slutsky. *Annu. Rev. Phys. Chem.*, 50, 279 (1999).

[73] S. Denys and M. E. Hendrickx. *J. Food Sci.*, 64, 709 (1999).

[74] S. Zhu, H. S. Ramaswamy, M. Marcotte, *et al.* *J. Food Sci.*, 72, E49 (2007).

[75] B. Kruppa and J. Straub. *Exp. Therm. Fluid Sci.*, 6, 28 (1993).

[76] M. Kubásek, M. Houška, A. Landfeld, *et al.* *J. Food Eng.*, 74, 286 (2006).

[77] A. Landfeld, J. Strohalm, M. Houska, *et al.* *High Pressure Res.*, 30, 108 (2010).

[78] A. Landfeld, J. Strohalm, J. Stancl, *et al.* *High Pressure Res.*, 31, 358 (2011).

[79] J. J. C. for Guides in Metrology/Working Group 1. Evaluation of measurement dataguide to the expression of uncertainty in measurement. JCGM 100:2008. GUM 1995 with minor corrections. Bureau International des Poids et Mesures. Sevres, France (2008).

[80] B. N. Taylor and C. E. Kuyatt. Guidelines for the evaluation and expression of uncertainty in NIST measurement results. NIST Technical note 1297. Gaithersburg, Germany (1994).

[81] W. Wagner and A. Pruss. *J. Phys. Chem. Ref. Data*, 31, 387 (2002).

[82] J. D. Bernal and R. H. Fowler. *J. Chem. Phys.*, 1, 515 (1933).

[83] P. W. Mirwald. *J. Chem. Phys.*, 123 (2005).

[84] Y. Koga, P. Westh, K. Yoshida, *et al.* *AIP Adv.*, 4 (2014).

[85] C. W. Lin and J. P. M. Trusler. *J. Chem. Phys.*, 136, 094511 (2012).

[86] B. Guignon, C. Aparicio, and P. D. Sanz. *J. Chem. Eng. Data*, 55, 3017 (2010).

[87] A. Marco Saitta and F. Datchi. *Phys. Rev. E*, 67, 202011 (2003).

[88] F. Li, Q. Cui, Z. He, *et al.* *J. Chem. Phys.*, 123 (2005).

[89] S. Fanetti, A. Lapini, M. Pagliai, *et al.* *J. Phys. Chem. Lett.*, 5, 235 (2014).

[90] E. Vikingstad, A. Skauge, and H. Høiland. *J. Colloid Interface Sci.*, 72, 59 (1979).

[91] E. Hidalgo Baltasar, M. Taravillo, P. D. Sanz, *et al.* *Langmuir*, 30, 7343 (2014).

[92] B. Guignon, C. Aparicio, P. D. Sanz, *et al.* *Food Res. Int.*, 46, 83 (2012).

[93] B. Guignon, I. Rey-Santos, and P. D. Sanz. *Food Res. Int.*, 64, 336 (2014).

[94] L. Kulisiewicz and A. Delgado. *Appl. Rheol.*, 20, 130181 (2010).

[95] V. M. Stippl, A. Delgado, and T. M. Becker. *High Pressure Res.*, 22, 757 (2002).

[96] C. P. Samaranayake and S. K. Sastry. *J. Phys. Chem. B*, 114, 13326 (2010).

[97] S. Min, S. K. Sastry, and V. M. Balasubramaniam. *J. Food Eng.*, 82, 489 (2007).

III

APPLICATIONS

CHAPTER 11

Microbiology under Pressure: How Microorganisms Are Affected by Pressure and How They May Cope with It

Jordi Saldo

MALTA Consolider Team and Animal and Food Science Department, CERPTA, Universitat Autònoma de Barcelona, Bellaterra, Spain

CONTENTS

11.1 INTRODUCTION

SOME of the least explored frontiers in the world are the high-pressure environments. Until the *HMS Challenger*'s expedition between 1872 and 1876 it was not even widely accepted that organisms can live under pressure far above the normal atmospheric pressure. And even much more recently we recently learned about the microorganisms withstanding not only the hydrostatic pressure but also the lithostatic pressure living within rocks. The development of a number of technical advances has been vital for acquiring relevant knowledge in this field.

Since ancient times, there have seen indirect clues about the existence of inhabitants of the deep ocean from observation of fishermen's catches and the anecdotal findings of abyssal "monsters". However, determining which organisms inhabited the dark, cold, deep ocean waters was always difficult because of hydrostatic pressure. Direct observation had to wait until the mid-20th century when access by bathyscaphes to depths lower than divers could reach extended research areas from a few to several thousand meters and increased the range of pressures that could be observed directly from 1 to 100 MPa. Metazoans were studied far earlier than deep-sea microorganisms.

This chapter presents the results of studies on the ocean floors and the isolation of microorganisms from several high-pressure environments. It will describe the peculiarities that allow these microorganisms to survive to high pressure and highlight characteristics shared by other creatures living under extreme conditions. This chapter will also review of the effects of high pressure on the cellular elements, explaining how high pressure can be lethal to living things and describe how some have developed adaptations to such environments. Some of these adaptations exhibit promising biotechnological potential. The final section deals with the application of high pressure to foods and other products to remove microorganisms and also explains how variables, such as growth stage of the microorganism or the pH levels of media may affect the lethality caused by high-pressure treatment.

11.2 EARLY HISTORY

The scientific study of microorganisms living under pressure was hindered by a theory that suggested that the absence of light and the presence of intense cold and high pressure should prohibit the development of life forms. This azoic theory became popular during the second half of 19th century. Its champion, Edward Forbes, claimed in 1843 that the ocean level below 300 fathoms ($\sim$550 m) were barren by extrapolating observations of the decrease in the number of organisms captured as dredges were equipped to dig deeper [1]. This theory was questioned when Louis F. de Pourtales of the United States Coast Survey Service reported in 1853 a review of the reports and results from depths to 2000 m, finding signs of life in all of them. This review led to a more systematic study and Portuales conducted a seabed dredging of the southern coast of Florida in 1863, finding abundant animals below the presumed azoic limit of 300 fathoms. The evidence against the azoic theory became overwhelming after the dredgings made by Charles Wyville Thomson, who in 1868 found life forms to depths of 2400 fathoms (~4400 m).

In December 1882, the *HMS Challenger* began a 4-year voyage around the world, a journey that followed the lead of the voyage of *HMS Beagle* 50 years earlier. The *Challenger*, equipped with biological and chemical laboratories, started mapping the deep ocean including trenches and studying samples obtained using improvements of survey techniques. From the 362 sampling sites, 25 dredges explored depths over 4500 m and down to 5700 m. The expedition described 4717 new species. At the same time, *Travaillieur* and *Talisman* (1882-1883) collected samples of sediment and water at depths to 5000 m (50 MPa). These samples were examined by A. Certes, who found bacteria in most samples. Certes indicated that the bacteria survived at high pressure and live in a state of suspended animation [2]. These studies marked the beginning of high-pressure microbiology.

A few years later the idea of suspending microbial activity by means of pressure found its application in the pioneer studies of B. H. Hite, who avoided acidification of milk by microorganisms by using pressure rather than heat, destroying 99.99 % of microorganisms by a pressure treatment at 700 MPa for 10 min [3].

The study of the effects of pressure on microorganisms led to the development of a classification into three categories, depending on their response to this factor (Figure 11.1). Pressure-sensitive microorganisms decrease their vitality continuously as pressure increases. This group is designated *piezosensitive* (from the Greek piezos = pressure) and replaced the earlier *barosensitive* (from the Greek baros = load) term, that was inaccurate. *Piezotolerant* microorganisms can withstand some pressure stress, having the same growth pattern at atmospheric pressure at higher pressure. The real piezophilic microorganisms exhibit their maximum growth above atmospheric pressure. Within this group we find microorganisms that are unable to grow at atmospheric pressure, but they can grow at higher pressures and are known as extreme piezophilic.

Both the experimental designs for high-pressure application and the sampling methods to study the mechanisms affecting the sensitivity of bacteria to pressure took a long time to be developed. The development of the equipment

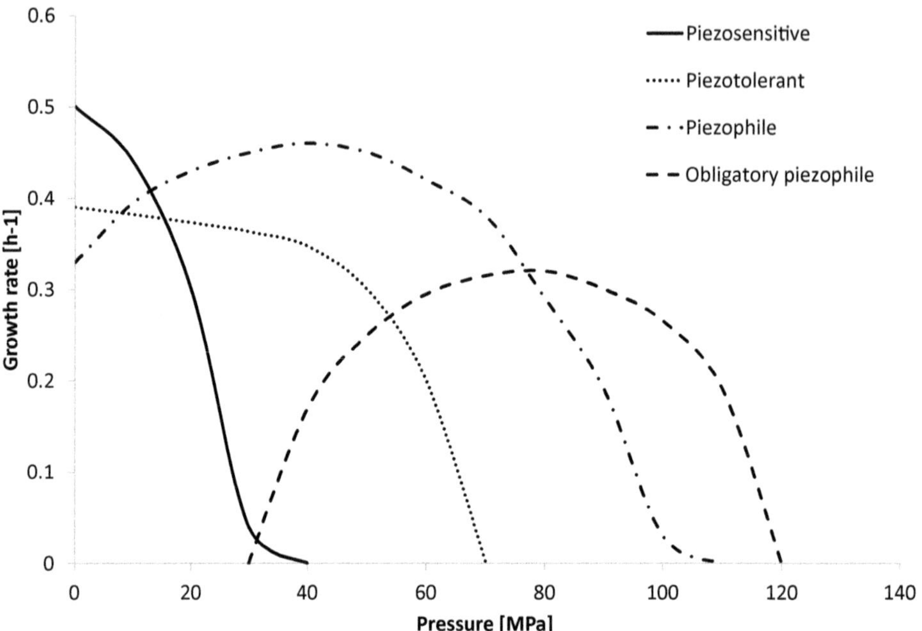

Figure 11.1 Microorganisms can be classified according to their behavioral responses to pressure. Adapted, with permission, from reference 4.

required to culture microorganisms under pressure and observe their metabolic changes was accelerate in the late 20th century.

11.3 PIEZOPHILIC MICROORGANISMS. BACTERIA AND ARCHAEA

Microbiologists have been able to isolate living things from various extreme environments that threaten the development of life because of temperature, chemical composition or pressure. Microorganisms adapted to high pressures are known as piezophiles. Some of the microorganisms isolated from these environments exhibited characteristics that differed from those of known bacteria. Thus, the classification of living things was rearranged into three main domains: *eukaryota*, *eubacteria* and *archaean*. Since microbial taxonomy is one of the scientific disciplines subjected to change and revision, the earlier domains were later re-named *bacteria* and *archaea* [5].

Many differences exist among the three groups of living things and we have to rethink the diversity and complexity of the evolution of living beings. Biologists have drawn new evolutionary trees, but because of lateral flows of genes between organisms some authors prefer to speak of evolving networks. In spite of their appearance (small size, unicellular structure, lack of internal organelles) we can not consider archaea and bacteria as more similar to each other than eukaryotes, The fact is that the three groups represent the main branches that evolved from the origin of life on Earth. The hypothetical Last Universal Common Ancestor (LUCA) would share some common characteristics, but some would have appeared or disappeared during evolution. Some similarities between archaea and eukaryota suggest they shared some common history after separating from bacteria and before starting their independent evolution.

Archaea possess some unique features that help them to be the predominant living things in extreme environments. Their cell membrane has important peculiarities in composition and structure. The major components of the cell membrane are phospholipids, in which a phosphate group and two fatty acids are linked to glycerol. Archaea's membrane lipids are derived from isoprene, with methyl groups on each side at every four carbons, while eukaryotic and bacterial membrane lipids are linear fatty acids. These side groups allow cyclizations to occur, and even link their non-polar ends creating transmembrane structures with two phosphoglycerol polar lipid groups. Another important feature is that fatty acids bind to glycerol by an ether linkage instead of the usual ester bond. This membrane is more resistant to chemicals, and transmembrane structures reinforce their integrity. Lipid side groups hinder lipid crystal packing and allow lipid membranes to remain fluid even at high pressures and low temperatures. Another special feature of the archaea cell membranes is that they contain sulfolipids, in addition to the usual phospholipid [6].

Archaea DNA is stabilized by histones, as happens in eukaryote DNA, but like bacteria, archaea's DNA is circular. This similarity can be considered a homology in terms of function and evolution. Only bacteria have cell walls made of peptidoglycan. In most archaea the cell wall is made of multiple layers. Archaea peptidoglycan is really unique, containing sugars which are linked together by $\beta 1 \rightarrow 3$ bond, which is insensitive to lysozyme. Peptidoglycan containing talasoaminouronic acid binds to oligopeptides having L-amino acids by a transpeptidase insensitive to penicillin. These peculiarities in archaea walls make them insensitive to many antibiotics affecting bacteria, although some are sensitive to antibiotics affecting eukaryotes.

The three types of living organisms (eukaryotes, bacteria and archaea) have photosynthetic representatives, but archaea are unique in this. Photosynthetic archaea belong to the halobacteria group in which the bacteriorhodopsin pigment operates directly as a proton pump without using water as an electron donor, and therefore not releasing oxygen.

Adaptations to high-pressure environments differ depending on the associated temperature. For instance, the bacteria that live at high pressure or low temperature must modify their membrane composition to include lipids of lower melting point and maintain the fluidity of the cover, as both factors tend to induce a higher order and favor crystal structures. A higher temperature can offset the effect of pressure.

11.4 NATURAL ENVIRONMENTS WITH MICROORGANISMS LIVING UNDER HIGH PRESSURE

11.4.1 High Pressure and Low Temperature: Ocean Trenches

The lowest oceanic depth (10 911 m) in the Mariana Trench in the Pacific Ocean was named Challenger Deep in honor of the research vessel; it was first reached by Picard and Walsh on the *Triestre* in 1960. Expeditions of the Japan Agency for marine-Earth Science and Technology (JAMSTECH) utilizing the remotely operated *Kaiko* vehicle took images of and samples from the area from 1995 to 1998. In 2009, the Woods Hole Oceanographic Institution (WHOI) replaced *Kaiko* with *Neraeus.*

Animals groups documented by the expeditions include fish, crustaceans, and echinoderms. Microorganisms adapted to life in this high-pressure and low-temperature environment were isolated from this and other sampling sites. Table 11.1 shows pressure and temperature gradients for extremophilic organisms. A few genera of bacteria found in abyssal samples are similar to types living in surface environments.

In 2012, a second manned expedition with J. Cameron at the control of the *Deepsea Challenger* reached the Challenger Deep. The vehicle took sea-floor samples and produced a high definition three-dimensional film that has been shown in theaters since 2014. Results of the analysis will be released soon by WHOI.

Table 11.1 Pressure and temperature gradients of potential environments for thermophilics

	Pressure	Temperature
Oceans	10 MPa · km^{-1}	2-3 °C when below thermocline
Continental crust	300 MPa · km^{-1}	25 °C · km^{-1}

The ocean depths receive very little organic carbon and are dependent on photosynthesis in the surface photic layer. Most of the energy is consumed in the first meters of the water column, a very small fraction reaching greater depths. The scarcity of nutrients, low oxygen saturation, and low temperature cause the creatures that inhabit these environments to be oligotrophic, and have slow metabolisms. This stable environment preserved from the radiation reaching the surface, could have been the environment in which the first living beings evolved, even if the energy source was not linked to photosynthesis. While some organisms originated in the deep evolved to live on the surface and some inhabiting the surface developed adaptations to live in abyssal conditions, both communities are now connected. Ocean circulation brings oxygen into the deep waters and minerals to the photic zone, and can also allow microorganisms to reach new habitats.

The distribution of microorganisms along the water column has a number of gradients. In the first 1000 m, bacteria tend to be more prolific, probably reflecting the availability of organic carbon and nitrogen. The composition of these communities also have differences. In surface and subsurface samples (<100 m), the archaea diversity is low; they reach up to 20 to 30 % of prokaryotes, and in deep water they outnumber bacteria in samples obtained at depths exceeding 1000 m [7]. Yeast populations decrease when distance from coast and depth of sampling are increased, suggesting a terrestrial origin. In contrast, filamentous molds have been isolated from abyssal marine waters, and yeasts such *Rhodotorula* and *Sporobolomyces* have also been isolated from deep ocean sediments.

11.4.2 High Pressure and Large Temperature Gradients: Ocean Ridge Smokers

In the deep ocean we have also found microorganisms adapted to hydrostatic pressure associated with hydrothermal vents. Oceanic ridge vents sustain entire ecosystems which are independent from photosynthesis - a feature also shared with the communities that live deep in the continental crust.

From the hydrothermal vents, the fumaroles have provided with the most astonishing findings. The fumaroles occur when seawater soaks the oceanic crust and reaches a subsurface deposit of magma. There the water is heated to supercritical conditions and dissolves large amounts of minerals, among which are magnesium, iron, and sulfides. In some cases the water is cooled before re-emerging, leading to warm springs in which temperature is 4 to 50 °C. In the hot springs, water emerges at temperatures up to 400 °C, and precipitation of minerals dissolved in the plume of hot water causes the image that lead to fumarole designation. The pressure in these cases is more moderate (25 to 30 MPa) but the temperature conditions favor the presence of other extremophiles, piezophiles, and hyperthermophiles. These microorganisms do not have counterparts on the surface. Biomass associated with these springs is much greater than could be explained by contributions from the upper photic zone, indicating primary production not dependent on photosynthesis. It has been found that these ocean oases depend on the activity of chemolithotrophic microorganisms, many of them archaea. Many microscopic eukaryotes (ciliates, flagellates, nematodes, that prey on microorganisms) are observed. Some of the larger animals that characterize these ecosystems, such as tubeworms and mussels, have virtually no digestive systems and host huge numbers of chemolithotrophic microorganisms as endosymbionts [8].

11.4.3 High Pressure and High Temperatures: Life within Earth's Crust

Microorganisms within petroleum deposits 600 m deep were first isolated in the 1920s. The discoverers attributed the presence of these microorganisms to oil evolution from microorganisms that were buried with sediment 340 million years ago. This discovery aroused skepticism from the scientific community, which attributed the presence of life on deposits to pollution produced during drilling from surface materials. The discussion led to the design of drilling techniques and controls to ensure that the samples obtained were from materials *in situ* and not artifacts produced by drilling.

Although studies claimed to observe microorganisms isolated from sediments of the ocean, doubts arose about the possible contamination of samples through the drilling process. A prospecting program was organized in 1985 to resolve the debate over the existence of life in the rocks, using anti-pollution drilling processes. Thus, it was possible to sample communities of bacteria and archaea from depths of 500 m in South Carolina. Applying these precautions developed for drilling techniques to oil drilling allowed us also to find new species at depths of 2.7 km in 1993 [9].

The microorganisms obtained from the depths of the earth proved to be quite different from those which inhabit in the surface, definitively ruling out the possibility of contamination. Unlike marine habitats and high pressures environments in which the ocean circulation contacts the surface in cycles lasting from years to centuries, underground environments have been isolated from the surface for much longer periods. The sediment samples obtained in 1993 were isolated for 340 million years, since the sedimentary rocks were formed. After rocks were buried, they underwent high pressures and temperatures (Table 11.1) as did the microorganisms they contained. Microorganisms live in rock pores, colonizing spaces with very tinny communications of even less than a tenth in size of containing bacteria and archaea. The organisms get their energy and nutrients from the hydrocarbons and compounds dissolved in water, and are strict anaerobes. Nutrient scarcity has forced these microorganisms to maintain low metabolic rates, and remain in a kind of suspended animation. The very low rate of division has made these microorganisms evolve very slowly during the long time they have remained trapped [9].

In addition to the sedimentary rock samples from which microorganisms were isolated, we now have gathered evidence of the presence of microorganisms in other environments such as from igneous rocks in Sweden. These organisms use hydrogen as an energy source in a system completely independent of photosynthesis. The presence of microorganisms embedded in granite cannot be explained as entrapment during the formation of the rock, and must be explained by active habitat colonization through groundwater flowing through cracks in the rocks.

These findings have extended the biosphere far beyond that expected a few years ago. On a drilling conducted during gold mining in Mooneng (South Africa) temperatures reached 60 °C at depth of down 3.5 km. Some thermophilic microorganisms were isolated from these very extreme conditions.

11.4.4 Environments Subject to Remote Observation

One exiting exotic environment is Vostok Lake, East Antarctica. This lake was discovered from a seismological study conducted between 1955 and 1964, and confirmed by eco-radio survey between 1973 and 1975. This freshwater lake is among the largest in the world and is divided into two basins. It covers about 14000 km^2 of surface and depths exceeded 1100 m. It is a lake of liquid water under about 4000 m of solid ice. The pressure on the lake is around 40 MPa. This pressure, along with heating inside of the Earth, allows the water to maintain a relatively warm temperature of −2.65 °C. This environment should maintain microorganism groups that existed when the lake was isolated from the surface about 15 million years ago [10].

Vostok Lake has proven to be the most interesting of a series of lakes occluded by Antarctic ice. One hundred forty-five lakes have been identified to date. They have not been drilled and will remain undisturbed until they can be sampled securely without compromising measurements. Until then, all data concerning compositions, dissolved gases, and geothermal energy sources will come from indirect observations. Sampling of the accretion ice high above the Antarctic subglacial lakes is the most direct observation possible to date.

It is believed the waters of Vostok Lake are supersaturated in gas, with dissolved gases in equilibrium within a solid gas hydrate known as clathrate (see Chapter 16). The oxygen concentration can be 50 times higher than that found dissolved in the water in equilibrium with air at 0.1 MPa. Organisms living in the water of these deep lakes are expected to possess very active catalase, peroxidase, and superoxide dismutase systems to compensate for oxidative stress problems [10].

Despite the extreme conditions of the waters of these subglacial lakes, there is no doubt that living organisms, most probably microorganisms, will be found in it. Despite the continuous exchange of materials from the cover ice cap and the lake waters, we can expect a community that has remained active for a very long period without contact with the rest of the biosphere.

If these microorganisms have maintained sufficient activity and reproduction they may serve as an interesting experiment of separate evolution. At the Russian Vostok station drilling has been done to explore the underlying lake. The most relevant is the 5G hole that allows us to explore the frozen ice formed from the free water of the lake. To prevent the freezing, the hole has been filled with a mix of kerosene and freon, but extreme care has been taken to prevent any possible contamination [11].

Another attraction of Vostok Lake is that it can serve as a model for one of the most accessible exobiology studies. Europa is one of the satellites of Jupiter, with a liquid ocean 50 to 100 km deep and covered with a layer of ice 3 to 4 km thick. Considering the gravity of the satellite, the pressure at the bottom of the ocean would be around 150 MPa, 50 % higher than pressure at Challenger Deep [12]. These conditions do not seem incompatible with life, according to Earth standards. After these Antarctic studies, the scientific community will be capable of undertaking studies of Europa. If life were to be found there, it would be one of the most exciting discoveries of decades (see also Chapter 16).

11.5 MICROORGANISM CHANGES CAUSED BY HIGH PRESSURE

11.5.1 Main Effects of High Pressure on Microbial Biochemistry

According to the principle of Le Châtelier, all pressure effects which are accompanied by changes in volume of a system are favored or hindered, depending on the sign of the change in volume. One of the major changes is an increase in the dissociation of water, since most biochemical processes occur in aqueous phase, with a negative volume variation −21 mL·mol^{-1}. Similarly some weak acids, such as carbonic and phosphoric,trend towards dissociation under pressure at −26 mL·mol^{-1}. These acids are very important in the homoeostasis of cytoplasm, as parts of buffer systems. The internal environment of the microorganisms tends to be more acidic when weak acids are subjected to high pressures, because of the equilibrium shift.

It has been possible to recover microorganisms adapted to all high-pressure environments studied, although the hostility of environmental factors limits the densities achieved. It has been found that above 70 to 80 MPa there is no significant RNA synthesis, which seems to indicate that the limit of active life under pressure may be around this figure. Some cells may be in a form of suspended animation above this pressure, recovering their activity upon return to more favorable conditions.

11.5.2 Cell Motility

Flagella loss in mesophilic bacterial species under high pressure was first observed by Claude ZoBell in 1970. In subsequent studies with *Escherichia coli* the flagellar syntheses, that is one of the most pressure-sensitive cellular processes, was suppressed at 20 MPa. In studies on the flagella of a strain of *Salmonella enterica*, a volume increase of 340 mL·mol^{-1} of polymerization of flagellin monomer was measured [13], indicating that polymerization cannot occur under pressure and even causes depolymerization of existing flagella.

11.5.3 Cell Membranes

The membranes of eukaryotes and bacteria are made of phospholipids, sterols, and proteins, while the membrane lipids of archaea are different, as discussed above (Section 11.3). Their function is to separate the internal media from the external environment and regulate their exchanges. To maintain functionality,their membranes must remain in lamellar form, since when the liquid crystal transition occurs the membranes lose their physiological function. This ordering of the structure of membrane components can be produced by decreasing the temperature and increasing the pressure, and microorganisms have evolved homeoviscous adaptations to handle both phenomena. To maintain the same viscosity, the membrane can vary the degree of saturation and the lengths of the fatty acids formed. There are more long chain polyunsaturated fatty acids in seabed microorganisms when they are cultured at high pressure. Significant amounts of EPA (C20:5) and DHA (C22:6) are produced [14]. Those lipids are of interest as their consumption is associated with the prevention of cardiovascular risks in humans, but are unusual in bacteria and are found only in microorganisms isolated in low temperature environments or ocean depths. One of the genes involved in both high-pressure and low-temperature adaptations is the one encoding the enzyme that synthesizes cis-vaccenic acid (β-ketoacyl ACP synthase II) and allows the synthesis of polyunsaturated fatty acids. When the microorganisms are exposed to high pressure, the expression of genes associated with isoprene synthesis is increased. These genes are also activated in the cold stress response, together with a general activation of the biosynthesis of fatty acids [15].

Cholesterol also increases cell membrane viscosity, but at higher concentrations causes the membrane crystal transition to disappear. Other sterols seem to be important in the regulation of membrane fluidity and hydrophobicity of organisms, allowing membranes to withstand drastic changes in temperature and/or pressure while maintaining their functionality.

11.5.4 Genetic Information

It is possible to study the effect of pressure on a microbial population by studying the changes found in the survivors of a sublethal treatment. Among these are changes in the levels of expression of various genes in a way that can allow us to extrapolate the mechanisms of adaptation to high pressure presented by other microorganisms in their natural environment. Elevated pressures cause some damage in cell membranes and dissociation of complex and oligomeric proteins and macromolecules. Ribosomes and translational apparatus are specially sensitive. When pressure is high enough, as for instance the level used for food preservation (Section 11.7), even monomeric proteins can be denatured. In some experiments, the sensitivity of microorganisms to high pressures depended on the physiological state of the cells prior to application of pressure. The more resistant microorganisms are those in stationary growth state or those previously exposed to stressful conditions. The medium in which the microorganisms are subjected to pressure also affects their sensitivity to treatment. Media with high osmotic pressures somewhat counteract the effects of high pressure.

The most pressure-sensitive cellular process is the synthesis of DNAs, and initiation is more sensitive than polymerization. The double helix structure of DNA is reinforced because of the stabilizing effect of high pressure on hydrogen bonds and the access of proteins responsible to start replication to DNA chain becomes more difficult. Since the cell membrane is necessary to start DNA replication, these structure changes also significantly affect this process. In contrast to the difficulties found on DNA replication, RNA synthesis is relatively insensitive to pressure and can occur at pressures up to 65 MPa. Bear in mind that RNA is always found as a monofilament and the initiation process may not be hindered by the stabilizing effect of pressure on hydrogen bonds.

Ribosomes undergo dissociation at pressures above 60 MPa, preventing protein synthesis, but if they are associated with mRNA or tRNA they become stabilized against pressure and that makes them resistant up to 100 MPa. This increased stability has no physiological implications; it does not facilitate protein synthesis as the active sites are already occupied.

One bacterial response to high hydrostatic pressure is the increased expression of genes related to DNA repair and DNA polymerase, to compensate for the effects of pressure on the genetic material. In some bacteria, the increase in production of histone-like proteins has been observed. Pressure tends to cause a compaction of chromosomes, while histones organize DNAon a more open form. These histone-like proteins could act to compensate pressure induced compaction of chromosome. Exposure to elevated pressures induces proteins like those found in response to thermal shock, but it affects both cold and hot shock proteins synthesis. The induced cold shock proteins are related to RNA and transcription, tending to correct the decrease in RNA synthesis and transcription stalling. Dissociative effects of high pressure on ribosomes induce a response through increased expression in genes encoding ribosomal proteins and genes associated to translation [14].

11.5.5 Proteins: Transporters and Enzymes

One of the effects of high pressure is the dissociation of oligomeric proteins and protein complexes. Proteins are participants in many aspects of biological processes, as for instance as enzymes in metabolic pathways or associated with the cell membrane as conveyors. Conformational changes in quaternary structure may compromise oligomeric protein functionality. The high pressure tends to cause the protein to adopt a more compact structure, although the loss of secondary structure occurs only at extreme pressures.

In bacteria exposed to high pressure, we can find a number of proteins whose synthesis is increased in survivors as compared to control microorganisms kept at atmospheric pressure. This reaction shows that many of the bacteria can develop responses to the high pressure, and the non-specificity of the response shows that piezotolerance is widely distributed in nature. The mechanisms of protection against high pressure are no different from those which bacteria exhibit when faced with other stressors [16].

Adaptations to withstand high pressures and low temperatures act in a different direction, since increased rigidity to resist compression and high flexibility to adapt to low temperature are necessary. The difficulty of finding a compromise solution helps explain the low rates of growth of abyssal psycropiezophilics. Similarly, it can be found that the optimal growth pressure for these organisms increases at temperatures above those at which they were isolated [17].

11.5.6 Cell Division

When cell cultures are high-pressure treated, it is possible to observe the development of an unusual filamentous growth. An increase in the synthesis of cell wall polymers occurs, with increased expression of genes associated with the production of peptidoglycan and lipoprotein transporters, that are used to assemble various components of cell membranes, such as peptidoglycan and teichoic acids.

One of the effects on cell wall is the difficulty of process of chromosome segregation and formation of the septal ring. An essential protein for cell division in bacteria is FtsZ, which has a significant homology with eukaryotic tubulins.

In eukaryotic cells, tubulin form a sort of cytoskeleton when polymerizing in the shape of microtubules, directing the chromatids to two sibling cells during division. High pressure interferes with polymerization of FtsZ protein [14], which in addition to directing the two copies of chromosomes to the sibling cells, forms a separation mechanism by a structure known as a Z ring [18]. Since the septal ring is an essential part of division, any damage to this system leads to a loss of ability to form colonies, while restoring this capability allows the recovery of cell proliferation.

11.6 BIOTECHNOLOGICAL APPLICATIONS OF PIEZOPHYLIA

The biotechnological potential of extremophiles was shown clearly Kary Mullis, then working for Cetus Corporation, developed the polymerase chain reaction that earned him a Nobel Prize in 1993. The key to the reaction is the use of DNA polymerase derived from the hyperthermophilic *Thermus aquaticus*, discovered by T.D. Brock in 1965 and designated *taq* polymerase. Brock's and Mullis' discovery enabled many applications that extended the field of biotechnology and generated millions of euros in royalties. Biotechnological applications of piezophilia are not as advanced as applications of other extremophilic organisms, primarly because of the difficult growth requirements [19].

Some of the existing applications work by maintaining enzymatic reactions under pressure. Although proteins that are stabilized by electrostatic interactions are destabilized by pressure, those stabilized by hydrophobic interactions may experience increased thermostability with increasing pressure [17]. A hydrogenase of thermophilic *Methanococcus jannaschii*, obtained at 2 600 m, shows a marked stabilization to temperatures of 70 to 90 °C at 50 MPa, a temperature that causes denaturation at 0.1 MPa. This stabilization by pressure seems more related to thermophilia and would not be exclusive to microorganisms from oceanic hydrothermal vents.

The molar volumes of different reaction products or intermediates can allow reactions to be more specific at high pressures. Thus porcine pancreatic α-amylase using maltose as a substrate produces maltotriose, maltotetraose, and maltohexaose in the same proportion when the reaction occurs at 0.1 MPa. When the reaction is carried out between 100 and 300 MPa, the equilibrium shifts to the preferential production of maltotriose [4]. A Pseudomonas-like microorganism isolated from the deepest point of the Mariana Trench exhibiting optimal growth at 20 MPa can produce a α-maltotetraohydrolase when cultured at pressures up to 100 MPa [20]. The enzyme production is greater when cultured under high pressure than at ambient pressure. This enzyme is of relevance for the pharmaceutical and food industries because it can produce maltotetraose from starchy substrates.

Studies of microorganism responses to pressure and differential expression of genes (Section 11.5.4) may lead to interesting applications. A biotechnological tool is the promoter activated by growth at 50 MPa discovered in *Shewanella benthica* and *S. violacea* that can be used to activate differential gene expression as a function of pressure. This promoter regulates the ORF1 and ORF2 genes in these organisms; their function is unknown but they are unique to piezophilic microorganisms [19]. It is expected that these genes can be used as markers to identify piezophilia in microbial isolates.

Manipulation of *Saccharomyces cerevisiae* to induce piezophilia will allow us to introduce genes encoding products of industrial interest; their expression will be improved by culturing under pressure thanks to the advantages listed above. Piezophilic mutants of *Actinomyces species* antibiotic producers have also been obtained. It is expected that new antibiotics produced under high pressure can present interesting clinical applications.

11.7 MICROORGANISM DESTRUCTION FOR FOOD PRESERVATION

In the late 19th century, the first experiments of preserving food without heating were conducted using high pressures. Hite published a report on experiences in Virginia in 1899 [3]. These studies were not continued, and very few studies were conducted during the first half of the 20th century due to limited availability of the necessary equipment.

The microbial lethality achieved by a high-pressure treatment varies considerably, depending on the growth phase of the microbial culture. Microbial cells in stationary growth phase are more resistant to treatment by high pressure. High osmotic pressure is also protective, especially from sodium chloride, metal cations and various organic compounds. Unfavorable growth conditions may cause more resistance to high pressure. In food, we frequently find that microorganisms are in exponential phases of development and high levels of organic acids that cause low pH in the product, favoring pressure inactivation. Other factors would enhance the resistance of target microorganisms, such as high concentrations of low molecular weight solutes raising osmotic pressure. High concentrations of taurine and monoamine oxidase inhibitors found in deep sea microorganisms seem to counteract the effects of pressure [21].

Inactivation of microorganisms may not be total when food is preserved by high-pressure treatment (see Chapter 12). High-pressure treatments cause high rates of sublethal damage to microorganisms. These cells have difficulty proliferating after treatment and do not form colonies in culture media.

However, they can recover their ability to reproduce if they have enough time to recover in a non-aggressive medium,

Immediately after high-pressure treatment at 600 MPa of *Escherichia coli* in Tris buffer at pH 6.5, counts were reduced by more than 2 orders of magnitude. After the samples were refrigerated for 18 h, plate counts increased. This indicated that 10 % of survivors of high-pressure treatment were stressed but recovered easily [22].

Counts in non-selective media were maintained when the sample was refrigerated for two weeks. However, in selective media such as violet red bile glucose (VRBG) agar, bacterial cells suffering sublethal damage had difficulties proliferating and forming colonies. After high-pressure treatment, colony forming numbers were below the detection limit, but the microorganisms gradually repaired their metabolic systems. After two weeks of cold storage, the number of colonies on selective media increased almost a thousandfold and eventually achieved the same growth levels in both selective and non-selective media. In contrast, no recovery of *E. coli* from the initial stress or subsequent sublethal damage was observed in an orange juice medium with a distinctly acidic pH. High-pressure treated cells slowly lost their viability based on decreased counts in non-selective media; untreated cells were not affected.

Sublethal damage in microorganisms caused by high pressure has been studied by using fluorescent dyes in flow cytometry. This technique allows has the identification of metabolically active cells—even those that cannot proliferate and are thus undetectable on plate count.

Bibliography

[1] R. Kunzig. *Science*, 302, 991 (2003).

[2] G. Demazeau and N. Rivalain. *Appl. Microbiol. Biotechnol.*, 89, 1305 (2011).

[3] B. Hite. *Bull. W. V. Univ. Agric. Exp. Stn.*, 58, 15 (1899).

[4] F. Abe and K. Horikoshi. *Trends Biotechnol.*, 19, 102 (2001).

[5] B. Speers and B. Waggoner. *Introduction to the Archaea, Life's Extremists.* The University of California Museum of Paleontology (2001).

[6] M. Kate, D. J. Kushner, and A. T. Matheson. In *The Biochemistry of Archaea*, Volume 26, New comprehensive biochemistry series, 261–295. Elsevier, Amsterdam (1993).

[7] K. Horikoshi. *Frontier Research System for Extremophiles.* Japan Agency for Marine-Earth Science and Technology, Yokosuka, Japan (2004).

[8] Z. Minic, V. Serre, and G. Hervé. *C. R. Biol.*, 329, 527 (2006).

[9] R. Monastersky. *Sci. News*, 151, 192 (1997).

[10] G. di Prisco. In N. Glansdorff and C. Gerday, eds., *Physiology and Biochemistry of Extremophiles*, 145–154. American Society of Microbiology, Birmingham, AL (2007).

[11] V. V. Lukin and N. I. Vasiliev. *Ann. Glaciol.*, 55, 83 (2014).

[12] S. A. Bulat, I. A. Alekhina, D. Marie, *et al.* *Adv. Space Res.*, 48, 697 (2011).

[13] D. H. Bartlett, F. M. Lauro, and E. A. Eloe. In N. Glansdorff and C. Gerday, eds., *Physiolology and Biochememistry of Extremophiles*, 333–348. American Society of Microbiology, Birmingham, AL (2007).

[14] D. H. Bartlett. *Biochim. Biophys. Acta*, 1595, 367 (2002).

[15] Y. Yano, A. Nakayama, K. Ishihara, *et al.* *Appl. Environ. Microbiol.*, 64, 479 (1998).

[16] J. P. Bowman, C. R. Bittencourt, and T. Ross. *Microbiology*, 154, 462 (2008).

[17] D. Demirjian, F. Morís-Varas, and C. Cassidy. *Curr. Opin. Chem. Biol.*, 5, 144 (2001).

[18] A. Ishii, T. Sato, M. Wachi, *et al.* *Microbiology*, 150, 1965 (2004).

[19] B. van den Burg. *Curr. Opin. Microbiol.*, 6, 213 (2003).

[20] H. Kobayashi, Y. Takaki, K. Kobata, *et al.* *Extremophiles*, 2, 401 (1998).

[21] P. M. Oger and M. Jebbar. *Res. Microbiol.*, 161, 799 (2010).

[22] Q.-A. Syed, M. Buffa, B. Guamis, *et al.* *High Pressure Res.*, 33, 64 (2013).

CHAPTER 12

Food Science and Technology: Preservation and Processing below 50 MPa and up to 1 GPa

Bérengère Guignon

MALTA Consolider Team and Departamento de Procesos. Instituto de Ciencia y Tecnología de Alimentos y Nutrición (ICTAN–CSIC), Madrid, Spain

CONTENTS

12.1 INTRODUCTION

"A balanced diet promotes good health". This is indeed an old idea. What is more recent is the collective awareness of it. Scientific proofs of what has been intuitive for centuries were progressively provided and intensive researches are still going on to elucidate the connections between diet and health. Media development has made the diffusion of information easier, faster and more efficient. Food industry scandals, obesity statistics, discovery of non-healthy components as a result of processing, food alerts are reported around the world in seconds. Beyond the social and human importance of nutrition, this collective awareness has also an economic dimension. In developed countries, nutritional diseases which originate in food excesses, deficiencies, intoxications, and allergies represent significant health costs. Therefore, American, European, and other governments have included specific initiatives in their policies to deal with these issues. All this has led consumers to increase their requirements regarding foods. Consequently, putting aside pleasure foods, current consumers' preferences are for natural-taste foods. Moreover, they want foods that can be stored for the longest time possible for convenience reasons (*e.g.*, limited time for shopping and cooking). Food has even added value if it can offer some health benefit. The ideal products are safe, with high nutritional values and providing relatively long shelf lives and good sensory quality at the same time.

Food industry has responded to this demand with minimally processed foods (such processes exert minimal impacts on food characteristics), foods with low levels of sugar, salt, fat, and foods free of additives or preserving agents. Manufacturers have also seen the opportunity to develop markets with a new range of products and the food industry has entered the era of (nutri-) functional foods that, together with nutraceutics, help intestinal transit, cholesterol regulation, immune system stimulation and so on. The preservation of such nutritional properties for long periods is often a complicated task. Prehistoric humans already salted fishes, sun dried fruits, and froze meat. Later, in the 19th century, Pasteur and Appert invented thermal technologies of pasteurization and sterilization, respectively. However, none of these technologies is totally satisfactory because foods rapidly lose their organoleptic and nutritional qualities, and microbial activity may not be completely eliminated. The quest for innovative solutions is one of the current challenges of the food industry. The improvement of food processing and preservation is the main rationale for the use of high pressure in this field. High pressure technology fulfills the current requirements for food safety and quality.

High-pressure processing efficiently prolongs the shelf lives of food products while having a limited impacts on their sensory and nutritional characteristics. High-pressure technology is not limited to the food preservation objective. It offers numerous other applications, from breakfast cereal manufacture to non-contaminant food cutting and oyster opening. Some of these techniques, usually those which employ the lowest pressure levels (< 50 MPa), were developed more than two centuries ago. Thanks to technological progress starting in the 1950s, the pressure levels achievable in modern industrial production are close to gigaPascal level and efforts to design processes that will achieve even higher pressures are ongoing. In the history of food processing by high pressure, the most significant advance was undoubtedly the discovery of high hydrostatic pressure effects on foods. Hite (1899) showed how pressurization extends the shelf life of milk. Bridgman submitted egg whites to 700 MPa pressures and noted coagulation. These findings were at the root of diverse developments that led to the first pressurized commercial products. Jams were introduced in the Japanese market in 1991, followed by avocado puree in the U.S. and fruit juices in Europe. Worldwide production of foods treated by high hydrostatic pressure processes is now estimated at half a billion kilograms annually.

This chapter is divided into three principal parts. In the first one, the effects of pressure on the main food components are described for pressures up to 1 GPa. In the second part, a short overview of the different uses of pressure in the food industry and research is given based on the pressure levels employed. The final section focuses on high hydrostatic pressure processing as one of the most modern and versatile techniques. Information is provided about current industrial applications and research.

12.2 PRESSURE EFFECTS ON MAIN FOOD COMPONENTS

12.2.1 Le Châtelier's Principle

Food component state and stability depend on environmental conditions such as temperature, pressure, water activity, and pH. Modifications of the state of a component and interactions between different components can also take place as a function of time under constant conditions. All these changes originate from physical, chemical and biological reactions. The equilibrium of the reaction in which a component is involved is governed by Le Châtelier's principle (1884). This principle establishes that if a system is perturbed by a change in temperature, pressure or concentration in one of its components, it will displace its equilibrium position to counteract the effect of the perturbation. In the case of pressure, the main effect of this perturbation is a volume reduction. Consequently, any biochemical reaction, conformational change, or phase transition which is accompanied by a volume decrease will be favored at high pressure, while reactions and changes that involve a volume increase will be inhibited.

For comparison, when a food is heated, its volume increases. Temperature acts on volume contrarily to pressure. Thus, different effects arise from pressure and temperature change in foods during processing. The combination of compression and heat gives rise to antagonist actions whose impact on food is usually different from the action of a single factor. Covalent bond formation leads to a volume decrease. Therefore, pre-existing bonds are maintained through compression, and biochemical reactions with creation of new bonds are favored under pressure. This explains why the majority of food components remain intact after pressure treatment. However, foods are complex systems and several reactions usually occurs simultaneously or successively. The prediction or explanation of a given pressure effect often requires the identification of the limiting step(s) in the chain of reactions. Moreover, the structures of food components and their interactions also involve non-covalent bonds. Such bonds are sensitive to pressure and they account for the main effects observed in food after pressure treatment. In Table 12.1, the general effects of pressure on several types of bonds present in foods are summarized. In Reference [1], the effects of pressure on the reaction rates are reviewed for relevant phenomena in foods.

Table 12.1 Pressure general effects on types of bonds present in foods [2, 3]

Bond	Case	Activation volume ΔV (mL·mol^{-1})	Pressure effect
Covalent	Formation	−10	Favorable
	Rupture	+10	Unfavorable
	Bond and angle change	~ 0	Protection up to 1 GPa
Hydrogen	Formation / Rupture	~ 0	System-dependent
	Modification	~ 0	Stabilizing (in general)
Ionic	Formation	10 to 20	Unfavorable
	Charged group hydration	−10	Favorable
Hydrophobic	Interaction formation	1 to 20	Destabilizing (up to ~100 MPa for proteins)

12.2.2 Effect of Pressure on Food Water

Pressure modifies water properties in a peculiar way. It causes a greater dissociation of molecules and reduces the freezing point, both phenomena of importance in the food domain. Greater molecule dissociation leads to pH

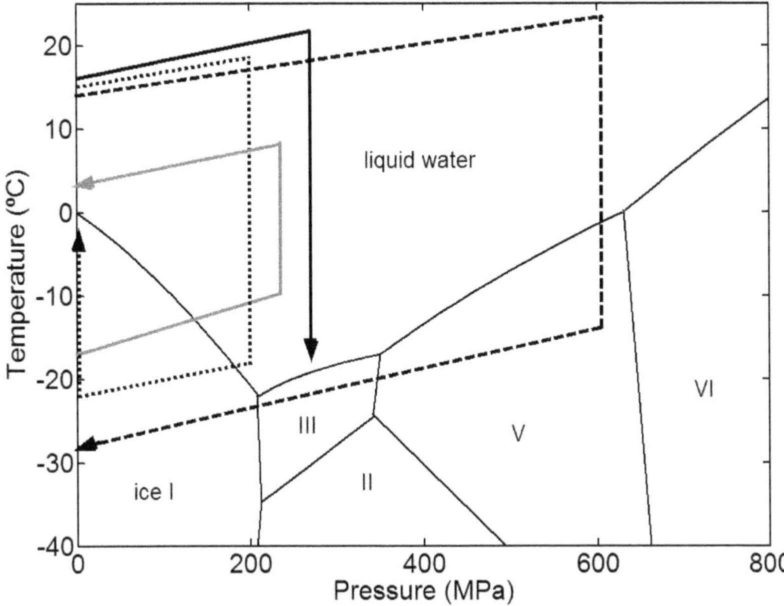

Figure 12.1 Examples of food processes at high pressure and low temperature on the water phase diagram: pressure-induced thawing (grey solid line), subzero preservation in liquid sate (black solid line), pressure-shift freezing (dotted line), and freezing to different ice-modifications (dashed line).

decrease with pressure. Depending on the pH sensor, food, and pressure level, this pH decrease is more or less pronounced. For example, at 25 °C, pH decreases of 0.2 to 0.3 [4, 5] and 0.77 [6] were reported in distilled water upon compression to 100 MPa. The pH change was only significant above 100 MPa in fruit juices and milk, amounting to about 0.3 between 500 and 800 MPa. Consequently, microorganism inactivation and protein state which are strongly pH-dependent will be affected. A given microorganism that barely survives in acidic medium under room conditions will be more easily inactivated. Any protein that reaches its isoelectric point could irreversibly precipitate, giving rise to the formation of a gel. The decrease of the freezing temperature with pressure increment allows innovative food processing [7]. For example, food items can be stored at sub-zero temperature in an unfrozen state, or quickly frozen and/or thawed by a suitable pressure-temperature management around the liquid-ice phase transition curve (see Figure 12.1).

12.2.3 Effect of Pressure on Food Lipids

Contrary to the case of water, lipid volume decreases when it freezes. Therefore, with pressure, lipid crystallization is favored. Thus, the melting points of triglycerides are raised by about 10 °C per 100 MPa increment. Soya oil, which contains a great variety of triglycerides, tends, for example, to solidify under pressures of 250 to 300 MPa at room temperature [8]. One of the possible applications of this phenomenon in food processing is the use of pressure to better control crystallization and melting of cacao fats (chocolate tempering) and thus improve the sensory qualities of smoothness and lightness. Another consequence of lipid crystallization under pressure is its contribution to microorganism inactivation. Microorganisms cell membranes consist of phospholipids that, when solidified are going to modify the membrane structure, fluidity, and permeability, leading to cell dysfunction and, finally, microorganism inactivation. Another pressure effect on lipids is an indirect effect: lipids suffer from oxidation upon metallic ion release in a food under pressure (catalysis effect in meat and fish) [9].

12.2.4 Effect of Pressure on Food Proteins

Proteins possess a common complex structure. They are made of amino acids chains arranged in a three-dimensional way. Primary (amino acid sequence), secondary (α-helix, β-sheet and turns), tertiary (spatial organization), and quaternary structures (subunits) can be distinguished within the same protein. Protein primary structure is due to covalent bonds while the others are stabilized by non-covalent interactions (hydrogen bonds, hydrophobic effects, or salt bridges). Since pressure does not affect covalent bonds (at the levels employed in food treatments), protein primary structure remains unchanged after treatment. Protein secondary, tertiary, and quaternary structures can suffer from some irreversible conformational changes as a result of non-covalent interaction modification under pressure (see Table 12.1). β-sheets seem to be the first secondary structures to be lost [10]. The tertiary structure appears as a molten globule at pressures below 200 MPa and unfolds partially or totally upon further compression depending on the pressure level. In the case of a protein complex, subunits dissociate under pressure.

Since protein functionality is related to its ternary structure, its functional activity may also change with pressure. In general, pressure effects on proteins are reversible at moderate pressures ($<$ 400 MPa). Nevertheless, depending on the volume change (ΔV) with denaturation and on the balance between stabilizing-destabilizing interactions, pressure is able to induce permanent structural modifications at lower pressures. The observable consequences after treatment are enzyme inactivation, color change, cooked appearance for fish and meat, and gel formation. Temperature also plays an important role in the result of high pressure treatment because of its antagonist action as mentioned above. In Figure 12.2, the typical protein state is represented as a function of pressure and temperature conditions. The stability frontier

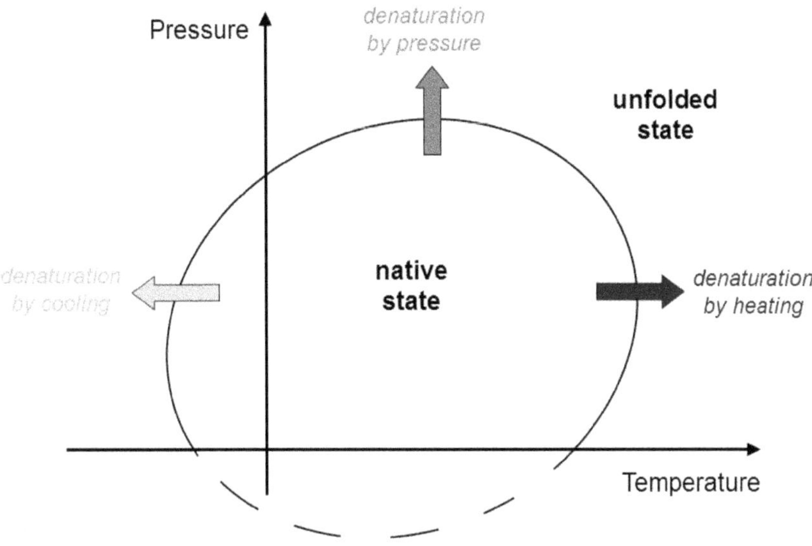

Figure 12.2 Typical stability diagram of proteins as a function of pressure and temperature.

between the native and denatured states usually looks like part of an ellipse [11]. Based on this diagram, a suitable management of both temperature and pressure parameters makes possible the control of protein activity [2, 6].

12.2.5 Effect of Pressure on Food Carbohydrates

Carbohydrates are classified according to the number of units in their structures as: monosaccharides, disaccharides, oligosaccharides, and polysaccharides. Monosaccharides (for example: glucose, fructose) and disaccharides (for example, sucrose which is formed by one glucose linked to one fructose molecule via an O-glycosidic bond) only involve covalent bonds and therefore are largely unaffected by pressure. No pressure effect has been reported on those simple sugars under the usual high-pressure processing conditions but pressure levels above 4.8 GPa give rise to hydrogen bond formation and polymorphs in the case of sucrose [12, 13]. Besides, pressure has an influence on Maillard reactions which involve carbohydrates [14]. More evident are the changes caused by pressure on starch structure. Starch is a polysaccharide which contains two polymers of glucose: amylase (mainly a lineal structure) and amylopectin (ramified). A starch granule is organized in layers of amylopectin between which amylose is intercalated. Amylopectin has both amorphous (ramifications) and crystalline (chains) regions. Starch granule

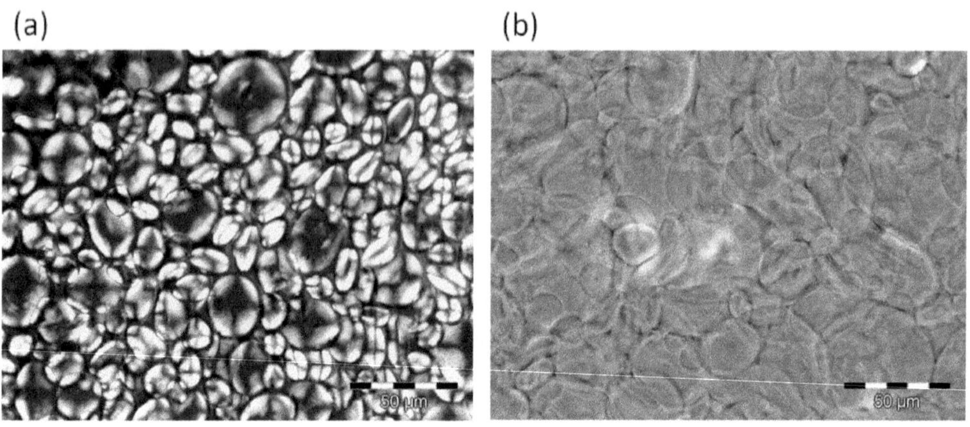

Figure 12.3 Observation under polarized light microscopy of whey starch granules before (a) and after (b) high-pressure treatment at 400 MPa and 40 °C for 20 min.

structure is globally crystalline and characterized by the birefringence phenomenon which appears as "Maltese cross" within each starch granule when observed under a polarized light microscope (Figure 12.3 (a)). This structure is lost with high pressure and the crosses disappear (Figure 12.3 (b)) but granule edges remain visible, indicating that part of the structure remains. When the starch gel is produced by heating, the granule is completely destroyed. This is the reason gel properties (viscosity, color, taste) obtained by high pressure treatment are different from those obtained by heat. The pressure from which the gel is formed depends on the starch source, temperature, and treatment time [15]. The mechanisms through which pressure produces starch gelatinization are not fully clear, but they are related to the influence of pressure on Van der Waals forces and on hydrogen bonds that ensure starch molecular stability. Compared to temperature effect, it is suggested that pressure favors granule hydration but protects the helicoidal structure of the crystalline part of amylopectin through hydrogen bond stabilization [16].

12.2.6 Effect of Pressure on Other Food Components

Pressure effect on other food components such as vitamins or mineral salts depends on the bonds that stabilize their structures and on possible reactions with other components present in food. As a general rule, as far as covalent bonds are concerned, their structure is preserved. This presents a great advantage because thermo-labile components are kept intact at high pressure while they are rapidly destroyed during traditional thermal processing. Hence, the preservation of nutritional and organoleptic quality characteristics is greatly improved. The stability of vitamins B1, B6, and C under pressure has been demonstrated in milk [17]. In fruits and vegetables, vitamins and antioxidants were also found to be pressure-stable [18–20]. Even when high temperatures were used in combination with high pressures, the preservation of such nutrients was enhanced compared to the traditional sterilization but some degradations were unavoidable [21, 22].

12.3 HIGH–PRESSURE TECHNIQUES FOR FOOD PROCESSING AND PRESERVATION

The technical difficulty affecting the implementation of high-pressure processes at industrial scale grows with the pressure level. This explains why processes above 50 to 100 MPa have been achieved in the food industry only recently. In this section, the different techniques based on the use of pressure to process or preserve foods are reviewed according to the level currently applied.

12.3.1 Processes at Pressures below 50 MPa

12.3.1.1 Pressure Cooking

The first time that pressure was employed in the food field was at the end of the 17th century with the invention by Denis Papin (1679) of the steam digester as a forerunner of the pressure cooker. The pressure inside the device could reach almost 2.5 MPa. The commercial version of the pressure cooker appeared in 1940 to 1950 and represented an important advance to cook foods more quickly. The food is placed in the cooker with water or broth and heated. When water boils, vapor formation in the closed container produces a pressure increase and the boiling temperature rises near 121 °C. A safety valve keeps pressure at a maximum of 0.2 MPa to avoid cooker explosion risks. The cooking time is diminished and aromas, colors, vitamins, and mineral salts are better preserved than after a traditional cooking process [23, 24]. Current industrial pressure cookers have a capacity exceeding 5000 L and reach 0.7 MPa. They are used to process marmalades, jellies, chocolate, sweets, meats, sausages, and other products.

12.3.1.2 Extrusion

Extrusion is another old food process involving pressure. The first extruders appeared between 1869 and 1930. Today, this is a common process in the food industry due to its versatility, high productivity, low cost, and energetic efficiency. The process consists of conveying and kneading a paste mass inside a cylindrical vessel by pushing it between a screw and the cylinder wall or between twin screws. This creates shear forces and pressure which can be used in combination or not with heat to modify the physico-chemical properties of the paste. Pressure is released at the exit of the machine. The basic equipment works in the continuous mode with two inputs: one for dry ingredients (flour, grains, sub-products) and the other for liquids (water, sugar, fat, colorant, meat, etc.). These components are introduced at the initial part of the barrel and they heat up and mix as they pass through the barrel. The resulting semi-solid mass (generally below 30 % in water content) is forced to pass throughout a small orifice (die) designed to give the final shape to the product. The simultaneous cooking involves temperatures up to 250 °C, whereas the residence time is relatively short (1 to 2 min). The mechanical and thermal actions generate a pressure of 1 to 10 MPa. At the exit of the machine, the cooked product expands upon water evaporation and a typical aerated texture is obtained. Thank to this technology, new products were launched on the market: snacks, breakfast cereals, candies, and animal foods. Extrusion applied to foods with higher water content (40 to 80 %) led to applications such as meat processing (sausages, surimi) or the creation of meat substitutes from vegetable proteins (soya) [25]. Recent developments aim to enrich extrudates with dietary fiber, resistant starch, antioxidants, inulin, and dairy proteins while ensuring acceptable taste and texture [26]. Other developments involve the use of extrusion for biodegradable packaging production including food components [27].

12.3.1.3 Supercritical Fluid Extraction

The use of supercritical fluids in the food industry began in the 1970s with the decaffeination of coffee. This ecological and reliable separation process produces high quality extracts and treated products. It usually uses carbon dioxide, nitrogen, or argon in a supercritical state. Such state is achieved at pressures between 10 and 40 MPa and temperatures between 30 and 60 °C. In the case of carbon dioxide, the critical point is 31.1 °C and 7.38 MPa. A supercritical fluid is able to diffuse through a solid like a gas and dissolve materials like a liquid. The extraction operation consists of putting into contact the raw material with the fluid in supercritical state. The process is usually discontinuous in the case of solid food, but specific designs make it possible to work in continuous mode, in particular when dealing with liquids (counter-current pumping) [28]. It is applied to the extraction of spices and plant aromas, hops for beer production, of healthy fatty acids from animal fats and oils, bioactive compounds from plants, food by-products, and algae [29]; it also permits extraction of caffeine from coffee and tea, elimination of cholesterol in products from animal sources, and removal of alcohol from wine, cider, and beer [3, 30]. Another objective of supercritical fluid application is micronization: particle sizes below 20 μm are obtained, simplifying their incorporation in a matrix (case of immiscible substances). For example, chocolate can be added to ice cream without a previous melting step; product viscosity is increased but granules cannot be detected by the tongue. Phytosterols which are liposoluble can be mixed directly as functional ingredients during food elaboration, without previous dissolution of fat: the anti-cholesterol effect of phytosterol is complemented by a lower fat intake [31]. Supercritical CO_2 is also efficient for microorganism inactivation and it is going to be introduced in industry [32, 33] for this purpose. Due to the large investment required for extraction equipment, the treated product must provide adequate added-value for the process to be profitable.

12.3.2 Processes at Pressures above 50 MPa

12.3.2.1 Water-Jet Cutting

High-pressure water jets are employed to cut foods under optimal hygienic conditions (no cross-contamination). The water jet is usually set at pressures between 200 and 400 MPa and can reach 800 MPa. It is applied to cut all kinds of foods: watermelon [34], potato chips, pizzas, sandwiches, fish, meat, chocolate, frozen products, and pastries as shown on the website of one equipment manufacturer (KMT). The technique is also valid for peeling fruits and vegetables [35]. Clean cuts are achieved and complex shapes can be programmed, for example, chicken nuggets with animal shapes. It can also be used for meat carcass cleaning reducing production costs and labor risks for the employer [36].

12.3.2.2 High–Pressure Homogenization

Pressure is also used to homogenize liquid products. At the beginning of the 20th century, standard homogenizers (below 60 MPa) were primarily used in the dairy industry to avoid creaming of whole milk. Later, with technological progress, high-pressure homogenizers were able to work at 150 to 200 MPa and today the ultra-high pressure homogenizers can reach nominal pressures of 400 MPa [37]. Homogenization is achieved by forcing a liquid to pass through a small orifice inside a pipe: the valve gap. Different valve designs exist but the phenomena leading to colloidal stabilization are the same: cavitation, torsion and shearing, impingement, and turbulence. Suspended droplets or particles are broken into smaller ones when flowing through the valve, which delays phase separation after processing and during storage. At the same time, microorganisms are inactivated and the product leaves the equipment in a sterile state for aseptic packaging. Ultra-high pressure homogenization improves emulsion texture and aroma characteristics and provides a longer shelf life [38]. Functional properties of proteins and polysaccharides are modified in addition to properties of the emulsions. The nutritional and sensory characteristics of vegetable milks and fruit juices are improved [39, 40]. Ultra-high pressure homogenizers are just starting to appear in the food industry. This high-pressure technique is also sometimes called "dynamic" high-pressure to differentiate it from food treatment by high hydrostatic pressures (see next paragraph). In comparison, the homogenization process allows the main advantages to be continuous and reach sterilisation objectives (no need for refrigeration), but it does not allow treatment of solid foods.

12.3.2.3 High Hydrostatic Pressure Processing

High hydrostatic (or isostatic) pressure processing is the treatment of foods by compression in a fluid medium (usually water). Pressure is fully transmitted in all the directions through the fluid to the food according to Pascal's principle (1663). Thank to this uniform pressure transmission, the products suffer almost no change in shape whereas if the pressure was unidirectional, the food would be squashed. Therefore, high hydrostatic pressure treatment is not going to depend on food size or geometry, which is an advantage for process scaling. The adiabatic heat generated during compression (temperature increase by about 3 °C per 100 MPa in water) is problematic. It can cause non-uniform treatment if preventive measures are not taken to avoid risk of non-uniform microorganism inactivation or over-processing.

Among all the available techniques, this method uses the highest pressure levels. Technological advances make it possible for laboratory equipment to reach more than 1000 MPa but current industrial equipment works at most at 600 MPa. These pressure levels are possible without too many risks for the operator because liquids are much less compressible than gases. Hence, any fluid leak involves a lower volume change than occurs with gas and, thanks to this, it is easier to control (no explosion as happens with gas). The main objective of high hydrostatic pressure processing is to extend food shelf life. Processing can be carried out by direct compression and immediate aseptic packaging (liquid foods), or by indirect compression which is the most common method for liquid or solid packaged items. Food products are isolated from the pressure transmitting fluid to avoid diffusion by vacuum packaging (although sometimes a small head space can be left). For processing, the packaged items are placed in baskets and introduced in the high pressure vessel. The vessel consists of a hollow cylinder with thick stainless steel walls able to withstand the pressure. The vessel is closed and completely filled with the pressure transmitting fluid. Air is evacuated through a purge valve. The valve is closed and the compression starts by means of a hydraulic pump. The generated pressure is multiplied by using an intensifier that also separates the hydraulic fluid from the fluid in contact with the packaged food. This fluid, usually water, surrounds the food and applies the pressure. In general, there is no regulation system for temperature and treatments are performed at room temperature. Once the required pressure for microorganism inactivation is reached, pressure is held for 1 to 15 min. Then it is released and the treated product is removed.

This is a discontinuous process. In order to increase the productivity, several vessels can be arranged to work in parallel, constituting a semi-continuous process. In addition to pressure level and temperature and pressure holding time, other processing factors are pressurization and depressurization rates, as well as the filling ratios of the vessel (filling ratio between 40 and 70 % are typical). To be profitable, a treatment time around 3 min is suitable. From 200 to 3000 kg of product per hour can be processed, depending on the equipment capacity and operation conditions (updated data can be found on the web sites of equipment manufacturers such as Hiperbaric or Avure). Investment for equipment is heavy: around one million euros based on the installation size and automation level. However, the energy requirement is lower than for thermal processes and is regarded as an environmentally friendly process (reusable water, no waste). Prices of pressure-treated products are low enough to be competitive. Although process costs are highly variable from one country to another and along time, some estimations can be found in the literature [41, 42]. High-pressure processing is acceptable to consumers who are sometimes reluctant to face novel technologies. A great variety of products are available on the market: sliced ham and tapas (Espuña in Spain), pre-cooked chicken (Tyson in the USA), spreadable products (Zwanenberg in Holland), ready-to-eat meals (MRM in Spain), fruit juices and smoothies (Macè in Italy, CJ in Korea), sandwich fillings (Rodilla in Spain), fruit salads (Chic Foods in China), snacks (Deli24 in the UK), salad dressings (Bolthouse in the USA), and seafood (Cinq Degrés Ouest in France, Mitsunori in Japan). This is a popular technique and the number of units installed keeps increasing: from 10 machines in 2000 to more than 250 in 2015. This success is due to the unique feature of high-pressure processing for the preservation of organoleptic and nutritional qualities of food. The intrinsic potential for innovation using pressure as a tool to design products is a considerable competitive advantage.

12.4 CHARACTERISTICS OF FOODS TREATED BY HIGH HYDROSTATIC PRESSURE

Among all the high pressure techniques employed in the food field, high hydrostatic pressure is the most versatile. Today pressure is seen as a tool providing real opportunities for innovation. This section focuses on the most outstanding effects of this process on food through examples drawn from research and from the industry.

12.4.1 Safety Aspects

12.4.1.1 Inactivation of Microorganisms: Yeasts, Molds, Bacteria, and Spores

High pressure enables the inactivation of microorganisms responsible for food quality degradation and foodborne diseases. For example, the shelf lives of cold meat products (ham, sausage, duck liver, etc.) are extended while avoiding the development of acidic tastes and undesirable aromas. This is the case for cooked ham for up to 60 days after treatment for 6 to 10 min at 400 or 600 MPa. *Listeria* is a common pathogen in meat products and ready-to-eat meals; it is efficiently inhibited by such treatment. A lot of published research work exists because microorganism inactivation strongly depends on family, species, and strain. For instance, Gram-negative bacteria are more sensitive to pressure than Gram-positive bacteria. Moreover, microorganism inactivation strongly depends on the food matrix: solid or liquid, pH, interactions with other components, etc. It also involves processing conditions (pressure and temperature levels, pressure release rate, number of pressure cycles, vessel dimensions, and so on). Kinetic aspects are relevant due to adiabatic heat generation during compression and the development of thermal non-uniformities during the pressure holding step. Kinetic studies (determination of reaction rate constant) and modeling are useful to deal with process optimization from the view of inactivation [43–45]. In general, yeasts and molds are inactivated from 200 or 300 MPa. Bacteria vegetative forms are easily inactivated between 400 and 600 MPa but their spores require the highest pressure levels (at least 500 MPa) and temperatures above 60 °C. Therefore, food products treated by high pressure still need refrigeration after processing to ensure their microbiological stability for the longest time possible. For spore destruction, even pressures of 1500 MPa at room temperature may not be enough to ensure food safety [46]. By raising the temperature of treatment to 80 °C, this constraint can be reduced to pressure levels currently more viable for industrial equipment (< 700 MPa). The benefit, compared to the traditional sterilization, is that the food organoleptic and nutritional characteristics are better preserved with pressure than without it. The so-called pressure-assisted thermal process (abbreviated PATP) for food sterilization has been and still is the object of intensive investigations. The Food and Drug Administration (FDA) approved its use in 2009 (tests on mashed potatoes). Semi-industrial equipment of 55 L capacity already exists on the market for such an application. High pressures act on microorganisms by interrupting their cellular functions. High pressure damages the microbial membrane and causes its rupture. Membrane phospholipids can solidify with pressure so the membrane loses its fluidity; it turns more porous under pressure and this interferes with exchanges of cell nutrients and wastes. Proteins and key enzymes of the cell are denatured and this can cause cellular dysfunction. Ribosome disintegration in subunits and cell internal acidification equally contribute to the cell impairment [6]. It seems that the accumulation of all these injuries leads to microorganism inactivation. The mechanisms involved in spore inactivation by high pressure are still under investigation but some information is already available [47]. Spore germination is induced at low pressures (about 200 MPa). A second pressure cycle or a moderate heat treatment after this pressure-induced germination was proposed as a strategy to inactivate spores. However, results were not fully satisfactory and more research is needed to solve this problem [48]. Other strategies for enhancing microorganism inactivation after high-pressure processing include hurdle techniques: successive minimal treatments, combination with preservation factors such as CO_2, natural antimicrobials, and even the return of food additives [42]. More characteristics of microorganism behavior under pressure can be found in the previous chapter.

12.4.1.2 Inactivation of Other Pathogens: Prions, Viruses, and Parasites

High pressure is also able to inactivate prions, viruses and parasites. The mechanisms by which they are inactivated are more complex than those for bacteria and they are still under study. Prions require pressures on the order of 1000 MPa or temperatures above 120 °C for their inactivation. In the case of the infectious agent of transmissible spongiform encephalopathy, treatments at 800 MPa and 60 °C for 30 min or at 800 MPa and 80 °C for 5 min can efficiently decrease the prion infectivity [49]. For viruses (*e.g.*, hepatitis A virus, norovirus), pressures between 200 and 500 MPa are enough, depending on the kind of virus and treatment duration [50]. Parasites such as *Anisakis simplex* in fish can be inactivated at pressures below 200 MPa [51]. Parasite destruction by high-pressure treatment between 100 and 400 MPa was also shown for *Toxoplasma gondii*, *Cryptosporidium parvum*, *Trichinella spiralis*, and *Ascaris* [52].

12.4.2 Enzyme-Related Quality Aspects

Enzymes are proteins characterized by strong catalytic power and specificity. In foods, enzymes catalyze biochemical reactions often linked to food quality. Since pressure is able to modify protein structure, a given enzyme can be activated (usually at pressures below 200 MPa) or inactivated (above 300 MPa). The pressure level required for achieving enzyme activation or inactivation depends on its intrinsic properties (category, origin), temperature, time spent under pressure, pH, water content and food matrix, among other factors [6]. Because they are present in a great variety of foods, pectinmethylesterase, polyphenol oxidase, lipoxygenase, peroxidase, lipases, and proteases are the most frequently studied enzymes. For instance, orange juice naturally contains pectinases enzymes that catalyze reactions that lead to a decrease in juice viscosity and turbidity. The juice separates into a dense phase and a transparent one, a feature rejected by consumers. When the juice is treated at a pressure above 300 MPa, most pectinases are irreversibly denatured and the juice quality loss is somewhat delayed [53]. Moreover, juice color and aromas are almost indistinguishable from those of natural orange juice [54]. Another example of enzyme inactivation is polyphenol oxidase and lipoxygenase inactivation in guacamole (avocado puree). These enzymes catalyse reactions responsible for browning and undesirable taste in avocado. With treatment of four cycles of 5 min at pressures close to 700 MPa, guacamole (an avocado puree) maintains its characteristic green color and natural taste longer [55]. Conversely, the activation of enzymes by low pressures can be used to clarify apple juices in less time than the traditional clarification process [56].

12.4.3 Other Pressure-Induced Changes in Food Quality Characteristics

High hydrostatic pressure enables food preservation and also constitutes a novel tool to transform food. In particular, unique effects on texture are observed. Its use as a pre-treatment or in combination with other technologies paves the way to a wide range of opportunities. Some illustrative examples are provided below.

Red meat turns initially firmer (and acquires a rose-colored appearance), but once cooked, it is more tender and juicy than unprocessed meat, with no need for additives [57]. Flans can be cooked at room temperature [58]. It is possible to produce marmalades within a reduced time and also at room temperature, preserving the natural fruit color and taste [59]. By treating milk at high pressure, its techno-functional properties are modified: gels formed at atmospheric pressure from high-pressure treated milks are creamier and cheese yield is higher [60]. Many other ways exist for structuring dairy products using high pressure as described [61]. Another unique effect of pressure is the opening of bivalves such as oysters and the separation of lobster meat from the shell [62].

High-pressure treatments can also be supports or alternatives to other processing technologies to improve food quality characteristics. High pressures can precede or replace blanching, osmotic dehydration, rehydration, solid-liquid extraction, frying, etc. For example, high-pressure treatment of potatoes before frying reduces the quantity of oil absorbed by 40 % during the frying step.

A thorough review of all these applications can be found [63]. The combination of high pressure with other emergent technologies such as electric pulses and ultrasounds would improve the preservation effects of each method while minimizing impacts on organoleptic and nutritional qualities.

12.5 CONCLUSIONS

High pressure constitutes a technology for both preservation and transformation of food products. The diversity of techniques enables a great variety of industrial applications. In all cases, pressure has become a promoter of safety, sensory, and nutritional quality. It is seen as a unique tool to obtain tailored characteristics such as a specific texture. However, many questions remain to be answered, especially regarding the newest techniques such as high hydrostatic pressure processing. Studies are necessary to complete the knowledge of bio- and physico-chemical phenomena under high pressure, to achieve better control of the process and optimize pressure effects. Lately, several strategies to solve non-uniform processing concerns have been proposed, including the development of pressure-temperature-time indicators [64].

The commercial success of this technology is in part due to the excellent acceptance of high-pressure treated products by consumers and the great potential to market new products with high added values. Production costs should also decrease if the prices of high-pressure equipment continue decreasing. Moreover, high pressure techniques are usually regarded as ecologically beneficial. This is an additional advantage which is becoming increasingly relevant in the context of international environmental policies and sustainability issues. The storage of foods under low pressures (below 200 MPa) at room temperature has recently been proposed as an ecological alternative to refrigeration [65]. With technological improvements and new research results, important advances are expected. There is no doubt that the enthusiasm for this technology in the food industry will still increase in the coming years.

Bibliography

[1] S. Martinez-Monteagudo and M. Saldaña. *Food Eng. Rev.*, 6, 105 (2014).

[2] V. Mozhaev, K. Heremans, J. Frank, *et al.* *Proteins: Struct., Funct., Genet.*, 24, 81 (1996).

[3] M. Raventos Santamaría. *Industria Alimentaria: Tecnologías Emergentes.* Universitat Politècnica de Catalunya, Barcelona (2005).

[4] P. Gervais. *J. Phys. Chem. A*, 103, 1785 (1999).

[5] C. Samaranayake and S. Sastry. *Innov. Food Sci. Emerg.*, 17, 22 (2013).

[6] E. Palou, A. Lopez-Malo, G. Barbosa-Cánovas, *et al.* *Handbook of food preservation.* CRC Press, Boca Raton, FL (2005).

[7] G. Urrutia Benet, O. Schlüter, and D. Knorr. *Innov. Food Sci. Emerg.*, 5, 413 (2004).

[8] A. LeBail, L. Boillereaux, A. Davenel, *et al.* *Innov. Food Sci. Emerg.*, 4, 15 (2003).

[9] H. Ma, D. Ledward, A. Zamri, *et al.* *Food Chem.*, 104, 1575 (2007).

[10] H. Imamura, Y. Isogai, and M. Kato. *Biochemistry*, 51, 3539 (2012).

[11] L. Smeller. *Biochim. Biophys. Acta, Protein Struct. Mol. Enzymol.*, 1595, 11 (2002).

[12] G. Ribeiro, T. Costa, A. Pereira, *et al.* *Vib. Spectrosc.*, 57, 152 (2011).

[13] E. Patyk, J. Skumiel, M. Podsiadło, *et al.* *Angew. Chem., Int. Ed.*, 51, 2146 (2012).

[14] H. Jaeger, A. Janositz, and D. Knorr. *Pathol. Biol.*, 58, 207 (2010).

[15] B. Bauer and D. Knorr. *J. Food Eng.*, 68, 329 (2005).

[16] D. Knorr, V. Heinz, and R. Buckow. *Biochim. Biophys. Acta,Proteins Proteomics*, 1764, 619 (2006).

[17] I. Sierra, C. Vidal-Valverde, and R. López-Fandiño. *Milchwissenschaft*, 55, 365 (2000).

[18] Y. Nuñez-Mancilla, M. Pérez-Won, E. Uribe, *et al. LWT Food Sci. Technol.*, 52, 151 (2013).

[19] J. Vázquez-Gutiérrez, L. Plaza, I. Hernando, *et al. Food Function*, 4, 586 (2013).

[20] D. Keenan, C. Rößle, R. Gormley, *et al. LWT Food Sci. Technol.*, 45, 50 (2012).

[21] I. Oey, I. Van der Plancken, A. Van Loey, *et al. Trends Food Sci. Technol.*, 19, 300 (2008).

[22] Z. Escobedo-Avellaneda, M. Pateiro-Moure, N. Chotyakul, *et al. CyTA J. Food*, 9, 351 (2011).

[23] F. Natella, F. Belelli, A. Ramberti, *et al. J. Food Biochem.*, 34, 796 (2010).

[24] R. Rocca-Poliméni, D. Flick, and J. Vasseur. *J. Food Eng.*, 107, 393 (2011).

[25] H. Akdogan. *Int. J. Food Sci. Technol.*, 34, 195 (1999).

[26] V. Obradović, J. Babić, D. Šubarić, *et al. J. Food Nutr. Res.*, 53, 189 (2014).

[27] V. Hernandez-Izquierdo and J. Krochta. *J. Food Sci.*, 73, R30 (2008).

[28] C. Pronyk and G. Mazza. *J. Food Eng.*, 95, 215 (2009).

[29] M. Herrero, J. Mendiola, A. Cifuentes, *et al. J. Chromatogr. A*, 1217, 2495 (2010).

[30] B. Machado, C. Pereira, S. Nunes, *et al. Sep. Sci. Technol.*, 48, 2741 (2013).

[31] E. Weidner. *J. Supercrit. Fluids*, 47, 556 (2009).

[32] L. Garcia-Gonzalez, A. Geeraerd, S. Spilimbergo, *et al. Int. J. Food Microbiol.*, 117, 1 (2007).

[33] M. Perrut. *J. Supercrit. Fluids*, 66, 359 (2012).

[34] W. McGlynn, D. Bellmer, and S. Reilly. *J. Food Qual.*, 26, 489 (2003).

[35] R. Carreño-Olejua, W. Hofacker, and O. Hensel. *Food Bioprocess Technol.*, 3, 853 (2010).

[36] M. Alitavoli and J. McGeough. *J. Mater. Process. Technol.*, 84, 130 (1998).

[37] E. Dumay, D. Chevalier-Lucia, L. Picart-Palmade, *et al. Trends Food Sci. Technol.*, 31, 13 (2013).

[38] A. Zamora and B. Guamis. *Food Eng. Rev.*, 1–13 (2014).

[39] I. Tahiri, J. Makhlouf, P. Paquin, *et al. Food Res. Int.*, 39, 98 (2006).

[40] F. Campos and M. Cristianini. *Innov. Food Sci. Emerg.*, 8, 226 (2007).

[41] F. Sampedro, A. McAloon, W. Yee, *et al. Food Bioprocess Technol.*, 7, 1928 (2014).

[42] D. Bermúdez-Aguirre and G. Barbosa-Cánovas. *Food Eng. Rev.*, 3, 44 (2011).

[43] V. Serment-Moreno, G. Barbosa-Cánovas, J. Torres, *et al. Food Eng. Rev.*, 6, 56 (2014).

[44] A. Delgado, C. Rauh, W. Kowalczyk, *et al. Trends Food Sci. Technol.*, 19, 329 (2008).

[45] H. Mújica-Paz, A. Valdez-Fragoso, C. Samson, *et al. Food Bioprocess Technol.*, 4, 969 (2011).

[46] D. Wilson, L. Dabrowski, S. Stringer, *et al. Trends Food Sci. Technol.*, 19, 289 (2008).

[47] E. Georget, S. Kapoor, R. Winter, *et al. Food Microbiol.*, 41, 8 (2014).

[48] Y. Shigeta, Y. Aoyama, T. Okazaki, *et al. Food Sci. Technol. Res.*, 13, 193 (2007).

[49] P. Heindl, A. Garcia, P. Butz, *et al. Innov. Food Sci. Emerg.*, 9, 290 (2008).

[50] T. Norton and D.-W. Sun. *Food Bioprocess Technol.*, 1, 2 (2008).

[51] A. Molina-García and P. Sanz. *J. Food Prot.*, 65, 383 (2002).

[52] E. Rendueles, M. Omer, O. Alvseike, *et al. LWT Food Sci. Technol.*, 44, 1251 (2011).

[53] M. Bull, K. Zerdin, E. Howe, *et al. Innov. Food Sci. Emerg.*, 5, 135 (2004).

[54] A. Polydera, E. Galanou, N. Stoforos, *et al. J. Food Eng.*, 62, 291 (2004).

[55] E. Palou, C. Hernández-Salgado, A. López-Malo, *et al. Innov. Food Sci. Emerg.*, 1, 69 (2000).

[56] B. Tomlin, S. Jones, A. Teixeira, *et al. J. Food Eng.*, 129, 47 (2014).

[57] M. Hendrickx and D. Knorr. *Ultra high pressure treatments of foods*. Kluwer Academic, Dordrecht (2001).

[58] A. Ibarz, E. Sangronis, G. Barbosa-Cánovas, *et al. Food Sci. Technol. Int.*, 5, 191 (1999).

[59] J. Gimenez, P. Kajda, L. Margomenou, *et al. J. Sci. Food Agric.*, 81, 1228 (2001).

[60] R. López-Fandiño. *Int. Dairy J.*, 16, 1119 (2006).

[61] A. Devi, R. Buckow, Y. Hemar, *et al. J. Food Eng.*, 114, 106 (2013).

[62] M. Cruz-Romero, A. Kelly, and J. Kerry. *Innov. Food Sci. Emerg.*, 8, 30 (2007).

[63] N. Rastogi, K. Raghavarao, V. Balasubramaniam, *et al. Crit. Rev. Food Sci. Nutr.*, 47, 69 (2007).

[64] T. Grauwet, C. Rauh, I. Van der Plancken, *et al. Trends Food Sci. Technol.*, 23, 97 (2012).

[65] K. Segovia-Bravo, B. Guignon, A. Bermejo-Prada, *et al. Innov. Food Sci. Emerg.*, 15, 14 (2012).

CHAPTER 13

Biotechnological Sciences and Industrial Applications

Oscar R. Montoro
MALTA Consolider Team and Departamento de Química Física I, Universidad Complutense, Madrid, Spain

Nadia A. S. Smith
Mathematics and Modelling Group, National Physical Laboratory, Teddington, United Kingdom

CONTENTS

13.1 INTRODUCTION: ULTRA HIGH-PRESSURE PROCESSING

THE aim of this chapter is to show the enormous potential of high-pressure technology, and its applications to different areas, focusing mainly on biotechnology and industrial applications.

The advantageous use of high-pressure (HP) in food technology has been discussed in the previous chapter. These innovative advances in the food industry are beginning to be extrapolated to the cosmetic industry, and to the medical and pharmaceutical sciences [1, 2].

Several applications of HP in industry have already been implemented, and many others are still under development. All of these applications are studied in the following sections, which are divided according to application areas, as shown in the contents list above.

High hydrostatic pressure processing (HHPP) is a method of treating products (such as food, pharmaceutical, and cosmetic products) to which pressure is applied to inactivate enzymes and pathogenic microorganisms (bacteria, viruses, yeasts, and fungi) contained in the product. Pressure can be applied with or without a thermal treatment. Hydrostatic pressure is isostatic, applied via a liquid medium such as water. If the absolute value of the applied pressure is the same in all spatial directions, the pressure is isostatic. This type of high pressure applied at moderate temperatures is sometimes called pascalization because of its correlation with pasteurization (application of high temperatures at atmospheric pressure). Two technical terms for this process are high-pressure processing (HPP) and ultra high-pressure processing (UHPP) [3–5].

In UHPP, products are exposed to pressures of 400 to 1000 MPa (4000 to 10000 atm), with exposure times ranging from a few seconds to over 30 min, depending on the microbial inactivation goal (the compressing is instantaneous and results in homogeneous sterilization with low energy consumption). UHPP is a type of cold sterilization that does not destroy the effective component [6].

UHPP technology has been widely implemented for food processing, and is also used in biotechnology. The effect pressure has on proteins, enzymatic reactions, and the lipid and protein components of microorganism membranes, explains how HP leads to inactivation of all of these elements. UHPP diminishes the fluidity of biomembranes, which in turn reduces the diffusion and active transport between intracellular and extracellular media.

To portray hydrostatic HPP more clearly, the mechanism of an industrial unit is shown in Figure 13.1. This unit was manufactured by Hiperbaric, a Spanish company whose industrial machines are distributed worldwide.

13.2 STERILIZATION AND DECONTAMINATION IN MEDICINE AND GALENIC FORMULATIONS

Based on the known effects of HPP, a promising application is sterilization of materials used for implants, protheses, surgical and medical materials, and endoscopic tools. The process also shows potential for treating hospital wastes.

Inactivation of bacterial spores using HHP is usually more efficient than inactivation at atmospheric pressure conditions, and it can be improved further by using a chemical agent or applying heat. For $P = 280$ MPa, soda 2N, processing time 1 hour at 45 °C, the inactivation results were better than at atmospheric pressure [8].

Pascalization allows implementation of new protocols for sterilization of therapeutic molecules that are sensitive to high temperatures and ionizing radiation that are commonly used in processing. Scientific work proves that organisms cited by the *European Pharmacopoeia* (*Candida albicans*, *Pseu-*

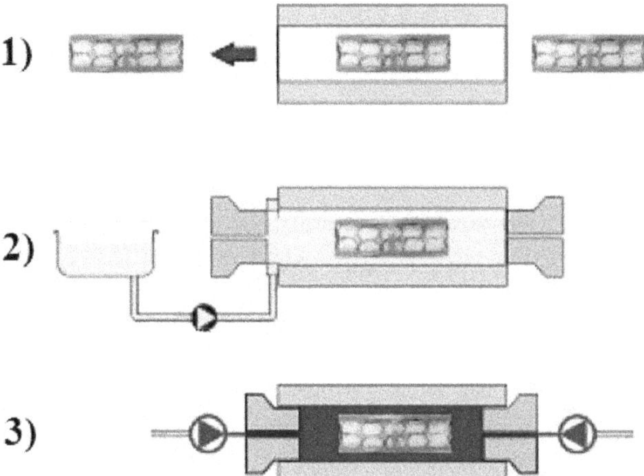

Figure 13.1 1) The product to be processed is placed inside a container in the processing chamber. 2) The container is sealed and the chamber is filled with water at low pressure. 3) Once the vessel is full and hermetically closed, the HP intensifiers are activated, pumping more water into the chamber until the desired pressure is reached [7].

domonas aeruginosa spores, and *Bacillus subtilis* spores) can be inactivated. The thermodynamic conditions of pressure and temperature ($P < 500$ MPa and $T < 37$ °C) retain the physical and chemical integrity of therapeutic molecules such as peptides, insulin, monoclonal antibodies [2].

Sterilization protocols can also be applied to the more modern galenic formulations such as nanodispersed systems made of biodegradable polymers (spherulites, nanocapsules, nanoshperes, and biodegradable liposomes) [2].

Spherulites have been found to be unalterable up to 500 MPa, where the granular spread of the components of the membranes and the encapsulation rates, for example, of coloring agents (E-124, cochineal, or amaranth), remain unchanged.

Studies of the behavior of seven different types of nanoparticles can be found in the literature. These nanoparticles do not show significant changes in the assembly of their dispersive system, after treatment at several pressures (200 MPa, 300 MPa, 400 MPa and 500 MPa) [9].

Little research has been done on sterilization of blood products and their treatment using HP or the effect on biological activity of the different components of blood. However, the effect that HP has on commercially prepared blood plasma is well known. It deforms platelets but does not seem to affect gamma globulins, thrombin, antithrombin, or factor IX; the only exception is factor VIII (an essential blood-clotting protein, also known as anti-hemophilic factor) [10]. This concept may be very useful, for example, to treat sera used to fight hemophilia B with HP, as its coagulation system does not activate correctly if factor IX is altered. However it would not work for hemophilia A, in which factor VIII does not activate.

More recent work using pressurizing and depressurizing cycles at temperatures of 0 °C down to -40 °C achieved the inactivation of viruses with envelopes (HIV and HSV) and those without envelopes (parvoviruses) without altering the coagulation factors or the immunoglobulins. This method is known as pressure cycling technology (PCT) [11].

It is possible that HP will be used in the future to produce plasma and other blood-derived products that contain no viruses. Another possibility is using HP as a direct blood treatment (similar to dialysis) to diminish viral loads in patients who suffer from severe viral diseases.

Another interesting work [12] describes a decellularization method using HHP technology (> 600 MPa). The HHP disrupts the cells inside tissues. Cell debris can be eliminated with a simple washing process and the result is clean decellularized tissue. Porcine aortic blood vessel were decellularized by HHP.

13.3 INACTIVATION AND EFFECTS OF HIGH PRESSURE IN PRODUCTS OF BIOLOGICAL ORIGIN

The protein denaturation due to HP produces intermediate states which in the case of viral proteins entails loss of their infectious character but preserves their immunogenicity. Thus, HPP can be considered an alternative to the typical chemical methods used for vaccine preparation.

Basset and collaborators from the Pasteur Institute in Paris were the first to research the use of pressure for vaccine preparation. In the middle of the 20th century, a vaccine for poliomyelitis was prepared using HP, but due to economic reasons it was not launched in the pharmaceutical market. A few years later the Sabin vaccine reached the market [13, 14].

In recent years, the interest in pressure for the preparation and development of vaccines against viruses and malignant cells has increased. For example, the proteins of the Rift Valley fever virus treated at 225 MPa, at 25 °C, for 30 min are less infectious and more immunogenic. Such treatment immunized mice against the virus [15].

The proteins present in viral capsids are susceptible to dissociation by HP (see Figure 13.2). Recent studies have been performed to inactivate pathogenic viruses in humans and other species. The HP treatment conditions used to treat a series of viruses are as follows:

- Bacteriophage T4 virus (T: 5 °C to 80 °C, $P > 600$ MPa) [16]
- Sindbis virus ($P = 174$ MPa) [17]

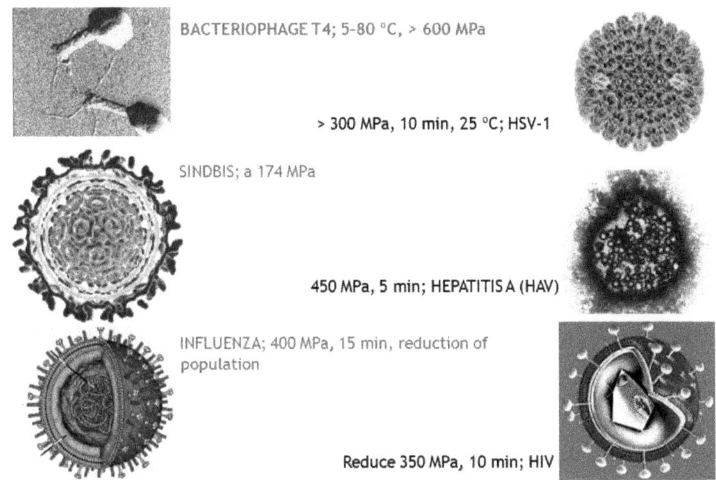

Figure 13.2 HP inactivation of a virus. Inactivation conditions applied to a series of virus.

- Flu virus (P =260 MPa for 12 h to reduce the population) [18]
- Herpes simplex virus 1 (HSV-1) ($P > 300$ MPa, $T = 25$ °C for 10 min) [19]
- Hepatitis A virus (HAV) ($P = 450$ MPa for 5 min) [20]
- HIV virus ($P = 350$ MPa, $T = 25$ °C to reduce the population) [10]

Prions or prion proteins are supramolecular aggregates with amyloid folds consisting of tightly packed beta sheets (glycoproteins). They are misfolded protein molecules that may propagate by transmitting a misfolded protein state. If a prion enters a healthy organism, it induces properly folded proteins to convert into the disease-associated, misfolded prion form; the prion acts as a template to guide the misfolding of more proteins into prion form. They can produce diseases called transmissible spongiform encephalopathies (TSEs) that affect the central nervous system (CNS). Creutzfeldt–Jakob disease is a TSE. Amyloids are peptides of 30 to 40 amino acids that are traditionally associated with Alzheimer's disease, but are not unique to it.

For the irreversible inactivation of prions by HP, values higher than 1 GPa are needed. However, pressures above 500 MPa inactivate the prion responsible for the spongiform encephalopathy of the Syrian hamster [14].

Misfolded protein structures or similar aggregates (such as amyloids), are parts of the cells involved in diseases like Alzheimer's, Parkinson's, spongiform encephalopathies, and even several cancers. Their isolation could allow developing antagonists to counteract these phenomena [21, 22].

Amyloids are peptides of 30 to 40 amino acids that are associated with Alzheimer's disease. They build up and form plaques in the brain samples of patients with this disease. Similar plaques appear in some variants of Lewy body dementia and in inclusion body myositis (a muscle disease).

Protein denaturation and aggregation-disaggregation mechanisms induced in amyloid plaques by pressure are complex processes that depend on the pressure used. Pressures under 200 MPa inactivate enzymes without causing major changes to the three-dimensional structure of the protein, although in some cases protein aggregates and even fibres have been formed [23].

The technique could be successfully applied to protein recombination for pharmaceutical purposes. The biological impact of protein aggregation in these diseases results from the deposition of insoluble protein aggregates in cells.

Paradoxically, in some cases there is evidence that at moderate pressures conformational transitions to partially folded states called molten globules (MGs) have been induced [24].

HP is considered a new strategy for revealing the formation of amyloids. Amyloid fibrils are formed by normally soluble proteins, which assemble to form insoluble fibers that are resistant to degradation. Their formation can accompany diseases, and each disease is characterized by a specific protein or peptide that aggregates. Well known examples of amyloid anomalies include Alzheimer's disease, Parkinson's disease, the spongiform encephalopathies, and even a number of cancers. The amyloid fibrils are deposited extracellularly in the tissues and are thought to exert pathogenic effects. The isolation of these abnormal fibrous proteinaceous structures would allow the development of antagonistic molecules capable of preventing or blocking this effect.

Recently it has been proposed that the proteins from polypeptide fibrils could produce new useful nanomaterials such as scaffolding to support conductive nanowires. The relevant properties are the high resistance to physical and chemical perturbations, including heat and pressure. A recent analysis of the mechanical properties of insulin fibrils showed that their hardness and stiffness levels are comparable to those of steel and silk [25]. Jansen *et al.* [26] show the formation of amyloid insulin of circular morphology under hydrostatic HPP, and Govers *et al.* [27] report the disassembly and reassembly of protein aggregates in *Escherichia coli.*

The application of HP to the processes of biopurification and bioelaboration of a product is of interest. The action of HP entails antigen-antibody dissociation without denaturing either entity. This observation gives way to a new immunoaffinity support and to processes of bioseparation that produce compounds of the high degree of purity required for therapeutic applications [28, 29].

HP is also effective for reducing the allergenic activities of foods. One of the three allergenic structures of celery root can be destroyed at 600 MPa and 20 °C. The Japanese company, Echigo–Seika, has commercialized pre-cooked hypo-allergenic rice, and this patented process could be extended to other cereals such as wheat and barley [30, 31].

Currently the wine industry is one of the most innovative from a research view. The main goal is to find new techniques to optimize the traditional processes. The use of HP to inactivate microbes or avoid the addition of sulfites for the reduction of the microbial population, is just another example of the use of HPP in industry. Sulfites are banned due to the high sensitivity of allergic and asthmatic people. In several studies of HPP applied to producing white and red wines, pressures around 200 MPa were enough to inactivate fungi, yeasts and lactic acid bacteria [32].

Another method worth mentioning in this section is the application of pressure to aging wine stocks. Natural aging takes place under low pressure near the surfaces of seas. The wine industry's use of pressure is an attempt to emulate the preservation of wine in vessels known as amphorae in ancient Greece and Rome [33].

13.4 PHARMACEUTICAL AND COSMETIC INDUSTRY: CHEMICAL EQUILIBRIUM

High-pressure homogenization (HPH) is used widely in the cosmetics and pharmaceutical industries to produce homogeneous and stable emulsions. HPH allows the reduction of particle sizes down to nanometres. High pressure forces particles through a special homogenizing valve and they are processed by the actions of shear force, turbulence, void effect, acceleration, and impact. Current industrial, pilot, and laboratory-scale HP homogenizers are equipped with plunger-type pumps and valves or nozzles made from abrasive-resistant ceramics or hard gemstones (see Figure 13.3). Particle size reduction and efficient dispersion can avoid phase separation, distribute product homogeneously, and intensify color [34].

In the cosmetic industry, HPH is used to improve the quality and stability of nail varnishes, lotions, beauty creams, tooth pastes, shampoos, and emulsions containing oils. In the pharmaceutical industry, HPH is also called micronization. This process is used for reducing the particle size of pharmaceutical products, and increasing the stability and effectiveness of medicaments.

The phenomena of polymorphism and solvate formation in the pharmaceutical industry are particularly important for bioavailability (most drugs are administered in solid form and rely on dissolution), processibility, and storage (crystallization, milling, freeze drying, wet granulation, mixing with

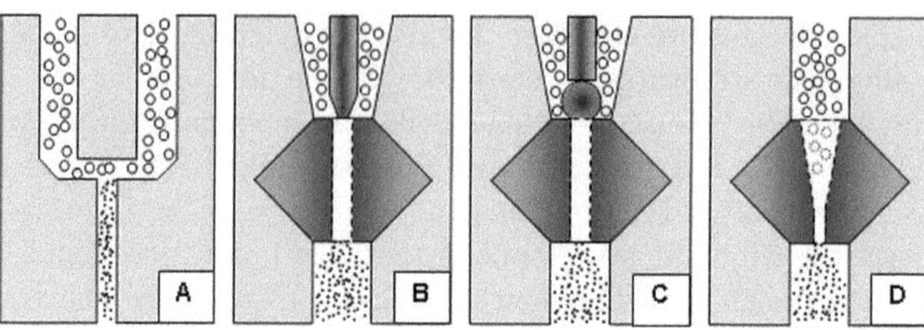

Figure 13.3 Common high-pressure homogenization valves: (A) microfluidic; (B) ceramic needle and seat; (C) ceramic ball and seat; (D) diamond, sapphire, or ruby nozzle [35].

excipients, and tabletting) [36]. To avoid sudden phase changes resulting in decrease of bioavailability in a marketed product, understanding the phase behavior of an active pharmaceutical ingredient in a drug formulation is fundamental.

In relation to polymorphism, the direct compression of paracetamol (analgesic) [37], direct compression of amino acids [38], *in situ* growth of crystals at high pressure (ketones, alcohols, carboxylic acids, amines, chloalkanes, chlorosilanes and mineral acids) [36], benzocaine (a local anesthetic), and L-citrulline polymorphism [39, 40], may be studied.

The formation under pressure of homogeneous protein gels, polysaccharide gels, and mixed proteins, is of great interest in the cosmetic and pharmaceutical industries, as seen in the previous section. However, its use can also be extrapolated to dentistry. Pressurized gels and flocculated emulsions of proteins can be used for their rheological properties or as transporters of active components that can be slowly released. Similarly, it is possible to rapidly obtain perfectly homogeneous gels from excipients used in galenic form such as alginates or carrageenans [41, 42].

HP can be used to enhance chemical reactions that create proteins. The synthesis of chiral compounds in optically pure form is a crucial requirement in modern organic chemistry. Currently HP is used for diastereoselective and enantioselective reactions. The reactions described below can be used to chemically modify proteins and for selective bioimmobilizations and/or biologically active molecules. Such substrates could be used as mechanisms to distribute medicaments or as biological compounds in biosensors.

Due to Le Châtelier's principle, several organic reactions may be enhanced with pressure. The main one is the Diels–Alder reaction (see Figure 13.4) which is a [4+2] cycloaddition between diene and dienophile [43]. This process occurs rapidly in aqueous media, is selective, and can occur at room

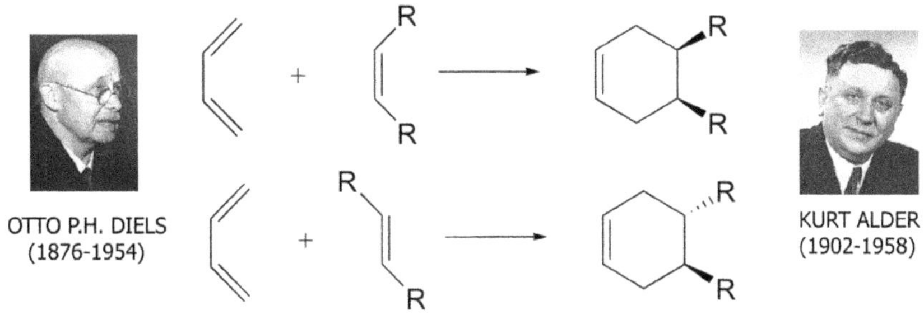

Figure 13.4 In 1928 Otto Paul Hermann Diels and Kurt Alder documented the Diels–Alder reaction for the first time, and in 1950 they were awarded the Nobel Prize in Chemistry for their work.

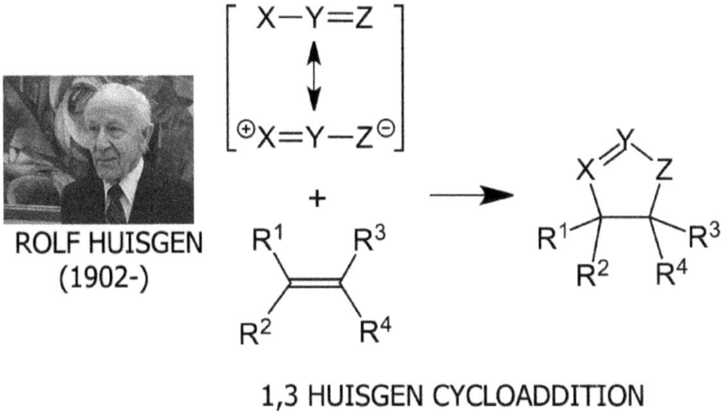

Figure 13.5 Rolf Huisgen, creator of the 1,3-dipolar cycloaddition reaction.

temperature. Therefore, it is ideal for selective modification, for example, by adding fluorophores [44], unnatural amino acids, or different types of peptides to proteins.

Other organic reactions, potentially enantioselective, enhanced by HP are 1,3-dipolar cycloaddition reactions, also known as Huisgen reactions (see Figure 13.5), additions to double and triple bonds, and Friedel–Crafts alkylation and acylation reactions.

It is worth noting that the formation of a brown melanoid is inhibited by pressure [14]. The formation of such molecules is related to the Maillard reaction, technically called non-enzymatic glycosylation, which is the result of covalent bonding of a protein or lipid molecule with a sugar molecule without the controlling action of an enzyme that occurs when heating food products. This reaction also occurs in aging processes.

Other reactions susceptible to optimization by HP are enzyme catalysis reactions. The synthesis of medicaments catalyzed by enzymes under mild

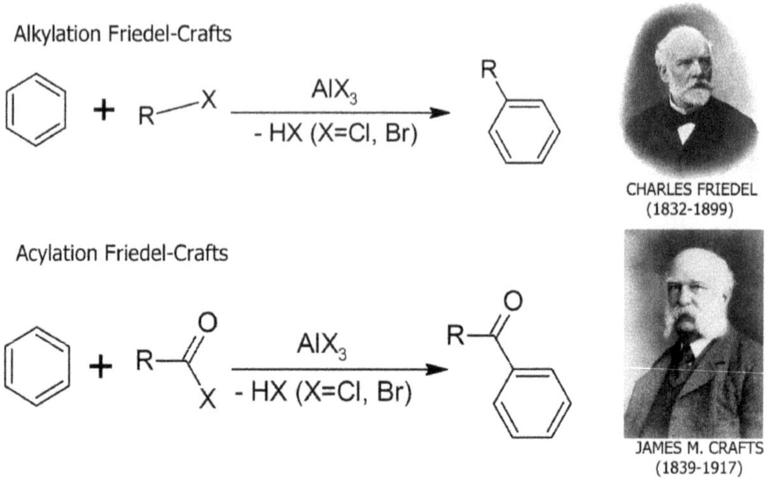

Figure 13.6 HP-aided enzyme catalysis reactions: Friedel–Crafts alkylation and acylation.

conditions is possible. Particularly, the enantioselective synthesis of esters such as ibuprofen, lipase esterification, synthesis of pharmacologic peptides by thermolysin, polyols, and carbohydrates are of great interest [45].

These reactions take place in bioreactors at pressures below 200 MPa which does not alter the stability and functionality of the catalytic enzymes. If one uses stabilizing co-solvents in the reactions or carries them out in an organic medium, the thermostability of the industrial enzymes may be increased [46].

13.5 MISCELLANEOUS: FREEZING OF LIVE TISSUE, ANESTHESIA, AND NARCOSIS

The characteristics of the solid-liquid phase diagram of water should be pointed out to enhance understanding of the applications based on its polymorphism: water remains liquid up to −20 °C at pressures of 200 MPa (2 kbar); there is an increase of volume in the transition from liquid to ice, which makes the density of ice lower than that of water; if ice is pressurized, water stabilizes back to liquid, thus at room temperature (25 °C) and at a few thousand atmospheres of pressure water remains liquid.

This characteristic allows us to freeze food in an innovative way. For example, food can be preserved at HP in liquid phase at temperatures below 0 °C; or fast thawing can occur by pressurizing a frozen food until the curve of change of phase is reached [47].

Freezing a product at temperatures between −10 °C and −20 °C at pressures between 100 and 400 MPa is nearly instantaneous once the pressure is

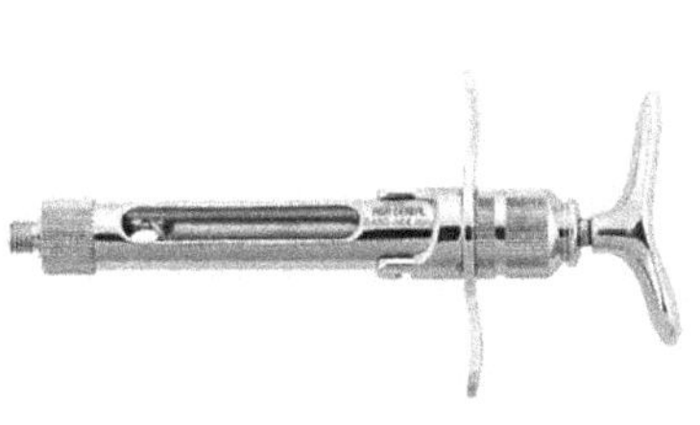

Figure 13.7 Interest in the effects of high pressure in anesthesia and sub-aquatic sports.

released. This could potentially allow fragile biopharmaceutical products to be frozen without structural damage that occurs when frozen at −30 °C or with liquid nitrogen at atmospheric pressure.

Several applications based on the effect of HP on the phase transition of water from liquid to ice are used in food technology. For example, in HP freezing such as the process described in [48], small uniformly distributed crystals are formed, avoiding the tissue damage that occurs with larger heterogeneous crystals.

More recently HP has been used to preserve pharmaceutical biological materials derived from blood and cells. In future, HP may even be used to preserve organs for transplants [49].

There is growing interest in determining the effects of HP on biological functions. Studies of brain processes under hyperbaric conditions can yield a better understanding of phenomena such as inert gas anesthesia, nitrogen narcosis and reversal of the effects of anesthetic and narcotic agents (of special interest in neurology and sub-aquatic sports); see Figure 13.7. Such research may provide an insight into the actions of anesthetics, which remain poorly understood. Various studies have established the behavioral responses of organisms to hyperbaric conditions in the presence or absence of anesthetic agents. The lipid theory based on the Meyer–Overton correlations suggests that the neuronal lipid bilayer is the primary target site of anesthetic action [50].

A topical cream developed by Astra AB in the early 1980s, consisting of a eutetic mixture of local anesthetics (EMLA), was studied under a range of temperatures and pressures. The components of the cream are lidocaine, prilocaine, and water. Pressure stabilized the solid state of lidocaine more than it stabilized prilocaine. The solid-liquid equilibrium of prilocaine was slightly less influenced by pressure with respect to temperature (the study range was 0 to 250 MPa) [51].

AP **Ammonium perchlorate**, $[NH_4]^+[ClO_4]^-$

ADM **Ammonium dinitramide**, $[NH_4]^+[O_2N\text{-}N\text{-}NO_2]^-$

AN **Ammonium nitrate**, $[NH_4]^+[NO_3]^-$

PETN (pentaerythritol tetranitrate), $[C(CH_2O\text{-}NO_2)]$

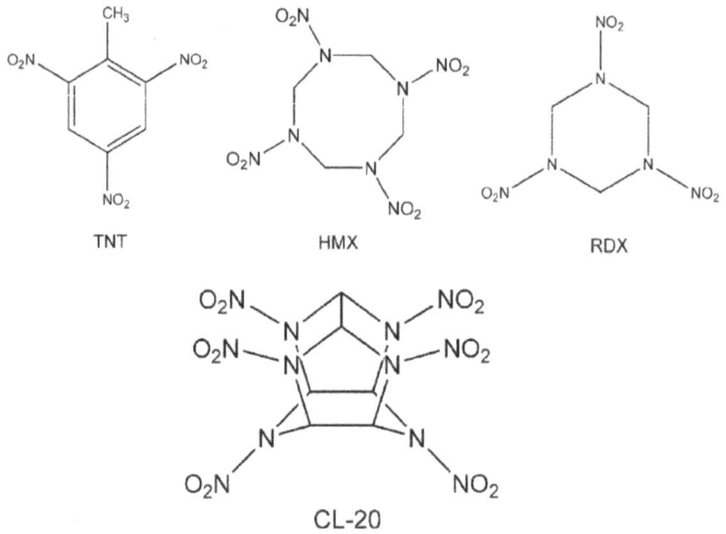

TNT HMX RDX

CL-20

Figure 13.8 Energetic materials studied at HP: ammonium perchlorate (AP), ammonium dinitride (ADM), ammonium nitrate (AN), PETN, HMX, RDX, and CL-20.

Energetic materials release heat and/or gaseous products at a high rate when stimulated by heat, impact, shock, or spark. They can be classified broadly as explosives, propellants, gas generators, and pyrotechnics. Polymorphism and phase transitions can alter the sensitivity and performance of an energetic material in a complex manner, and HP can play a fundamental role [36].

During the detonation of an explosive material, the shock wave may produce pressures up to 50 GPa and temperatures up to 5500 °C. These extreme conditions result in polymorphic transitions and the initiation of chemical reactions. Examples of compounds that have been studied under HP are ammonium perchlorate (AP), ammonium dinitride (ADM), ammonium nitrate (AN), PETN HMX, RDX, CL-20 (see Figure 13.8).

Bibliography

[1] A. Aertsen, F. Meersman, M. E. G. Hendrickx, *et al. Trends Biotechnol.*, 27, 434 (2009).

[2] Y. Rigaldie and G. Demazeau. *Ann Pharm Fr.*, 62, 116 (2004).

[3] E. Dumay, D. Chevalier-Lucia, L. Picart-Palmade, *et al. Trends in Food Science and Technology*, 1, 13 (2013).

[4] P. Paquin. *Int Dairy J.*, 9, 329 (1999).

[5] J. Yuste, M. Capellas, and R. Pla. *J Rapid Methods Autom Microbiol.*, 9, 1 (2001).

[6] H. Li. *J Guangdong. AIB Polytechnic College*, 24, 4 (2008).

[7] www.hiperbaric.com.

[8] H. Delacour. *Intérêts des hautes pressions hydrostatiques dans l'inactivation des spores bactériennes.* Ph.D. thesis, Lyon (2000).

[9] Y. Rigaldie. *Sur l'impact des traitements sous hautes pressions dans la décontamination et la stérilisation de formes pharmaceutiques renfermant des molecules thérapeutiques sensibles aux precédés énergétiques.* Ph.D. thesis, Bordeaux (2002).

[10] T. Shigehisa, T. Nakagami, H. Ohmo, *et al. High Pressure Biosci. Biotechnology*, 273–278 (1996).

[11] M. Manak. *Blood product safety and transmissible spongiform encephalopathies. Cambridge Healthtech Institute's Eighth Annual Meeting, Washington* (2002).

[12] S. Funamoto, K. Nama, T. Kimura, *et al. Biomaterials*, 31, 3590 (2010).

[13] J. Basset, P. Lépine, and L. Chaumont. *Ann Inst Pasteur*, 575–596 (1956).

[14] P. Masson, C. Tonello, and C. Balny. *J Biomed Biotechnology*, 85–88 (2001).

[15] P. Y. Perche, C. Clero, M. Bouloy, *et al. Am J Trop Med Hyg.*, 3S, 256 (1997).

[16] P. Gross and H. Ludwig. In C. Balny, R. Hayashi, K. Herremans, *et al.*, eds., *High pressure and biotechnology*, Volume 224, 57–59. John Libbey Eurotext, London (1992).

[17] P. Burz, B. Habison, and H. Ludwig. In C. Balny, R. Hayashi, K. Herremans, *et al.*, eds., *High Pressure and Biotechnology*, Volume 224, 61–64. Colloque inserm. John Libbey Eurotext, London (1992).

[18] J. L. Silva, P. Luan, M. Glaser, *et al. J Virol Methods*, 66, 2111 (1992).

[19] T. Nakagami, T. Shigeshisa, T. Ohmori, *et al. J Virol Methods*, 38, 255 (1992).

[20] D. H. Kingsley, D. G. Hoover, E. Papafragkou, *et al. Innov Food Sci Emerg.*, 2, 95 (2001).

[21] I. V. Baskakov, G. Legname, M. A. Baldwin, *et al. J Biol Chem.*, 227, 21140 (2002).

[22] S. T. Ferreira and F. G. de Felipe. *FEBS Letters*, 498, 129 (2001).

[23] L. Séller. *Biochim Biophys Acta*, 1595, 11 (2002).

[24] P. Masson and C. Cléry. In C. Balny, R. Hayashi, K. Herremans, *et al.*, eds., *High pressure and biotechnology*, Volume 224, 117–126. John Libbey Eurotext, London (1992).

[25] J. F. Smith, T. P. J. Knowles, C. M. Dobson, *et al. Proc Natl Acad Sci. USA*, 103, 15806 (2006).

[26] R. Jansen, S. Grudzielanek, W. Dzwolak, *et al. J Mol Biol.*, 338, 203 (2004).

[27] S. K. Govers, P. Dutre, and A. Aertsen. *J Bacteriol.*, 196, 2325 (2014).

[28] P. Lemay. *Biochim. Biophys. Acta*, 1595, 357 (2002).

[29] P. Lemay, L. Estevez-Burugorri, A. Largeteau, *et al. Effects des pressions hydrostatiques sur les couples antigène-anticorps: application à un procédé de bioséparation.* Actes du colloques "La pression pourquoi? Ses effects sur la Matière, des atomes aux systèmes complexes", Banyuls-sur-mer (2000).

[30] T. Inoue and Y. Kato. Japan Patent H6-7777; International Patent WO92/11772 (1990).

[31] A. Yamazaki. *Foods Food Ingredients J Jpn.*, 210, 1 (2005).

[32] D. Bermudez-Aguirre and G. V. Barbosa-Canovas. *Food Eng Rev.*, 3, 44 (2011).

[33] http://www.elmundo.es/elmundo/2013/01/30/andalucia_sevilla/1359554041.html.

[34] http://www.chinahomogenizers.com/cosmetic.html.

[35] http://web.utk.edu/~fede/high%20pressure%20homogenization.html.

[36] F. P. A. Fabbiani and C. R. Pulham. *Chem Soc Rev.*, 35, 932 (2006).

[37] E. V. Boldyreva. *J Mol Struc.*, 647, 159 (2003).

[38] S. A. Moggach, D. R. Allan, S. J. Clarck, *et al. Acta Crystallogr. Struct. Sci.*, 61, 449 (2005).

[39] H. Allouchi, B. Nicolaï, M. Barrio, *et al. Cryst Growth Des.*, 14, 1279 (2014).

[40] I. Gana, M. Barrio, B. Do, *et al. Int J Pharm.*, 456, 480 (2013).

[41] M. Schwertfeger. In Ludwig, ed., *Advances in high pressure bioscience and biotechnology*, 337–340. Proceedings of International Conference on High Pressure Bioscience and Biotechnology, Heidelberg (1998).

[42] B. Steyer, F. Béra, C. Massaux, *et al.* In Ludwig, ed., *Advances in high pressure bioscience and biotechnology*, 352–356. Proceedings of International Conference on High Pressure Bioscience and Biotechnology, Heidelberg (1998).

[43] C. Ménard-Moyon, F. Dumas, E. Doris, *et al. J Am Chem Soc.*, 128, 14764 (2006).

[44] A. D. Araújo, J. M. Palomo, J. Cramer, *et al. Angew Chem Int Ed.*, 45, 296 (2006).

[45] S. Kunugi. *Ann. NY Acad Sci.*, 672, 293 (1992).

[46] R. V. Rariy, N. Bec, N. Klyachko, *et al. Biotechnol. Bioengn.*, 57, 552 (1998).

[47] N. A. S. Smith, S. S. L. Peppin, and A. M. Ramos. *Proc Roy Soc A*, 468, 2744 (2012).

[48] N. A. S. Smith, V. Burlakov, and A. M. Ramos. *J Phys Chem B*, 117, 8887 (2013).

[49] V. V. Mozhaev, K. Heremans, J. Frank, *et al.* *Trends Biotechnol.*, 12, 493 (1994).

[50] A. Wlodarczyk, P. F. McMillan, and S. A. Greenfield. *Chem Soc Rev.*, 35, 890 (2006).

[51] I. B. Rietveld, M. A. Perrin, S. Toscani, *et al.* *Mol Pharmaceutics*, 10, 1332 (2013).

CHAPTER 14

High-Pressure and High-Temperature Conditions as Tools for Synthesis of Inorganic Materials

Miguel A. Alario-Franco

Departamento de Química Inorgánica I, Universidad Complutense de Madrid, Madrid, Spain

Antonio J. Dos santos-García

Departamento de Ingeniería Mecánica, Química y Diseño Industrial, Universidad Politécnica de Madrid, Madrid, Spain

Emilio Morán

Departamento de Química Inorgánica I, Universidad Complutense de Madrid, Madrid, Spain

CONTENTS

14.1 INTRODUCTION

SINCE the middle of the 19th century, when thermodynamics was established as one of the most rigorous and profound areas of chemistry and physics, it is well known that the classical variables for describing the behavior of any system are composition, temperature and pressure. Nevertheless, when dealing with solid state synthesis -the so-called ceramic method- because most processes are carried out at high temperatures under ambient pressure, in most phase diagrams only changes in composition associated with temperature are considered and pressure-related data are scarce. Thus, this third dimension, pressure, remains to be fully explored for the synthesis of ceramic materials. Some unique aspects of pressure as a thermodynamic variable have already been discussed in former chapters. From a quantitative point of view, the pressure range is extremely wide: from the 10^{-32} bar existing in interstellar space to the 10^{32} bar at which some processes take place inside some stars and from the 10^{-6} bar attained in ultra-high vacuum chambers, common in condensed-matter laboratories, to the 10^{6} bar that many materials can handle without problem at room temperature. We should mention that what we call ambient pressure (1 atmosphere $\sim$1 bar) is an exceptional situation in the universe, where most of the existing matter, because of gravity, supports very high pressures close to 9 GPa (9×10^4 bar) or higher. The Earth is a huge and dynamical high pressure and high temperature laboratory and very important physico-chemical changes take place deep in its interior. Finally, energy changes associated with pressure changes in a system are small and directly related to the system's compressiblity. This explains why pressure is a relevant thermodynamic parameter in geology, astronomy, physics, chemistry, materials science, architecture, biology, and food science and technology.

Among the milestones in the development of the high pressure science and technology, we should start by mentioning brilliant geologists such as De Senarmont, Spezia and, others who proposed in the middle of the 19th century that magmatic rocks and mineral were formed under high pressure (HP) and high temperature (HT) conditions. The formation of others minerals such as some sulphides and silicates (e.g., emeralds), requires aqueous media; the process is known as hydrothermal mineralogenesis. These ideas led to experiments that attempted to reproduce such extreme HP and HT conditions in the laboratories. The goals were to confirm the early hypothesis and produce new materials with qualities superior to those of existing materials. The greatest difficulty was the rapid deterioration of mechanical properties of most steels and alloys at high temperatures; these materials can handle high pressures at room temperature but not at increased temperatures.

The pioneering work of P. W. Bridgman at Harvard University in 1908 is

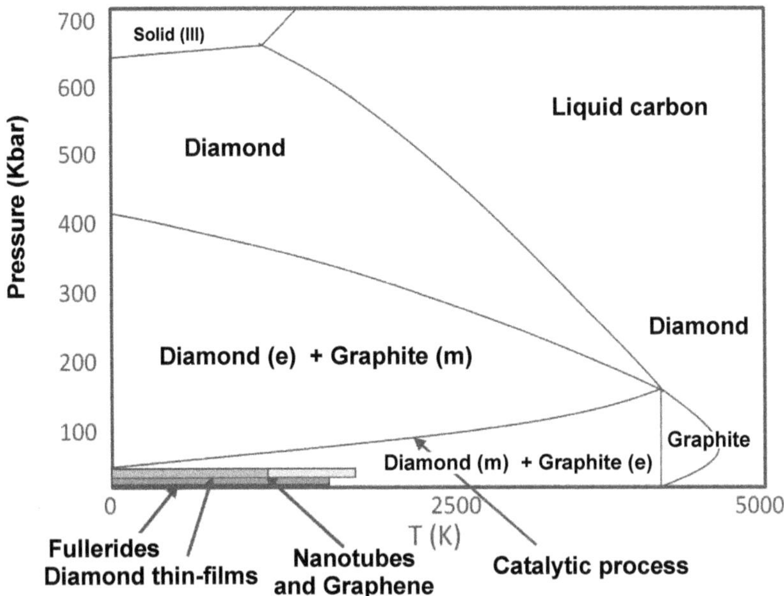

Figure 14.1 Carbon phase diagram.

worth of mention. Bridgman designed new systems and used innovative steels and alloys able to resist simultaneous high temperatures and high pressures. An ingenious inverse closing device that tightens as pressure increases still bears his name. His systems made important contributions to inorganic chemistry, condensed matter physics and materials science; he was awarded with the Nobel Prize in Physics in 1946 [1]. Contributions of others were also relevant to various scientific disciplines. Haber and Bosch (1913) made possible the synthesis of ammonia from atmospheric nitrogen and hydrogen gas [2]. They used a catalytic procedure utilizing high pressure and moderate temperature that required a special reactor. The procedure is used routinely in chemical industries.

The most relevant event in the high pressure synthesis of materials was the fulfilling of an old dream: producing diamonds from graphite. Chemists pursued the idea in the late 19th century but the dream did not become reality until 1955. General Electric Company in the United States and Almänna Electriska Aktiebolaget (ASEA, now ABB Group) invented the process around the same time; General Electric made the first announcement. Based on the carbon phase diagram (Figure 14.1), graphite should transform into diamond around 1200 °C and 10^5 bar (10 GPa) in oxygen-free conditions.

Hannay heated petrol in closed steel vessels. The vessels exploded and his experiment failed. In France, Moissan, who won a Nobel Prize in Chemistry, produced a matarial called silicon carbide (SiC), also known as carborundum and moissanite. The hardness of SiC is close to that of diamond and it is used in cutting tools and electrical resistance elements. Moissan's system was unable to reach the very high pressure (10 GPa) required for diamond synthesis.

Bridgman achieved 10 GPa pressure and attempted to use it for diamond synthesis. However, without suitable catalysts and sufficiently high temperatures, he also failed. Finally, Tracy Hall, a General Electric Company engineer, designed a belt-type apparatus to generate HP and HT conditions. He used molten metals as solvents for graphite and for generated reproducible diamonds crystals when the pressurized solutions were cooled down.

The procedure currently generates 80 % of global diamond production. Synthetic diamonds are used as abrasives in industry [3]; they are not of gem quality. The Tracy Hall Foundation maintains an excellent Web page that describes diamond production and other pressure-related subjects [4].

Another important milestone in the high-pressure synthesis of materials was the hydrothermal synthesis of α-quartz, the only variety of silica among many showing piezoelectricity. This type of quartz (also called rock crystal) is of hydrothermal origin and was known as early as the mid-19th century. Mineralogists De Senarmont (1851), Spezia (1905), and Nacken (1942) succeeded in producing small quantities of the quartz in their laboratories. In 1961, Bob Laudise of Bell Labs established precise conditions for the reproducible synthesis of the material at the industrial scale [5]. The discovery revolutionized watch production and quartz watches are popular all over the world.

Another extraordinary material also prepared at industrial by hydrothermal procedures is chromium dioxide (CrO_2). This compound is ferromagnetic and acts as a metallic conductor -a unique combination of transport properties [6]. This oxide can be prepared as acicular magnetic nanoparticles (a single particle is a bit with a value of 0 or 1 depending on whether it is oriented or disoriented). The use of CrO_2 in audio and video tapes and hard disks revolutioized magnetic information storage.

Hydrothermal synthesis has attracted a great deal of attention in recent years and it is used to produced zeolites and related materials. Zeolites are porous aluminosilicates. Acle Cronstedt, Swedish mineralogist, proposed the *zeolite* name in 1756; it is based on the Greek phrase for "boiling stones". Some zeolites are of mineral origin but the number of zeolites produced by hydrothermal synthesis continues to increase. They have open structures and topologies and are widely used as adsorbers, catalysts, and water softeners [7]. Template synthesis and building-block techniques allowed new porous materials to be produced by hydrothermal synthesis. Not all are aluminosilicates; some tungstates and vanadates are produced in the form of nanotubes.

High-temperature superconductors (HTSCs) are mixed oxides of copper and other elements such as alkaline-earths and/or lanthanides. The mercury cuprate superconductors (general formula $HgBa_2Ca_{n-1}Cu_nO_y$; $2 \leq n \geq 8$) first prepared by HP and HT synthesis in the early 1990s represented an important advance. The high pressure avoids decomposition of HgO used as a reactant and the material with n = 3 shows critical temperature T_c = 135 K which can be raised to 164 K if resistivity and magnetic susceptibility are also measured under applied pressure (25 GPa) [8–10]. Also under extreme conditions of pressure and temperature (6 GPa, 1000 °C), another family of superconductors was prepared. They are called cuprocarbonates and have a general formula $(Cu/C)Ba_2Ca_{n-1}Cu_nO_y$; $3 \leq n \geq 7$; the maximum T_c = 120 K was reached for the n = 4 member [11]. In recent years the coexistence of magnetic order and superconductivity in ruthenocuprates has been under study since both properties are antagonists and superconductors should show perfect diamagnetism when cooled below T_c. The first material of this family, also known M-1212 is $RuSr_2GdCu_2O_8$, it is ferromagnetic at 150 K and a superconductor below 35 K [12, 13]. Although this material is prepared at room pressure, replacing gadolinium with other rare earths always requires HP and HT conditions, for lanthanides bigger or smaller than gadolinium [14]. High pressures are also needed to replace Ru with other transition elements [15–18] and, interestingly, Cr is stabilized in this structure as Cr^{+4} in octahedral coordination [19].The pressures required to form the 1212-type structure depend on the cationic sizes (either on the M or RE positions) and, even more strikingly the plot of experimental P values needed for a specific M-1212 family versus the rare earth sizes seems to follow a Gaussian relation [17].

In a different scenario, the properties of di-hydrogen compressed under extraordinarily high pressures ($\geq$ 100 GPa = 1 Mbar) are of great interest, both from the experimental and theoretical views, as it has been predicted that this very simple molecule become a metallic fluid under such high pressures and this would account for the magnetic fields found in stars or gigantic planets such as Jupiter where di-hydrogen is the main component. Dynamic experiments have proved that conductivity increases remarkably for compressed di-hydrogen [20]. We can state that the research on the uses of high pressure a a tool for synthesizing new materials and modifiying their properties continues worldwide. Specialized academic and industrial laboratories conduct experiments and publish their work. Scientific societies such as the European High Pressure Research Group (EHPRG) meet regular to exchange ideas [21]. Journals such as *High Pressure Research* publish research results and foster further work. Highly specialized areas of research continue to search for new compounds containing metals that have unusual oxidation states. Attempts to develop super-hard materials such as carbon nitride (C_3N_4) are ongoing. Other topics of study include synthesis of metastable phases, *in situ* study of phase transformations, growing of crystals under HP and HT conditions, and measurements of physicochemical properties under pressure. A recent and spectacular result was achieving a new critical superconducting temperature for sulfur di-hidride. H_2S and its D_2S congener are both superconducting under very high pressures (around 200 GPa). They show clear BCS [22] superconducting behavior at maximum T_c levels around 191 and 70 K, respectively. More work is needed to establish the real nature of superconductors. The literature contains several excellent reviews on the subject [23–29].

14.2 PRESSURE: UNITS, RANGES, AND WAYS TO PRODUCE IT

Due to the different behavior of pressure, the units used are also diverse: atmospheres, pounds per square inch (psi), pascal, and bars. The pressure ranges utilized in solid state sciences are also very different. The high-pressure term used in the literature refers to experiments performed above room pressure, but the range of pressures varies widely. The authors propose that pressures be clasified based on the values attained:

Moderate (autogenous): 1–100 bar Intermediate: 100–10,000 bar (0.1–10 Kbar) High: 10–100 Kbar (1–10 GPa) Ultra High: above 10 GPa.

There are several methods and experimental setups for producing pressures based on working temperatues and samples sizes [23, 24]. The following are the most relevant:

1. For very small(μ-sized samples, the usual devices are diamond anvil cells (DACs) in which two diamonds of special cut are mechanically compressed against each other. A metal gasket is used to support the sample between the two diamonds. In spite of the small size of the setup (15 to 20 cm high), very high pressures can be reached, on the order of Mbar [30]. The maximum achieved today is 640 GPa in a double stage DAC at the Argonne National Laboratory. Most of the experiments are carried out at room temperature and laser heating through the diamonds has to be used if high temperatures are needed. A great advantage of these setups is that they are portable, allowing their use at synchrotron and neutron source facilities.

2. If the target is the synthesis of milligrams of material and GPa pressures and 1000 to 1500 °C or higher temperatures are required, piston-cylinder, belt-type, multi-anvil or similar devices are needed and all of them require hydraulic presses to compress small samples (Figure 14.2).

   In these setups the sample is placed inside a metallic capsule (usually made of Au,Pt, Ta, Re, Cu or stainless steel); the capsule is inserted in a graphite cylinder which acts as the furnace (instead of graphite Mo or Ta may be used); the cylinder is surrounded by pyrophillite which acts

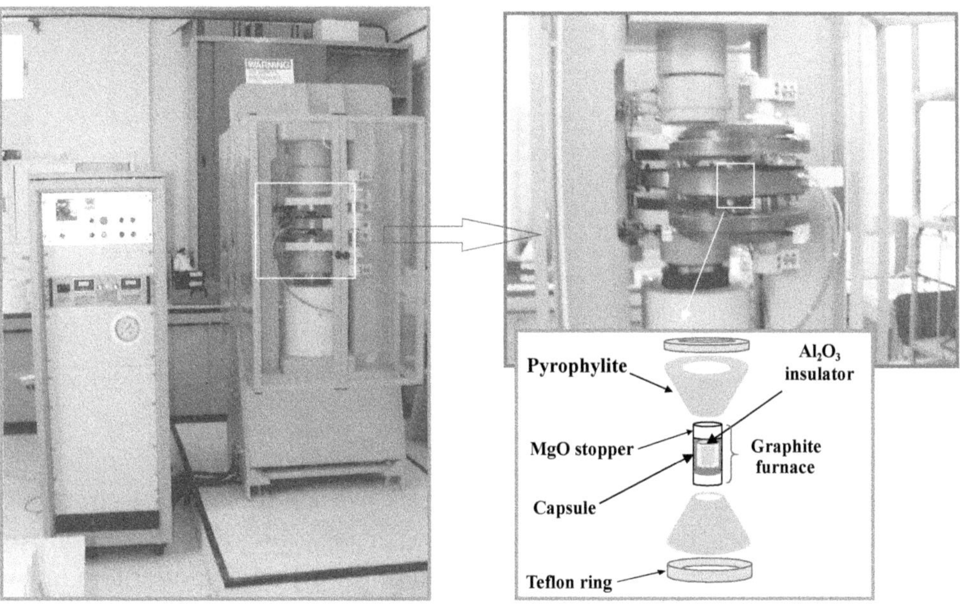

Figure 14.2 Belt-type press used at Laboratorio Complutense de Altas Presiones in Madrid (Spain).

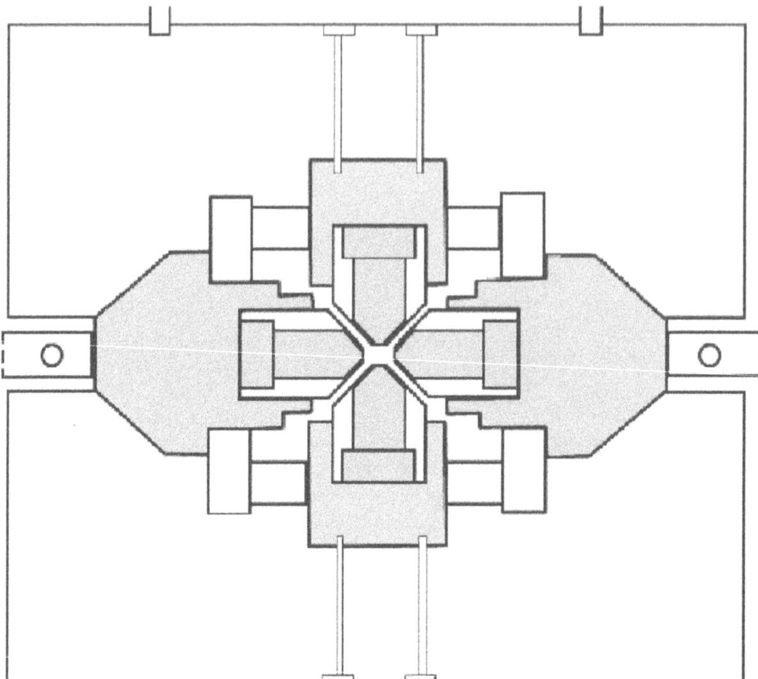

Figure 14.3 Multi-anvil design. Adapted with permission from Preparative Methods in Solid State Chemistry. Copyright 1972 ACADEMIC PRESS, INC [23].

a pressure transducer (this phyllosilicate behaves as a fluid under pressure). The cylindrical and conic pistons are made of tungsten carbide. The belt apparatus provides homogeneous pseudo-hydrostatic conditions and higher pressures (up to 10 GPa) than the piston cylinder one but the amount of sample produced is much smaller (~100 mg). Higher pressures out smaller samples are provided by the multianvil setups, also designed by Tracy Hall. Different pieces are assembled and pressed against each other in a tetrahedral, octahedral, cubic configuration, depending on the case and the capsule is placed at the center (Figure 14.3).

3. High pressures of a specific gas can be generated by different means in the laboratory. An usual way is to compress and heat an industrial gas in a closed furnace or vessel (made from special alloys such as Inconel for high pressures of oxygen) which allows application of only moderate pressures (i.e., 200 to 300 atm at 1000 °C for O_2) to a large amount of sample (grams). Much higher gas pressures (GPa) can be reached using a piston cylinder or a belt-type apparatus with *in situ* gas generation coming from a decomposition reaction of a reactant. For instance, by adding small amounts of potassium chlorate or perchlorate to the sample,

oxygen gas and potassium chloride are produced at 600 °C. The external pressure retains the gas inside the capsule and, the oxygen has to be considered as reactant, maybe the most important. Potassium chloride free sample can be obtained after wash it in deionized water.

4. Sometimes the high pressure is not the most relevant aspect factor. Many liquids when heated above their boiling points become much better solvents. This is the principle for solvothermal synthesis and the (moderate) pressure is generated by heating a specific liquid (often water) up to a desired temperature in an autoclave containing the sample (see Section 14.5).

5. When ultrahigh pressures (millions of atmospheres or TPa) and temperatures up to 4000 K are required, these extreme conditions can be produced by shock waves associated with the impact of a tantalum ball fired at speeds of ~7 km/s against a metallic plate that transmits pressure and temperature to the sample placed in between two monocrystalline alumina disks [31]. In a procedure involving less control, some researchers place samples and dynamite inside special reactors and detonate them underground.

14.3 HIGH PRESSURE: THERMODYNAMIC ISSUES

The most obvious effect of pressure is to increase the density of any solid system by decreasing its volume. This can be achieved through phase transitions (to produce more compact structures) or chemical reactions (the average molar volume of the products has to be lower than those of the reactants). Some other effects are less evident but no less important: interatomic distances decrease, higher oxidation states may be stabilized, coordination numbers usually increase and some metal-metal bonds can be formed. We will, in what follows, review some examples for the different cases.

14.3.1 Formation of More Compact Structures

The paradigm for this case is the phase transformation of graphite (density = 2.266 g/cm^3) into diamond (density = 3.514 g/cm^3). The transformations is also accompanied by an increase in the coordination number of C from 3 to 4 and an impressive change in all the physicochemical properties. One of the carbon polymorphs known as C_{60} or fullerene whose molecular structure is a truncated icosahedron (molecular cage with plenty of internal space), transforms into diamond when treated under high pressures [32]. This behavior is

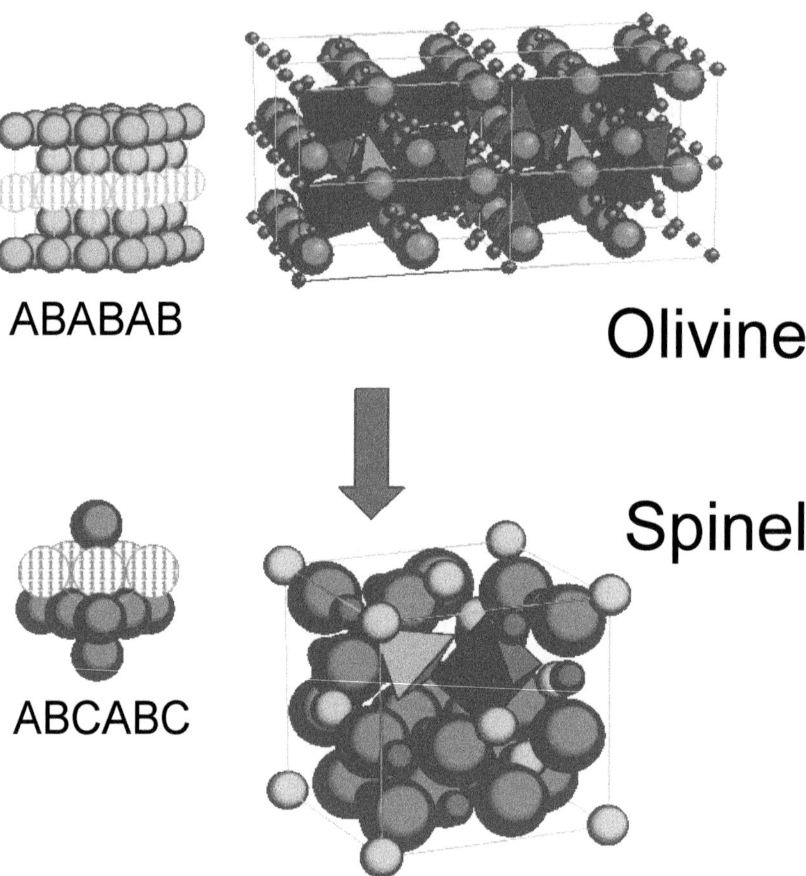

Figure 14.4 A high-pressure phase transformation from olivine to spinel takes place at the lithosphere when tectonic plates meet. It may be responsible for the genesis of deep earthquakes.

also extended to other covalent species such as boron nitride (BN) an artificial material which adopts the layered graphitic structure; its density is 2.1 g/cm^3 when prepared at room pressure but it transforms into the zincblende (diamond-like) structure when treated under HP and HT conditions and has a much higher density, 3.45 g/cm^3.

Another case, relevant because of its geological implications, is the transformation of olivine [$(Mg,Fe)_2SiO_4$] into another material with identical composition but a different structure: the spinel (Figure 14.4).

In this phase transition, the coordination numbers remain the same: silicon in tetrahedra and magnesium or iron in octahedra, but the anionic compact layer sequence changes from hexagonal ABABAB to cubic ABCAB. Although this process does not involve much molar energy [33], as it may happen spontaneously in the friction zones between tectonic plates, the total amount of energy released suddenly is gigantic and this could cause earthquakes. This olivine-to-spinel or-phenacite transformation is common if sufficient pressure and temperature are provided. We have studied the transformation of olivine-$LiCoXO_4$ (X = P and As) materials to the spinel high-pressure polymorph but another with the related Na_2CrO_4 structure appeared [34].

In this connection, high pressures applied to oxides favour the formation of compact and highly symmetric structures such as the ABX_3 perovskite whose coordination numbers are high (6 for the B cation and 12 for the A in the ideal cubic structure). Symmetry-lowering distortions are common when the sizes of A and B cations are not perfectly matched (Figure 14.5). This size match can be estimated from the Goldschmidt tolerance factor (TF):

$$TF = \frac{r_A + r_O}{\sqrt{2}(r_B + r_O)}, \tag{14.1}$$

where r_A, r_B and r_O are the ionic radii of A and B cations and oxygen, respectively [35] and TF takes the value of unity for the ideal cubic perovskite structure.

Perovskites are particularly common when the B cation is a transition metal ion (i.e, $CaTiO_3$, $CaMnO_3$, etc.) but also when B is a representative element (i. e., $LaAlO_3$, $BaPb_{1-x}Bi_xO_3$, $NaSbO_3$, etc.). As an example, cubic distorted perovskites $MSeO_3$ (M = Mn, Co, Cu, Zn) have been formed at 6 GPa and 1100 °C. This is remarkable because of the very small size of Se^{+4} entering the A position (bearing an inert pair of electrons) and also the Goldschmidt tolerance factor far below the usual room pressure values [36, 37]. Interestingly, there are few reports of perovskites containing Sb^{+5} and Bi^{+5} unlike the oxides of their d^0 transition metal counterparts Nb^{+5} and Ta^{+5}. In fact, structures with edge-sharing octahedra such as ilmenite and corundum derivatives seem to be the most common for antimonates and bismuthates. However, high-pressure and high-temperature reactions can help in converting ilmenite to perovskite or even to the polar $LiNbO_3$ structure [38–40].

Regarding perovskites, high pressure may help to stabilize other inert pair-containing metastable materials and some new advanced and functional materials known as multiferroics may be formed. An excellent review is provided by E. Gilioli and Lars Ehm [41]. Multiferroism (MF) can generally be defined as a phenomenon in which two or more of the so-called ferroic order parameters (ferroelectricity, ferromagnetism and ferroelasticity) simultaneously coexist in a single phase material. Among them, the most interesting ones are the magnetoelectric materials in which magnetization can be induced or modified by an electric field or electric polarization may be affected by a magnetic field. One of the ways to promote MF in the perovskite ABO_3 structure, is to place a magnetic ion in the B position (to induce magnetic order) and a lone pair

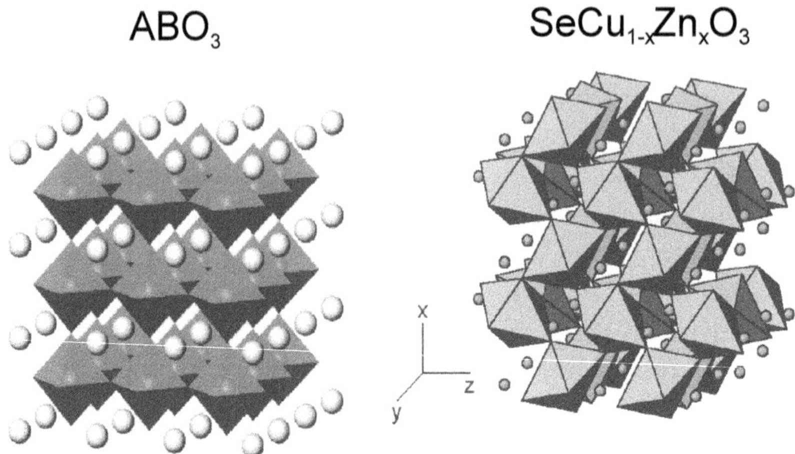

Figure 14.5 Comparison between the ideal cubic perovskite structure (left) and a strongly distorted one (right).

cation in the A position (such as Bi^{+3} or Pb^{+2} to induce polarization); the structure is non-centrosymmetric. There are many examples requiring HP and HT conditions for their synthesis including perovskites such as $BiMO_3$ (M = Mn, Cr, etc.) [42, 43] or $PbVO_3$ [44] containing a lone pair at the A-site and perovskites with small A-site cations [42]. The theoretical prediction made in 2005 by N. Spaldin et al. [45] claiming that Bi_2FeCrO_6 would show a giant MF effect if Fe^{+3} and Cr^{+3} were ordered promoted a worldwide interest in these materials. This perovskite was recently prepared under HP and HT conditions [46] but the B cations were not ordered and this hindered the onset of ferromagnetism. Ferroelectric behavior associated with lattice distortions rather than inert pair effect is also being investigated in many HP materials such as $AMnO_3$ (A = Sc, Y) and more complex stoichiometries such as those of quadruple perovskites ($AA'_3B_4O_{12}$) are also under scrutiny. Other interesting results are hexagonal perovskites transforming into corresponding cubic forms.

Similarly, MOOH oxihydroxides with bohemite structure transform into the InOOH structure which is similar to the rutile one with the addition of hydrogen bonds between every two oxygens. If, rather than phase transitions, we deal with solid state reactions carried out under pressure, a rule of thumb to predict whether a process such as $A_{solid} + B_{solid} \rightarrow C_{solid}$ will or will not take place is to calculate the respective volumes per formula unit (V/Z). Only if the volume of the product C is smaller than the sum of the volumes of A and B, the reaction will happen. As an example let us recall that the synthesis of manganites of divalent transition metals $MMnO_3$ (M= Mg, Co,

Zn) with the ilmenite structure is only possible when performed under HP and HT conditions while MnO_2 and M_2MnO_4 spinels are easily obtained by the ceramic route at room pressure.

A systematic series of high pressure phase transformations have been also observed for ABO_4 compounds: molybdates, vanadates, chromates, silicates, germanates, niobates, and tantalates. [47]. In many cases, the crystal structures of the polymorphs are related via simple crystallographic operations. For instance, ABO_4 compounds that consist of BO_4 tetrahedra and AO_8 bidisphenoid units can be related with MO_2 octahedral structures [48]. ABO_4 zircon types transform to scheelite types under high pressure (with no change in cation coordination numbers but reducing ~10 % the unit cell volume across the transition [49–51]). MO_2 rutile types transform to fluorite types by increasing the pressure. This fact allow us to draw a parallel between the high-pressure structural behaviors of ABO_4 and MO_2 (MMO_4) compounds. This common trend was first observed by Fukunaga and Yamaoka in the late 1970s [52] and revisited by Bastide in the late 1980s [53]. Errandonea and Manjon reviewed the results obtained by other studying groups the phase behaviors of different ABO_4 materials and picture a devised phase diagram that improves our understanding of these high-pressure phase transformations [47].

14.3.2 Increase of Coordination Numbers

We already noted the graphite (coordination number, c.n = 3) to diamond (c.n. 4) transformation but there are many other examples, for instance, silicon with the diamond structure adopts, when treated at very high pressures, a new polymorph with the β-Sn structure (c.n 6). A really remarkable case is silica, SiO_2: at very high pressures it transforms into a new variety known as stishovite named after Stishov who prepared it for the first time (1961). This new phase, rather than showing the usual tetrahedral coordination for silicon (c.n. 4, also usual in silicates), adopts the octahedral characteristic of the rutile structure (c.n. 6). This produces an extraordinary increase (50 %) in the density from 2.95 g/cm^3 of coesite (also a high-pressure polymorph formed at 30 Kbar) to 4.28 g/cm^3 for stishovite. This new form of silica was found in craters created by the impacts of giant meteorites. Canyon Diablo was found by S. Coe, of coesite fame. Note that the relationship between P and T for this phase transition is almost linear. Another example relevant to geophysics is the transformation under HP and HT conditions of some feldspars (3D silicates with Si in tetrahedra) into materials showing the hollandite structure where all cations including Si randomly occupy octahedra:

$$K^{IV}Al^{IV}Si^{IV}{}_3O_8 \text{ (feldspar) } 120 \text{ kbar, } 900\ ^\circ C \rightarrow K^{VI}Al^{VI}Si^{VI}{}_3O_8$$

It is also interesting to recall that due to the effect of cationic sizes for similar germanates, these phase transitions take place at much lower pressures, on the order of 30 Kbar. Many high-pressure transitions of the lithosphere in silicates may be replicated in the laboratory at more moderate conditions by using germanates instead. Regarding ionic bonding considerations, it is easy to understand the effect that high pressure has on cationic coordination numbers. As is well known, anions are much softer and easily polarizable than cations and also more compressible. Therefore, the ratio r_{cation}/r_{anion} increases which makes the coordination numbers for cations also increase. A remarkable example is zinc oxide (ZnO) which at room pressure and temperature shows the wurtzite structure (Zn^{+2} in tetrahedra, hexagonal close packing of oxide ions) but when treated at HP and HT transforms into a material with the rock salt or halite structure (Zn^{+2} in octahedral, cubic close packing of anions). In similar cases such as calcium oxide (CaO) the coordination number may still increase with pressure changing from the rock salt structure to the cesium chloride structure. The coordination number for Ca is 8 and the cubic lattice is primitive and not face centered (fcc). In oxides, similar examples are abundant, especially for elements such as aluminum or gallium which can be found in tetrahedral or octahedral coordination in binary and more complex oxides as well. Thus, an oxide such as $InGaO_3$-I (In in octahedra, Ga in tetrahedra) transforms under HP and HT conditions first into a new polymorph, $InGaO_3$-II, whose gallium sits in a square-pyramid position (c.n. 5) and at higher pressures transforms into $InGaO_3$-III where both cations occupy octahedral positions in a disordered ilmenite arrangement equivalent to the corundum. A similar case is the high pressure form of V_2O_5 where V^{+5} adopts octahedral coordination instead of the square-pyramid. The HP polymorph shows very good electrochemical performance as a cathode for lithium intercalation, better than the room pressure polymorph and, to our knowledge is the first example of a high pressure-material suitable for these purposes that usually require open structures [54].

14.3.3 Stabilization of High Oxidation States for Metallic Elements

High pressures of strongly oxidizing gases (oxygen and fluor) are used to produce new compounds characterized by the unusually high oxidation states of some metals. High pressure induces greater overlapping between the orbitals of cations and anions, thus increasing the covalent character of the bonding. The higher the oxidation state becomes, the shorter the bond distances are and this corresponds to stronger bonds increasing covalency. For instance, a variety of mixed oxides containing Fe (IV), Fe(V), Co(IV), Ni(III) and Cu(III) in the first transition series, [55, 56] and Rh(IV) and Ir(IV) in the second. Especially interesting is the work of Jansen *et al* on new mixed oxides of noble metals, in particular Ag and Au, which are synthesized by using liquid oxygen heated in special autoclaves at supercritical conditions. Sometimes high pressure has the effect of preventing dismutation of a particular high oxidation state as happens, for instance with the rare Cr^{+4}.

As is well known, the solid state chemistry of transition metal oxides is one of the basic sources of useful materials and given the high number of such elements present in the Periodic Table and their varied oxidation states and plural coordination numbers, it is far from unexpected that reactivity under HP and HT of transition metal oxides with other metal oxides is the subject of many HP studies. As one would expect, the chemistry of Cr(IV) becomes richer and structurally more complex when CrO_2 is reacted with other chemicals to make mixed oxides. We have been working for some time on the synthesis of perovskite-related materials by reacting at HP and HT mixtures of CrO_2 with different oxides. The study of transition-metal oxides with perovskite and related structures has attracted much interest in recent years, due to the wide range of interesting properties observed in these compounds. Figure 14.6, shows in chronological order the evolution of complexities of the crystal structural of various mixed oxides we have obtained under HP and HT conditions using CrO_2 as a starting material [57].

Some of these oxides, all of which are high pressure phases, were known but not fully characterized. Most are however, novel and recently produced by our group. The materials include the simple basic perovskites $ACrO_3$ (A(II) = Ca, Sr, Pb), the more elaborated misfit layer compound $[Sr_2O_2][CrO_2]_{1.85}$ [58], and the spinel $CdCr_2O_4$ [59] including interesting chromium oxides such as the family Cr-1212 [19], $Bi(Cr_{0.5}Ni_{0.5})O_3$ [43] or $(Bi_{0.5}Pb_{0.5})CrO_3$ [60], several Ruddlesden–Popper phases and the $(Ca/Sr)CrO_3$ solid solution. Another interesting example of a Cr^{+4} material is the half-metallic ferromagnet $K_2Cr_8O_{16}$ with the hollandite structure that can only be obtained under specific HP and HT conditions. In a second step, its Curie temperature can be driven close to room temperature upon K^+ extraction [61]. Starting from the bottom left of Figure 14.6 we see the apparent cubic perovskites $SrCrO_3$ and $PbCrO_3$, of which only the first is really cubic. $PbCrO_3$, only appears isometric under powder x-ray diffraction observations. When scrutinized with electron microscopy and diffraction, it shows a much more complex structure, in which, as shown at top center in Figure 14.6, the A (lead) positions are only partially occupied in a rather extensive ($\sim a_p x 3a_p x 14\text{-}18a_p$, depending on the crystals) incommensurate modulated mode which also shows phase displacement along the vertical axis of $3a_p$; this composition-based superlattice is due to lead

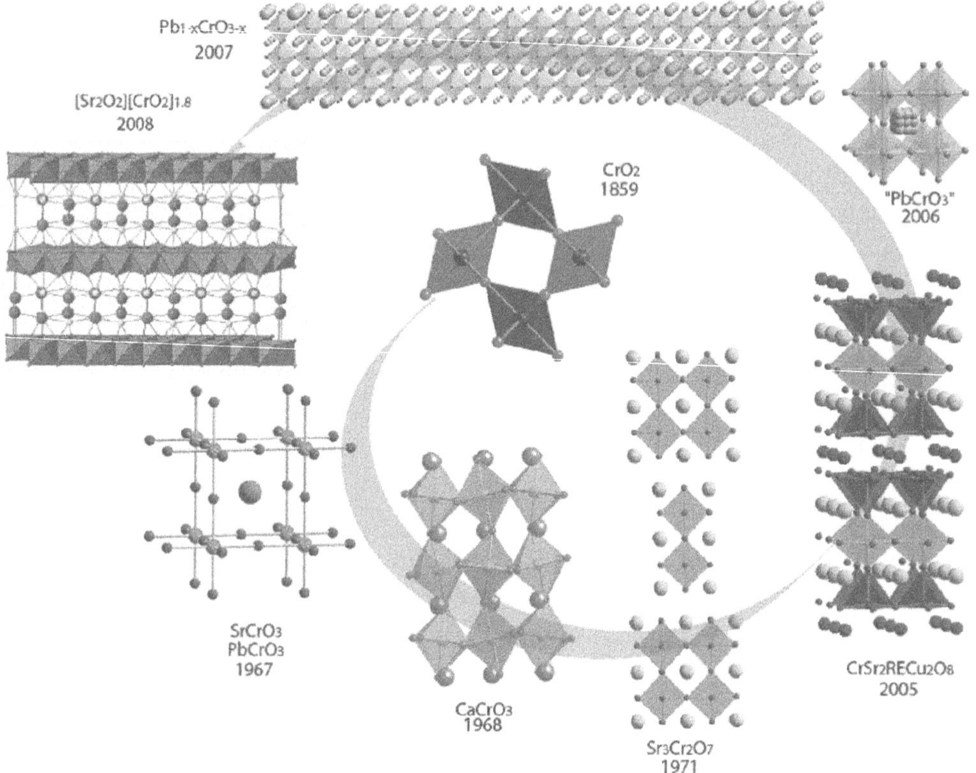

Figure 14.6 Chronological evolution as a function of structural complexity of Cr(IV)-containing oxides. Adapted with permission from Increasing the Structural Complexity of Chromium(IV) Oxides by High-Pressure and High-Temperature Reactions of CrO_2. Copyright (2008) American Chemical Society [57].

deficiency [62, 63]. Furthermore, this modulated structure is randomly distributed in the three space directions yielding a network of microdomains. When Sr is replaced by calcium, the symmetry lowers to orthorhombic due to the relatively small radius of the Ca(II) that increases the covalent character of the Ca-O bond inducing a tilting of the [Cr-O_6] octahedra. $(Sr_{1-x}Ca_x)CrO_3$ shows a progressive evolution of properties in parallel with the structural changes that show a tetragonal intermediate phase at x = 0.5, detected only with electron diffraction. In this interesting case [64] the idea of a solid solution or kind of random distribution between two dissimilar oxides is certainly somewhat naive as electron microscopy and diffraction show microdomains in the whole compositional range: that is, as soon as calcium enters the structure, one sees satellite reflections and three sets of domains differing in the relative orientations of the diagonal cell of the perovskite superstructure. There is evidence of Jahn–Teller distortion associated with the tilting in the orthorhombic phases. Interestingly, the domain size seems to be coupled to the composition and domains are larger near to the end members than at the middle of the range. Obviously, when we modify the molar ratio between Cr and Sr we obtain more materials. We obtained the well known Ruddlesden–Popper (RP) phases $A_{n+1}MnO_{3n+1}$, one of the types of the so-called layered perovskites, which are among the best known perovskite derivatives. Their structure is more commonly described as an intergrowth of single rock salt (RS) type AX layers and increasingly thick perovskite AMX_3 blocks and can then be formulated as $nAMX_3$ + AX (e.g., $Sr_3Cr_2O_7$ for n = 2 in Figure 14.6) [65]. These RP materials have received a great deal of attention for their metallic, ferromagnetic, colossal magnetoresistance, superconducting and thermoelectric properties. We have been able to obtain n = 1, 2 and ∞ members as nearly single phases. When pressure-synthesis is 8 GPa, new compounds appeared. The presence of n = 3 and a hexagonal layered perovskite (15-R) was observed by transmission electron microscopy and electron diffraction. The observed $SrCrO_3$-15R phase is the first example of a hexagonal perovskite in the Sr-Cr-O system; such phases were already well known in the Ba-Cr-O system. The interesting properties are reported elsewhere [65]. We recently reported the preparation and characterization of 11 members of the $CrSr_2RECu_2O_8$ family with RE = La, Pr, Nd, Eu, Gd, Tb, Dy, Y, Ho, Er, and Lu [17]. This family was synthesized in the pressure range of 6 to 8 GPa at 1300 °C for 30 minutes. In M-1212 type oxides, there is, in fact, a very interesting Gaussian relation between the optimum synthesis pressure and the rare earth ion size, which does not follow the well known lanthanide contraction (Figure 14.7).

This is discussed elsewhere [17] and correlates well with similar cases observed for cupro-ruthenates and cupro-iridates. We have also found in the Sr-Cr-O system new phases that adopt a misfit layer structure (MLS). These

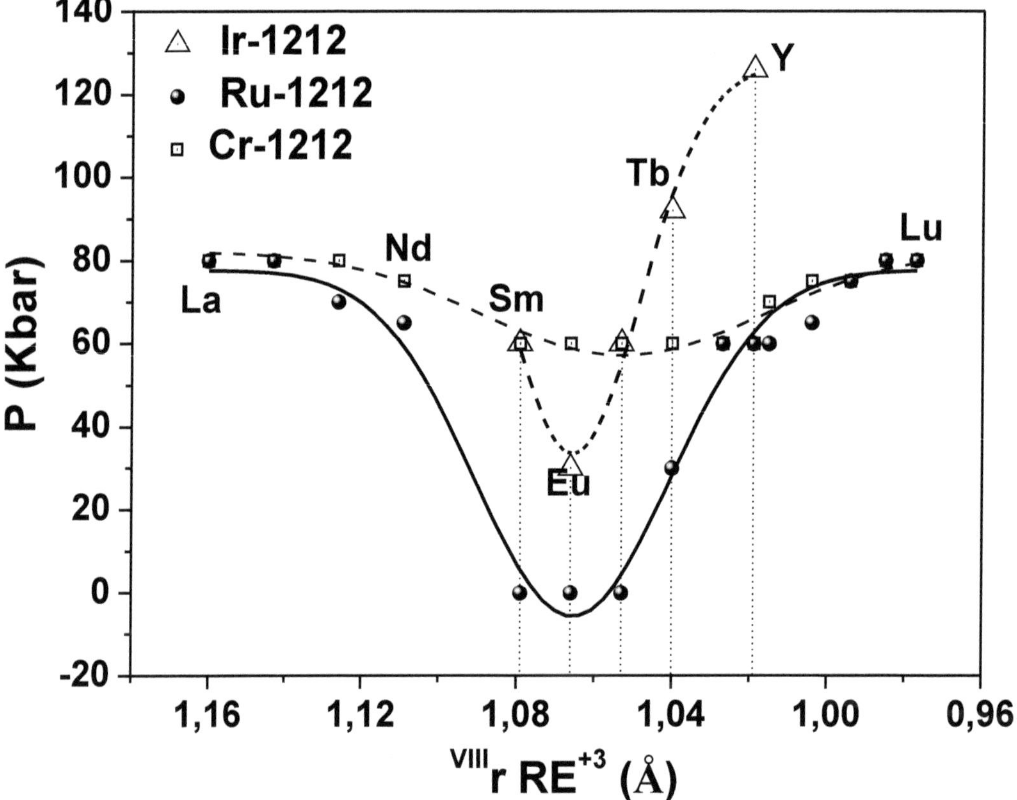

Figure 14.7 Optimum pressure for the synthesis of M-1212 family. Adapted with permission from Ahead of the Lanthanide Contraction; Pressure and Ionic Size in the Synthesis of $MSr_2RECu_2O_8$ (RE = Rare Earth, M = Ru, Cr, Ir): a Gaussian Relation. Copyright (2008) American Chemical Society [17].

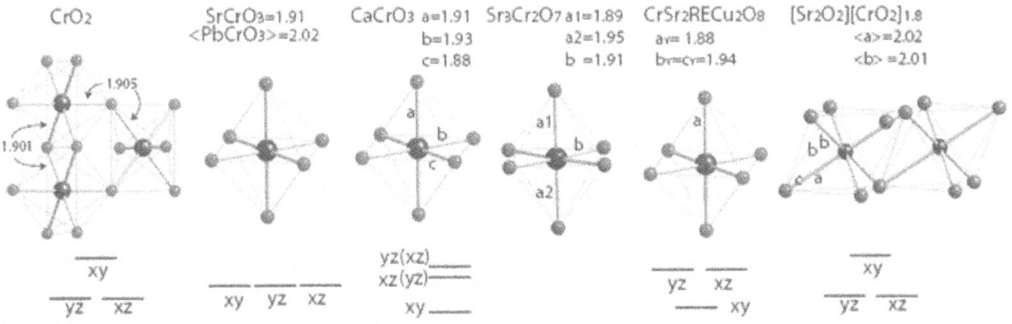

Figure 14.8 Irregular octahedral [Cr-O_6] geometries observed in the chromium compounds. Adapted with permission from Increasing the Structural Complexity of Chromium(IV) Oxides by High-Pressure and High-Temperature Reactions of CrO_2. Copyright (2008) American Chemical Society [57].

structures are formed by two different layered structures, having in common a lattice parameter along one of the layer directions and a different lattice parameter in the other layer direction; the third one indicates the stacking. We have observed that ML phases in the Sr-Cr(IV)-O system form at a wide range of pressures and temperatures above 3.5 GPa and at higher pressures they can coexist with the perovskite and RP n = 2 phase. However, no intergrowth of these phases in the same crystal is observed. In particular, $[Sr_2O_2][CrO_2]_{1.85}$ [58] seems to be the first example of misfit layer compounds among the chromium oxides and could eventually lead to research line in thermoelectrics as occurred with the similar $(Ca_{1-x}Bi_x)_3Co_4O_9$.

Based on the structural variety of high pressure phases, it is no surprise that the coordination of chromium - otherwise always octahedral - presents different types and degrees of distortions. Figure 14.8 shows the rich variety of irregular octahedral [Cr-O_6] geometries observed in the chromium compounds studied.

The corresponding qualitative splitting of the t_{2g} energy levels is also shown below each figure. Details of this work can be found in Reference [57].

14.3.4 Metal–Metal Bonding

Sometimes pressure makes elements more closer along a particular crystallographic direction and this may cause an orbital overlapping that would generate metal-metal bands and give rise to metallic bonding and electronic conductivity if partially filled. This is the case in covalent elements for hydrogen and iodine which become metallic under very high pressures or boron which becomes a superconductor under pressure at critical temperatures of ~10 K [66]. For oxides of transition metals, provided that the *d* orbitals are sufficiently large and the structure flexible enough (with perovskite) the same effect may happen.

14.3.5 Pressure Effects on Crystal Field and Spin Transitions

Pressure makes Co_2O_3 undergo a subtle transition: neither the oxidation state (Co^{+3}) nor the coordination number (c.n = 6, octahedral) nor the structure (corundum) are changed but the magnetic properties change completely (from paramagnetic to diamagnetic) and the volume reduction is noticeable (6.7 %). To explain this, one has to take into account the effects of pressure on the crystal field, taking oxide ions as the ligands and Co^{+3} as the central ion with its 3d orbitals split in two sets, a triplet of t_{2g} symmetry and a doublet e_g. Obviously, pressure makes the ligands to get closer to the cation orbitals and since the interaction is repulsive, the crystal field (△ or 10Dq) increases and the low spin configuration (LS) is more stable. Therefore, when the oxide is prepared under pressure (from cobalt fluoride and sodium peroxide at 60 Kbar and 700 °C), the electronic configuration is $(t_{2g})^6$. The LS configuration in which the six electrons are paired in the lower energy levels of the triplet and make this oxide diamagnetic. If this compound is heated to $\geq$ 400 °C at room pressure, it readily transforms into the high spin form (HS) with electronic configuration $(t_{2g})^4$ $(e_g)^2$ where four electrons are unpaired. The remarkable difference between cationic sizes for both configurations (Shannon and Prewitt ionic radii for Co^{+3} in octahedral coordination are 0.54 Å for the LS configuration and 0.61 Å for the HS one), results in a remarkable difference between the respective unit cell volumes [67]. Variations in the crystal field may be important even in the absence of external pressure: we refer to the so-called internal pressure effect induced by a crystal lattice on substitutional species inside it. To illustrate, consider the ruby, a crystalline material characterized by a beautiful and intense red color, used as a gem stone and also as a source for laser radiation. This color is due to the internal pressure induced by isolated Cr^{+3} cations when substitute Al^{+3} in the corundum Al_2O_3 structure. The bigger cation-size of Cr^{+3} split *d* orbitals allowing further electronic transitions with visible light emission (pure Al_2O_3 is colorless, Cr_2O_3 is green while $Al_{2-x}Cr_xO_3$ is red for very small values of x). Actually, this crystal field effect is used to calibrate pressures in DACs. A tiny ruby piece is introduced in the device and the change in the frequency of the light emitted is used to measure the applied pressure because the P versus frequency relationship is linear.

14.3.6 Decomposition Reactions under Pressure

In the search for new materials one tends to think that for a given stoichiometry, a change under high pressure would be just a phase transition to a more compact structure. This is often the desired effect but we should keep in mind that the main effect of pressure is densification. If the average density of a mixture of phases is higher than that of a given phase, the mixture would certainly decompose, especially if temperatures are high enough to overcome kinetic barriers. A very interesting case, again with geophysical implications, starts deep in the Earth in the so-called 670 km discontinuity where pressures of 26 GPa and temperatures of 2000 K are already reached and deeper 2900 km where HP and HT conditions become more extreme. We already mentioned that olivine $(Mg_{2-x}Fe_x)SiO_4$ transforms into spinel and this may cause earthquakes but at higher pressures the spinel phase decomposes to yield perovskite and wustite as follows:

$$(Mg_{2-x}Fe_x)SiO_4 \rightarrow (Mg_{1-y}Fe_y)SiO_3 + Mg_{1-z}Fe_zO$$

This implies an important increase in coordination numbers (Si occupies the octahedral B site, c.n. = 6, of the perovskite structure ABO_3, while Mg and Fe go to the cubooctahedral A position, c. n. = 12), far from the room pressure condition.

A similar situation can be experimentally found at lower pressures for other materials, i.e.:

$$M_2TiO_4 \text{ (spinel, M = divalent)} \rightarrow MTiO_3 \text{ (ilmenite)} + MO \text{ (rock-salt)},$$

$$Y_3Fe_5O_{12} \text{ (garnet)} \rightarrow 3\ YFeO_3 \text{ (perovskite)} + Fe_2O_3 \text{ (corundum)}$$

14.3.7 Pressure-Induced Amorphization

High pressure synthesis is usually perfomed under high temperature conditions needed to enhance diffusion rates. To prevent possible hardware damage, experiments are carried out over short times (in our laboratory 6 GPa experiments usually require 30 min). Sometimes, under these conditions very dense, crystalline materials are obtained and exhibit with physicochemical features different from those of glasses (which are obtained from a molten solid) and amorphous (and porous) materials obtained by other means. This phenomenon was reported first in 1938 for silicates [68] and in 1984 for the hexagonal polymorph of ice [69] but many questions remain unsolved. This method can be used at temperatures low enough to avoid solid to solid phase transitions in order to obtain amorphous materials when a particular system.

Hardly vitrifies this is the case, for instance, of zircon wolframate (ZrW_2O_8) which is amorphized at room temperature under high pressures; this material is unique since it undergoes contraction upon heating instead of the usual expansion. One of the materials referred in the literature as a negative thermal expansion (NTE) should have been designated a thermal contraction [70, 71]. Likewise, new non-oxide glasses such as Li-Ca-P nitrides are prepared under HP [72].

14.4 SOLVOTHERMAL SYNTHESIS

Several techniques of heating a liquid above its boiling point in a closed vessel are grouped under this name. Obviously this procedure originates a moderate pressure (autogenous) higher than atmospheric and, if the liquid is just water, the term hydrothermal synthesis is used in the literature. Nevertheless, other liquid media such as organic solvents, liquid ammonia and hydrazine are increasingly used [73]. By these means a great variety of new materials have been prepared and a great number of minerals have been formed in the lithosphere in a pressurized and hot aqueous media and often heterogeneous in composition. Unlike previously described methods, solvothermal synthesis requires dissolution of the reactants. The reaction proceeds by heating the solution above the room pressure boiling point of the solvent. Various low solubility species such as silicas, aluminosilicates, titanates, and sulfides have been dissolved and made reactive. It has been estimated that at 600 °C liquid water dissociates to a much higher degree than at room conditions (to produce H_3O^+ and OH^- ions) and K_w becomes 10^{-6} instead of 10^{-14}. Because of the amphoteric character of this solvent, it will act both as a high acid and a high base simultaneously, and therefore will perform acid-base reactions in a very aggressive way. On the other hand, some species known as mineralizers may be added (basic materials such as carbonates or hydroxides, acidic compounds such as nitric or hydrochloric acids or ammonium salts, and redox or complexing agents) that will enhance the dissolving power of water in one or another direction. Under these conditions, water acts as a reducing agent which means that the oxidation states of some transition metal cations may be lowered; for instance, hydrothermal synthesis conditions applied to Fe^{+3} solutions usually yield magnetite Fe_3O_4 in which the iron oxidation state is a mixed Fe^{+2} Fe^{+3}. No less important is the addition of molecular template species to perform as nucleating agents and be removed afterwards to generate pre-designed cavities or porosity. In this connection, organic cations with specific geometries such as tetramethyl-ammonium (TMA) and others are successfully used in the synthesis of new species [74, 75]. If liquid ammonia rather than water is used, the acid and basis ions are NH^{4+} and NH^{2-}, respectively, and the

Table 14.1 Examples of materials synthesized by hydrothermal method

Compound	T (°C)	P (bar)	t (days)
$La_{1-x}M_xMnO_3$ (M = Ca, Sr, Ba)	240	autogenous	3–6
$A_xV_6O_{16} \cdot nH_2O$ [1]	250–280	autogenous	2–4
$KMg_2AlSi_4O_{12}$	700–800	500–1000	2–3
BaY_2F_8	240	autogenous	3
C_3N_4 (graphitic)	250	1300	-

[1] *A* stands for an alkaline element.

reaction products will be nitrides but not oxides. The reaction parameters are the nature of the solvent, pressure, temperature, pH, relative amounts of solvent and reactants that determine effective concentrations, additives, and thermal gradients (matter transport is by convection). It is important to know whether the temperature of the solvent (374.1 °C and 218.3 atmospheres for water). Above that point, the solvent becomes a supercritical fluid which is neither a liquid nor a vapour. The pressure increases exponentially above this temperature.

We might state that solids with open or cage structures (i.e., zeolites) are produced under supercritical conditions; under supercritical conditions, the huge pressure increase produces compact structures such as $BaTiO_3$ and other perovskites. In aqueous solutions, subcritical conditions lead to solids with different amounts of water of crystallization while anhydrous compounds are produced under supercritical conditions. Solvothermal methodology, although not effective at very high pressures and high temperatures, certainly shows many advantages as it can be applied to the preparation of: (i) materials decomposing at high temperature, (ii) materials with low solubility and/or low reactivity, (iii) low temperature polymorphs, (iv) species with unusual oxidation states (by adding oxidizing or reducing agents), and (v) single crystals (such as α-quartz).

By these means we produced a new family of silicon-free hydrogarnets [76], rhodium dioxide and oxihydroxide [77], and other materials. Table 14.1 lists relevant examples of materials synthesized by this method. Regarding the apparatus suitable for solvothermal synthesis, below 700 °C and 3 Kbar, autoclaves are commonly used.

Depending on the working conditions, these autoclaves may be quite different (type of alloy, design, wall thickness, internal sleevings); they can be totally or partially introduced in the furnace (which produces different thermal gradients) and pressure can be just autogenous (produced by temperature) or increased upon external pressurizing. Once the vessel is closed, the pressure will be determined by the degree of filling and the presence of volatile species

Table 14.2 Examples of hydrothermal synthesis crystal growth.[1]

Compound	Solvent	$T_{solution}$	T_{growth}	P (MPa)
α-SiO_2	NaOH 0.5-1 M	350-400	325-335	80-150
Al_2O_3	Na_2CO_3 2-3 M	460-535	390-450	110-160
$Y_3Fe_5O_{12}$	Na_2CO_3 2-3 M	420	685-765	130
$Y_3Fe_5O_{12}$	KOH 20 M	420	380	100-120
MS [2]	HCl 1 M	410	430	240
$AlPO_4$	H_3PO_4 6.1 M + $NH_4H_2PO_4$ 3.8 M	-	250-400	150

[1] Temperatures (T) in °C.
[2] S= Cd, Hg, Pb.

(many times produced in situ) and all this has to be carefully designed to avoid dangerous explosions. If the autoclave is connected to a system with external control of pressure, the final pressure can be enhanced by completely filling the autoclave and pressurizing it prior to heating: the pressure is set to a fixed value and automatic valves are needed to open the system if the pressure exceeds it. A common problem is corrosion and to avoid it, the reactions are carried out in sealed tubes (made from noble metals) so that the surrounding liquid only acts as a pressure transmitter. An alternative is to use jacketed autoclaves in which an internal sleeve (made of gold or silver, for instance) is used to perform the reactions inside. Provided that the temperature of synthesis is below 220 °C, Teflon liners can be used conveniently. Sealed pyrex tubes can also be used for temperatures below 500 °C but basic conditions have to be avoided and internal and external pressure carefully equilibrated since these tubes may break down. By these means, a great variety of materials such as zeolites, silicates, phosphates, vanadates, and others have been synthesized. For hydrothermal synthesis at very high pressures (above 1 GPa), the equipment to be used is the same as that used for solid state synthesis (i.e., the belt type or similar) and hydrothermal conditions are achieved by adding a few drops of water or solution to the reactants to wet them. A new variety of iron oxihydroxide, δ-FeOOH, has been prepared at 6 GPa and 900 °C [78]. This methodology can be applied to the growth of single crystals such as α-quartz which is produced on industrial scale this way. To get large crystals (centimeter-sized) special conditions are needed: large autoclaves (dm^3 internal volume or higher), two well defined temperature zones (one for dissolving the nutrient and another for growing a previously placed seed), and very long times (sometimes several months). Table 14.2 shows the conditions for growing crystals of some important materials.

14.5 IN SITU STUDIES

Most of the synthesis and phase transformations at HP and HT are carried out on a trial-and-error basis mainly because the required extreme conditions do not allow to a researcher to "get inside the system" and observe the sample. In essence, the system behaves as a "black box". Obviously this is inconvenient because the researcher has no control of and cannot observe the processes that take place during such experiments. If phase transformations induced by pressure are reversible, the high-pressure polymorphs cannot be observed at room pressure and temperature after the experiment. This is the case for monoclinic zirconia which transforms into a tetragonal polymorph at 40 GPa or manganese difluoride (rutile type, c.n. 6 for Mn at room pressure) which transforms into a fluorite polymorph (c.n. 8) at 10 GPa and into a third with the α-PbO_2 structure (c.n. 6) when releasing pressure. To circumvent these problems, new setups are being designed and used where radiation gets into the system while it is working to provide spectroscopic or diffractometric information. The most common devices are the aforementioned DACs because, as the diamond anvils are transparent, UV-visible laser radiation can pass through and be used for a Raman analysis in order to follow a phase transition in situ. In the same way and, taking benefit from portability, the set-up can be taken to a Synchrotron facility for x-ray diffraction experiments. If both high pressures and temperatures are needed, the situation becomes more difficult. A high-pressures and high-temperature setup similar to the belt device with a window that allows radiation to pass through is needed. We noted the formation and subsequent decomposition of a new infinite-layer cuprate containing Sr and Ca at 2 GPa and 1200 °C [79].

In recent years, great effort has been devoted to neutron diffraction using modified Paris–Edinburgh chambers at facilities such as Institut Laue–Langevin (ILL) in Grenoble where one neutron beam is dedicated to such experiments.

The authors are gratefully indebted to many co-workers and colleagues who contributed to HP/HT studies: J. C. Joubert, R. Argoud, J. J. Capponi, J. Chenavas and C. Bougerol from the "Laboratoire de Crystallographie" in Grenoble and to Prof. G. Demazeau in Bordeaux, all at the CNRS (France). At the Universidad Complutense de Madrid (Spain), Dr. E. Castillo-Martínez and Dr. A. M. Arévalo-López for their contribution to the work here described; Dr. J. M. Gallardo-Amores is acknowledged for performing most of the experiments and Prof. V. García-Baonza for Raman experiments and helpful discussions.

Bibliography

[1] P. Bridgman. *Collected experimental papers.* Harvard University Press. Cambridge (1964).

[2] J. Mellor. *Comprehensive treatise on organic, inorganic and theoretical chemistry, Volume 5.* Longmans, Green & Co., London (1924).

[3] R. Hazen. *The new alchemists.* Times Books, New York (1993).

[4] H. Tracy Hall Foundation. http://www.htracyhall.org.

[5] R. Laudise and J. Nielsen. Volume 12 of Solid state physics, 149–222. Academic Press, New York (1961).

[6] B. Kubota. *J. Phys. Soc. Jpn.*, 15, 1706 (1960).

[7] R. Barrer. *Hydrothermal chemistry of zeolites.* Academic Press, New York (1982).

[8] S. N. Putilin, E. V. Antipov, O. Chamaissem, *et al. Nature*, 362, 226 (1993).

[9] A. J. Dos santos-García, M. Á. Alario-Franco, and R. Sáez-Puche. *in D. A. Atwood, ed., The Rare Earth Elements: Fundamentals and Applications.* Wiley, New York (2012).

[10] M. Nuñez-Regueiro, J. L. Tholence, E. V. Antipov, *et al. Science*, 262, 97 (1993).

[11] M. Á. Alario-Franco. *Advanced Materials*, 7, 229 (1995).

[12] C. Bernhard, J. L. Tallon, C. Niedermayer, *et al. Phys. Rev. B*, 59, 14099 (1999).

[13] A. C. McLaughlin, W. Zhou, J. P. Attfield, *et al. Phys. Rev. B*, 60, 7512 (1999).

[14] R. Ruiz-Bustos, J. Gallardo-Amores, R. Sáez-Puche, *et al. Physica C*, 382, 395 (2002).

[15] A. J. Dos santos-García, M. H. Aguirre, E. Morán, *et al. J. Solid State Chem.*, 179, 1296 (2006).

[16] A. J. Dos santos-García, R. Ruiz-Bustos, Á. M. Arévalo-Lopez, *et al. High Pressure Res.*, 28, 525 (2008).

[17] M. Á. Alario-Franco, R. Ruiz-Bustos, and A. J. Dos santos-García. *Inorg. Chem.*, 47, 6475 (2008).

[18] S. Marik, A. J. Dos santos-García, E. Morán, *et al. J. Phys. Condens. Matter*, 25 (2013).

[19] R. Ruiz-Bustos, M. H. Aguirre, and M. Á. Alario-Franco. *Inorg. Chem.*, 44, 3063 (2005).

[20] B. Edwards and N. Ashcroft. *Nature*, 388, 652 (1997).

[21] The European High Pressure Research Group. Http://www.ehprg.org.

[22] A. Drozdov, M. I. Eremets, and I. A. Troyan. *arXiv:1412.0460* (2014).

[23] C. Rooymans. In P. Hagenmuller, ed., *Preparative methods in solid state chemistry*, 71–132. Academic Press, New York (1972).

[24] J. Goodenough, J. Kafalas, and J. Longo. In P. Hagenmuller, ed., *Preparative methods in solid state chemistry*, 1–69. Academic Press, New York (1972).

[25] M. Takano, Y. Takeda, and O. Ohtaka. *High pressure synthesis of solids.* Wiley, New York (2006).

[26] G. Demazeau. *High Pressure Res.*, 18, 203 (2000).

[27] P. F. McMillan. *Curr. Opin. Solid State Mater. Sci.*, 4, 171 (1999).

[28] E. Takayama-Muromachi. *Chem. Mater.*, 10, 2686 (1998).

[29] C. N. R. Rao and J. Gopalakrishnan. *New directions in solid state chemistry.* Cambridge University Press, 2nd ed., cambridge edition (1997).

[30] A. Jayaraman. *Rev. Mod. Phys.*, 55, 65 (1983).

[31] W. J. Nellis, A. C. Mitchell, M. van Thiel, *et al. J. Chem. Phys.*, 79, 1480 (1983).

[32] B. Wei, J. Liang, Z. Gao, *et al. J. Mater. Process. Tech.*, 63, 573 (1997).

[33] U. Amador, J. Gallardo-Amores, G. Heymann, *et al.* *Solid State Sci.*, 11, 343 (2009).

[34] A. Navrotsky. *J. Solid State Chem.*, 6, 21 (1973).

[35] V. M. Goldschmidt, T. Barth, G. Lunde, *et al.* *Mat.Nat. Kl.*, 2, 117 (1926).

[36] K. Kohn, K. Inoue, O. Horie, *et al.* *J. Solid State Chem.*, 18, 27 (1976).

[37] R. Escamilla, J. M. Gallardo-Amores, E. Morán, *et al.* *High Pressure Res.*, 22, 551 (2002).

[38] A. J. Dos santos-García, C. Ritter, E. Solana-Madruga, *et al.* *J. Phys. Condens. Matter*, 25, 206004 (2013).

[39] G. Bazuev, B. Golovkin, N. Lukin, *et al.* *J. Solid State Chem.*, 124, 333 (1996).

[40] A. M. Arévalo-López and J. P. Attfield. *Phys. Rev. B*, 88, 104416 (2013).

[41] E. Gilioli and L. Ehm. *IUCrJ*, 1, 590 (2014).

[42] A. A. Belik, Y. Matsushita, M. Tanaka, *et al.* *Chem. Mater.*, 24, 2197 (2012).

[43] Á. M. Arévalo-López, A. J. Dos santos-García, J. R. Levin, *et al.* *Inorg. Chem.*, 54, 832 (2015).

[44] R. V. Shpanchenko, V. V. Chernaya, A. A. Tsirlin, *et al.* *Chem. Mater.*, 16, 3267 (2004).

[45] N. A. Spaldin and M. Fiebig. *Science*, 309, 391 (2005).

[46] Y. Zheng, C. H. Woo, B. Wang, *et al.* *Appl. Phys. Lett.*, 90, 092905 (2007).

[47] D. Errandonea and F. J. Manjón. *Progr. Mater. Sci.*, 53, 711 (2008).

[48] B. G. Hyde and S. Anderson. *Inorganic crystal structures*. Wiley, New York (1989).

[49] E. Climent-Pascual, J. Romero de Paz, J. M. Gallardo-Amores, *et al.* *Solid State Sci.*, 9, 574 (2007).

[50] E. Climent Pascual, J. M. Gallardo Amores, R. Sáez Puche, *et al.* *Phys. Rev. B*, 81, 174419 (2010).

[51] A. J. Dos santos-García, E. Climent-Pascual, J. M. Gallardo-Amores, *et al. J. Solid State Chem.*, 194, 119 (2012).

[52] O. Fukunaga and S. Yamaoka. *Phys. Chem. Miner.*, 5, 167 (1979).

[53] J. Bastide. *J. Solid State Chem.*, 71, 115 (1987).

[54] M. A. y de Dompablo, J. Gallardo-Amores, U. Amador, *et al. Electrochem. Commun.*, 9, 1305 (2007).

[55] J.-H. Choy, G. Demazeau, J.-M. Dance, *et al. J. Solid State Chem.*, 109, 289 (1994).

[56] G. Demazeau, D. Y. Jung, A. Largeteau, *et al. Rev. High Pressure Sci. Technol.*, 7, 1025 (1998).

[57] E. Castillo-Martínez, A. M. Arévalo-López, R. Ruiz-Bustos, *et al. Inorg. Chem.*, 47, 8526 (2008).

[58] E. Castillo-Martínez, A. Schonleber, S. van Smaalen, *et al. J. Solid State Chem.*, 181, 1840 (2008).

[59] Á. M. Arévalo-López, A. J. Dos santos-García, E. Castillo-Martínez, *et al. Inorg. Chem.*, 49, 2827 (2010).

[60] I. Pirrotta, R. Schmidt, A. J. Dos santos-García, *et al. J. Solid State Chem.*, 225, 321 (2015).

[61] I. Pirrotta, J. Fernandez-Sanjulian, E. Morán, *et al. Dalton Trans.*, 41, 1840 (2012).

[62] Á. M. Arévalo-López and M. Á. Alario-Franco. *J. Solid State Chem.*, 180, 3271 (2007).

[63] Á. M. Arévalo-López, A. J. Dos santos-García, and M. Á. Alario-Franco. *Inorg. Chem.*, 48, 5434 (2009).

[64] E. Castillo-Martínez, A. Durán, and M. Á. Alario-Franco. *J. Solid State Chem.*, 181, 895 (2008).

[65] E. Castillo-Martínez and M. Á. Alario-Franco. *Solid State Sci.*, 9, 564 (2007).

[66] M. I. Eremets, V. V. Struzhkin, H.-k. Mao, *et al. Science*, 293, 272 (2001).

[67] R. D. Shannon and C. T. Prewitt. *Acta Crystallogr. B*, 25, 925 (1969).

[68] J. W. Greig and T. F. W. Barth. *Am. J. Sci.*, 35A, 93 (1938).

[69] O. Mishima, L. D. Calvert, and E. Whalley. *Nature*, 310, 393 (1984).

[70] J. M. Gallardo-Amores, U. Amador, E. Morán, *et al. Int. J. Inorg. Mater.*, 2, 123 (2000).

[71] C. A. Perottoni and J. A. H. Da Jornada. *Science*, 280, 886 (1998).

[72] T. Grande, J. Holloway, P. McMillan, *et al. Nature*, 369, 43 (1994).

[73] G. Demazeau. *J. Mater. Chem.*, 9, 15 (1999).

[74] T. Chirayil, P. Y. Zavalij, and M. S. Whittingham. *Chem. Mater.*, 10, 2629 (1998).

[75] P. Zavalij, J. Guo, M. Whittingham, *et al. J. Solid State Chem.*, 123, 83 (1996).

[76] E. Morán-Miguelez, M. Á. Alario-Franco, and J. C. Joubert. *Mater. Res. Bull.*, 21, 107 (1986).

[77] E. Morán-Miguelez and M. Á. Alario-Franco. *Thermochim. Acta*, 60, 181 (1983).

[78] M. Pernet, J. Chenavas, J. Joubert, *et al. Solid State Commun.*, 13, 1147 (1973).

[79] J. García-Jaca, X. Turrillas, S. M. Clark, *et al. Physica C*, 341- 348, Part 2, 779 (2000).

CHAPTER 15

Structure of Earth's Interior

Fernando Aguado

MALTA Consolider Team and DCITIMAC, Universidad de Cantabria, Santander, Spain

David Santamaría

MALTA Consolider Team and Departamento de Física Aplicada, Universidad de Valencia, Valencia, Spain

CONTENTS

15.1 INTRODUCTION

The Earth's interior remains unexplored to a great extent. Only a small number of constituents from different levels can be extracted and analyzed at ambient conditions. Therefore, our understanding of the composition of the Earth by direct observation is quite limited and subject to other indirect methods leading to a changing knowledge, which needs to be constantly updated in view of new evidence. There are three fundamental approaches providing information to be compared and complemented to establish our view of the materials and processes inside the Earth. The first one arises from the seismic data which reveal the pressure distribution, density and elasticity moduli as a function of depth. There are also thermal models, influenced by different thermodynamic parameters, which play a fundamental role in establishing correlations between temperature and pressure of minerals at different depths. Finally, compositional models give us information about the distribution of constituents in minerals for particular pressure and temperature conditions. These depend on thermal and seismological models [1].

Whereas the common research protocol in materials science is to explore the structure and properties of a given compound, in geophysics the procedure is frequently the opposite, that is, to look for suitable candidates matching given physical and chemical properties (constrained by indirect evidence). Mineral physics is a primary source for the knowledge of the Earth's deep interior. Research on the structural behavior of compounds under high pressure and high temperature conditions allows us to correlate seismic features with structural evolution of main constituents at different levels. In this way, seismic discontinuities observed in the inner Earth can be well correlated with pure structural changes (phase transitions) or abrupt compositional variations at those levels, e.g. at the core-mantle boundary (CMB). Since it is crucial to precisely determine the behaviour of materials under conditions mimicking the Earth's interior, many high-pressure experimental developments are frequently introduced by scientists in the field of geophysics.

The Earth can be divided in three fundamental parts: a thin solid crust (only 65 km thick at most), the mantle, a solid area made of silicate minerals, and the metallic core, subdivided into an outer molten layer and an inner solid region. The mantle extends to approximately 2900 km depth, which corresponds to 135 GPa and temperatures around 4000 K. This is a heterogeneous layer of the Earth, subdivided in three main parts and delimited by two seismic discontinuities at 410 and 660 km.

This chapter includes non-exhaustive descriptions of materials and processes in the mantle and Earth's core, where most of the research activity of the geophysics community is currently focused. Numerous references on different research topics from both zones have been incorporated to provide the reader with a starting point for further investigation. The structural properties and phase transitions associated with mantle's constituents will be explained in relation to the seismic features delimiting different regions. Special attention will be paid to the lower mantle dominated by perovskite-type compounds and, particularly, the research activity on the CMB. Interest in this region has increased enormously since the discovery of a new structural phase transition providing new insights into long-standing issues associated with the seismic features in the so-called D" layer. Finally, the structural properties of the

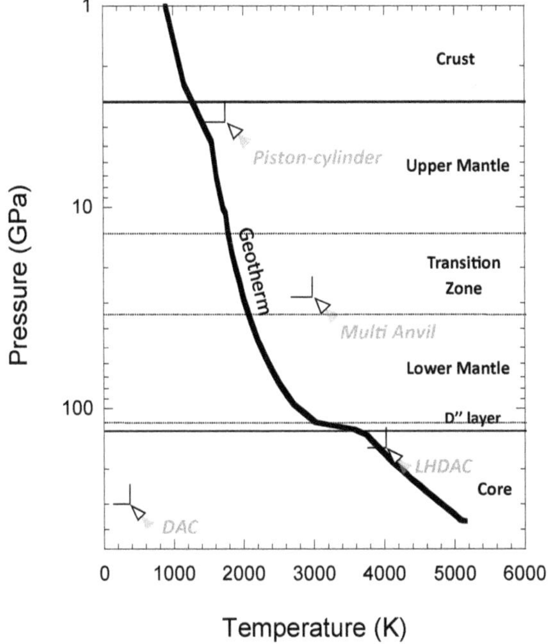

Figure 15.1 Pressure-temperature diagram of the Earth's interior together with the accessibility shown by different pressure devices. Solid curve represents the geotherm according to models.

core, its main seismic features and some outstanding problems corresponding to that area will be presented. First, methods for generation of extreme conditions corresponding to zones of geophysical interest and their adaptation to standard structural characterization techniques will be briefly introduced.

15.2 HIGH-PRESSURE TECHNIQUES

In this section, the devices used in high-pressure experiments will be briefly described, with focus on structural studies of compounds of interest in mineral physics under extreme conditions. A detailed explanation of high-pressure generation devices can be found in Chapter 6, whereas fundamentals of techniques for structural characterization (synchrotron radiation) are explained in more detail in Chapter 7. In addition, a number of textbooks cover the basics of these topics, some of which will be referenced in this section.

15.2.1 Generation of High-Pressure and High-Temperature Conditions

Devices for generating high static pressures can be roughly separated into two groups: the large volume cells, which reach moderate maximum pressures and are frequently employed for synthesis (see Chapter 14), and those devices based on small anvils (mainly diamond anvil cells or DACs [2]) which allow work under wider pressure and temperature ranges, but compromise the amount of sample available. Another group based on dynamical pressure generation (shock experiments) to allow access to higher pressures and temperatures simultaneously during short times is especially interesting for describing structures belonging to the Earth's core and other planetary interior structures. This technique is also described in Chapter 6 in more detail.

The first group includes devices for material synthesis, like piston-cylinder apparatuses (generating mostly uniaxial pressures up to 4 GPa) and multi-anvil and Paris–Edinburgh devices that nowadays can achieve pressures over 30 GPa at temperatures of thousands degrees Celsius. Some of these can be incorporated into x-ray or neutron beamlines to perform *in situ* studies. In terms of Earth's interior accessibility, this group of devices can reproduce easily conditions corresponding to the upper mantle and the transition zone, as well as part of the lower mantle (Figure 15.1). However, to work comfortably at conditions corresponding to the lower mantle (D" layer), DAC-type devices must be employed. This drastically reduces sample volumes (typically 10^{-7} cm^3), conditioning techniques and instruments for characterization under such conditions. In spite of the inconvenience associated with technical requirements, the pairing of new generation synchrotrons and DACs, together with modern laser heating techniques, has been a real catalyst for explaining the physical and chemical processes within the Earth. From the first prototypes [3, 4] to the most recent devices, maximum pressures have raised an order of magnitude in 50 years (reaching the Mbar range). Translated into Earth's language, this means we can routinely reproduce conditions corresponding not only to the lowermost mantle but also the core.

15.2.2 Structural Characterization under Extreme Conditions

In order to gain knowledge of the processes in the Earth's interior, it is vital to determine the crystal structures of its constituents under pressure and temperature conditions. Therefore, standard characterization techniques must be adapted for work in such sample environments. Among the choices, x-ray diffraction is by far the technique most commonly employed for determination of structural evolution under extreme conditions (see Chapter 7). Other techniques such as different types of spectroscopies can also provide valuable information. In any case, it is very important to be able to obtain such information in real time (*in situ* measurements), instead of after applying P/T cycles, given that some mineral phases are stable only under such conditions (unquenchable). However, this fact implies issues for the control of P and T and also their precise determination. In addition, in many cases it is difficult to perform a full structural characterization by powder diffraction (the most common procedure) on a crystalline sample, i.e., applying the Rietveld method [5, 6], since that strongly relies on the quality of diffraction intensities. Single crystal diffraction is an alternative technique yielding full structural information up to megabar level but also shows considerable limitations [7]. These studies require diamonds with small flat surfaces (culets) and their size relates to the maximum pressure achieved. Typically, culet diameter ranges between 400 ($P_{max} \sim 30$ GPa) and 40 ($P_{max} \sim 2$ Mbar) microns and requires a preparation protocol to achieve the ideal ratio between cavity and culet diameters [8]. In addition, to get a clean signal from diffraction it is necessary to reduce the x-ray beam focus, frequently monochromatic, to a few microns. Nevertheless, high-brilliance synchrotron sources together with modern bidimensional detectors allow acquisition of good quality patterns in just a few seconds under these working conditions. Specific DACs are used for studies reproducing lower mantle conditions (over 1 Mbar). Smaller diamonds (which can be beveled or double-beveled) and harder materials for gaskets (e.g., rhenium) need to be used for these experiments. In addition, high-temperature experiments require additional preparation. Although it is possible to get access to considerable temperature inside a cell through resistive heating (both in large volume and DACs [9]), it is more convenient to employ laser heating for studies at lower mantle conditions where temperatures in excess of 2000 K are to be achieved. Additional equipment and specially prepared DACs (LHDACs) are required. Usually a double-side heating geometry is employed to reduce temperature gradients over the sample, which can drastically affect the results. Powerful infrared lasers such as CO_2 ($\lambda = 10.6$ μm) or YLF or Nd-doped YAG ($\lambda = 1.06$ μm) can be focused on both sides of the DAC.

Figure 15.2 shows the experimental setup for a high P/T x-ray diffraction experiment with double-side laser heating [10]. Concerning the DAC, the cavity must be loaded with the sample and some insulating material, due to the high conductivity of diamond anvils. In case of near infrared lasers, a piece of inert metal foil is needed to effectively absorb the laser energy and then transfer it to heat the mineral sample. The spectroradiometric method allows us to estimate temperature inside the sample chamber [11]. Determination of temperature (and pressure) under such conditions is a topic of continuous analysis within the scientific community.

15.2.3 Pressure and Temperature Determination: Pressure Scales

Whereas for static experiments at room temperature the ruby scale (emission shift) is widely employed for pressure determination, typical mineral physics experiments under high pressure and high temperature require the utilization

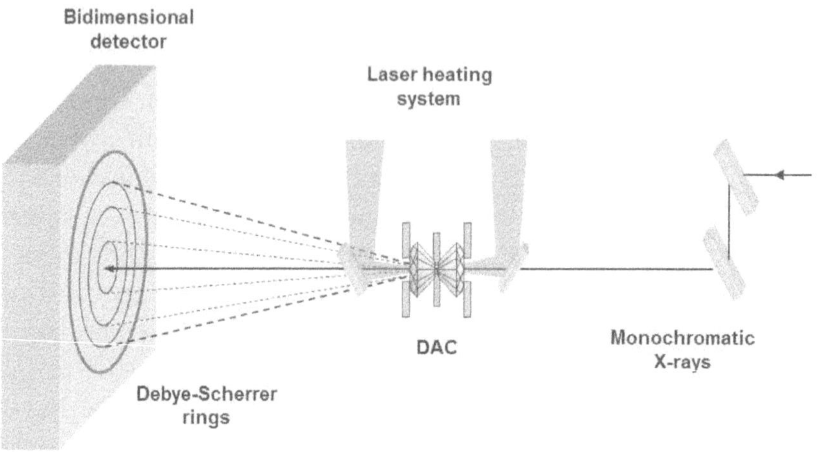

Figure 15.2 Angle-dispersive x-ray diffraction (ADXRD) setup for structural characterization under high pressure and high temperature conditions. Temperature is applied by a double-side laser heating system focused on the sample chamber inside the diamond anvil cell (LHDAC).

of $P-V-T$ equations of state of specific and well known compounds used as standards. It is important to note that small pressure differences from standard to standard are translated into uncertainties of several kilometers within the Earth. Therefore, there is an open debate about the interpretation of different phenomena in the mantle (particularly the lower mantle) due to discrepancies introduced from different pressure scales. The most important scales for pressure determination are gold [12], Speziale's MgO [13], Pt from Jamieson [14] and Holmes [15]. In particular, pressure accuracy is very important to determine the suitability of structural phase transitions to explain seismic features found in the inner Earth. Irifune *et al.* determined, based on Anderson's gold scale [16], that postspinel phase transition was not consistent with the 660 km seismic discontinuity [17]. Later reports based on different scales led to the opposite conclusion [18]. Fei *et al.* eventually demonstrated that differences of approximately 2 GPa were attained from different scales and that the phase transition accounts for such discontinuity based on the MgO scale [19]. Actually, differences between scales increase over 100 GPa, reaching up to 15 GPa for pressures around 120 GPa (which corresponds to D" layer depth). In spite of several comparative studies, there is still some controversy on which scale is the most appropriate for structural characterization experiments. Concerning the mantle, MgO scale is probably the best suited for reconciling the phase transitions with seismic observations along this region.

Temperature is independently determined by the black body emission in high $P-T$ diffraction experiments [20]. Such estimation can imply a considerable associated uncertainty due to the assumption of wavelength-independent emissivities. Some enhancements have been recently introduced to improve accuracy in temperature determination [21, 22].

15.3 STRUCTURE OF EARTH'S MANTLE

As previously stated, the inner Earth can be partially investigated through several methods, from the direct analysis of samples coming from depths up to a few hundred kilometers to the study of meteorites. Models are most commonly developed using seismologic data in combination with experiments conducted on suitable candidates under extreme conditions. Thus, structural and elastic properties can be determined and validated against seismic models, complemented with different kinds of calculations and simulations.

Our view on the mantle is continuously challenged and modified by new experimental results. Thus, recent discoveries have helped us to improve our knowledge of this region, i.e., the transition from the perovskite structure to a new denser conformation at lower mantle conditions is potentially able to explain most of the seismic features observed at the D" layer (above the CMB). In addition, spin crossover transitions in Fe-bearing minerals forming the mantle and the presence of important amounts of water in the transition region (which would yield conciliation between minerals phase boundaries and seismic discontinuities) are other examples of outstanding topics from the ongoing mantle research, with many recent contributions [23].

This section covers some general aspects of the structure of Earth's mantle, including its basic composition and the structural changes in minerals which explain the main seismic discontinuities dividing different regions. The lowermost mantle and the CMB will be treated in the next two sections. Some of the outstanding problems, particularly those related to the lower mantle, will be introduced in the final section.

15.3.1 Mantle Composition

The most common element (in weight) in the mantle is oxygen (*ca.* 44 %), followed by magnesium and silicon (23 % and 21 %, respectively) [25]. However, other less-common elements like iron (6 %) or aluminum (2 %) can play a decisive role in some geophysical processes. In particular, they are able to modify some properties of host minerals significantly, even at very low concentrations (e.g. effects of Fe in $MgSiO_3$ perovskite [26]). Although the mantle is a massive heterogeneous region subdivided in three distinct parts, only five elements are the main constituents (96 % in weight) of all the minerals in this layer. The upper mantle consists basically of four minerals according to the so-called pyrolitic composition (see Figure 15.3): olivine $(Mg,Fe)SiO_4$ (ca. 60 % at

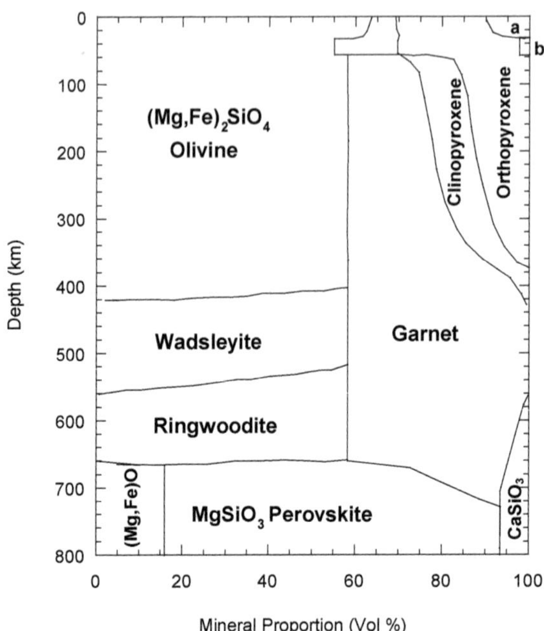

Figure 15.3 Mantle composition (pyrolite) as a function of depth. (a) and (b) represent spinel and anorthite (plagioclase feldspar), respectively. Adapted, with permission, from reference 24.

low pressures), orthopyroxenes $(Mg,Fe)SiO_3$ and clinopyroxenes $CaMgSi_2O_6$ (30 %), and garnet $(Mg,Fe,Ca)_3Al_2Si_3O_{12}$ (10 %). Alternative compositional models can yield considerable differences in phases and mineral proportions along mantle regions. Thus, different SiO_2 phases representing *ca.* 10 % would be present in the entire mantle when the MORB (mid ocean ridge basalt) compositional model is considered [27]. Nevertheless, phase transitions in olivine are believed to represent boundaries in the mantle since they are mostly consistent with seismic observations.

15.3.2 Phase Transitions at 410 km and 660 km

Two fundamental questions in geophysics are: (i) to determine whether the compositional model remains the same at different depths and (ii) to establish whether it correctly represents the observed seismic features. To answer these questions it is necessary to combine laboratory results on structural behavior of minerals with available seismic data. The Preliminary Reference Earth Model (PREM) [28] establishes the dependence of the density and the *P*- and *S*-wave velocities (V_P and V_S, respectively) as a function of depth (one-dimensional profiles), as shown in Figure 15.4. It can be inferred from these data that there are at least two important global discontinuities around 410 and 660 km (14

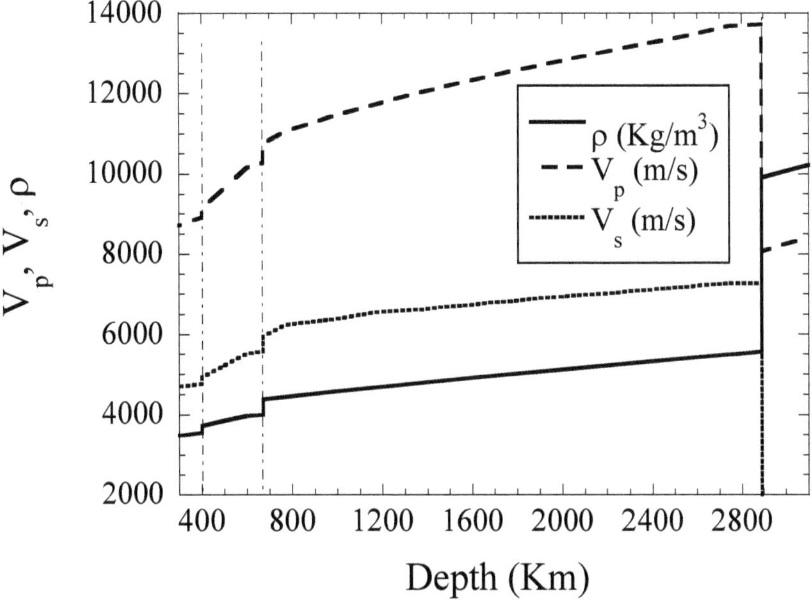

Figure 15.4 Variation of seismic P- and S-wave velocities (V_P and V_S, respectively), and density (ρ) as a function of depth according to PREM [28]. Vertical lines indicate the most important discontinuities in the Earth's mantle.

GPa and 24 GPa, approximately), corresponding to boundaries of the three main mantle regions. These features are common to other models (e.g., AK135 [29]) that provide slightly different profiles since they use different seismic data and can also employ other constraints from known mineral properties. Also, there are other weaker discontinuities not present in most models, but actually corresponding to known mineral phase transitions, like the non-global complex discontinuity at 520 km [30].

The structures of most silicates at the upper mantle can be described on the basis of SiO_4 tetrahedra arranged in related flexible structures, almost equivalent from the energetic point of view. As the depth (pressure) increases, the densification process favors structures with different SiO_4 connectivities. However, SiO_6-based structures become more stable at pressures above 24 GPa, mostly giving rise to perovskite-related structures. The end members of olivine, forsterite Mg_2SiO_4 and fayalite Fe_2SiO_4, both present orthorhombic symmetry (space group *Pbnm* or *Pnma*). The SiO_4 units are isolated within the structure, with Mg and Fe occupying two distinct octahedral sites, M1 and M2 [31] (Figure 15.5).

High-pressure olivine polymorphs can be stabilized at different pressures (depths), depending on factors like the temperature or Fe content. However, olivine-spinel and spinel-perovskite are generally sharp structural phase

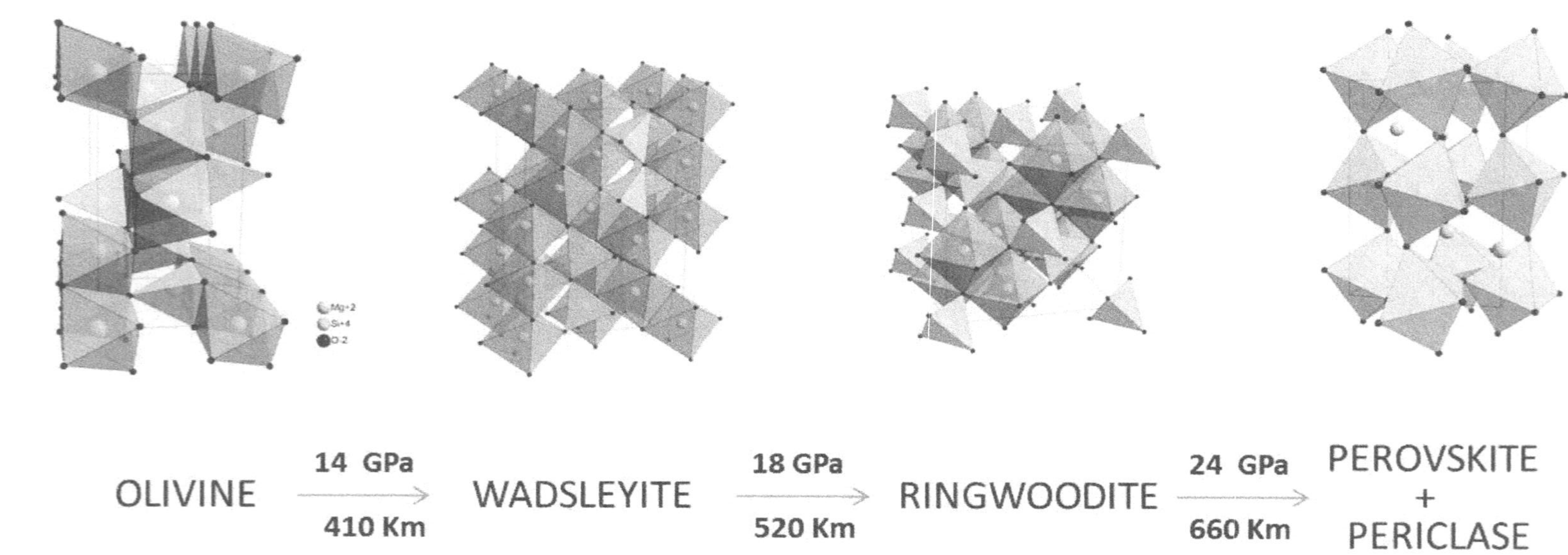

Figure 15.5 Pressure-induced structural phase transition sequence in $(Mg,Fe)_2SiO_4$ (olivine), which is believed to define the main mantle regions. Light grey tetrahedra correspond to SiO_4 units, whereas Mg and Fe are located at octahedral sites (dark grey polyhedra). Above 24 GPa (660 km) olivine decomposes to form $(Mg,Fe)SiO_3$ perovskite and $(Mg,Fe)O$ ferropericlase.

transitions defining region boundaries within the mantle. Several experiments have determined that the transition taking place at about 13 to 14 GPa from olivine to wadsleyite (β-spinel) is the origin of the first seismic discontinuity within the mantle. The structure of wadsleyite (orthorhombic symmetry, S.G. $Imma$) consists of Si_2O_7 isolated units with three different sites for cations in octahedral coordination (Figure 15.5). The idea of the transition as an explanation for seismic discontinuity was introduced in the 1930s [32], although experimental evidences appeared in the 1960s through studies on $(Mg,Fe)_2SiO_4$ by using large volume devices for pressure generation under high temperatures achieved by resistive heating [33]. In spite of initial discrepancies concerning pressures, at the beginning of the 1970s the structural phase transition in the $(Mg,Fe)_2SiO_4$ solid solution to a modified spinel structure was established as the origin of the corresponding seismic features at 410 km [34]. When technical developments made it possible to achieve higher pressures and temperatures, the scientific community's interest turned to a deeper challenge, that is, to investigate the abrupt discontinuity at 660 km. Early studies were unable to unambiguously characterize structures at such extreme conditions [35]. Different hypotheses were used to investigate the post-spinel phase transition, from candidate structures within the same stoichiometry to decomposition processes. When laser heating in DACs became available, the structural behavior of upper mantle constituents was investigated and in all cases the final stable phase was shown to be perovskite [36]. The debate continued about the transition pressures, as previously explained. Figure 15.5 shows the structural evolution of olivine under high pressure. There is an additional phase transition, from the wadsleyite to the ringwoodite structure (cubic symmetry, S.G. $Fd-3m$). The latter consists of SiO_4 isolated tetrahedra, with Mg and Fe occupying octahedral sites. Wadsleyite and ringwoodite structures can both incorporate water (hydrous phases) constituting more than 2 % in weight, which implies that the transition zone could be a very important water reservoir inside the Earth [23]. It is important to stress once again that the transition pressures can be modified considerably when the effect of impurities or water on the structure is considered. In any case, the associated phase transition is believed to represent the seismic discontinuity at 520 km, although a complex seismic structure with several discontinuities in some areas has been detected [30], indicating the heterogeneous character of this zone. Above 24 GPa, the transformation of $(Mg,Fe)_2SiO_4$ into perovskite and periclase is globally observed (matching the discontinuity at 660 km). Nonetheless, more complex scenarios at the base of the transition zone have been recently proposed, although they are also consistent with phase transformations within a pyrolitic composition [37].

15.3.3 Lower Mantle Composition: Perovskite Structure

The lower mantle is dominated by the perovskite structure, independently of the initial compositional model. The most abundant mineral is the ferrosilicate perovskite (Mg,Fe)SiO_3, which can incorporate trivalent Al as well. Another important mineral in the lower mantle is ferropericlase (or magnesiowustite) (Mg,Fe)O, which is believed to be the second major mineral. Whereas MgO shows a very stable structure (NaCl-type, S.G. $Fm-3m$) all along the mantle [38], FeO undergoes phase transitions under extreme conditions [39], similarly to some solid solution members [40, 41]. $CaSiO_3$ perovskite is also present in smaller amounts, although its structural behavior at this level is still controversial. Some basic concepts about the perovskite structure will be briefly introduced before we discuss structural properties of both silicate systems.

Perovskite structure is very common among ternary compounds in solid state. Apart from the obvious interest from the field of geophysics, interesting applications including giant and colossal magnetoresistance and superconductivity are closely connected to this type of structure. After the discovery of $CaTiO_3$ by G. Rose in 1839 these compounds are known as Perovskites. The term is also used to name different families of compounds of similar structure with different stoichiometries. The basic ABX_3 structure can be rationalized as a three-dimensional network of corner-sharing BX_6 octahedra, with A cations occupying 12 coordinated voids in the ideal non-distorted case (Figure 15.5). In spite of being a very stable structure from the thermodynamical view (this structure is favored at conditions comprising most of the lower mantle), atomic internal displacements inducing different types of distortions in the parent structure or aristotype (cubic symmetry) are commonly present. Thus, several hettotypes or low symmetry structures can be induced by different deformation mechanisms, which can be divided in three groups: tilting of BX_6 octahedra (with or without internal deformation), regular octahedron distortion due to electronic effects (e.g., Jahn–Teller) or stress, and off-center shifts from symmetric positions of both the B and A cations. The latter mechanism is responsible for interesting properties like ferroelectricity [42]. In many cases, more than one type of distortion can be present in a given perovskite structure.

Many efforts have been made aiming to rationalize the perovskite structure. Historically, the first attempt to determine the formability of such a structure was exclusively based on steric considerations, that is, the relationship of ionic radii through a tolerance factor (Goldsmith, 1926 [43]):

$$TF = \frac{r_A + r_X}{\sqrt{2}(r_B + r_X)}, \tag{15.1}$$

where r_A, r_B and r_X determine whether an ABX_3 compound will be stable in the perovskite conformation. $TF = 1$ would correspond to ideal cubic perovskites, whereas values below that would imply a size mismatch accommodated by structural deformations, mainly octahedral tilting. $TF > 1$ corresponds to different types of hexagonal structures made of octahedra that share faces or edges (as well as corners). In spite of being a very simple criterion, it is still used due to its efficiency to predict the structures of novel compounds. An interesting property in the perovskite family is that chemical pressure from cationic substitution seems to be equivalent to the application of extreme conditions, as observed in studies of fluorides and oxides [44]. Another commonly used structural criterion is the ratio of AX_{12} and BX_6 polyhedron volumes, which must be $V_A/V_B = 5$ for the ideal case. Distorted perovskites show lower values and it has been recently shown that there is a value defining the stability limit of tilted perovskites: $V_A/V_B=3.8$ [45]. This is an important result revealing possible destabilization mechanisms in the structure. For the prediction of structural evolution of distorted perovskites, several methods have been introduced. More significant are those based on the bond valence theory [46, 47], which have been used to predict the degree of distortion (variation of the tilting angles) undergone by pressure.

The crystallography of the perovskite structure is well known and has been analyzed in detail in numerous works [48–50]. The cubic aristotype (S.G. $Pm-3m$) is related to several hettotypes of lower symmetry involving tilting distortions, being the following group-subgroup sequence observed most frequently in phase transitions [51]:

$$Pm-3m \rightarrow P4/mbm \rightarrow Pbnm \rightarrow P21/m.$$

In fact, orthorhombic $Pbnm$ (or $Pnma$) is the most common group among perovskites and its distortion can be described as a function of two independent tilting angles. This is the case of $MgSiO_3$ at lower mantle conditions, presenting out-of-phase tilting through two directions and in-phase tilting through the other direction ($a^-a^-c^+$, following Glazer's notation [52]). On the contrary, it is generally believed that $CaSiO_3$ shows a high stability at lower mantle conditions, with no tilting distortions (S.G. $Pm-3m, a^0a^0a^0$).

The structural behavior of perovskites under external pressure is commonly opposite to the high temperature effect, that is, perovskites would evolve towards a higher degree of distortion as pressure increases. However, this behavior is not general in the perovskite family, as previously demonstrated [53]. This could have enormous implications in mantle geophysics, since elastic response depends strongly on the structure and is fundamental in the cases of $MgSiO_3$ and $CaSiO_3$ (representing up to 9 wt% [54, 55]). Actually, there is an open debate on the symmetry of $CaSiO_3$ under extreme conditions. The structure at room temperature is unanimously represented by the tetragonal symmetry. However, at high temperature, different possibilities have been proposed: cubic $Pm-3m$ in most of the cases [56], tetragonal $I4/mcm$ from calculations [57] and, more recently, orthorhombic $Pbnm$ or $Cmcm$ (transition zone conditions) [58]. Apart from experimental difficulties related to *in situ* structural studies at mantle conditions, the $CaSiO_3$ perovskite phase is unquenchable, hampering its unambiguous characterization. Concerning compressional behavior of $MgSiO_3$ perovskite under extreme conditions, contradictory results have been reported over the years. Some early studies indicated an increase of symmetry under pressure (reduction of tilting angles) [59], high structural stability with no significant tilting variation [60], and even a decomposition in simple oxides [61]. This controversy was settled a few years ago, when reliable diffraction experiments could be conducted, leading to the discovery of a new high-pressure phase [62, 63]. Later, a general model based on the bond valence theory applied to numerous orthorhombic perovskites was used to estimate the tilting variation in $MgSiO_3$ under pressure, although it yielded a large uncertainty [47]. In general, the low compressibility of perovskite silicates makes it difficult to model their distortion mechanisms. Therefore, studies on isostructural compounds can provide invaluable information.

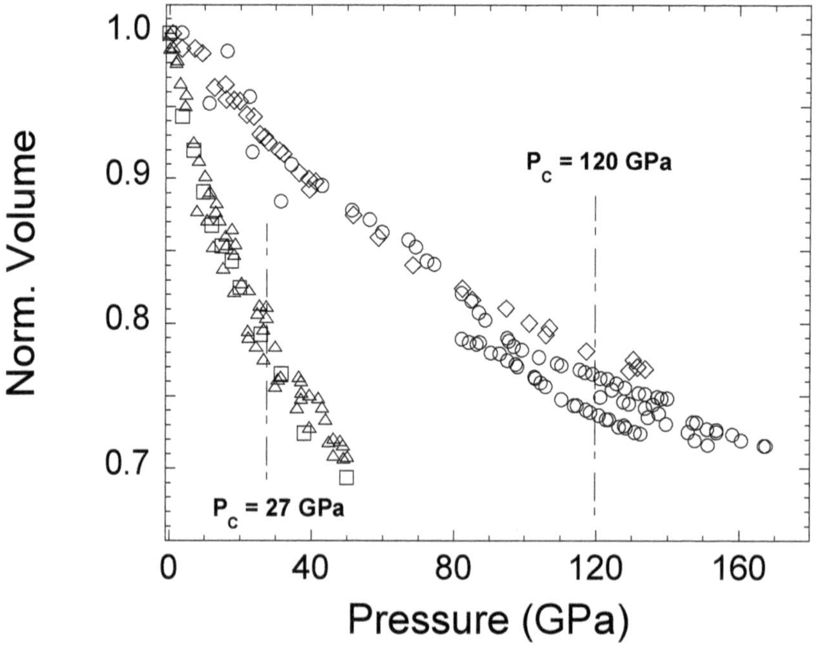

Figure 15.6 Variation of the normalized unit cell volume as a function of pressure for mantle perovskites $MgSiO_3$ [64] (circles) and $CaSiO_3$ [65] (diamonds), as well as the associated soft analogs, $NaMgF_3$ [66] (triangles) and $KMgF_3$ [67] (squares), respectively. The structural stability limit is considerably lower in halides.

Analogs are compounds studied to understand the structural properties of systems of interest in geophysics, that are hardly accessible through experiments at extreme conditions. The precise structural characterization of $MgSiO_3$ perovskite under D" layer conditions required challenging experimental setups until very recently and the associated high pressure phases are also unquenchable. Nevertheless, there are many other isostructural systems which have been investigated to shed light on its structure and compression mechanisms under more favorable experimental conditions. Thus, $AGeO_3$ compounds have been studied for a long time to investigate the elastic properties and structural limits of perovskite silicates, achieving similar compression rates at lower pressures [68, 69]. In addition, ABF_3 fluorides are even more compressible with bulk moduli being typically below 100 GPa, which facilitates the distortion of these structures and the induction of phase transitions. To illustrate the behavior of silicate analogs, the normalized unit cell volume as a function of pressure has been represented in Figure 15.6 for $NaMgF_3$ [66] and $KMgF_3$ [67], as well as for the isostructural oxides: $MgSiO_3$ [64] and $CaSiO_3$ [65], respectively. It can be noted that the volume reduction achieved in oxides at the maximum pressure is reached below 40GPa in both analogs. The stability limit of the $MgSiO_3$ perovskite is met at approximately 120 GPa, whereas for neighborite, $NaMgF_3$, such condition is observed at much lower pressures (P_C= 27 GPa). Above the critical pressure a phase transition towards a denser structure is observed in both compounds, the so-called post-Perovskite phase (PPv).

15.3.4 Post-Perovskite Phase Transition

In 2004 two independent studies by Murakami et al. [62] and Oganov and Ono [63] were published almost simultaneously, demonstrating the existence of a new structural phase transition in $MgSiO_3$ perovskite at lower mantle conditions. First principles calculations and x-ray diffraction experiments proved the stability of a denser structure above 127 GPa and temperatures exceeding 2500 K. This new phase could eventually explain a series of seismic properties observed at the D" layer, providing new insights into the geophysics processes at the lower mantle. Although this phase has been shown to be effective as an explanation for many long-standing problems, there are still some controversial points. This section introduces the most important contributions and some unresolved problems surrounding this fundamental phase transition in geophysics and some basic facts about its structural behavior will be presented.

The PPv structure (also $CaIrO_3$-type, S.G. *Cmcm*) is based on a layer conformation of SiO_6 octahedra, which share corners along one crystallographic direction (*c* axis) and edges along the other (see Figure 15.7). The Pv-PPv transition in $MgSiO_3$ is associated with a volume reduction of ca. 1 to 1.5

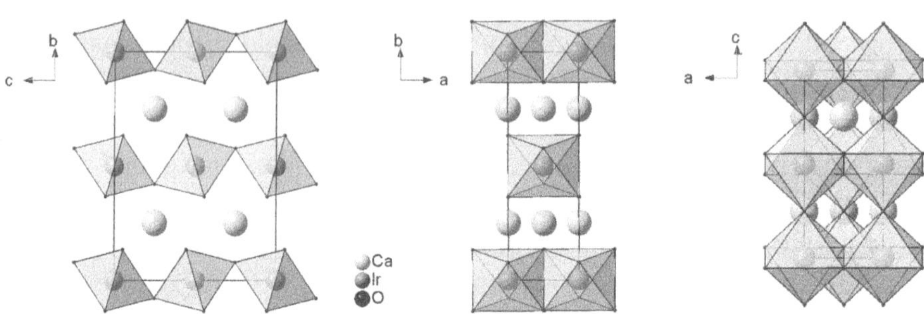

Figure 15.7 View of the post-perovskite (PPv) phase in $CaIrO_3$ along the three crystallographic axes. The structure is formed by layers of IrO_6 octahedra stacked along the b direction, sharing corners along the c axis and edges along the other direction (a axis).

%, with no substantial reduction of the SiO_6 polyhedron's volume. This reduction has also been observed in other geophysically interesting compounds like the (Mg,Fe)SiO_3 solid solution [64], different mantle compositions (e.g., KLB-1 peridotite [70], olivine, pyroxene and garnet [71–73]), as well as in MORB [72, 74]. In addition, many ABF_3 and ABO_3 analogs undergo the phase transition under extreme conditions. Other compounds with similar structures like A_2O_3 sexquioxides are also capable of presenting an equivalent phase (e.g., Fe_2O_3 [75]). Table 15.1 shows a summary of the structural properties of compounds in which a PPv $CaIrO_3$-type structure has been found. The reported data correspond to both *in situ* experiments at high P/T conditions and ambient pressure characterization on quenchable phases. The unique tilting angle associated with the PPv phase (B-X-B) is similar for most compounds while the BX_6 octahedra are quite regular (BX_1/BX_2). The number of systems presenting this phase below 1 Mbar is still scarce, especially in comparison with the total number of known perovskites. At the moment of the discovery of the PPv phase transition there was only one oxide known to retain the same structure at ambient conditions, i.e., $CaIrO_3$. However, later some other Ca-bearing oxides were stabilized in the same phase by synthesis in multivianvil apparatuses, like $CaRuO_3$ [76] or $CaRhO_3$ [77]. Although many oxides and halides present the same layer conformation as $MgSiO_3$ PPv with similar distortion rates (like tolerance factor, lattice parameters ratios and tilting angles), it is not easy to predict whether the high pressure response can be extrapolated from analogs. Thus, in spite of being the prototypical PPv structure, $CaIrO_3$ shows unexpected elastic and mechanical properties [78]. Conversely, $NaMgF_3$ has proven to be a good analog through structural studies under pressure [45, 66] and by first principles calculations [79].

Table 15.1 Structural data for ABX_3 and A_2X_3 compounds crystallizing in a post-perovskite ($CaIrO_3$-type) structure.

Compound	P	a	b	c	V	TF	<B-X>	$B\text{-}X_1/B\text{-}X_2$	B-X-B	Reference
$MgSiO_3$	125	2.456	8.042	6.093	120.34	0.905	1.657	0.988	135.8	[62]
$MgGeO_3$	63	2.613	8.473	6.443	142.65	0.843	1.723	0.987	141.2	[80]
$MnGeO_3$	60	2.703	8.921	6.668	160.79	0.869	-	-	-	[68]
$ZnGeO_3$	126	2.574	8.245	6.385	135.52	0.846	-	-	-	[81]
$CaSnO_3$	56	2.854	9.343	7.090	189.1	0.856	-	-	-	[82]
$NaMgF_3$	54	2.716	8.381	6.849	155.92	0.869	1.812	0.971	149.0	[83]
$NaZnF_3$	19	2.951	9.141	7.236	195.21	0.861	2.028	0.991	135.5	[84, 85]
$NaNiF_3$	10^{-4}	3.025	10.054	7.399	225.01	0.883	2.005	0.995	135.7	[86]
$NaCoF_3$	10^{-4}	3.070	10.129	7.466	232.20	0.859	-	-	-	[86]
$NaFeF_3$	10^{-4}	3.138	10.209	7.459	238.92	0.844	2.073	0.986	130.55	[87]
$CaIrO_3$	10^{-4}	3.144	9.862	7.297	226.30	0.884	2.023	0.962	135.4	[88]
$CaRuO_3$	10^{-4}	3.115	9.827	7.296	223.34	0.887	2.008	0.955	139.0	[76]
$CaPtO_3$	10^{-4}	3.123	9.912	7.346	227.41	0.884	2.017	1.000	131.2	[89]
$CaRhO_3$	10^{-4}	3.101	9.856	7.264	222.03	0.896	1.972	0.972	137.3	[77]
Fe_2O_3	68	2.640	8.544	6.386	144.04	-	-	-	129.1	[75]
Al_2O_3	163	2.402	7.863	5.989	113.12	-	-	-	-	[90]
Mn_2O_3	10^{-4}	2.721	8.901	6.728	162.95	-	-	-	133.5	[91]

Notes: First column indicates the pressure condition (in GPa) for the structural characterization. Unit cell volume and lattice parameters correspond to the *Cmcm* space group (units of Å are used). TF is the tolerance factor for the corresponding perovskite phase calculated from Equation (15.1) and Shannon's radii [92]. The average octahedral bond length, <B-X>, octahedral elongation, $B\text{-}X_1/B\text{-}X_2$, and the B-X-B tilting angle (in degree) are included when available.

A systematic exploration of the transition among perovskite families will eventually provide additional keys to explain the mechanisms governing the transition. Parameters affecting the perovskite destabilization are not fully understood at the moment, so it is not possible to predict conditions for the transition to occur. The tolerance factor and other size-related factors are not well correlated with the transition for all the systems, although some general trends can be observed.

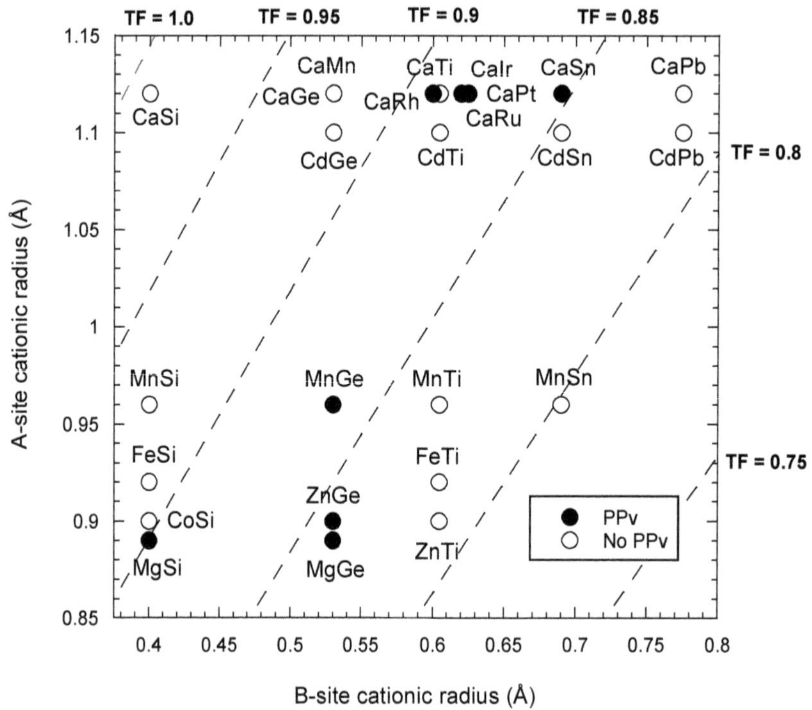

Figure 15.8 ABO_3 perovskite and post-perovskite diagram as a function of A and B ionic radii (according to Shannon [92] for 8 and 6 coordinations, respectively). Diagonal broken lines represent tolerance factors (TF). Compounds presenting post-perovskite structures are represented by solid circles. Note that most of them have TF in the range of 0.85 to 0.9. Detailed information on post-perovskite structures can be found in Table 15.1.

Figure 15.8 shows results of Pv-PPv phase transition studies on ABO_3 perovskites, plotted as a function of ionic radii of A and B cations. Considering only steric factors, it can be seen that most of the compounds in which the transition has been observed present TF values between 0.85 and 0.9. Some studies have revealed a general condition associated with the perovskite instability: the polyhedral volume ratios $V(AX_{12})/V(BX_6)$ together with the anion-anion repulsion have been used to define the stability limits among Pv and PPv phases [45]. The transition found in several ABX_3 systems under

P/T conditions has been accounted for by means of those factors. However, the transition mechanism needs further investigation since ambiguous reports can hamper developments of appropriate models for the transition. Furthermore, it is difficult to extrapolate results to real systems where complex compositions are present. Thus, recent results on (Mg,Fe)SiO_3 solid solution under extreme conditions show symmetries other than $Cmcm$ for particular compositions [93]. Therefore, partitioning of Fe and Al in the Pv and PPv phases plays an important role in the structural properties, affecting the elastic response [94]. Changes in Fe spin state are also fundamental factors, not only in the perovskite, but in other Fe-bearing phases, e.g., (Mg,Fe)O [95]. Finally, it is also necessary to consider different structural scenarios like decomposition that can take place in some perovskites at extreme conditions [96]. The general problem becomes even more intricate when considering seismic evidence concerning the lower mantle, which can be difficult to explain only on the basis of a structural phase transition. Thus, it is not trivial to reconcile seismic observations at a given depth with the physical properties observed under the corresponding pressure and temperature conditions. In any case, the discovery of the $MgSiO_3$-PPv phase led the scientific community to very important advances in the understanding of chemistry and physical processes in the lower mantle.

There are several anomalies associated with the CMB: the seismic discontinuity in D" layer, the polarization anisotropy of S-waves, the anticorrelation between the V_S anomaly and sound velocity V_ϕ , together with the existence of ultra low velocity zones (ULVZ), are the most important features observed. A detailed description of these anomalies is out of the scope of this chapter. However, an analysis of the effectiveness of the PPv transition as an explanation of these features can be found elsewhere [97, 98].

The most important anomaly is the discontinuity appearing between 2600 and 2700 km (119 to 125 GPa approximately), which is associated with an increment of V_P and V_S velocities from 2.4 % to 3 %, although in some zones is not observed for P waves. This has been historically attributed to thermal and chemical variations at that level and it could also be consistent with the PPv transition. However, there are two important issues to unambiguously determine whether the transition is suitable for explaining the seismic features: first, the deviations in pressure determination can imply 3 to 4 GPa or, equivalently, 30 to 75 km in associated depths. Second, variations in the dP/dT value (Clapeyron slope) can range from 5 to 11 MPa/K for $MgSiO_3$ for different pressure scales [99]. In any case, the large positive slope implies that the topography of the D" layer is consistent with the transition being deeper in warmer regions and shallower in colder zones [100].

The polarization anisotropy is a difference in the vertically and horizontally polarized S wave velocities (V_{SV} and V_{SH}, respectively) below the D" layer, ranging from 1 to 3 %. It is possible to correlate the structural changes at this level with this and the previous anomaly, as a result of the preferred orientation in the layered PPv structure. Although it is plausible to explain the polarization anisotropy (Pv structure would be incompatible), there is still an open debate on whether it would be enough alone to account for the observed differences based on possible deformation mechanisms [101, 102].

The anticorrelation between V_s and V_ϕ (bulk sound) velocities is only observed at the lower mantle. Originally this feature was explained by models based on the chemical heterogeneities in the mantle (not only temperature), but the PPv phase transition also offers a plausible explanation since it provides different velocities from those of the Pv structure [103]. Thus, the negative correlation could be explained by differences in Pv/PPv ratios at equivalent depths, which would be related to temperature variations at the lower mantle.

The ULVZs are regions less than 50 km thick showing a drastic reduction of wave propagation velocities. This phenomenon has been explained by partial melting of silicates in contact with the core [104]. An alternative explanation is the reduction of velocity when Fe is incorporated into the PPv structure [105]. In general, this is an intricate problem since the structure of the interface is supposed to be a complex region where different heterogeneities and chemical reactions can also take place. In this regard, it is important to validate thermal models, some of which even postulate a double crossing in the Pv-PPv transition, leading to stabilization of the former (Pv) at the base of the mantle. This would be compatible with a double seismic anomaly at the D" layer, which has been detected in some regions [106].

15.3.5 Unresolved Issues and Future Research

Current investigation of the mantle is associated with several unanswered questions, especially those related to structures and processes in the vicinity of the CMB. One outstanding issue that needs to be addressed is the unification of pressure scales for the precise and univocal determination of phenomena at the lower mantle. Concerning structural responses of materials of interest in the lowermost mantle, different topics need to be thoroughly investigated, starting with the Pv-PPv phase transition in more perovskites. The exploration of wide pressure and temperature ranges is necessary to gain deeper knowledge of the stability criteria and mechanisms governing the PPv transition among perovskites, facilitating the prediction of the structural phase transition at given conditions. This would help to shed light on the structure and mechanical properties of complex systems present at the CMB. In addition, some future technical developments could help to obtain valuable information, e.g., new designs of large volume cells to reach lower mantle pressures, and more uniform sample heating and precise temperature measurement inside the DAC. Other questions, like the partitioning of Fe and Al among several phases (Pv, PPv and periclase) considering different Fe spin states, possible disproportionation and decomposition in fundamental compounds, and the structural behavior of other minor mantle constituents under extreme conditions require also further investigation.

In spite of the significance of the PPv phase transition for the understanding of processes at the lowermost mantle, it is still essential to determine to what extent seismic data can be explained only by means of this structural transition [100]. It is necessary to determine the compatibility of observations corresponding to a given depth with the seismic features found in the mantle, which is proven to be a very complex zone from recent characterizations [107]. Some seismic anomalies, like the large low shear wave velocity provinces (LLSVPs) [108] or the ULVZs occurring at different depths need to be correctly interpreted considering not only the PPv but also possible compositional (and thermal) variations. Factors affecting anisotropy (which is considerably higher in PPv than Pv) and the effects of chemical heterogeneities on the seismic properties must be also analyzed with more detail.

15.4 COMPOSITION AND STRUCTURE OF EARTH'S CORE

The core comprises the innermost parts of the Earth and it is known to be formed by a solid inner core with an approximate radius of 1250 km surrounded by a liquid outer core that extents to a radius of 3400 km. Earth's core was discovered by R.D. Oldham in 1906 [109]. His paper on the transmission of seismic waves demonstrated the existence of the Earth's core as well as the value of Earthquake data for studying the structure of the deep interior of our planet. Further seismic analyses allowed us to deduce the core boundary [110], to demonstrate that part of the core is liquid [111], discover the existence of the inner core [112], and infer its solidity [113].

Geomagnetism and chemical considerations, like the comparison with chrondrite composition, indicate that iron should be the main constituent of the Earth's core. This fact was firmly established as a result of the analyses of mass density/sound-wave systematics [114]. However, the density difference between pure Fe and that suggested by seismic models requires the presence of elements lighter that iron in the core. The amounts of light elements in the outer core were first estimated by Birch to be 10 wt% [115]. This percentage of light elements is within the currently accepted 6 to 10 % range in the

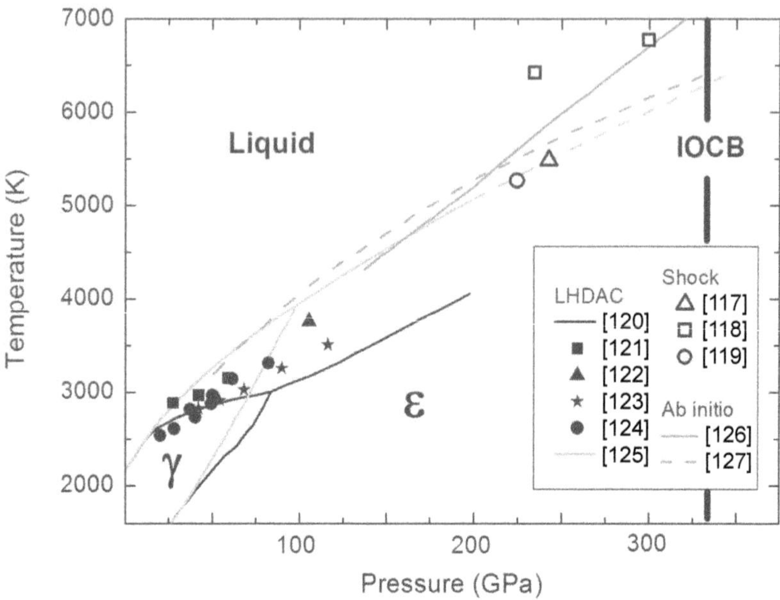

Figure 15.9 *P-T* phase diagram of iron. The melting curves of six LHDAC experiments, three shock studies and three *ab initio* melting calculations are shown. Based on these results (or corresponding extrapolations), the melting temperatures at the inner-outer core boundary could range from 5000 K to 7000 K.

outer core, while the solid inner core is thought to be slightly less dense than crystalline Fe, also containing minor amounts of Ni (~5 %) [116].

The following sections give an overview of our present knowledge of the Earth's core structure. In particular, we will focus on laboratory experiments at high-pressure and -temperature conditions which simulate how these thermodynamic variables affect the physical behavior of elemental iron and iron-rich alloys. The characterization of matter under these conditions and the results obtained from other indirect methods (seismic data, heat flux, etc.), seek to constraint the chemical composition and structure of the core of our planet. In-depth reviews of recent experiments and seismological observations are published elsewhere [117, 118].

15.4.1 Phase Diagram of Iron

Understanding the Earth's thermal gradient and thermal history, as well as the seismic density and structure of the core, requires a thorough knowledge of the phase behaviour, equation of state and melting temperature of Fe at the inner-outer core boundary (IOCB) at 330 GPa.

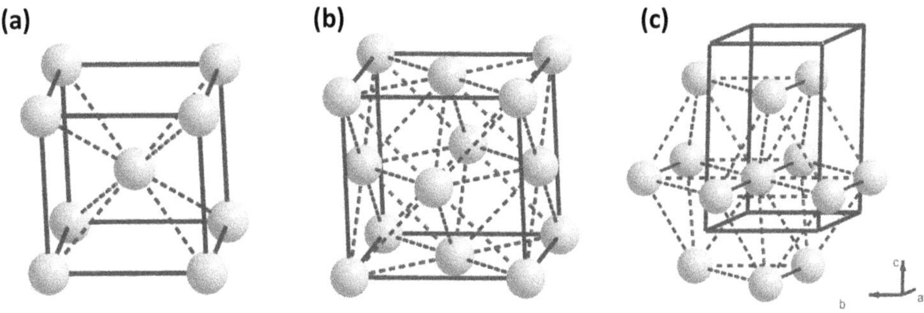

Figure 15.10 (a) Body-centered cubic (bcc, α) phase of iron, stable at ambient conditions. (b) Hexagonal close-packed (hcp, ϵ) phase, stable at IOCB conditions according to most studies. (c) Face-centered cubic (fcc, γ) phase of iron. Unit cells are depicted as black solid lines. To illustrate the coordination of Fe atoms in these structures (bcc = 8 Fe neighbors, hcp and fcc = 12 Fe neighbors), Fe-Fe contacts are represented as dashed lines.

The IOCB temperature is bracketed between the melting temperature of pure Fe and the liquidus temperature of the outer core iron-rich alloy (expected to be depressed ~900 K, as discussed below). Melting temperatures of pure Fe determined by shock velocity (dynamic) [119–121], laser-heated DAC measurements (LHDAC, static) [122–127], and thermodynamical modelling [128–130], do not present a complete consensus, with temperatures ranging between 5000 K [122] and 7000 K [113, 128] (see Figure 15.9). Differences among studies come from different uncertainty sources: (i) the possibility of superheating in shockwave experiments, (ii) pyrometry measurements and the criterion used to identify melting in LHDACs, and (iii) the underlying approximations and assumptions in theoretical calculations. More recent studies suggest a temperature of 6230 ± 500 K at the IOCB region [127, 130]. Moreover, the temperature of the CMB can be estimated using this value to be approximately 4100 K [127], the temperature at which partial melting of mantle material has been observed [131].

Regarding the solid part of the phase diagram of elemental Fe, only three phases are known. At ambient conditions, iron has a body-centered cubic (bcc, α) structure (Figure 15.10a). At 13 GPa and room temperature, it transforms into a hexagonal close-packed (hcp, ϵ) structure which is stable up to 300 GPa [132] (Figure 15.10b). At high temperatures and pressures below 70 GPa, the face-centered cubic phase (fcc, γ) is also observed (Figure 15.10c). The existence of a β phase at moderately high pressures and temperatures with an orthorhombic-distorted or doubled-hcp structures has also been reported [133, 134], but it is likely that this distorted phase is due to deviatoric stresses

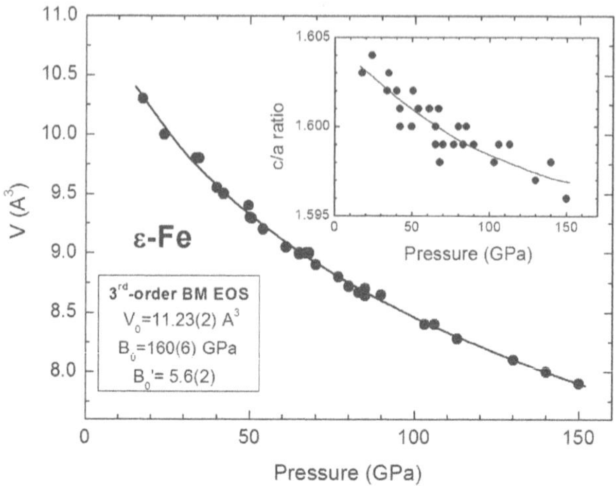

Figure 15.11 Pressure dependence of the volume per atom for hcp-Fe phase under compression. Experimental data from Reference [125] and the fitted third-order Birch–Murnaghan EOS are represented by black circles and solid line, respectively. The zero-pressure volume, the bulk modulus, and its first pressure derivative are collected in the bottom-left box. (Inset) Evolution of the c/a lattice ratio of the hcp-Fe phase with pressure. As can be seen, the contraction of the lattice parameters is slightly anisotropic.

in the DAC. Recently, Tateno et al. reported x-ray diffraction measurements of pure Fe up to 377 GPa and 5700 K, demonstrating that the hcp structure is the stable form of iron in the entire inner core *P-T* range [135]. On the other hand, theoretical calculations show very small energy differences among the hcp, bcc, and fcc phases at core conditions and some studies predict the stability of these three structural types [136–138]. Most recent calculations support Tateno's results [138].

15.4.2 Anisotropy of Solid Inner Core

Compressional seismic waves in the north-south or polar direction travel faster than waves in the equatorial plane [139]. Recent studies indicate up to 4.4 % and 1 % anisotropy in the western and eastern hemispheres, respectively, with an almost isotropic 60 km-deep region at the top of the inner core surrounding a more anisotropic region and, possibly, an innermost inner core with even different properties [118].

Seismic anisotropy observed at the global scale is the signature of single-crystal anisotropy of the solid material in the inner core, that is, pure iron plus small amounts of light elements, not clearly identified. The equation of state of pure iron reported by different research groups is in good agreement, giving a bulk modulus of ∼160 GPa [125, 140], as shown in Figure 15.11. These recent results also indicate that the compressibility of the lattice parameters of hcp-Fe at room temperature is slightly anisotropic, the c/a ratio decreasing with pressure [125, 140] (see inset in Figure 15.11). On the other hand, c/a increases with temperature [125, 128] and suggests that at the core P-T conditions this ratio is close to that at room conditions. In addition to the observed axial anisotropy, dynamics simulations predict a strong non-linear behavior of the elastic properties of ϵ-Fe at 360 GPa just before melting [141]. This strongly non-linear effect in Fe could provide enough anisotropy to match seismic observations of the inner core. Anisotropy at the single-crystal scale may also be increased by including the effect of elements other than Fe. Varying compositional stratification and solidification textures may make it possible to match seismic observations [142].

15.4.3 Light Elements in Core

As mentioned above, the addition of elements lighter than iron is needed to account for the observed seismic propagation in Earth's core [115]. On top of that, there exists a density jump across the IOCB as large as 640 kg/m^3, a value that cannot be reconciled only by the density difference between solid and liquid (∼220 kg/m^3) and therefore requires enrichment in light elements in the outer core relative to the inner core [117]. The list of potential light elements in the outer core includes S, Si, O, C, and H [116]. C and H are expected to be strongly depleted in Earth and are less likely to be incorporated in the core due to their high volatility. The possible presence of the other three elements is supported by (i) the metal-silicate element partitioning at the bottom of the mantle and (ii) the high content of S in planetesimal cores. Numerous high P/T studies on Fe-rich alloys have been performed in the past 20 years and some conclusions can be drawn. First, the core likely contains ∼6 wt% Si and ∼ 1–2 wt% S [143], data based on geochemical constraints. Secondly, from *ab initio* partitioning calculations, similar amounts of these two elements seem to be present in coexisting solid and liquid iron, indicating that neither of them can account for the density jump at the IOCB [143]. Finally, theoretical simulations point to a ∼4 wt% O in the outer core to explain such density contrast [143].

P-T phase diagrams on Fe-rich alloys could shed some light into the possible structure of the core, but literature shows some controversial data. Further studies are therefore needed. Hereafter, we will summarize most recent results obtained in high-pressure high-temperature static experiments:

- Fe - 10 wt% Ni alloy was reported to be stable in the bcc structure at 225 GPa and 3400 K [144] but later experiments did not observe this phase up to 340 GPa and 4700 K and supported the hcp structure for Ni content $\leq$10 % [145].

- Studies on Fe - 4 wt% to 10 wt% Si alloys suggested that they adopt the hcp structure at inner core conditions [146, 147].
- The solubility of S in hcp-Fe increases with pressure and points to the fact that a small percentage in weight of this element could be accommodated in the inner core [148].

Ni, Si, and S form continuous solid solutions with iron in a wide compositional range, and liquid-solid partitioning indicates that similar amounts of these elements would be present at inner and outer core. Oxygen, in contrast, does not mix with iron at all. Recent experiments in the Fe + FeO mixture up to 197 GPa and 3600 K [149] showed that (i) they form a binary eutectic system in the whole P-T range, and (ii) Fe-richer compounds than FeO were not formed, this latter adopting the CsCl-type B2 structure above 240 GPa and 4000 K [150].

It must be highlighted that light element partitioning is responsible for the depression of about 900 K of the melting temperature of the core mixture with respect to the melting temperature of pure Fe [143], which would produce an IOCB temperature of about 5300 K, in a similar way as sea water freezing temperature decreases well below zero when salt is dissolved into it. More accurate experimental and theoretical studies are needed to determine the composition of light elements in the core and to determine the chemical and physical properties of the corresponding iron alloys at $P - T$ conditions of Earth's interior.

15.4.4 Remaining Questions and Future Challenges

A number of seismically observed features have yet not been explained. These include both the polar-equatorial and the suggested external-internal anisotropy (innermost region) of the inner core. The inner core displays lateral variations and the western and eastern hemispheres show different propagation velocities. Another interesting question to be answered is the superrotation rate of the inner core, if it exists. For more details on mechanisms proposed to explain core heterogeneity, the reader is referred to Reference [118]. To address the study of these and other uncertainties, improved viscosity measurements and compressional velocity measurements are required.

Bibliography

[1] J. Poirier. *Introduction to the Physics of the Earth's Interior.* Cambridge University Press (2000).

[2] A. Jayaraman. *Rev. Mod. Phys.*, 55, 65 (1983).

[3] J. Jamieson, A. Lawson, and N. Nachtrieb. *Rev. Sci. Instrum.*, 30, 1016 (1959).

[4] C. Weir, E. Lippincott, A. Van Valkenburg, *et al.* *J. Res. Natl. Bur. Stand. A*, 63, 55 (1959).

[5] V. Pecharsky and P. Zavalij. *Fundamentals of Powder Diffraction and Structural Characterization of Materials.* Springer (2005).

[6] R. Young and R. Young. *The Rietveld Method.* IUCr monographs on crystallography. Oxford University Press (1995).

[7] E. Boldyreva and P. Dera. *High-Pressure Crystallography: From Fundamental Phenomena to Technological Applications.* NATO Science for Peace and Security Series B: Physics and Biophysics. Springer (2010).

[8] W. Holzapfel and N. Isaacs. *High Pressure Techniques in Chemistry and Physics: A Practical Approach.* Practical approach in chemistry series. Oxford University Press (1997).

[9] C.-S. Zha and W. Bassett. *Rev. Sci. Instrum.*, 74, 1255 (2003).

[10] M. Akaishi and M. Z. Kenkyūjo. *Advanced Materials '96: New Trends in High Pressure Research: Proceedings of 3rd NIRIM International Symposium on Advanced Materials, Tsukuba, Japan, March 4-8, 1996.* The Institute (1996).

[11] M. Manghnani and Y. Syono. *High Pressure Research in Mineral Physics.* Geophysical monograph. Terra Scientific Publishing Company (1987).

[12] T. Tsuchiya. *J. Geophys. Res. B Solid Earth*, 108, ECV 1 (2003).

[13] S. Speziale, C.-S. Zha, T. Duffy, *et al. J. Geophys. Res. B Solid Earth*, 106, 515 (2001).

[14] S. Akimoto and M. Manghnani. *High-Pressure Research in Geophysics.* Advances in Earth and Planetary Sciences. Center for Academic Publications Japan (1982).

[15] N. Holmes, J. Moriarty, G. Gathers, *et al. J. Appl. Phys.*, 66, 2962 (1989).

[16] O. Anderson, D. Isaak, and S. Yamamoto. *J. Appl. Phys.*, 65, 1534 (1989).

[17] T. Irifune, N. Nishiyama, K. Kuroda, *et al. Science*, 279, 1698 (1998).

[18] S.-H. Shim, T. S. Duffy, and G. Shen. *Nature*, 411, 571 (2001).

[19] Y. Fei, J. Li, K. Hirose, *et al. Phys. Earth Planet. Inter.*, 143, 515 (2004).

[20] T. Watanuki, O. Shimomura, T. Yagi, *et al. Rev. Sci. Instrum.*, 72, 1289 (2001).

[21] J.-F. Lin, W. Sturhahn, J. Zhao, *et al. Geophys. Res. Lett.*, 31, L14611 1 (2004).

[22] S. Deemyad, A. Papathanassiou, and I. Silvera. *J. Appl. Phys.*, 105, 093543 (2009).

[23] E. Ohtani and T. Sakai. *Phys. Earth Planet. Inter.*, 170, 240 (2008).

[24] L. Stixrude and C. Lithgow-Bertelloni. *Earth Planet. Sci. Lett.*, 263, 45 (2007).

[25] W. McDonough and S.-S. Sun. *Chem. Geol.*, 120, 223 (1995).

[26] S. Dorfman and T. Duffy. *Geophys. J. Int.*, 197, 910 (2014).

[27] T. Irifune and A. Ringwood. *Earth Planet. Sci. Lett.*, 117, 101 (1993).

[28] A. Dziewonski and D. Anderson. *Phys. Earth Planet. Inter.*, 25, 297 (1981).

[29] B. Kennett, E. Engdahl, and R. Buland. *Geophys. J. Int.*, 122, 108 (1995).

[30] A. Deuss and J. Woodhouse. *Science*, 294, 354 (2001).

[31] J. Birle, G. Gibbs, P. Moore, *et al.* *Am. Mineral.*, 53, 807 (1968).

[32] V. M. Goldschmidt. *Nachrichten von der Gesellschaft der Wissenschaften zu Göttingen, Mathematisch-Physikalische Klasse*, 1, 184 (1931).

[33] S.-I. Akimoto and H. Fujisawa. *Earth Planet. Sci. Lett.*, 1, 237 (1966).

[34] S.-I. Akimoto. *Tectonophysics*, 13, 161 (1972).

[35] A. E. Ringwood. *Composition and Petrology of the Earth's Mantle.* McGraw-Hill (1975).

[36] L.-G. Liu. *Geophys. Res. Lett.*, 2, 417 (1975).

[37] A. Deuss, S. Redfern, K. Chambers, *et al.* *Science*, 311, 198 (2006).

[38] T. Duffy, R. Hemley, and H.-K. Mao. *Phys. Rev. Lett.*, 74, 1371 (1995).

[39] M. Murakami, K. Hirose, S. Ono, *et al.* *Phys. Earth Planet. Inter.*, 146, 273 (2004).

[40] J.-F. Lin, D. Heinz, H.-k. Mao, *et al.* *Proc. Natl. Acad. Sci. U.S.A.*, 100, 4405 (2003).

[41] Y. Fei, L. Zhang, A. Corgne, *et al.* *Geophys. Res. Lett.*, 34, L17307 (2007).

[42] H. Megaw. *Ferroelectricity in Crystals.* Methuen (1957).

[43] V. Goldschmidt. *Die Naturwissenschaften*, 14, 477 (1926).

[44] Y. Syono, S.-I. Akimoto, and K. Kohn. *J. Phys. Soc. Jpn.*, 26, 993 (1969).

[45] C. Martin and J. Parise. *Earth Planet. Sci. Lett.*, 265, 630 (2008).

[46] J. Zhao, N. Ross, and R. Angel. *Acta Crystallogr. B: Struct. Sci.*, 60, 263 (2004).

[47] J. Zhao, N. Ross, and R. Angel. *Acta Crystallogr. B Struct. Sci.*, 62, 431 (2006).

[48] A. Glazer. *Acta Crystallogr. B: Struct. Sci.*, 28, 3384 (1972).

[49] K. Aleksandrov and V. Beznosikov. *Phys. Solid State*, 39, 695 (1997).

[50] C. Howard and H. Stokes. *Acta Crystallogr. A Found. Crystallogr.*, 61, 93 (2005).

[51] C. Howard and H. Stokes. *Acta Crystallogr. B Struct. Sci.*, 54, 782 (1998).

[52] A. M. Glazer. *Acta Crystallogr. A Found. Crystallogr.*, 31, 756 (1975).

[53] R. Angel, J. Zhao, and N. Ross. *Phys. Rev. Lett.*, 95, 025503 (2005).

[54] T. Irifune. *Nature*, 370, 131 (1994).

[55] S. Kesson, J. Fitz Gerald, and J. Shelley. *Nature*, 393, 252 (1998).

[56] S. Ono, Y. Ohishi, and K. Mibe. *Am. Mineral.*, 89, 1480 (2004).

[57] L. Li, D. Weidner, J. Brodholt, *et al. Phys. Earth Planet. Inter.*, 155, 249 (2006).

[58] T. Uchida, Y. Wang, N. Nishiyama, *et al. Earth Planet. Sci. Lett.*, 282, 268 (2009).

[59] Y. Wang, F. Guyot, A. Yeganeh-Haeri, *et al. Science*, 248, 468 (1990).

[60] L. Stixrude and R. Cohen. *Nature*, 364, 613 (1993).

[61] S. Saxena, L. Dubrovinsky, P. Lazor, *et al. Eur. J. Mineral.*, 10, 1275 (1998).

[62] M. Murakami, K. Hirose, K. Kawamura, *et al. Science*, 304, 855 (2004).

[63] A. Oganov and S. Ono. *Nature*, 430, 445 (2004).

[64] S.-H. Shim. *Annu. Rev. Earth Planet. Sci.*, 36, 569 (2008).

[65] H. K. Mao, L. Chen, R. Hemley, *et al. J. Geophys. Res. Solid Earth*, 94, 17889 (1989).

[66] H.-Z. Liu, J. Chen, J. Hu, *et al. Geophys. Res. Lett.*, 32, 1 (2005).

[67] F. Aguado, F. Rodriguez, S. Hirai, *et al. High Pressure Res.*, 28, 539 (2008).

[68] S. Tateno, K. Hirose, N. Sata, *et al. Phys. Chem. Miner.*, 32, 721 (2006).

[69] A. Kubo, B. Kiefer, S.-H. Shim, *et al. Am. Mineral.*, 93, 965 (2008).

[70] M. Murakami, K. Hirose, N. Sata, *et al.* *Geophys. Res. Lett.*, 32, L03304 (2005).

[71] W. L. Mao, G. Shen, V. B. Prakapenka, *et al.* *Proc. Natl. Acad. Sci. U.S.A.*, 101, 15867 (2004).

[72] B. Grocholski, K. Catalli, S.-H. Shim, *et al.* *Proc. Natl. Acad. Sci. U.S.A.*, 109, 2275 (2012).

[73] S. Tateno, K. Hirose, N. Sata, *et al.* *Geophys. Res. Lett.*, 32, L15306 (2005).

[74] K. Hirose, N. Takafuji, N. Sata, *et al.* *Earth Planet. Sci. Lett.*, 237, 239 (2005).

[75] S. Ono and Y. Ohishi. *J. Phys. Chem. Solids*, 66, 1714 (2005).

[76] H. Kojitani, Y. Shirako, and M. Akaogi. *Phys. Earth Planet. Inter.*, 165, 127 (2007).

[77] Y. Shirako, H. Kojitani, M. Akaogi, *et al.* *Phys. Chem. Miner.*, 36, 455 (2009).

[78] L. Miyagi, N. Nishiyama, Y. Wang, *et al.* *Earth Planet. Sci. Lett.*, 268, 515 (2008).

[79] K. Umemoto, R. Wentzcovitch, D. Weidner, *et al.* *Geophys. Res. Lett.*, 33, L15304 (2006).

[80] K. Hirose, K. Kawamura, Y. Ohishi, *et al.* *Am. Mineral.*, 90, 262 (2005).

[81] H. Yusa, T. Tsuchiya, M. Akaogi, *et al.* *Inorg. Chem.*, 53, 11732 (2014).

[82] S. Tateno, K. Hirose, N. Sata, *et al.* *Phys. Earth Planet. Inter.*, 181, 54 (2010).

[83] C. Martin, W. Crichton, H. Liu, *et al.* *Am. Mineral.*, 91, 1703 (2006).

[84] S. Yakovlev, M. Avdeev, and M. Mezouar. *J. Solid State Chem.*, 182, 1545 (2009).

[85] M. Akaogi, Y. Shirako, H. Kojitani, *et al.* *Phys. Earth Planet. Inter.*, 228, 160 (2014).

[86] D. Dobson, S. Hunt, A. Lindsay-Scott, *et al.* *Phys. Earth Planet. Inter.*, 189, 171 (2011).

[87] F. L. Bernal, K. V. Yusenko, J. Sottmann, *et al.* *Inorg. Chem.*, 53, 12205 (2014).

[88] S. Hirai, M. Welch, F. Aguado, *et al.* *Zeitschr. Kristallogr.*, 224, 345 (2009).

[89] Y. Inaguma, K.-I. Hasumi, M. Yoshida, *et al.* *Inorg. Chem.*, 47, 1868 (2008).

[90] A. Oganov and S. Ono. *Proc. Natl. Acad. Sci. U.S.A.*, 102, 10828 (2005).

[91] J. Santillan, S.-H. Shim, G. Shen, *et al.* *Geophys. Res. Lett.*, 33, L15307 (2006).

[92] R. D. Shannon. *Acta Crystallogr. A Found. Crystallogr.*, 32, 751 (1976).

[93] T. Yamanaka, K. Hirose, W. Mao, *et al.* *Proc. Natl. Acad. Sci. U.S.A.*, 109, 1035 (2012).

[94] Z. Mao, J.-F. Lin, J. Yang, *et al.* *Earth Planet. Sci. Lett.*, 403, 157 (2014).

[95] K. Fujino, D. Nishio-Hamane, T. Nagai, *et al.* *Phys. Earth Planet. Inter.*, 228, 186 (2014).

[96] L. Zhang, Y. Meng, W. Yang, *et al.* *Science*, 344, 877 (2014).

[97] J. Wookey, S. Stackhouse, J.-M. Kendall, *et al.* *Nature*, 438, 1004 (2005).

[98] K. Hirose, J. Brodholt, T. Lay, *et al.* *Post-Perovskite: The Last Mantle Phase Transition*, Volume 174. John Wiley & Sons (2013).

[99] K. Hirose, R. Sinmyo, N. Sata, *et al.* *Geophys. Res. Lett.*, 33, L01310 (2006).

[100] K. Hirose, S.-I. Karato, V. Cormier, *et al.* *Geophys. Res. Lett.*, 33, L12S01 (2006).

[101] A. Oganov, R. Martonak, A. Laio, *et al.* *Nature*, 438, 1142 (2005).

[102] S. Merkel, A. Kubo, L. Miyagi, *et al.* *Science*, 311, 644 (2006).

[103] R. M. Wentzcovitch, T. Tsuchiya, and J. Tsuchiya. *Proc. Natl. Acad. Sci. U.S.A.*, 103, 543 (2006).

[104] M. Gurnis. *The Core-Mantle Boundary Region*, Volume 28 of Geodynamics series. American Geophysical Union (1998).

[105] W. Mao, H.-K. Mao, W. Sturhahn, *et al. Science*, 312, 564 (2006).

[106] J. Hernlund, C. Thomas, and P. Tackley. *Nature*, 434, 882 (2005).

[107] X. Shang, S.-H. Shim, M. de Hoop, *et al. Proc. Natl. Acad. Sci. U.S.A.*, 111, 2442 (2014).

[108] E. J. Garnero and A. K. McNamara. *Science*, 320, 626 (2008).

[109] R. D. Oldham. *Q. J. Geol. Soc.*, 62, 456 (1906).

[110] B. Gutenberg. *Phys. Z.*, 14, 1217 (1913).

[111] H. Jeffreys. *Geophys. J. Int.*, 1, 385 (1926).

[112] I. Lehmann. *Publ. Bur. Centr. Seism. Int. A*, 14, 87 (1936).

[113] A. Dziewonski and F. Gilbert. *Nature*, 234, 465 (1971).

[114] A. F. Birch. *Am. J. Sci.*, 238, 192 (1940).

[115] F. Birch. *J. Geophys. Res.*, 69, 4377 (1964).

[116] J.-P. Poirier. *Phys. Earth Planet. Inter.*, 85, 319 (1994).

[117] K. Hirose, S. Labrosse, and J. Hernlund. *Annu. Rev. Earth Planet. Sci.*, 41, 657 (2013).

[118] A. Deuss. *Annu. Rev. Earth Planet. Sci.*, 42, 103 (2014).

[119] J. M. Brown and R. G. McQueen. *J. Geophys. Res. Solid Earth*, 91, 7485 (1986).

[120] C. Yoo, N. Holmes, M. Ross, *et al. Phys. Rev. Lett.*, 70, 3931 (1993).

[121] J. H. Nguyen and N. C. Holmes. *Nature*, 427, 339 (2004).

[122] R. Boehler. *Nature*, 363, 534 (1993).

[123] G. Shen, V. B. Prakapenka, M. L. Rivers, *et al. Phys. Rev. Lett.*, 92, 185701 (2004).

[124] Y. Ma, M. Somayazulu, G. Shen, *et al. Phys. Earth Planet. Inter.*, 143, 455 (2004).

[125] R. Boehler, D. Santamaría-Pérez, D. Errandonea, *et al. J. Phys. Conf. Series*, 121, 022018 (2008).

[126] J. M. Jackson, W. Sturhahn, M. Lerche, *et al. Earth Planet. Sci. Lett.*, 362, 143 (2013).

[127] S. Anzellini, A. Dewaele, M. Mezouar, *et al. Science*, 340, 464 (2013).

[128] A. B. Belonoshko, R. Ahuja, and B. Johansson. *Phys. Rev. Lett.*, 84, 3638 (2000).

[129] D. Alfè, G. Price, and M. Gillan. *Phys. Rev. B*, 65, 165118 (2002).

[130] D. Alfè. *Phys. Rev. B*, 79, 060101 (2009).

[131] G. Fiquet, A. Auzende, J. Siebert, *et al. Science*, 329, 1516 (2010).

[132] H. Mao, Y. Wu, L. Chen, *et al. J. Geophys. Res. Solid Earth*, 95, 21737 (1990).

[133] S. Saxena, L. Dubrovinsky, P. Häggkvist, *et al. Science*, 269, 1703 (1995).

[134] D. Andrault, G. Fiquet, M. Kunz, *et al. Science*, 278, 831 (1997).

[135] S. Tateno, K. Hirose, Y. Ohishi, *et al. Science*, 330, 359 (2010).

[136] A. B. Belonoshko, R. Ahuja, and B. Johansson. *Nature*, 424, 1032 (2003).

[137] A. S. Mikhaylushkin, S. Simak, L. Dubrovinsky, *et al. Phys. Rev. Lett.*, 99, 165505 (2007).

[138] L. Stixrude. *Phys. Rev. Lett.*, 108, 055505 (2012).

[139] A. Morelli, A. M. Dziewonski, and J. H. Woodhouse. *Geophys. Res. Lett.*, 13, 1545 (1986).

[140] A. Dewaele, P. Loubeyre, F. Occelli, *et al. Phys. Rev. Lett.*, 97, 215504 (2006).

[141] B. Martorell, L. Vočadlo, J. Brodholt, *et al. Science*, 342, 466 (2013).

[142] A. Lincot, R. Deguen, S. Merkel, *et al. C. R. Geosci.*, 346, 148 (2014).

[143] D. Alfè, M. Gillan, and G. Price. *Contemp. Phys.*, 48, 63 (2007).

[144] L. Dubrovinsky, N. Dubrovinskaia, O. Narygina, *et al. Science*, 316, 1880 (2007).

[145] S. Tateno, K. Hirose, T. Komabayashi, *et al. Geophys. Res. Lett.*, 39, 12305 (2012).

[146] H. Asanuma, E. Ohtani, T. Sakai, *et al. Geophys. Res. Lett.*, 35, 12307 (2008).

[147] Y. Kuwayama, T. Sawai, K. Hirose, *et al. Phys. Chem. Miner.*, 36, 511 (2009).

[148] S. Kamada, H. Terasaki, E. Ohtani, *et al. Earth Planet. Sci. Lett.*, 294, 94 (2010).

[149] H. Ozawa, K. Hirose, S. Tateno, *et al. Phys. Earth Planet. Inter.*, 179, 157 (2010).

[150] H. Ozawa, F. Takahashi, K. Hirose, *et al. Science*, 334, 792 (2011).

CHAPTER 16

Interiors of Icy Moons from an Astrobiology Perspective: Deep Oceans and Icy Crusts

Olga Prieto-Ballesteros

MALTA Consolider Team and Centro de Astrobiología, INTA-CSIC, Torrejón de Ardoz, Spain

Victoria Muñoz-Iglesias

MALTA Consolider Team and Centro de Astrobiología, INTA-CSIC, Torrejón de Ardoz, Spain

Laura J. Bonales

MALTA Consolider Team and CIEMAT, Madrid, Spain

CONTENTS

16.1 INTRODUCTION

THE study of the interior of the icy moons of our solar system is stimulated by the evidence of large aqueous reservoirs buried tens or hundreds of kilometers below their surfaces. These deep reservoirs may form global oceans between two layers of different ice phases, constitute local aqueous chambers within the icy crust, or be in contact with the rock layers underneath. All these scenarios represent promising habitable environments that ought to be considered in future astrobiological exploration.

The emergence of deep aqueous environments extends the classical concept of the habitable zone around a star, which, in principle, takes into account just the planetary surfaces that may retain liquid water (H_2O). Today, the requirements for a habitable environment, established by the astrobiology community, are the presence of(1) liquid water, (2) essential chemical elements (carbon, hydrogen, oxygen, nitrogen, phosphorus, and sulfur), and (3) energy available for life. Due to this fact, the assessment of the habitability of icy moons is highly prioritized in the roadmaps of the main space agencies. This is illustrated by the recent selection of the JUpiter and ICy moons Explorer (JUICE) mission by the European Space Agency (ESA), which will be launched in 2022 [1]. The exploration of deep environments is challenging because they are technologically inaccessible for direct sampling, at least at present. While we wait for new data from future planetary missions, high-pressure experiments on ices and aqueous systems are necessary to make progress in the understanding of the hidden environments within these planetary bodies.

Our present knowledge about the interiors of the icy moons is based on the information provided by three sources: space missions and ground telescopes (e.g., gravity field measurements), laboratory experiments, and theoretical modeling. Direct information about the moons of the outer solar system mainly comes from the space missions Voyager 1 and 2, Galileo, and Cassini-Huygens. They provide geological and geophysical evidence that icy moons such as Europa, Ganymede, Callisto, Enceladus, and Titan may have the three presumed requisites of habitability. All the aforementioned satellites of Jupiter and Saturn might harbor global aqueous layers below the icy crusts, but each one has different physical, chemical, and geological characteristics, which constrain their potential for habitability (Figure 16.1; see the following section).

Spectral and geophysical data confirm that ices compose the surfaces and crusts of these satellites. It does not mean that there is just water, which is the standard chemical definition of ice. The planetary definition of ice includes solid phases of any compound with low melting and boiling point, e.g., carbon dioxide (CO_2), methane (CH_4), ammonia (NH_3), and nitrogen (N_2). In

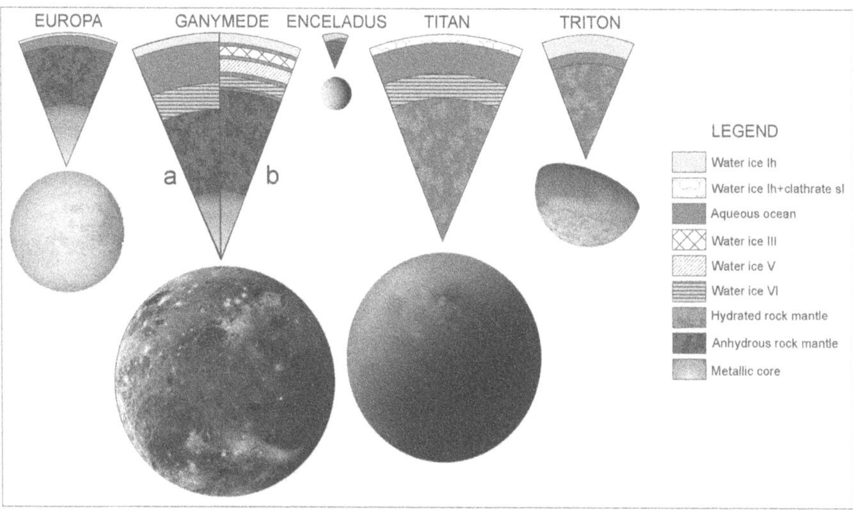

Figure 16.1 Geophysical models of the main icy satellites of the solar system that may harbor global oceans: Europa, Ganymede, Enceladus, Titan, and Triton (see the explanation in the main text). Ganymede shows two models in support: a) one aqueous ocean, b) several aqueous solutions with different densities between layers of water ice phases. (*Sources*: Planetary images courtesy of National Aeronautics and Space Administration and Jet Propulsion Laboratory, California Institute of Technology, Pasadena. Geophysical models courtesy of H. Hussmann *et al. Icarus*, 185, 258 (2006); F. Sohl *et al. Space Sci. Rev.*, 153, 485 (2010); C. Sotin and G. Tobie. *Science*, 320, 1588 (2008); S. Vance *et al. Planet. Space Sci.*, 96, 62 (2014); L. Iess *et al. Science*, 344, 78 (2014). With permission.)

addition, some salts such as magnesium sulfates ($MgSO_4$) are observed on the surfaces of the Jupiter satellites. All these molecules form multi-compound systems with water. Different phases of these systems take place at high pressures depending on their thermal states and depths within the moons, including aqueous solutions, gases, and crystalline and glassy solids.

The research field of high pressure is now a spotlight for astrobiology because the governing conditions at deep environments also affect the principal habitability requirements. It is recognized that high pressure determines environmental attributes like the solubility of gases in aqueous solutions, the physical properties of minerals and fluids, the equilibrium of the aqueous system, the availability of organic matter, the pathways in cycling essential elements, and the reactivity of simple organics and stability of biomolecules [2–5]. Despite this strong influence, some studies indicate that pressure is not an insurmountable obstacle for life. Indeed, we aware that some terrestrial organisms have adapted to live at the highest pressure limits of the oceans [6].

Here, we revise some contributions from high pressure science to the understanding of the interiors of those moons of the outer solar system that have deep aqueous reservoirs. We pay special attention to the applications to Europa, Ganymede, Enceladus, and Titan. We start by describing the general characteristics of the deep environments. After that, the main chemical systems relevant to the aqueous environments are examined. We assume that the moons evolve chemically and structurally to their current states from the initial compositions of the condensates of the solar nebula. We do not explain the details about the endogenous and exogenous processes that contribute to the planetary crust evolution. An exhaustive discussion about geochemical evolution of the icy satellites can be found in References [7–9]. For specific details regarding the properties of the solid phases that we discuss here, we refer the reader to other monographic manuscripts such as References [10–13].

16.2 BULK COMPOSITION OF ICY MOONS

Our solar system is 99 % hydrogen, helium, carbon, and oxygen, which are the building compounds of molecules such as H_2O, CO_2, CH_4, and NH_3. Most of these volatile molecules were transported to the colder outer parts of our planetary system, beyond the asteroid belt by the solar wind during its early formation stage. Then they condensed as solid grains and accreted with dust particles, growing into icy objects like comets or the moons of the giant planets. Primordial condensed ices were distributed in the solar system following a gradient of freezing temperature (see Table 16.1), thus the distance to the sun. A secondary batch of ices ensued from the chemical evolution of the moons, e.g., from surface photolysis and other reactions acting over the original compounds [8].

After 4.6 gigayears from the origin of the solar system, the icy moons appear diverse in their physical and chemical properties. Some are tiny, but others exceed the size of the planet Mercury: Ganymede, Callisto, and Titan. This means that pressure in their interiors is sufficient to yield phase transitions in some materials, preferentially the ices. Space missions have measured geophysical parameters such as the momentum of inertia and the gravity field, which help us know the nature of the interiors of the icy satellites. The mean density of the planetary object is used to estimate its bulk composition. The relationship between the radius and the bulk density of the moons allows us to assess their rock/water ice ratios. Taking this into account, Hussmann *et al.* [14] organized the icy objects in different groups: large, small, transient, and dense satellites (Figure 16.2). Saturn's and Uranus' families, except for Titan, constitute the small moons group. However, Triton and the trans-neptunian Pluto are transient bodies that we call border objects. The Jupiter family

Table 16.1 Freezing temperature (FT) of ices of the solar system and the icy moons where they are detected.

Compounds	H_2O	CO_2	SO_2	NH_3	CH_4	O_3	CO	N_2	O_2
FT (K)	273	215	200	195	91	80	68	63	55
Jupiter–Europa	Y	Y	Y	-	-	Y	-	-	Y
Jupiter–Ganymede	Y	Y	-	-	-	Y	-	-	Y
Jupiter–Callisto	Y	Y	Y	-	-	-	-	-	-
Saturn–Small moons	Y	Y	-	-	-	Y	-	-	-
Saturn–Enceladus	Y	Y	-	plume	-	Y	-	-	-
Saturn–Titan	Y	-	-	-	Y	-	-	-	-
Uranus–Small moons	Y	Y	-	-	-	-	-	-	-
Neptune–Triton	Y	Y	-	-	Y	-	Y	Y	-
Pluto	Y	-	-	-	Y	-	Y	Y	-
Comets	Y	Y	-	Y	Y	-	Y	Y	-

Y = yes.

exhibits peculiar density distribution based on the noticeable variation among the Galilean satellites. This is explained by the intense geological activity and loss of volatiles in Europa and Io, which, strictly speaking, are not considered icy. Besides, the compressional response on the ices constrains the geophysical models because of the reduction of porosity and higher density of the high-pressure phases.

The differentiation of the icy bodies in compositional layers occurs only if they accumulate enough energy. There are four main energy sources triggering planetary evolution: radiogenic isotopes (e.g., uranium, potassium) stored in the silicate component; accretional energy as a remnant from the formational processes; gravitational energy during core formation; and tidal energy. The heating is sufficient in some icy bodies to melt the ices, generating liquid layers and activating the geological processes. Cryovolcanic features are the observable geological consequences that warn us about endogenous activity that leads to differentiation. The geomorphology on the surface resulting from the melting activity is comparable to features of terrestrial volcanism (e.g., cones, lava flows, clastic deposits), but here the materials are cryogenic and made of silicates.

Geophysical models show that large moons like Ganymede and Titan are highly differentiated, while Callisto's differentiation seems incomplete. Water-rich layers may extend deeper than 800 km, which means that the pressures are higher than 1.5 GPa at the bases of these layers. Eventually, the thickness of the water layer depends on the dehydration rate of the silicate layer and its differentiation pathway in addition to the available energy. If the silicate

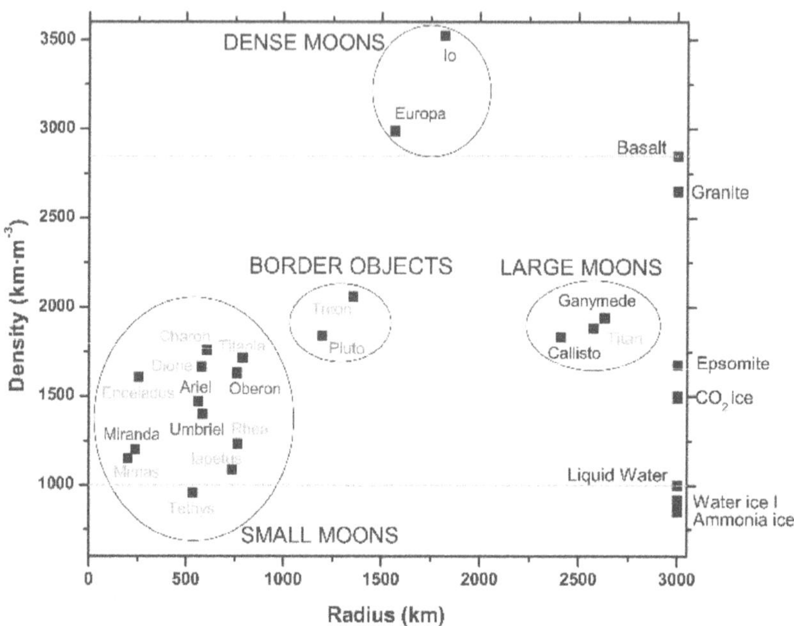

Figure 16.2 Relationship between radius and density of icy moons of the solar system. Modified from H. Hussmann *et al.* Icarus, 185, 258 (2006).

cores are still hydrated, then the water-rich layer could be thinner. As will be discussed later, global oceans may form between solid ice layers down to 0.9 GPa. On the other hand, small moons have low radiogenic heat and thus they differentiate only if tidal energy operates efficiently. They may have internal oceans at tens of megapascals if some refrigerant like ammonia is added to the water system [13]. While differentiation runs or when the liquid layers are already established, exchanges of energy and materials between fluids and minerals happen and modify the composition of the aqueous system (e.g., [8] and references therein). Light volatiles like hydrogen rise upward. They may be dissolved in the aqueous fluids and react with them or escape from the object. These exchange processes at deep conditions regulate the physical and chemical properties of the aqueous system (e.g., oxidation state, acidity) and the potential for habitability.

In the aim to reveal the moon interiors, we find that direct compositional data about planetary objects come only from the surfaces. Nonetheless, the identification of materials coming from endogenous processes like cryovolcanism provides hints to the interior state. Spectroscopy and imaging data supply the chemistry and the geological context in order to help reveal the origins of the materials. In order to select the candidates to model the interior, we should take into account that exogenous alteration (e.g., sputtering or photolysis) may modify the surface. Thus, the choice is tricky: while the surface materials are clues to determine the interior composition, they could be modified later from their original state.

16.3 DEEP AQUEOUS ENVIRONMENTS INSIDE ICY MOONS

Considering the groups that were previously proposed, brief descriptions of the representative moons with potential habitable environments are shown below and describe the high-pressure aqueous systems.

16.3.1 Dense Moons: Europa

Europa is the main target of astrobiology, competing with Mars. Geophysical measurements evidence a differentiated structure in several layers. They include a metallic core, a silicate mantle, and a water-rich ocean and crust (see Figure 16.1). The melting and transport of the inner materials to the present configuration is supported by the tidal and radiogenic energy of the body. The release of this energy also promotes the geological activity on the surface, which generates features such as fractures and other chaotic terrains and lenticulae areas. The resurfacing has occurred until recent times and even may be happening at present if the crater counting analysis is taken into account [15] and the recent plume discovery is confirmed [16]. The presence of temperature depressants would facilitate the melting of the H_2O-rich ice crust. The collapsed structure of Thera Macula, for instance, could have occurred by the rising of a fluid which developed at shallow depths. Partial melting of the crust may happen when warm material rises through and thaws it, forming a new upper reservoir. The interaction of the deep water with indigenous fluids in the crust may involve both the heat transfer and the chemical mixing, which would favor emerging habitable environments. Aqueous solutions may also include organic compounds, volatiles, and metals delivered from the rocky mantle or from an exogenous source. Environmental properties of the deeper liquids in contact with the silicate layer may be assimilated or mixed in the secondary fluid and be recorded in materials that reach the surface. The chemical composition of the liquid solution and cryovolcanic materials finally exposed to the surface in solid state including the concentration of essential elements for life will be the result of these cryomagmatic events.

The remote spectroscopy data at Galileo's resolution reveal that the surface composition is dominated by bright water ice and dark materials which are other volatile ices (e.g., CO_2, SO_2) and a combination of hydrated salts, mainly sulfates of magnesium and sodium, and hydrates of sulfuric acid [17, 18]. These materials are often associated with tectonic and/or cryovolcanic features, which indicate their possible endogenous origin [19].

The magnetic field signatures of Europa, Ganymede, and Callisto reveal a liquid layer inside them. These signatures are induced by the movement of electrolytic ions in a global ocean, like the sulfates observed on the surface. It is hypothesized that the global aqueous layer of Europa is currently in contact with the rocky mantle at pressures approximately from 20 to 150 MPa. This scenario produces the key difference with respect to the large satellites. The rocky substrate is probably geothermally active, which stimulates the interaction between water and mineral and may provide both the nutrients and chemical energy sources to support the habitability of Europa. The terrestrial analogue environment generally suggested is the seafloor at hydrothermal vents, which is mostly chemosynthetically sustained and biologically exuberant.

16.3.2 Large Moons: Ganymede and Titan

In the larger icy moons, the ocean is under a thicker crust, sandwiched inside different phases of water ice at pressures around 1 GPa. As mentioned, internal differentiation is significant in Ganymede, which has a metallic core, a silicate mantle, and a water layer (see Figure 16.1). Ganymede has an intrinsic magnetic field in addition to the induced one previously mentioned. Indeed, the magnetosphere interaction between Ganymede and Jupiter is unique in the solar system.

The chemistry on the surface observed by the Galileo spacecraft is mainly composed of water ice, carbon dioxide and sulfur dioxide ices, and some silicates and organic materials. Salt hydrates are also present on the surface of Ganymede. Although an oceanic origin is also proposed for these salts [19], the link between the liquid layer and the surface is not obvious because the current icy crust is too thick and the cryovolcanic features are infrequent. Only local indirect evidence of cryovolcanism, including caldera-like features and smooth bright lanes, have been identified. The source of the oceanic solutes of the larger satellites is presumably the rocky fraction, as in Europa. The solute enrichment took place during a period when the structure of the interior was less evolved or resulted from melted brine plumes from the rock-water interface. The most obvious resurfacing occurs by faulting. The relatively young grooved terrain formed 2 to 3.6 gigayears ago. It probably happened during periods of high tidal flexing activity and heating, which resulted in the visible dichotomy of dark and bright terrains. During those heating periods, the rise of the deep melt brine plumes in Ganymede's ocean were produced. Early in this period, cryovolcanic extrusions could arise but they were mostly eroded by tectonics or mass wasting. Certainly, the lack of evidence for cryovolcanism in Galileo's coverage does not mean that it does not exist or that deep cryomagmatism could not occur. Another way to access to deep layers may be during early resurfacing and excavation by meteoritic impacts. If cryoxenoliths reached the surface by tectonics, they might have incorporated the endogenic salts and other phases from the high-pressure conditions of the aqueous layers.

Another exciting large icy world is Titan, which is commonly recognized by the great assortment of organic compounds detected in its atmosphere and surface. The deep aqueous ocean in Titan is apparently rich in ammonia and lies between ice phases with different densities [20]. Below the high-pressure ices is the silicate core whose hydration state is still debated [21]. The interaction between the silicate layer and the high-pressure ice may occur at 0.9 GPa. Above the aqueous layer, the crust is probably enriched in non-water volatiles forming clathrate hydrate phases (see Figure 16.1). Titan is an exceptional satellite because it has an atmosphere of 0.16 MPa dominated by nitrogen and methane. Data from the Cassini-Huygens mission show that organic species are diverse in the atmosphere and on the surface, ranging from simple molecules like methane to tholins. The origins of some of these molecules are endogenous. Cryovolcanic features are observed on the radar images of the surface. Materials coming from the interior are strongly affected by the exogenous radiation and are altered in short periods of time [22]. In addition to photolysis, the interaction between the surface and the atmosphere modifies the geological substrate as well. It produces weathering, erosion and transport features such as rounded pebbles, dunes, lakes, and channels similar to those on Earth, but made of ices and hydrocarbon fluids that characterize the geochemistry of this moon.

16.3.3 Small Moons: Enceladus

Some small moons may have developed internal aqueous reservoirs below the ice layer. One special case is Enceladus, where the Cassini spacecraft detected active water-rich plumes jetting from localized fractures at the South Pole. They seem to be the sources of the E ring particles around Saturn. The composition of these plumes suggests that a habitable environment inside this moon is possible because of volatiles containing essential chemical elements (e.g., H_2O, CO_2, CH_4, N_2). The tidal friction generated by Saturn and its orbital eccentricity forced by Dione support the power for the observed activity. The measured density indicates that the body has a relatively high rock fraction for a small moon (see Figure 16.1). A recent model based on the Cassini gravity measurements shows that the density of the core is low and probably still hydrous [23]. It also indicates the presence of a negative mass anomaly in the south polar region, which is compatible with the presence of a regional liquid reservoir at depths of 30 to 40 km and extending up to south latitudes of about 50°. The aqueous activity may be supported by compounds that depress the melting point of the aqueous system, such as ammonia [24] or chloride salts [25]. Two opposite hypothesis are proposed to explain the water ice-rich geysers. Both relate the tectonic activation of the differentiated icy crust, with the tiger stripes of the South Polar circle as conductors for the rising materials. One of the models agrees with the presence of the regional ocean at shallow depths and under pressures up to 3 to 4 MPa [26, 27] below the South Pole. The other predicts the formation of the plumes by the dissociation of gas clathrates when the crust is fractured [28]. The deep aqueous reservoirs are not required in this hypothesis unless it is assumed that clathrates are initially formed from gas-rich aqueous solutions.

16.3.4 Border Moons: Triton

Voyager 2 also detected plume activity in Triton. Some dark jets around 8 km high were observed on the south hemisphere of this Neptune satellite [29]. On the surface, more than a hundred dark streak features are found in the same area indicating that this activity occurred recently in the past. Resurfacing also seems to occur by cryovolcanic flows, originating plains and the distinctive Cantaloupe terrain [30]. The composition of the Triton surface is mainly ices of N_2, CH_4, H_2O, CO and CO_2, [31, 32]. Some of these ices could change to gas like N_2 seasonally by insulation. The geological activity shown on the surface reflects the internal activity and differentiation. Probably Triton has a rock-metal core, and a multi-volatile mantle and icy crust [33]. Some authors believe that Triton still retains an ocean. However, there are many uncertainties about its distribution inside or even its mere existence due to the lack of data on gravity and composition. Hussmann *et al.* [13] calculate that the ice crust is 140 to 200 km thick, and suggest that there is a water-ammonia ocean lying below 150 km (see Figure 16.1). Meanwhile, Ruiz [34] estimates an ice layer of only 20 km. In any case, the deep liquid reservoirs need an efficient temperature depressant like NH_3.

16.4 AQUEOUS SYSTEMS UNDER HIGH PRESSURE

We see that some compounds dominate the chemistry of the deep icy layers of the moons. Water is the common molecule of the chemical systems in the crust of these planetary bodies, similar to what silica does in the crusts of terrestrial planets. In this section, we examine some of these governing chemical systems, and describe the phase relationships and changes at high pressure relevant to planetary science. The results of the high-pressure studies of those chemical systems successfully explain some characteristics of the moons, including the potential habitability.

16.4.1 H_2O: Everywhere

Water is the main volatile of crusts and surfaces of the moons of the outer solar system. Although, there are some other materials accompanying the water, it is relevant to examine the mono-compound system in detail due to its predominance and special attributes at high-pressure. In our planet, high-pressure aqueous environments are located at the seafloor. As an example, the pressure reaches 108 MPa at the Mariana Trench, where liquid water is at temperature from 274 to 278 K and its density is about 5 % higher than water at the surface. This condition is mild compared to conditions within the giant satellites. Even at this terrestrial range of pressure, this variable affects chemical systems, e.g., fluid critical points, acidity, and oxidation, among others. It is especially intriguing that pressure alters the activity coefficient of the water, which might require a reformulation of the accepted concept of pH and other electrochemical coefficients [4]. The critical point of water is at 647 K and 22.1 MPa (locally feasible at hydrothermal vents on the seafloor). At supercritical conditions, the unique properties of aqueous solutions alter the synthesis and stability of important small biomolecules or proto-biomolecules [3]. The common state of water ice at temperatures and pressures of the Earth's surface is the hexagonal structure (ice Ih). Nevertheless, water has over 15 known crystalline and amorphous phases [9, 35], most of which are only known to occur naturally at very high pressures in the interiors of the large planetary moons. The hydrogen bonding is responsible for the peculiar properties, which convert water into an essential substance for life as we know it. Again, one manifest example is the density, which is 9 % less than that of the liquid phase, causing ice Ih to float (see Figure 16.3).

The phase diagram of water shows several crystalline phases in contact with the liquid at low (ice Ih), medium (ice phases III and V), and high pressures (ice phases VI, VII, and X). The hydrogen bonding is disordered, in contrast with bonding at lower temperature phases (II, XI, VIII) further away from the liquid field. Each polymorph has different properties, so the changes with depth produce discontinuities in the moon structures, and in some cases mechanical layer decoupling. As in the case of the transformation from Ih to water, the decrease in volume when ice Ih changes to ice II ($\Delta V = 3.92$ cm^3 mol^{-1}), ice II to ice V ($\Delta V = 0.7$ cm^3 mol^{-1}), ice V to ice VI ($\Delta V = 0.7$ cm^3 mol^{-1}), and ice VI to ice VII ($\Delta V = 1.05$ cm^3 mol^{-1}) is noticeable. Several crystalline phases can be present inside the icy moons, including liquid water, depending on the total water layer thickness and thermal regime. Small and dense moons are restrained to hold ice Ih, while large moons may stabilize denser phases up to ice VII. Density relationships among the solids and the liquids produce the

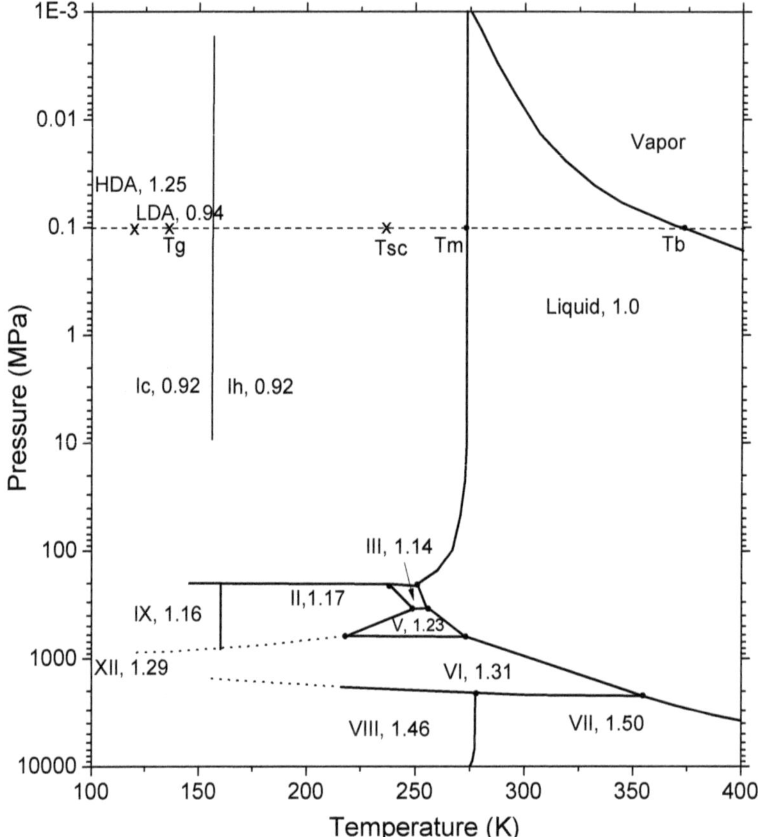

Figure 16.3 H_2O phase diagram in the range from 10^{-3} to 10^4 MPa, which shows the stability field of crystalline, liquid, and vapor phases of water. The number near the name of each phase indicates its respective density (after M. Choukroun and O. Grasset, *J. Chem. Phys.*, 127, 124506 (2007) and references therein). Dashed line points, 0.1 MPa location of the boiling (Tb), melting (Tm), supercooling (Tsc), and glass temperatures. The temperature at which low (LDA) and high (HAD) density amorphous phases appear are from P. G. Debenedetti and H. E. Stanley, *Phys. Today*, 40 (2003).

alternating distribution of layers in large moons, and impede the buoyancy of aqueous cryomagmas in ice Ih-rich crusts (Figure 16.1, see Ganymede).

Water also displays polyamorphism. Indeed, amorphous water ice is the most ordinary phase in the universe [36]. Amorphous ice phases appear if water is cooled to its glass transition temperature with no crystal nucleation. This is usual under vacuum, but possible at high pressure as well. There are three documented amorphous phases of water, different in both density and formation conditions. Low density amorphous ice (LDA) is naturally formed by condensation of water molecules over dust particles and cold surfaces at less than 120 K and 0.1 MPa. Glass transition to ice Ic is around 140 to 210 K. Otherwise, if temperature is colder than 30 K, high density amorphous ice (HDA) condenses [37]. HDA also forms by compressing Ih to 1.6 GPa at 140 K or LDA to 0.6 GPa. The very high density amorphous ice (VHA) forms when HDA is warmed to 160 K between 1 and 2 GPa. Amorphous water is detected on the surfaces of Europa, Ganymede, and Enceladus. Its presence in these moons is probably induced by radiation; however flash freezing of cryovolcanic fluids arising from the interior also impedes the mineral crystallinity, similarly to the formation of terrestrial volcanic glasses. If supercooling affects the cryomagmatic fluids, they can persist transiently in the crystalline domain of stability. Supercooled water happens in metastable equilibrium, and minor perturbations favor the solidification to the glass state.

16.4.2 H_2O and Other Volatiles: Retention, Escape, and Cycling Essential Elements

Systems containing water and other simple volatiles, such as CO_2, CH_4, SO_2, H_2S, CO, O_2, and H_2, are representative of different scenarios of the icy moons. The most familiar scenario among the terrestrial aqueous environments is that occurring in the shallow ocean of Europa and some small moons, where relative warm temperature and moderate pressure are possible. Volatiles may dissolve and dissociate in liquid water, producing several chemical species [6, 7, 38, 39].

Depending on the physical chemical conditions of the aqueous fluid, gases such as CO_2 can react and form carbonates or bicarbonates that buffer the acidity of the system. Pressure, temperature, and salinity constrain the carbonate compensation depth (CCD) at which these minerals dissolve in terrestrial oceans; thus the sedimentation of carbonates and formation of organism shells are prevented. At present, the CCD is at 4.5 km below the terrestrial sea level. At extreme pressures higher than 3.5 MPa and around 273 K, CO_2 is in liquid state, which is denser than water. This condition is reported in some exotic areas of the terrestrial seafloor (e.g., the Okinawa Trough), where microbial communities emerge above the pools of liquid CO_2 [40]. At low temperatures, water ice Ih and separated volatile solid phases form. However, if

Table 16.2 Structures and cavities of clathrate hydrates.

Structure	I		II		H		
Crystal system	Cubic		Cubic		Hexagonal		
Crystal system	Pm3m		Fd3m		P6/mmm		
Cavity	D	T	D	H	D	D'	E
Description	5^{12}	$5^{12}6^{2}$	5^{12}	$5^{12}6^{4}$	5^{12}	$4^{3}5^{6}6^{3}$	$5^{12}6^{8}$
Cavities/unit cell	2	6	16	8	3	2	1
Cavity radius (Å)	3.95	4.33	3.91	4.73	3.94	4.04	5.79
Water molecules/cavity	20	24	20	28	20	20	36

D=pentagonal dodecahedron, T=tetrakaidecahedron, H=hexakaidecahedron, D'=irregular dodecahedron, E=icosahedron. Adapted, with permission, from Reference 41.

sufficient gas phase is in contact with water, volatile molecules distort the structure of water in order to find their place until a new crystal phase is formed. H_2O forms clathrate hydrates at specific conditions for each volatile: relatively low temperature and high pressure.

In clathrate hydrates, there is no covalent bond between the water shell and the guest molecule, but Van der Waals bonds attach them. Still, the structure is unstable until the guest molecules occupy a determined percentage of the cavities. Clathrate hydrates are part of the carbon biogeochemical cycle of the Earth, in addition to the carbonates. In fact, large amounts of hydrocarbons are stored where conditions may promote the formation of these minerals, which occur mostly in seafloor sediments. It is likely that in the potential oceans of icy satellites, clathrates constitute the dominant sink for many non-polar molecules like CH_4 and CO_2. There are several types of packaging in clathrate hydrates [41]. At relatively low pressure, structures are cubic (sI or sII), and hexagonal (sH). The difference between them is the number of water molecules that constitute the unit cell and the amounts and types of polyhedral cavities generated. Each cavity is able to accommodate a determined size of the guest molecule. Molecules whose diameters are between 4.3 and 5.3 Å (e.g., CH_4, CO_2) form sI structures. They configure cavities type D and T (see Table 16.2 for a description). Molecules with diameters smaller than 4.3 Å, such as N_2 or H_2, and those larger than 6 Å (e.g., C_3H_8), form sII structures, filling cavities type D and H. The sH structure hosts larger molecules and mixtures of small and large molecules filling D, D' and E cavities. At pressures in the range of gigapascals, clathrates undergo transitions to other tetragonal (sT) and hexagonal (sH) phases, and finally form filled ice structures (e.g., CH_4 clathrates in this phase are orthorhombic, sO, and stable up to 42 GPa [42]).

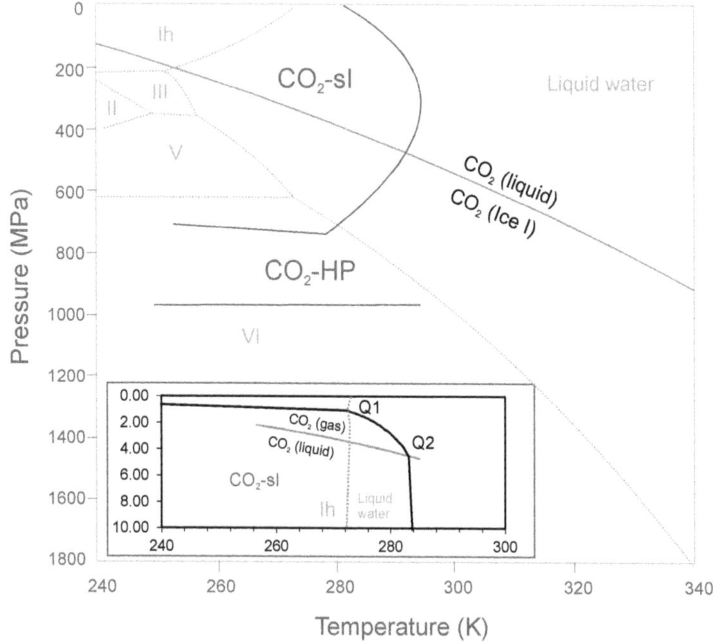

Figure 16.4 Phase diagram of CO_2 clathrates in the range of pressures from 0 to 1800 MPa (modified from H. Hirai *et al. J. Chem. Phys.*, 133, 124511 (2010); O. Bollengier *et al. Geochim. et Cosmochim. Acta*, 119, 322 (2013); and C. A. Tulk *et al. J. Chem. Phys.*, 141, 174503 (2014)). Black solid line distinguishes the sI phase (CO_2-sI) to high-pressure phase (CO_2-HP), which has been identified as filled water ice. Grey solid line marks the stability field of liquid and solid CO_2-I. The dotted line shows the crystalline phases of water. The inset is the phase diagram of the system H_2O-CO_2 in the range of pressures from 0 to 10 MPa.

We take CO_2 clathrates as examples for describing the phase relationships in the clathrate-forming systems (see [41] and references therein). The phase diagram of CO_2 clathrates exhibits two quadruple points at relative low pressure: Q1 (272 K, 1.05 MPa) where the clathrate hydrate coexists with liquid water, water vapor, and water ice, and Q2 (283.1 K, 4.5 MPa) where it coexists with liquid water, water vapor, and liquid CO_2. At high pressure, CO_2 clathrate transitions are poorly studied. It is known that at 700 MPa the sI CO_2 hydrate transforms directly to a filled water ice structure [43–45] (Figure 16.4). Neutron diffraction data show that the filled water ice phase can be recovered to ambient pressure (0.1 MPa) at 96 K, and recrystallization to sI hydrate occurs upon subsequent heating to 150 K, possibly by first forming low density amorphous ice. Meanwhile, other gas clathrates reach this phase though sII and sH states. This is the case of CH_4 clathrates, which have two documented phase transitions, one at 0.8 to 1 GPa and other at 2 GPa [46–49].

It is suggested that clathrate hydrates are present in the icy satellites, preferentially in the subsurface [50]. Among others, methane, carbon dioxide, and oxygen clathrates have special interest for planetary science. Large amounts of volatiles may have been trapped in clathrates during planetary accretion in the outer solar system (e.g. [51, 52]). In some cases, they may be retained in the planetary interior [46, 53] or alternatively reprocessed during thermal evolution, triggering the dissociation and degasification of the body. If they are still exposed and retained on the surface, their detection by traditional remote sensing is not easy because the IR reflectance spectrum is similar to that of water ice, except for the absorption bands of the guest molecule. Oancea *et al.* [54] obtained the IR spectra of CO_2 clathrates at vacuum conditions and low temperature, roughly resembling planetary conditions. Signatures at 2.71 and 4.28 microns identified the CO_2 clathrate. However, the spectral search on icy moons has not been successful until now. One explanation is the amorphization of the exposed ices by the radiation. On Europa, SO_2, CO_2, and O_2 clathrates are predicted to form from endogenous volatiles [6, 50, 55], or alternatively by radiolytically produced oxidants [56]. All of them are stable in the crust and in the ocean but their buoyancy depends on the type of the guest molecule and the occupancy of the clathrate cavities. Crystallization from aqueous fluids removes the volatiles and concentrates the solutions in the rest of the solvent. This results in the differentiation of the planetary fluids. Due to the thermodynamics associated to the formation and dissociation of the clathrate hydrates and the low thermal conductivity, the fluids may be responsible for some geological activity [50, 57]. In this regard, it is suggested that layers of clathrates in the icy crust may isolate the oceans, preventing them from freezing. Besides, the disruption of the surface to form chaotic terrains is attributed to clathrate destruction. CH_4 clathrates may exist at several depths inside Titan and play a relevant role in its planetary activity. It is suggested that they are the current sources releasing the methane detected in the atmosphere. sI CH_4 clathrates may constitute global layers in the crust that are buoyant over the water ice Ih [22]. Keeping in mind that high-pressure phases of CH_4 clathrates are stable below the deep aqueous ocean, some investigators [46, 58] claim that they also trap the volatiles originated from the alteration of the still hydrous core in the icy mantle. Similarly, the volatiles from the geysers of Enceladus are explained by the dissociation of clathrate layers (see Section 4.2). Apart from the formation of clathrate deposits, Fortes [59] postulates that dragged xenoliths of clathrates may locally rise to the upper crust and dissociate, driving explosive eruptions.

16.4.3 H_2O and Salts: Exchange Consequences

The origin of the salts detected on the Galilean satellites and in Enceladus' plume is assumed to be the rock fraction of these planetary bodies. This hypothesis is based on the mineral products resulting from the aqueous alteration of chondrite meteorites [6, 60, 61]. The abundance and chemistry of the salts in the moon depend on the nature of the primordial accreted rocks and ices, the physical and chemical processing of the materials, and other details that we do not understand about the moon interiors [7]. For instance, if oxidizing conditions do not rule, sulfates might not form from the exchange processes between the silicate rock and water. Salts are important for life as we know it. Life forms utilize salts for different functions such as delivery of nutrients, transmission of nerve impulses, and protect against radiation. However, high salt concentration impacts an aqueous environment. Macromolecules can form from simple organic molecules under the low water activity of brines, but hypersalinity may inhibit the formation of some polymers and base pairs. On our planet, halophile organisms have found the mechanisms to survive under hypersalinity conditions from very ancient times, and it is suspected that these salts may have played an important role in the evolution of life.

As is stated in Section 16.2, magnesium sulfate is especially relevant in the Galilean moons, where a wide range of pressures may exist. Chlorides may also be present, but they have not been detected yet. Magnesium sulfates are known to form minerals with different hydration numbers. Epsomite ($MgSO_4{\cdot}7H_2O$) is the stable phase at terrestrial standard conditions, while meridianiite is the lower temperature state ($MgSO_4{\cdot}11H_2O$). This mineral may be present on Mars and icy satellites [62, 63]. Other hydration numbers are possible at higher temperatures (e.g., 1, 6) or in metastable state (e.g., 2, 4, 5). The hydration reaction is exothermic and involves a significant volume expansion, especially from epsomite to meridianiite. The eutectic point of the H_2O-$MgSO_4$ system at 0.1 MPa is 269.25 K, with 17.5 wt.% $MgSO_4$, while the peritectic is 274.95 K, and 21.6 wt.% $MgSO_4$. Laboratory experiments at high pressure show that the eutectic point shifts to lower content of salt when the pressure and temperature increase [62, 64], approaching 14 wt.% $MgSO_4$ at 2 GPa and 298 K. From this condition, up to 4 GPa, the epsomite coexists with water ice VI first, and with water ice VII later (Figure 16.5). Vance and Brown [65] report some thermodynamic properties of the brine solutions at 800 MPa (e.g., density, thermal expansion, sound speed, and specific heat). They show that the density of the brine changes under pressure, so it is buoyant between different phases of ice (e.g. VI-V or V-III). Using these results, they propose a multilayer model for the internal structure of Ganymede (see Figure 16.1, Ganymede). The influence of salinity on Ganymede's bulk structure is also

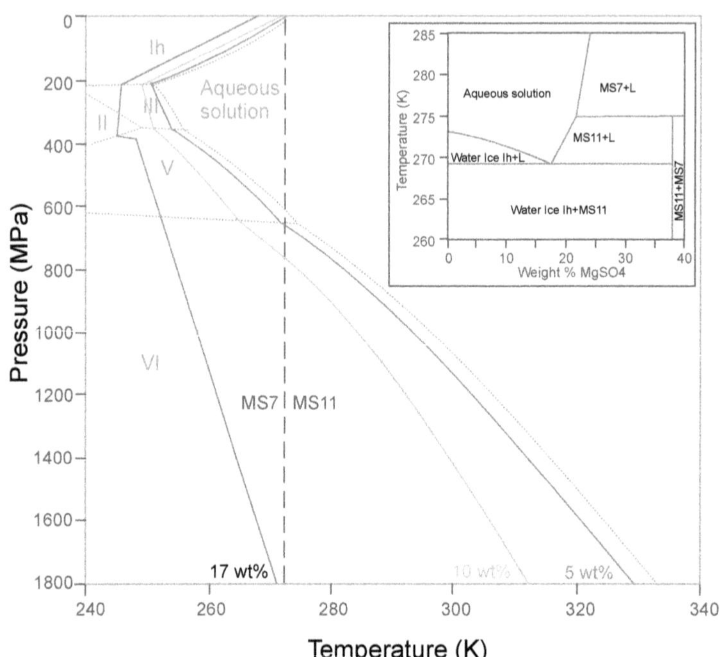

Figure 16.5 Phase diagram of the binary system H_2O-$MgSO_4$ in a range of pressures from 0 to 1800 MPa. Solid line indicates the liquidus at several concentrations of salt (after D. L. Hogenboom *et al. Icarus*, 115, 258 (1995); R. Nakamura and E. Ohtani. *Icarus*, 211, 648 (2012); S. Vance *et al. Planet. Space Sci.*, 96, 62 (2014)). Dotted line shows the phases of water. Dashed line divides the stability field of epsomite (MS7) from that of meridianiite (MS11). The inset shows the phase diagram at 0.1 MPa. Adapted, with permission, from reference 66 and 70.

assessed from the available phase equilibrium data of the water-salt system [66]. This work indicates that higher salinities imply small silicate fractions in the bulk structure in order to compensate for gravitational data. High-pressure studies on solids of this system are focused on epsomite. Bridgman [67] was the first to study this mineral at pressure up to 4 GPa, detecting two transitions, at 1 to 1.5 GPa and at 2.5 GPa. Latter works detect up to five phase transitions for this mineral [64, 68, 69]. There are few studies of meridianiite [21, 70] that show a transition relevant for the large icy moons (1 GPa, 270 K) where this mineral dehydrates to ice VI and an unidentified $MgSO_4$ solid, probably a phase with a hydration number between 7 and 11. This reaction is associated with a few percents of volume expansion. If the thickness of the sulfate layer is significant within large moons like Ganymede, then its dehydration causes global expansion in the body, which would be evidenced as extensional tectonism on the surface.

Likewise, the $NaSO_4$-H_2O system at 0.1 MPa has a eutectic point at 271.8 K and 4.15 wt.% $NaSO_4$, where solution crystallizes to ice Ih and mirabillite ($NaSO_4 \cdot 10H_2O$). Mirabillite decomposes incongruently in the peritectic point, 305.5 K with 33.2 wt.% $NaSO_4$. A metastable heptahydrate phase is detected at standard pressure and temperature [71]. This hydrate was also identified as a stable state at high pressure, later recognized as octahydrate [72]. The study of the system up to 800 MPa determines the shift in both the incongruent temperature in the same way as $MgSO_4$-H_2O and the stability fields of high-pressure phases.

Sulfuric acid hydrates are detected on the surface of Europa [18] and seem to be associated with the alteration by radiation of sulfates [73]. However, their presence within the crust or the ocean cannot be disregarded [6, 74, 75]. The binary system at 0.1 MPa reaches the eutectic at 210 K with 35.5 wt.% H_2SO_4,which shifts toward more water-rich compositions as pressure increases to 200 MPa. There are many hydrated phases possible, but the hemitriskaidecahydrate ($H_2SO_4 \cdot 6.5H_2O$) and the octahydrate ($H_2SO_4 \cdot 8H_2O$) are the relevant states to the icy moons (see Reference [76], and references therein). High-pressure studies about the solids of this system are scarce. They reveal the polymorphism of $D_2SO_4 \cdot 4H_2O$ up to 550 MPa by neutron powder diffraction [70].

Ammonia may play in the ionic state (NH_4^+) in the deep aqueous solutions to produce ammonium brines. Fortes *et al.* [77] suggest that ammonium brines occur in Titan's ocean and lie beneath the thick shell of methane clathrates and ice. However, the reducing condition evidenced by the presence of CH_4 in this moon contradicts the ammonium sulfate stability.

The detection of salts by different sensors on several structural levels within the moons hints that these layers might be linked. If brines stay at deep reservoirs in the crust or the ocean, the liquid state persistence is constrained by the thermal gradient. When they cool down, some minerals crystallize and separate depending on the density contrast with the remnant fluid. This fractionation process differentiates the deep fluids and results in several mineral deposits that produce heterogeneities in the interior. Theoretical modeling [78] and laboratory experiments up to 30 MPa [79], both using different initial compositions for the brine, confirm the changes in salinity and pH in the liquid during the cooling, proposing that these processes are applicable to Europa's crust. The transport of this material is commonly associated to cryovolcanic activity. However, brines above are denser than water ice Ih, and cannot raise to the surface if the crust is pure water. The problem of fluid buoyancy in the crust is solved if the density contrast is balanced by adding a light compound in the fluid or a heavy contaminant in the crust. The first option incorporates ternary systems with salts and volatiles as a possible answer. Bonales

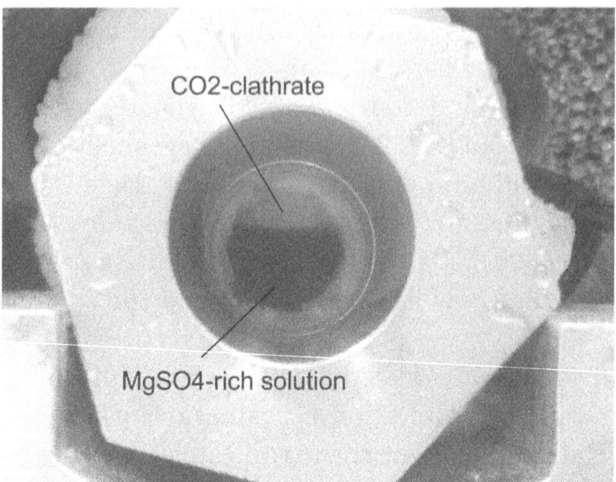

Figure 16.6 Picture of the interior of a high-pressure chamber during the formation of CO_2 clathrates from the ternary system $MgSO_4$-H_2O-CO_2 (3 MPa, 273 K). CO_2 clathrate hydrates form from the interphase of vapor-liquid. The density of the solution at 17 wt.% $MgSO_4$ is higher than that of the clathrate, so the solid stays buoyantly over the liquid.

et al. [80] report the solubility of CO_2 in the aqueous solution of $MgSO_4$ at different concentrations (5 and 17 wt.%), at a range of temperatures from 278 to 293 K and pressures up to 5 MPa. The results are applied to model both the ascending and degassing processes of brine cryomagma through the crust of Europa. Muñoz *et al.* [55] examined by Raman spectroscopy the CO_2 clathrate formation and dissociation in the presence of other phases crystallized from magnesium sulfate aqueous solutions at several concentrations, from 268 to 290 K and pressures up to 6 MPa. During the cooling process, the mineral assemblage of the system evolved differently, depending on the initial salt concentration. They found that the evolution of the presumed fluids and the changes in the mineral assemblages resulted in differentiation, layering (see Figure 16.6), and volume changes, which promote the generation of local stresses and the resurfacing of the moon.

16.4.4 H_2O and NH_3 in Extreme Conditions

Ammonia was cited as an important constituent of the outer solar system bodies in the early 1970s [81–85]. It is the prominent antifreeze agent in geophysical models of the icy moon interiors to accomodate liquid layers and explain the cryovolcanic features on their surfaces [86–90]. Ammonia has been observed only in the Enceladus plume and in the clouds of Jupiter and has not been detected on the surface of the moons.

Liquid ammonia is proposed as an alternative to pure water as a possible solvent for life. Ammonia dissolves many organic compounds and may be maintained in a liquid state at a wide range of temperatures that are even broader under pressure. However, this is not beneficial in the case of dissolution of hydrophobic organic molecules because it limits the possibility of compartmentalization in the liquid: liposomes do not work. Its high basicity is also troublesome (pH is 11.3 if the concentration of the aqueous system is 15 wt.% NH_3), although extreme alkalophile organisms have developed adaptation mechanisms to live comfortably under this condition on Earth (e.g., in lakes of the Rift valley).

The binary system at 0.1 MPa has a eutectic point at only 175 K (35 wt.% NH_3), very close to the peritectic at 176 K (32.1 wt.% NH_3) at which the ammonia dihydrate phase (ADH) melts incongruently [85]. There are two more hydrates at this pressure: ammonia monohydrate (AMH) composed of 50.25 wt.% NH_3 and ammonia hemihydrate (AHH) with 75 wt.% NH_3. The phase equilibria of this system at high pressures were extensively studied (Figure 16.7). The eutectic point moves to form water-rich liquids in the range from 20 to 300 MPa, but the shift is reversed below and above this interval of pressures. At 100 MPa, the eutectic is at 235 K; and at 450 MPa it is at 240 K. The peritectic temperature is higher from 400 MPa to 1 GPa, which has significance in the geophysics of large moons. ADH changes from phase I to II at 175 K and 450 MPa. The ADH IV phase appears above 600 MPa and is stable up to 6 GPa at temperatures below 170 K [91]. AMH changes from I to II at 195 K and 340 MPa [62, 86] and is also obtained in addition to water ice II by warming ADH II at 190 K and 550 MPa [92].

The ternary system H_2O-NH_3-CH_4 has interesting planetary applications for the Saturn satellites due to the inhibiting character of ammonia in forming clathrates. Experimental data show that ammonia at 10 wt.% decreases by 14 to 25 K the dissociation curve of the CH_4 clathrates at pressures above 5 MPa. The behavior of this system is explored up to 800 MPa [93, 94]. This effect is proposed to control the endogenous activity in Titan. It has been suggested that the dissociation of CH_4 clathrates is only possible by intense heating or in presence of large local amounts of ammonia.

Choukroun *et al.* [94] propose that dissociation of CH_4 clathrate layers happens if cryomagmatic fluids enriched in ammonia interact with the clathrate layers and induce their dissociation. In contrast, Shin *et al.* [95] recently demonstrated experimentally that ammonia molecules form clathrate hydrates as well. Changing the pressure and temperature to the surface conditions, prevents the inhibition of clathrate formation. Ammonia can be trapped alone and also participates synergistically in the formation of binary clathrates with methane and other volatiles at low temperatures from solutions and vapor

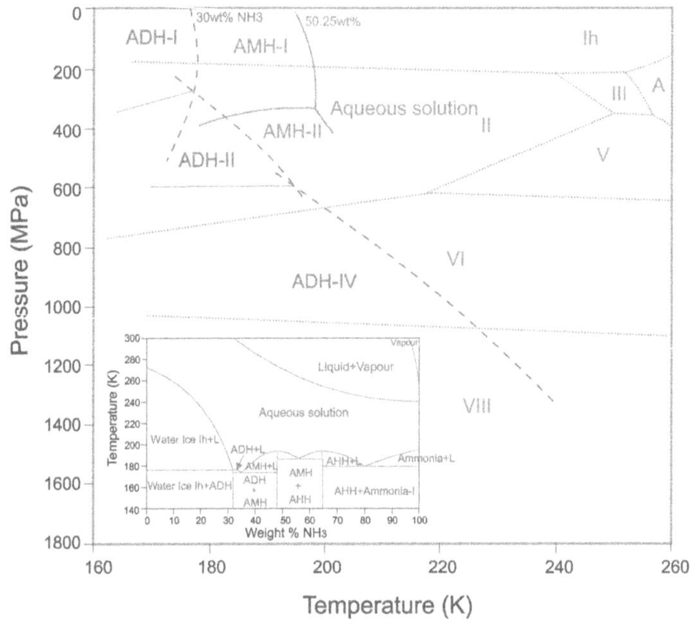

Figure 16.7 Phase diagram of ADH and NH_3-H_2O system at a range of pressures from 0 to 1800 MPa. ADH = ammonia dihydrate; AMH = ammonia monohydrate; AHH = ammonia hemihydrate. Dashed line points to ADH phases, while solid line is for the AHM high-pressure phases. Dotted line shows the phases of water. Inset figure is the composition versus temperature diagram at 0.1 MPa (after D. L. Hogenboom *et al. Icarus*, 128, 171 (1997); A. D. Fortes *et al. J. Appl.Cryst.*, 42, 846 (2009); and J. S. Loveday *et al. High Pressure Research*, 29, 396 (2015)).

deposition (at vacuum). This is conceivable since ammonia promotes defects in the lattice water ice and stabilizes the guest molecule from which the clathrate hydrate is formed. The plume of Enceladus, Titan's surface, and the primordial solar nebula would be environments where ammonia can play this role.

16.5 CONCLUSIONS

Planetary science is taking advantage of the improvements of high-pressure studies. The latest technology for simulation and chemical analysis at high pressures allows the discovery of new phases of ices and characterization of chemical systems that are stable at the extreme conditions of the interiors of the icy moons. The properties of water as a simple molecule or in association with other compounds produce many polymorphic transitions that trigger geological phenomena, as Bridgman stated many years ago [96]. The experimental support is required for designing geophysical models of the hidden interiors of the moons, where aqueous systems may develop. It helps us to approach the habitability of deep environments in the solar system. Future space exploration will provide data to confirm and enhance the current models.

Bibliography

[1] O. Grasset, M. K. Dougherty, A. Coustenis, *et al.* *Planet. Space Sci.*, 78, 1 (2013).

[2] M. Citroni, M. Ceppatelli, R. Bini, *et al.* *J. Chem. Phys.*, 123 (2005).

[3] R. M. Hazen, N. Boctor, J. A. Brandes, *et al.* *J. Phys. C Solid State Phys.*, 14, 11489 (2002).

[4] P. C. Michels and D. S. Clark. *Appl. Environ. Microbiol.*, 63, 3985 (1997).

[5] C. P. Samaranayake and S. K. Sastry. *J. Phys. Chem. B*, 114, 13326 (2010).

[6] K. Horikoshi. *Curr. Opin. Microbiol.*, 1, 291 (1998).

[7] J. S. Kargel, J. Z. Kaye, J. W. Head, *et al.* *Icarus*, 148, 226 (2000).

[8] M. Y. Zolotov and J. S. Kargel. *On the Chemical Composition of Europa's Icy Shell, Ocean, and Underlaying Rocks.* Space Science Series. University of Arizona Press, Tucson (2009).

[9] P. V. Hobbs. *Ice Physics.* Classic Texts in the Physical Sciences. Oxford University Press, Oxford (1974).

[10] F. Sohl, M. Choukroun, J. Kargel, *et al.* *Space Sci. Rev.*, 153, 485 (2010).

[11] J. Klinger, D. Benest, A. Dollfus, *et al.*, eds. *Ices in the Solar System*, Volume 156 of NATO ASI Series C: Mathematical and Physical Sciences. Springer, Heidelberg (1985).

[12] M. F. B. Schmitt, C. de Berg, ed. *Solar System Ices*, Volume 227 of Astrophysics and Space Science Library. Springer, Berlin (1998).

[13] A. D. Fortes and M. Choukroun. *Space Sci. Rev.*, 153, 185 (2010).

[14] H. Hussmann, F. Sohl, and T. Spohn. *Icarus*, 185, 258 (2006).

[15] K. Zahnle, L. Dones, and H. F. Levison. *Icarus*, 136, 202 (1998).

[16] L. Roth, J. Saur, K. D. Retherford, *et al.* *Science*, 343, 171 (2014).

[17] T. B. McCord, G. B. Hansen, F. P. Fanale, *et al.* *Science*, 280, 1242 (1998).

[18] R. W. Carlson, R. E. Johnson, and M. S. Anderson. *Science*, 286, 97 (1999).

[19] T. B. McCord, T. M. Orlando, G. Teeter, *et al.* *J. Geophys. Res. E Planets*, 106, 3311 (2001).

[20] C. Sotin and G. Tobie. *Science*, 320, 1588 (2008).

[21] A. D. Fortes, F. Browning, and I. G. Wood. *Phys. Chem. Miner.*, 39, 443 (2012).

[22] J. I. Lunine and S. K. Atreya. *Nat. Geosci.*, 1, 159 (2008).

[23] L. Iess, D. J. Stevenson, M. Parisi, *et al.* *Science*, 344, 78 (2014).

[24] J. H. Waite, M. R. Combi, W. H. Ip, *et al.* *Science*, 311, 1419 (2006).

[25] F. Postberg, S. Kempf, J. Schmidt, *et al.* *Nature*, 459, 1098 (2009).

[26] C. C. Porco, P. Helfenstein, P. C. Thomas, *et al.* *Science*, 311, 1393 (2006).

[27] F. Nimmo and R. T. Pappalardo. *Nature*, 441, 614 (2006).

[28] S. W. Kieffer and B. M. Jakosky. *Science*, 320, 1432 (2008).

[29] L. Soderblom, S. Kieffer, T. Becker, *et al.* *Science*, 250, 410 (1990).

[30] P. Schenk and M. P. A. Jackson. *Geology*, 21, 299 (1993).

[31] R. H. Brown, D. P. Cruikshank, J. Veverka, *et al.* In D. P. Cruikshank, ed., *Neptune and Triton*, Space Science Series. University of Arizona Press, Tucson (1995).

[32] D. P. Cruikshank, B. Schmitt, T. L. Roush, *et al.* *Icarus*, 147, 309 (2000).

[33] W. B. McKinnon and R. L. Kirk. In L. A. McFadden, P. R. Weissman, and T. V. Johnson, eds., *Encyclopedia of the Solar System, 2nd ed.* Academic Press, San Diego (2007).

[34] J. Ruiz. *Icarus*, 166, 436 (2003).

[35] P. W. Bridgman. *J. Franklin Inst.*, 177, 315 (1914).

[36] P. G. Debenedetti and H. E. Stanley. *Phys. Today*, 56, 40 (2003).

[37] P. Jenniskens and D. F. Blake. *Science*, 265, 753 (1994).

[38] D. J. Stevenson. *Nature*, 298, 142 (1982).

[39] G. D. Crawford and D. J. Stevenson. *Icarus*, 73, 66 (1988).

[40] F. Inagaki, M. M. M. Kuypers, U. Tsunogai, *et al.* *Proc. Natl. Acad. Sci. U.S.A.*, 103, 14164 (2006).

[41] E. D. Sloan and C. A. Koh. *Clathrate Hydrates of Natural Gases, 3rd ed.* CRC Press, Boca Raton, FL (2007).

[42] H. Hirai, T. Tanaka, T. Kawamura, *et al.* *J. Phys. Chem. Solids*, 65, 1555 (2004).

[43] C. A. Tulk, S. Machida, D. D. Klug, *et al.* *J. Chem Phys.*, 141 (2014).

[44] H. Hirai, K. Komatsu, M. Honda, *et al.* *J. Chem Phys.*, 133 (2010).

[45] O. Bollengier, M. Choukroun, O. Grasset, *et al.* *Geochim. Cosmochim. Acta*, 119, 322 (2013).

[46] J. S. Loveday, R. J. Nelmes, M. Guthrie, *et al.* *Nature*, 410, 661 (2001).

[47] H. Hirai, S. I. Machida, T. Kawamura, *et al.* *Am. Mineral.*, 91, 826 (2006).

[48] M. Choukroun, Y. Morizet, and O. Grasset. *J. Raman Spectrosc.*, 38, 440 (2007).

[49] J. S. Loveday and R. J. Nelmes. *Phys. Chem. Chem. Phys.*, 10, 937 (2008).

[50] O. Prieto-Ballesteros, J. S. Kargel, M. Fernandez-Sampedro, *et al.* *Icarus*, 177, 491 (2005).

[51] J. I. Lunine and D. J. Stevenson. *Icarus*, 52, 14 (1982).

[52] Y. Alibert, O. Mousis, and W. Benz. *Astrophys. J.*, 622, L145 (2005).

[53] J. S. Loveday, R. J. Nelmes, and M. Guthrie. *Chem. Phys. Lett.*, 350, 459 (2001).

[54] A. Oancea, O. Grasset, E. Le Menn, *et al.* *Icarus*, 221, 900 (2012).

[55] V. Munoz-Iglesias, O. Prieto-Ballesteros, and L. J. Bonales. *Geochim. Cosmochim. Acta*, 125, 466 (2014).

[56] K. P. Hand, C. F. Chyba, R. W. Carlson, *et al.* *Astrobiology*, 6, 463 (2006).

[57] V. Munoz-Iglesias, L. Jimenez Bonales, D. Santamaria-Perez, *et al.* *Spectrosc. Lett.*, 45, 407 (2012).

[58] A. D. Fortes. *Planet. Space Sci.*, 60, 10 (2012).

[59] A. D. Fortes. *Icarus*, 191, 743 (2007).

[60] K. Frediksson and J. F. Kerridge. *Meteoritics*, 23, 35 (1988).

[61] O. Prieto-Ballesteros and J. S. Kargel. *Icarus*, 173, 212 (2005).

[62] D. L. Hogenboom, J. S. Kargel, J. P. Ganasan, *et al.* *Icarus*, 115, 258 (1995).

[63] J. Dalton, O. Prieto-Ballesteros, J. Kargel, *et al.* *Icarus*, 177, 472 (2005).

[64] R. Nakamura and E. Ohtani. *Icarus*, 211, 648 (2011).

[65] S. Vance and J. M. Brown. *Geochim. Cosmochim. Acta*, 110, 176 (2013).

[66] S. Vance, M. Bouffard, M. Choukroun, *et al.* *Planet. Space Sci.*, 96, 62 (2014).

[67] P. Bridgman. *Proc. Natl. Acad. Sci. U.S.A.*, 76, 71 (1948).

[68] A. D. Fortes, I. G. Wood, M. Alfredsson, *et al.* *Eur. J. Mineral.*, 18, 449 (2006).

[69] E. L. Gromnitskaya, O. F. Yagafarov, A. G. Lyapin, *et al.* *Phys. Chem. Miner.*, 40, 271 (2013).

[70] A. D. Fortes, I. G. Wood, and K. S. Knight. *Phys. Chem. Miner.*, 35, 207 (2008).

[71] A. Hamilton, C. Hall, and L. Pel. *J. Phys. D Appl. Phys.*, 41 (2008).

[72] I. D. H. Oswald, A. Hamilton, C. Hall, *et al.* *J. Am. Chem. Soc.*, 130, 17795 (2008).

[73] J. B. Dalton III, T. Cassidy, C. Paranicas, *et al.* *Planet. Space Sci.*, 77, 45 (2013).

[74] R. T. Pappalardo and A. C. Barr. *Geophys. Res. Lett.*, 31 (2004).

[75] G. M. Marion. *Geochim. Cosmochim. Acta*, 66, 2499 (2002).

[76] K. D. Beyer, A. R. Hansen, and M. Poston. *J. Phys. Chem. A*, 107, 2025 (2003).

[77] P. M. Grindrod, A. D. Fortes, F. Nimmo, *et al. Icarus*, 197, 137 (2008).

[78] G. M. Marion, J. S. Kargel, D. C. Catling, *et al. Geochim. Cosmochim. Acta*, 69, 259 (2005).

[79] V. Munoz-Iglesias, L. J. Bonales, and O. Prieto-Ballesteros. *Astrobiology*, 13, 693 (2013).

[80] L. J. Bonales, V. Munoz-Iglesias, and O. Prieto-Ballesteros. *Eur. J. Mineral.*, 25, 735 (2013).

[81] J. S. Lewis. *Icarus*, 15, 174 (1971).

[82] J. S. Lewis. *Icarus*, 16, 241 (1972).

[83] J. S. Lewis and R. G. Prinn. *Astrophys. J.*, 238, 357 (1980).

[84] R. G. Prinn and B. Fegley. *Astrophys. J.*, 249, 308 (1981).

[85] J. S. Kargel, S. K. Croft, J. I. Lunine, *et al. Icarus*, 89, 93 (1991).

[86] D. L. Hogenboom, J. S. Kargel, G. J. Consolmagno, *et al. Icarus*, 128, 171 (1997).

[87] O. Grasset and J. Pargamin. *Planet. Space Sci.*, 53, 371 (2005).

[88] F. Sohl, H. Hussmann, B. Schwentker, *et al. J. Geophys. Res. E Planets*, 108 (2003).

[89] G. Tobie, O. Grasset, J. I. Lunine, *et al. Icarus*, 175, 496 (2005).

[90] G. M. Marion, J. S. Kargel, D. C. Catling, *et al. Icarus*, 220, 932 (2012).

[91] J. S. Loveday, R. J. Nelmes, C. L. Bull, *et al. High Pressure Res.*, 29, 396 (2009).

[92] A. D. Fortes, E. Suard, M.-H. Lemee-Cailleau, *et al. J. Am. Chem. Soc.*, 131, 13508 (2009).

[93] A. Kurnosov, L. Dubrovinsky, A. Kuznetsov, *et al. Z. Naturforsch., B Chem. Sci.*, 61, 1573 (2006).

[94] M. Choukroun, O. Grasset, G. Tobie, *et al.* *Icarus*, 205, 581 (2010).

[95] K. Shin, R. Kumar, K. A. Udachin, *et al.* *Proc. Natl. Acad. Sci. U.S.A.*, 109, 14785 (2012).

[96] P. W. Bridgman. *Geol. Soc. Amer. Bull.*, 62, 533 (1951).

Epilogue

BACK in 2007, I was gifted with the responsibility of leading the Matter at High Pressure project, now internationally known as MALTA-Consolider Team. Eight years later, my retrospective view is quite rewarding and I want to thank every person involved with or related to MALTA-Consolider for their illusion, dedication and abnegation to our common interests, and also for helping me to reach a decent outcome of our multidisciplinary effort. Overall, I learned a lot from all my MALTA fellows and this book somehow recalls the philosophy of MALTA-Consolider: gathering our efforts to benefit the project as a whole. Unfortunately, there is no room here, and perhaps this is not the right place, to write down a personal acknowledgement to every people involved in MALTA-Consolider, but I think necessary to dedicate a few words to thank some people that have shared the responsibility of leading the project.

Alfredo Segura has been, and always will be, our guiding light, a generous person, and the most respected scientist in our crew. Fernando Rodríguez, with his unique personality, has provided us with the required international links and visibility. Buenaventura Guamis has turned into the most enthusiastic PI for MALTA-Consolider, and his continuous support is invaluable. Alfonso Muñoz suggested the idea of applying for the Consolider call, he supports many groups with his computational knowledge, and he is (incredibly) always around. Pedro Sanz, our fellow neighbour at the Madrid Campus, always pushing our knowledge and scientific capacity to the limit. Olga Prieto is our link to the outer space, her success and recognition are ours. Víctor Lavín is our most enthusiastic and generous supporter (wide sense), a living creature of MALTA-Consolider's spirit. Javier Manjón is the most prominent example of MALTA-Consolider success, and his passionate scientific personality is second to none among us. Juan Andrés joined us when he had much more attractive opportunities, providing the theoretical-chemistry support that MALTA-Consolider lacked in its origins. Mercedes Taravillo, she is the project coordinator in the shadow, the most dedicated person to the project, prioritizing the successful development of the project over her own interests (even in this book). Óscar Rodríguez-Montoro, our project manager, has saved our health, lifestyle and families from catastrophe. José Manuel Menéndez, our computational manager, and much more, put his scientific visibility aside; he is a lovely guy. Jesús González, our high pressure "maestro", put all his wide knowledge to our service. Stefan Klotz, our honoris causa MALTA-Consolider fellow, who witnessed the scientific (and regional food) exhibition in Miraflores de la Sierra back in 2011; thanks for your continuous and uninterested engagement to our project. And last, but not least, José Manuel Recio (aka Michi) and I are close friends, thanks for being always here and there, and, on behalf of MALTA-Consolider, thanks for your effort for this book to see the light.

And for those who still wonder where the MALTA term comes from, just to say that Matter at High Pressure reads "Materia a ALTA Presión" in Spanish. The term was coined by Ángel Vegas and myself while we were filling up the proposal form for the Consolider call. The rest is now our history.

Valentín García Baonza
MALTA Principal Investigator
Madrid, May 2015

Index

Milton Keynes UK
Ingram Content Group UK Ltd.
UKHW050307111024
449327UK00043B/2100